Motivic Homotopy Theory and Refined Enumerative Geometry

CONTEMPORARY MATHEMATICS

745

Motivic Homotopy Theory and Refined Enumerative Geometry

Workshop on
Motivic Homotopy Theory
and Refined Enumerative Geometry
May 14–18, 2018
Universität Duisburg-Essen, Essen, Germany

Federico Binda
Marc Levine
Manh Toan Nguyen
Oliver Röndigs
Editors

AMS AMERICAN MATHEMATICAL SOCIETY

2010 *Mathematics Subject Classification.* Primary 14F42, 14N99.

Library of Congress Cataloging-in-Publication Data

Names: Workshop on Motivic Homotopy Theory and Refined Enumerative Geometry (2018 : Universität Duisburg-Essen), author. | Binda, Federico, 1988- editor. | Levine, Marc, 1952- editor. | Nguyen, Manh Toan, 1987- editor. | Röndigs, Oliver, 1973- editor.

Title: Motivic homotopy theory and refined enumerative geometry / Federico Binda, Marc Levine, Manh Toan Nguyen, Oliver Röndigs, editors.

Description: Providence, Rhode Island : American Mathematical Society, [2020] | Series: Contemporary mathematics, 0271-4132 ; volume 745 | Includes bibliographical references.

Identifiers: LCCN 2019039738 | ISBN 9781470448981 (paperback) | ISBN 9781470454555 (ebook)

Subjects: LCSH: Homotopy theory–Congresses. | Geometry, Enumerative–Congresses. | Motives (Mathematics)–Congresses. | AMS: Algebraic geometry – (Co)homology theory [See also 13Dxx] – Motivic cohomology; motivic homotopy theory [See also 19E15]. | Algebraic geometry – Projective and enumerative geometry [See also 51-XX] – None of the above, but in this section.

Classification: LCC QA612.7 .W67 2018 | DDC 516.3/5–dc23

LC record available at https://lccn.loc.gov/2019039738

Contemporary Mathematics ISSN: 0271-4132 (print); ISSN: 1098-3627 (online)

DOI: https://doi.org/10.1090/conm/745

∞ The paper used in this book is acid-free and falls within the guidelines
established to ensure permanence and durability.
Visit the AMS home page at `https://www.ams.org/`

10 9 8 7 6 5 4 3 2 1 25 24 23 22 21 20

Contents

Preface

Refined enumerative geometry, an emerging area of study, enriches our understanding of solutions to geometric problems. Instead of a mere integer counting such solutions, refined enumerative geometry produces a class in the Grothendieck-Witt ring of quadratic forms over a field, whose dimension is the integer that classical enumerative geometry provides. Such a perspective already occurs in real algebraic geometry, where signs are introduced in order to capture certain counting phenomena over real closed fields, corresponding to the signature of quadratic forms.

Motivic or $\mathbf{A}^1$-homotopy theory supplies a convenient foundational framework for this endeavour. A standard target for Euler characteristics is the endomorphism ring of the unit in a suitable tensor triangulated category, such as the $\mathbf{P}^1$-stable homotopy category over a field. Morel's theorem identifies this endomorphism ring with the Grothendieck-Witt ring of the field. The $\mathbf{P}^1$-stable homotopy category over a field also contains Voevodsky's motivic Eilenberg-MacLane spectrum representing motivic cohomology, and in particular Chow groups of smooth varieties. The latter invariant constitutes a classical tool for enumerative geometry. Between the theories of $\mathbf{P}^1$-stable homotopy and motivic cohomology sits the theory of Chow-Witt groups, represented by algebraic cycles with additional orientation data. Chow-Witt groups are rich enough to capture quadratic form information yet manageable enough to allow for successful computations.

This new viewpoint on enumerative geometry allows beautiful refinements of very classical results (such as counting lines on a cubic surface) but also opens up new possibilities in various other subjects (such as Gromov-Witten theory). In order to promote this viewpoint, the workshop "Motivic Homotopy Theory and Refined Enumerative Geometry" took place in Essen (NRW, Germany) from the 14th to the 18th of May 2018. Central to the workshop were three series of lectures held by as many experts:

- Denis-Charles Cisinski: Intersection theory and mixed motives with rational coefficients
- Jean Fasel: Introduction to Chow-Witt groups
- Marc Levine: Enumerative geometry via quadratic forms

In addition to these lectures, nine research talks on refined enumerative geometry, motivic homotopy theory, and algebraic K-theory were given by Alexey Ananyevskiy, Frédéric Déglise, Jens Hornbostel, Oliver Röndigs, Shuji Saito, Marco Schlichting, Matthias Wendt, Kirsten Wickelgren, and Paul Arne Østvær. The workshop was supported by the DFG Priority Programme "Homotopy theory and algebraic geometry". We would like to express our sincere gratitute to all lecturers, speakers, participants, and administrative personnel for contributing to the success of this workshop. Moreover, we would like to thank all authors for their contributions to these proceedings, as well as the expert referees and the staff at AMS (in particular Chris Thivierge) for their valuable efforts.

Federico Binda
Marc Levine
Manh Toan Nguyen
Oliver Röndigs

FIGURE 1. Workshop picture

Contemporary Mathematics
Volume **745**, 2020

List of talks

Lecture series

Denis-Charles Cisinski: Intersection theory and mixed motives with rational coefficients
Jean Fasel: Introduction to Chow-Witt groups
Marc Levine: Enumerative geometry via quadratic forms

Research talks

Alexey Ananyevskiy: SL-oriented cohomology theories
Frédéric Déglise: Characteristic classes in non oriented theories
Jens Hornbostel: Chow-Witt groups of classifying spaces
Oliver Röndigs: On very effective hermitian K-theory
Shuji Saito: Rigid analytic K-theory
Marco Schlichting: Nisnevich descent for Grothendieck-Witt groups
Matthias Wendt: Chow-Witt rings of Grassmannians and oriented Schubert calculus
Kirsten Wickelgren: An arithmetic count of lines through 4 general lines in $\mathbb{P}^3$
Paul Arne Østvær: A motivic Segal conjecture for the group of order two

List of participants

Alexey Ananyevskiy
St. Petersburg State University

Peter Arndt
Universität Düsseldorf

Ran Azouri
Universität Duisburg-Essen

Tom Bachmann
Universität Duisburg-Essen

Grégory Berhuy
Université Grenoble Alpes

Candace Bethea
University of South Carolina

Sai Somanjana Sreedhar Bhamidi
Korea Institute for Advanced Study

Sascha Biberhofer
University of Vienna

Federico Binda
Universität Regensburg

Denis-Charles Cisinski
Universität Regensburg

David Corwin
MIT

Frédéric Déglise
Institut de Mathématiques de Bourgogne

Elena Dimitriadis Bermejo
Université Paul Sabatier (Toulouse III)

Jean Fasel
Université Grenoble Alpes

Niels Feld
Université de Dijon

Kévin Francois
IMT Toulouse

Martin Frankland
Universität Osnabrück

Rahul Gupta
TIFR Mumbai

Gabriela Guzman
Universität Duisburg-Essen

Daniel Harrer
Universität Duisburg-Essen

Corey Harris
MPI Leipzig

Olivier Haution
LMU Munich

Jeroen Hekking
KTH Stockholm

Jens Hornbostel
Bergische Universität Wuppertal

Katharina Hübner
Universität Heidelberg

Thomas Hudson
Bergische Universität Wuppertal

Fangzhou Jin
Universität Duisburg-Essen

Wataru Kai
Tohoku University

Håkon Kolderup
University of Oslo

Jonas Irgens Kylling
University of Oslo

Tobias Lenz
MPI and Universität Bonn

Marc Levine
Universität Duisburg-Essen

Martin Lüdtke
Universität Frankfurt

Lorenzo Mantovani
Universität Duisburg-Essen

Alberto Merici
Universität Zürich

Alberto Navarro
Universität Zürich

Manh Toan Nguyen
Universität Osnabrück

Paul Arne Østvær
University of Oslo

Shusuke Otabe
Tohoku University

Sabrina Pauli
University of Oslo

Alina Pavlikova
Université de Genève

Rakesh Pawar
Universität Duisburg-Essen

Massimo Pippi
Institut de Mathématiques de Toulouse

Charanya Ravi
University of Oslo

Oliver Röndigs
Universität Osnabrück

Vivek Sadhu
IISER Bhopal

Shuji Saito
Tokyo Institute of Technology

Anand Sawant
LMU München

Marco Schlichting
University of Warwick

Pavel Sechin
Universität Heidelberg

Enzo Serandon
Universität Duisburg-Essen

Mikhail Shkolnikov
IST Austria

Markus Spitzweck
Universität Osnabrück

Vasudevan Srinivas
TIFR Mumbai and FU Berlin

Padmavathi Srinivasan
Georgia Institute of Technology

Rin Sugiyama
Japan Women's University

Tariq Syed
LMU München

Matteo Tamiozzo
Universität Duisburg-Essen

Fabio Tanania
University of Nottingham

Konrad Voelkel
Universität Osnabrück

Matthias Wendt
Universität Freiburg

Kirsten Wickelgren
Georgia Institute of Technology

Heng Xie
Bergische Universität Wuppertal

Maria Yakerson
Universität Duisburg-Essen

Xuan-Gottfried Yang
Independent researcher

Ferdinando Zanchetta
Warwick University

Heer Zhao
Universität Duisburg-Essen

Marcus Zibrowius
Universität Düsseldorf

Contemporary Mathematics
Volume **745**, 2020
https://doi.org/10.1090/conm/745/15020

SL-oriented cohomology theories

Alexey Ananyevskiy

ABSTRACT. We show that a representable motivic cohomology theory admits a unique normalized SL^c-orientation if the zeroth cohomology presheaf is a Zariski sheaf. We also construct Thom isomorphisms in SL-oriented cohomology for SL^c-bundles and obtain new results on the η-torsion characteristic classes, in particular, we prove that the Euler class of an oriented bundle admitting a (possibly non-orientable) odd rank subbundle is annihilated by the Hopf element.

1. Introduction

The setting of oriented cohomology theories in algebraic geometry [**Pan03**, **Sm07a**, **LM07**] is rather well developed, and currently the arising characteristic classes, pushforward maps and operations between oriented cohomology theories, governed by variants of Riemann–Roch theorem [**Pan04**, **Sm07b**], are understood in many details. Particular examples of oriented cohomology theories, such as Chow groups, motivic cohomology and Quillen K-theory proved to be extremely useful tools for studying various questions of algebro-geometric nature. In the recent years attention was drawn to the quadratic refinements of these classical cohomology theories, namely to Chow–Witt groups, Milnor–Witt motivic cohomology and hermitian K-theory. The latter theories are genuinely non-orientable, the projective bundle formula fails for them. Nevertheless these theories share a certain generalized orientation property that was introduced by I. Panin and Ch. Walter [**PW10a**, **PW10b**, **PW10c**].

One of the existing approaches to oriented cohomology theories is based on the existence of Thom isomorphisms for vector bundles [**Pan03**, **Sm07a**]. Roughly speaking, a cohomology theory $A^{*,*}(-)$ is oriented if for every vector bundle E over X one has a so-called Thom isomorphism $A^{*,*}(X) \cong A_X^{*+2n,*+n}(E)$, where $A^{*,*}(X)$ stands for the cohomology of X, $A_X^{*+2n,*+n}(E)$ is the cohomology of E supported on the zero section and $n = \operatorname{rank} E$. Additionally one asks for some natural properties of these Thom isomorphisms – functoriality and compatibility with direct sums of vector bundles. The notion of orientation may be generalized as it is done in [**PW10a**, **PW10b**] by postulating the existence of Thom isomorphisms

---

2010 *Mathematics Subject Classification.* Primary 14F42; Secondary 55P99, 55R40.

The research is supported by Young Russian Mathematics award, by "Native towns", a social investment program of PJSC "Gazprom Neft", and by RFBR grant 18-31-20044. A part of the work was done during the author's stay at the University of Oslo, the visit was supported by the RCN Frontier Research Group Project no. 250399 "Motivic Hopf equations".

only for vector bundles with a certain additional structure – symplectic bundles (symplectic orientation), vector bundles with trivialized determinant line bundle (SL-orientation), vector bundles with the chosen square root of the determinant line bundle (SL^c-orientation). One may also introduce the notion of G-orientation for a general family of group sheaves $G = \{G_n\}_{n\in\mathbb{N}}$ following Remark 2.9 and Definition 3.3 of the current paper, but at the moment we are mostly interested in SL^c and SL-orientations.

It is easy to see that an SL^c-oriented cohomology theory is naturally SL-oriented since if the determinant of a vector bundle is trivial then it clearly has a square root. But it is rather surprising that an SL-oriented cohomology theory admits Thom isomorphisms for vector SL^c-bundles as well:

THEOREM 1.1 (Theorem 4.3). *Let $A \in \mathcal{SH}(S)$ be an SL-oriented spectrum and $\mathcal{E}$ be a rank n vector SL^c-bundle over $X \in \mathcal{S}m_S$. Then there exists an isomorphism*

$$A^{*,*}(X) \xrightarrow{\simeq} A_X^{*+2n,*+n}(\mathcal{E}).$$

The reason for this is that for a line bundle L the Thom spaces $Th(L)$ and $Th(L^\vee)$ are naturally isomorphic which allows one to cancel a twist by the square of a line bundle in SL-oriented cohomology. The above theorem yields that SL^c and SL-orientations are closely related and because of that we do not loose much information focusing on SL-orientation.

In order to show that a cohomology theory is SL-oriented one usually has to come up with a construction of Thom classes for vector SL-bundles which in particular cases may be quite tricky. In the current paper we show that one always has a unique SL-orientation (and SL^c-orientation) for a commutative ring spectrum A provided that $A^{0,0}(-)$ is a sheaf. Namely, we prove the following theorem.

THEOREM 1.2 (Theorem 5.3 and Corollary 5.4). *Let $A \in \mathcal{SH}(S)$ be a commutative ring spectrum and suppose that $A^{0,0}(-)$ is a sheaf in the Zariski topology. Then A admits unique normalized SL^c and SL-orientations.*

Note that if the base scheme is the spectrum of a field then $A^{0,0}(-)$ is a sheaf in particular for the spectra representing sheaf cohomology with the coefficients in a homotopy module, Voevodsky's motivic cohomology and Milnor-Witt motivic cohomology. The proof of the above theorem roughly goes as follows. First we define Thom classes for trivial vector SL^c-bundles choosing an arbitrary trivialization and pulling back the appropriate suspension $\Sigma_{\mathrm{T}}^n 1$. Then we show that the class defined in such a way does not depend on the choice of a trivialization because the action of a matrix from $\mathrm{SL}_{\mathrm{n}}^{\mathrm{c}}$ (a matrix with the determinant being a perfect square) on the Thom space of a trivial bundle is locally homotopy equivalent to the trivial action. This allows one to patch the Thom class of a nontrivial vector SL^c-bundle from a trivializing cover.

We also improve some results on characteristic classes for SL-oriented cohomology theories obtained in [**An15**] where it was mostly assumed that the Hopf element η is inverted in the coefficients. Extending the proof of [**Lev17**, Proposition 7.2] to the general setting of SL-oriented cohomology theories we obtain the following theorem.

THEOREM 1.3 (Theorem 7.4). *Let $A \in \mathcal{SH}(S)$ be an SL-oriented spectrum and $\mathcal{E} = (E, \lambda)$ be a vector SL-bundle over $X \in \mathcal{S}m_S$. Suppose that there exists* *an isomorphism of vector bundles* $E \cong E_1 \oplus E_2$ *with* E_1 *being of odd rank. Then* $\eta \cup e(\mathcal{E}) = 0$.

As a corollary we obtain the following properties of the Borel and Pontryagin classes.

COROLLARY 1.4 (Corollary 7.9). *Suppose that* $S = \operatorname{Spec} k$ *for a field* k *of characteristic different from 2. Let* $A \in \mathcal{SH}(k)$ *be an SL-oriented spectrum. Then the following holds after inverting* η *in the coefficients.*

(1) $b_{2i+1}(\mathcal{H}(E)) = 0, i \geq 0$, *for the hyperbolic symplectic bundle* $\mathcal{H}(E)$ *associated to a vector bundle* E *over* $X \in \mathcal{S}m_k$.
(2) $p_t(E \oplus E') = p_t(E)p_t(E')$ *for vector bundles* E, E' *over* $X \in \mathcal{S}m_k$.
(3) $p_i(E) = 0$ *for a rank* n *vector bundle* E *over* $X \in \mathcal{S}m_k$ *and* $i > \frac{n}{2}$.

Here b_{2i+1}, p_i *and* p_t *denote the Borel, Pontryagin and total Pontryagin classes of the respective bundles, see Definitons* 7.6 *and* 7.7 *or* [**PW10a**, Definition 14.1], [**An15**, Definition 7] *and* [**HW17**, Definition 5.6].

This generalizes [**An15**, Corollary 3] where the same properties were obtained assuming the determinant of E to be trivial.

The paper is organized as follows. In Section 2 we recall the notions of symplectic, SL and SL^c-bundles and outline the general setting of vector bundles with a chosen reduction of the structure group. In the next section we give a uniform definition of G-oriented cohomology theories and prove some basic properties. In Section 4 we show that an SL-oriented cohomology theory admits Thom isomorphisms for vector SL^c-bundles. In Section 5 we construct a normalized SL^c and SL-orientations on a commutative ring spectrum provided that $A^{0,0}(-)$ is a sheaf and show that such orientations are unique. In the next section we recall some well-known facts on the Hopf map and related elements. Then in the last section we show that the Euler class of a vector SL-bundle admitting an odd rank vector subbundle is annihilated by η and derive consequences on the Pontryagin classes.

Throughout the paper we employ the following assumptions and notations.

S	a quasi-compact separated scheme
$\mathcal{S}m_S$	the category of smooth schemes over S
$\mathcal{SH}(S)$	the motivic stable homotopy category over S [**MV99, Voe98**]
$A^{*,*}(-)$	the cohomology theory represented by $A \in \mathcal{SH}(S)$, with the multiplicative structure given by [**An17**, Section 3]
$Th(E)$	the Thom space of a vector bundle E
$\mathbf{1}_X$	the trivialized line bundle over $X \in \mathcal{S}m_S$

2. Vector bundles with additional structure

DEFINITION 2.1. A *rank* $2n$ *symplectic bundle* (or *vector Sp-bundle*) over $X \in \mathcal{S}m_S$ is a pair $\mathcal{E} = (E, \phi)$ with E being a rank $2n$ vector bundle over X and ϕ being a symplectic form on E. An *isomorphism of symplectic bundles* $\theta\colon (E,\phi) \xrightarrow{\simeq} (E',\phi')$ is an isomorphism of vector bundles $\theta\colon E \xrightarrow{\simeq} E'$ such that the diagram

$$\begin{array}{ccc} E & \xrightarrow{\theta} & E' \\ \simeq\downarrow & & \downarrow\simeq \\ E^\vee & \xleftarrow{\theta^\vee} & (E')^\vee \end{array}$$

commutes, where the vertical isomorphisms are given by the symplectic forms. The *sum of symplectic bundles* is induced by the direct sum of vector bundles together with the orthogonal sum of symplectic forms,

$$(E,\phi)\oplus(E',\phi')=(E\oplus E',\phi\perp\phi').$$

For a vector bundle E over $X\in\mathcal{S}m_S$ the *associated hyperbolic symplectic bundle* (or *hyperbolization of* E) is the symplectic bundle

$$\mathcal{H}(E)=\left(E\oplus E^{\vee},\begin{pmatrix}0&1\\-1&0\end{pmatrix}\right).$$

The *trivialized* rank $2n$ symplectic bundle over $X\in\mathcal{S}m_S$ is

$$\mathbf{1}_X^{\mathrm{Sp},2n}=\mathcal{H}(\mathbf{1}_X)^{\oplus n}=\left(\mathbf{1}_X^{\oplus 2n},\begin{pmatrix}0&1\\-1&0\end{pmatrix}^{\oplus n}\right).$$

A *trivial* rank $2n$ symplectic bundle over $X\in\mathcal{S}m_S$ is a rank $2n$ symplectic bundle over $X\in\mathcal{S}m_S$ isomorphic to the trivialized rank $2n$ symplectic bundle $\mathbf{1}_X^{\mathrm{Sp},2n}$.

DEFINITION 2.2. A *rank n special linear vector bundle* (or *vector SL-bundle*) over $X\in\mathcal{S}m_S$ is a pair $\mathcal{E}=(E,\lambda)$ with E being a rank n vector bundle over X and $\lambda\colon\det E\xrightarrow{\simeq}\mathbf{1}_X$ being an isomorphism of line bundles. An *isomorphism of vector SL-bundles* $\theta\colon(E,\lambda)\xrightarrow{\simeq}(E',\lambda')$ is an isomorphism of vector bundles $\theta\colon E\xrightarrow{\simeq}E'$ such that $\lambda=\lambda'\circ\det\theta$,

$$\begin{array}{ccc}\det E & \xrightarrow[\simeq]{\det\theta} & \det E'\\ & {\scriptstyle\lambda}\searrow\simeq \quad \simeq\swarrow{\scriptstyle\lambda'} & \\ & \mathbf{1}_X & \end{array}$$

The *sum of special linear vector bundles* is induced by the direct sum of vector bundles together with the canonical isomorphisms $\det(E\oplus E')\cong\det E\otimes\det E'$ and $\mathbf{1}_X\otimes\mathbf{1}_X\cong\mathbf{1}_X$ which we omit from the notation,

$$(E,\lambda)\oplus(E',\lambda')=(E\oplus E',\lambda\otimes\lambda').$$

The *trivialized* rank n vector SL-bundle over $X\in\mathcal{S}m_S$ is $\mathbf{1}_X^{\mathrm{SL},n}=(\mathbf{1}_X,\mathrm{id})^{\oplus n}$. A *trivial* rank n vector SL-bundle over $X\in\mathcal{S}m_S$ is a rank n vector SL-bundle over $X\in\mathcal{S}m_S$ isomorphic to the trivialized rank n vector SL-bundle.

DEFINITION 2.3. A *rank n vector SL^c-bundle* over $X\in\mathcal{S}m_S$ is a triple $\mathcal{E}=(E,L,\lambda)$ with E being a rank n vector bundle over X, L being a line bundle over X and $\lambda\colon\det E\xrightarrow{\simeq}L^{\otimes 2}$ being an isomorphism of line bundles. An *isomorphism of vector SL^c-bundles* $\theta\colon(E,L,\lambda)\xrightarrow{\simeq}(E',L',\lambda')$ is a pair $\theta=(\underline{\theta}\colon E\xrightarrow{\simeq}E',\hat{\theta}\colon L\xrightarrow{\simeq}L')$ consisting of an isomorphism of vector bundles and an isomorphism of line bundles such that the following diagram commutes.

$$\begin{array}{ccc}\det E & \xrightarrow[\simeq]{\det\underline{\theta}} & \det E'\\ {\scriptstyle\lambda}\big\downarrow\simeq & & \simeq\big\downarrow{\scriptstyle\lambda'}\\ L^{\otimes 2} & \xrightarrow[\hat{\theta}^{\otimes 2}]{\simeq} & (L')^{\otimes 2}\end{array}$$

We will usually abuse the notation omitting the underline and denoting the isomorphism $\underline{\theta}\colon E\xrightarrow{\simeq}E'$ by θ itself.

The *sum of vector SL^c-bundles* is induced by the direct sum of vector bundles together with the canonical isomorphisms

$$\det(E \oplus E') \simeq \det E \otimes \det E', \quad (L \otimes L')^{\otimes 2} \simeq L^{\otimes 2} \otimes (L')^{\otimes 2}$$

which we omit from the notation,

$$(E, L, \lambda) \oplus (E', L', \lambda') = (E \oplus E', L \otimes L', \lambda \otimes \lambda').$$

The *trivialized* rank n vector SL^c-bundle over $X \in \mathcal{S}m_S$ is $\mathbf{1}_X^{\mathrm{SL}^c,n} = (\mathbf{1}_X, \mathbf{1}_X, \theta)^{\oplus n}$ with $\theta\colon \mathbf{1}_X \xrightarrow{\simeq} \mathbf{1}_X^{\otimes 2}$ being the canonical isomorphism. A *trivial* rank n vector SL^c-bundle over $X \in \mathcal{S}m_S$ is a rank n vector SL^c-bundle over $X \in \mathcal{S}m_S$ isomorphic to the trivialized rank n vector SL^c-bundle.

DEFINITION 2.4. Following the above notation we sometimes refer to rank n vector bundles over $X \in \mathcal{S}m_S$ as *rank n vector GL-bundles*. The *trivialized* rank n vector GL-bundle over $X \in \mathcal{S}m_S$ is $\mathbf{1}_X^{\mathrm{GL},n} = \mathbf{1}_X^{\oplus n}$. A *trivial* rank n vector GL-bundle over $X \in \mathcal{S}m_S$ is a vector bundle isomorphic to $\mathbf{1}_X^{\oplus n}$.

DEFINITION 2.5. The hyperbolization construction from Definition 2.1 gives rise to a morphism of groupoids

$$\{\text{vector bundles over X}\} \xrightarrow{\mathcal{H}} \{\text{symplectic bundles over X}\}.$$

Let (E, ϕ) be a symplectic bundle over $X \in \mathcal{S}m_S$. The Pfaffian of ϕ induces an isomorphism of line bundles $\lambda_\phi\colon \det E \xrightarrow{\simeq} \mathbf{1}_X$ (see, for example, the discussion above [**An16a**, Definition 4.5]) giving rise to the *associated SL-bundle* (E, λ_ϕ). This rule can be promoted in a canonical way to a morphism of groupoids

$$\{\text{symplectic bundles over X}\} \xrightarrow{\Phi} \{\text{vector SL-bundles over X}\}.$$

Let (E, λ) be a vector SL-bundle over $X \in \mathcal{S}m_S$. The canonical isomorphism $\theta\colon \mathbf{1}_X \xrightarrow{\simeq} \mathbf{1}_X^{\otimes 2}$ gives rise to the *associated SL^c-bundle* $(E, \mathbf{1}_X, \theta \circ \lambda)$. This rule can be promoted in a canonical way to a morphism of groupoids

$$\{\text{vector SL-bundles over X}\} \xrightarrow{\Psi} \{\text{vector SL}^c\text{-bundles over X}\}.$$

Let (E, L, λ) be a vector SL^c-bundle over $X \in \mathcal{S}m_S$. Forgetting about the additional structure (L, λ) we obtain a morphism of groupoids

$$\{\text{vector SL}^c\text{-bundles over X}\} \xrightarrow{\Theta} \{\text{vector bundles over X}\}.$$

LEMMA 2.6. *In the notation of Definition 2.5 there are canonical isomorphisms*

(1) $\mathcal{H}(E_1 \oplus E_2) \cong \mathcal{H}(E_1) \oplus \mathcal{H}(E_2)$ *for vector bundles* E_1, E_2,
(2) $\Phi(\mathcal{E}_1 \oplus \mathcal{E}_2) \cong \Phi(\mathcal{E}_1) \oplus \Phi(\mathcal{E}_2)$ *for symplectic bundles* $\mathcal{E}_1, \mathcal{E}_2$,
(3) $\Psi(\mathcal{E}_1 \oplus \mathcal{E}_2) \cong \Psi(\mathcal{E}_1) \oplus \Psi(\mathcal{E}_2)$ *for vector SL-bundles* $\mathcal{E}_1, \mathcal{E}_2$,
(4) $\Theta(\mathcal{E}_1 \oplus \mathcal{E}_2) \cong \Theta(\mathcal{E}_1) \oplus \Theta(\mathcal{E}_2)$ *for vector SL^c-bundles* $\mathcal{E}_1, \mathcal{E}_2$,
(5) $\Phi(\mathbf{1}_X^{\mathrm{Sp},2n}) \cong \mathbf{1}_X^{\mathrm{SL},2n}$,
(6) $\Psi(\mathbf{1}_X^{\mathrm{SL},n}) \cong \mathbf{1}_X^{\mathrm{SL}^c,n}$,
(7) $\Theta(\mathbf{1}_X^{\mathrm{SL}^c,n}) \cong \mathbf{1}_X^{\oplus n}$.

PROOF. Straightforward. □

LEMMA 2.7. *Let $G = Sp$, SL or SL^c. For a rank n vector G-bundle $\mathcal{E}$ over $X \in \mathcal{S}m_S$ and a point $x \in X$ there exists a Zariski open subset $U \subseteq X$ such that $x \in U$ and the restriction $\mathcal{E}|_U$ is a trivial rank n vector G-bundle.*

PROOF. It is sufficient to show that over a local ring every vector G-bundle is trivial. Let $W = \operatorname{Spec} R$ be the spectrum of a local ring. Recall that every vector bundle (without additional structure) over W is trivial.

$G = Sp$: follows from [**MH73**, Chapter 1, Corollary 3.5].

$G = SL$: we need to show that for a vector SL-bundle $(\mathbf{1}_W^{\oplus n}, \lambda)$ there exists an isomorphism of vector SL-bundles $f\colon (\mathbf{1}_W^{\oplus n}, \lambda) \xrightarrow{\simeq} \mathbf{1}_W^{\mathrm{SL},n}$. Put $\mathbf{1}_W^{\mathrm{SL},n} = (\mathbf{1}_W^{\oplus n}, \lambda_{\mathrm{triv}})$. One may take $f\colon \mathbf{1}_W^{\oplus n} \xrightarrow{\simeq} \mathbf{1}_W^{\oplus n}$ to be the isomorphism given by the diagonal matrix $\operatorname{diag}(\lambda_{\mathrm{triv}} \circ \lambda^{-1}, 1, 1, \ldots, 1)$.

$G = SL^c$: we need to show that for a vector SL^c-bundle $(\mathbf{1}_W^{\oplus n}, \mathbf{1}_W, \lambda)$ there exists an isomorphism of vector SL^c-bundles $f\colon (\mathbf{1}_W^{\oplus n}, \mathbf{1}_W, \lambda) \xrightarrow{\simeq} \mathbf{1}_W^{\mathrm{SL}^c,n}$. Put $\mathbf{1}_W^{\mathrm{SL}^c,n} = (\mathbf{1}_W^{\oplus n}, \mathbf{1}_W, \lambda_{\mathrm{triv}})$. One may take $f = (\underline{f}, \mathrm{id})$ with $\underline{f}\colon \mathbf{1}_W^{\oplus n} \xrightarrow{\simeq} \mathbf{1}_W^{\oplus n}$ given by the diagonal matrix $\operatorname{diag}(\theta^{-1} \circ \lambda_{\mathrm{triv}} \circ \lambda^{-1} \circ \theta, 1, \ldots, 1)$ with $\theta\colon \mathbf{1}_X \xrightarrow{\simeq} \mathbf{1}_X^{\otimes 2}$ being the canonical isomorphism. □

REMARK 2.8. Let SL_n^c be the kernel of the homomorphism

$$\mathrm{GL}_n \times \mathbb{G}_m \to \mathbb{G}_m, \quad (g,t) \mapsto t^{-2} \det g,$$

(cf. [**PW10b**, Definition 3.3]). It is easy to see that there are canonical equivalences of groupoids

$$\{\text{rank } 2n \text{ symplectic bundles over X}\} \xleftrightarrow{\simeq} \{\mathrm{Sp}_{2n}\text{-torsors over X}\},$$
$$\{\text{rank } n \text{ vector SL bundles over X}\} \xleftrightarrow{\simeq} \{\mathrm{SL}_n\text{-torsors over X}\},$$
$$\{\text{rank } n \text{ vector SL}^c\text{-bundles over X}\} \xleftrightarrow{\simeq} \{\mathrm{SL}_n^c\text{-torsors over X}\},$$

with the torsors considered in the Zariski topology. Under the above equivalences the sum of vector G-bundles corresponds to the canonical homomorphisms

$$\mathrm{Sp}_{2n} \times \mathrm{Sp}_{2m} \to \mathrm{Sp}_{2(n+m)}, \quad \mathrm{SL}_n \times \mathrm{SL}_m \to \mathrm{SL}_{n+m}, \quad \mathrm{SL}_n^c \times \mathrm{SL}_m^c \to \mathrm{SL}_{n+m}^c,$$

while the morphisms $\mathcal{H}$, Φ, Ψ and Θ from Definition 2.5 correspond to the canonical homomorphisms

$$\mathrm{GL}_n \to \mathrm{Sp}_{2n}, \quad \mathrm{Sp}_{2n} \to \mathrm{SL}_{2n}, \quad \mathrm{SL}_n \to \mathrm{SL}_n^c, \quad \mathrm{SL}_n^c \to \mathrm{GL}_n.$$

REMARK 2.9. Let G be a Zariski sheaf of groups on $\mathcal{S}m_S$ and $\rho\colon G \to \mathrm{GL}_n$ be a homomorphism. Following Remark 2.8 one can define the notion of a *vector bundle with the structure group reduced to* G or a *vector* G*-bundle* as a triple $(E, \mathcal{P}, \psi)$ where

(1) E is a vector bundle over X,
(2) $\mathcal{P}$ is a G-torsor over X,
(3) $\psi\colon \mathbf{1}_X^{\oplus n} \times_G \mathcal{P} \xrightarrow{\simeq} E$ is an isomorphism of vector bundles.

The case of $G = \mathrm{St}_n$ the Steinberg group seems to be of particular interest since, as the author learned from F. Morel and A. Sawant, St_n is closely related to the universal $\mathbb{A}^1$-covering group of SL_n whence the notion of a vector bundle with the structure group reduced to St_n may be related to the notion of Spin structure from classical topology.

3. Orientations and twisted cohomology

From now on we assume $G = \mathrm{GL}, \mathrm{Sp}, \mathrm{SL}$ or SL^c.

DEFINITION 3.1. Let $A \in \mathcal{SH}(S)$ be a spectrum and E be a rank n vector bundle over $X \in \mathcal{S}m_S$. Denote $A^{*,*}(-;E)$ the presheaf on the slice category $\mathcal{S}m_S/X$ (i.e. the category of morphisms of S-schemes $Y \xrightarrow{p} X$ with $Y \in \mathcal{S}m_S$) given by

$$[Y \xrightarrow{p} X] \mapsto A^{*+2n,*+n}(Th(p^*E)).$$

We refer to this presheaf as *A-cohomology groups twisted by E*. For a rank n vector G-bundle $\mathcal{E}$ we abuse the notation writing $A^{*,*}(-;\mathcal{E})$ for $A^{*,*}(-;E)$ with E being the vector bundle obtained from $\mathcal{E}$ forgetting the additional structure.

REMARK 3.2. For $A \in \mathcal{SH}(S)$ and $X \in \mathcal{S}m_S$ the suspension isomorphism

$$\Sigma^n_{\mathrm{T}}\colon A^{*,*}(-) \xrightarrow{\simeq} A^{*+2n,*+n}((-)_+ \wedge \mathrm{T}^{\wedge n})$$

of presheaves on $\mathcal{S}m_S$ induces an isomorphism

$$\Sigma^n_{\mathrm{T}}\colon A^{*,*}(-) \xrightarrow{\simeq} A^{*,*}(-;\mathbf{1}_X^{\oplus n})$$

of presheaves on the slice category $\mathcal{S}m_S/X$.

DEFINITION 3.3 (cf. [**Pan03**, Definition 3.1.1], [**Sm07a**, Definition 2.2.1], [**PW10a**, Definition 14.2], [**PW10b**, Definitions 5.1 and 12.1] and [**An15**, Definition 4]). A *normalized G-orientation* of a commutative ring spectrum $A \in \mathcal{SH}(S)$ is a rule which assigns to each rank n vector G-bundle $\mathcal{E}$ over $X \in \mathcal{S}m_S$ an element

$$th(\mathcal{E}) \in A^{0,0}(X;\mathcal{E}) = A^{2n,n}(Th(\mathcal{E}))$$

with the following properties:

(1) For an isomorphism $\theta\colon \mathcal{E} \xrightarrow{\simeq} \mathcal{E}'$ of vector G-bundles over $X \in \mathcal{S}m_S$ one has
$$th(\mathcal{E}) = \theta^A th(\mathcal{E}')$$
where $\theta^A\colon A^{0,0}(X;\mathcal{E}') = A^{2n,n}(Th(\mathcal{E}')) \to A^{2n,n}(Th(\mathcal{E})) = A^{0,0}(X;\mathcal{E})$ is the pullback induced by θ.

(2) For a morphism $f\colon Y \to X$ of smooth S-schemes and a vector G-bundle $\mathcal{E}$ over X one has
$$f^A th(\mathcal{E}) = th(f^*\mathcal{E})$$
where $f^A\colon A^{0,0}(X;\mathcal{E}) = A^{2n,n}(Th(\mathcal{E})) \to A^{2n,n}(Th(f^*\mathcal{E})) = A^{0,0}(Y;f^*\mathcal{E})$ is the pullback induced by the morphism of total spaces $f^*\mathcal{E} \to \mathcal{E}$.

(3) For vector G-bundles $\mathcal{E}$ and $\mathcal{E}'$ over $X \in \mathcal{S}m_S$ one has
$$th(\mathcal{E} \oplus \mathcal{E}') = q_1^A th(\mathcal{E}) \cup q_2^A th(\mathcal{E}')$$
where q_1, q_2 are the projections from $\mathcal{E} \oplus \mathcal{E}'$ to its factors.

(4) $G = \mathrm{GL}, \mathrm{SL}$ or SL^c:
$$th(\mathbf{1}_S^{G,1}) = \Sigma_{\mathrm{T}} 1 \in A^{0,0}(S;\mathbf{1}_S).$$
$G = \mathrm{Sp}$:
$$th(\mathbf{1}_S^{G,2}) = \Sigma^2_{\mathrm{T}} 1 \in A^{0,0}(S;\mathbf{1}_S).$$

We refer to the classes $th(\mathcal{E})$ as *Thom classes*. A commutative ring spectrum A with a chosen normalized G-orientation is called a *G-oriented spectrum*.

LEMMA 3.4. *Let $A \in \mathcal{SH}(S)$ be a commutative ring spectrum. Then normalized orientations of A for different structure groups induce each other as follows.*

$$GL\text{-oriented} \Rightarrow SL^c\text{-oriented} \Rightarrow SL\text{-oriented} \Rightarrow Sp\text{-oriented}.$$

PROOF. We will show that an SL^c-oriented spectrum A has a canonical normalized SL-orientation, the other cases are similar. Let $\mathcal{E}$ be an SL-bundle over $X \in \mathcal{S}m_S$ and put $th(\mathcal{E}) = th(\Psi(\mathcal{E}))$ for the associated SL^c-bundle $\Psi(\mathcal{E})$ (see Definition 2.5). Properties 1 and 2 of Definition 3.3 follow from the functoriality of Ψ, properties 3 and 4 follow from Lemma 2.6. □

LEMMA 3.5. *Let $A \in \mathcal{SH}(S)$ be a G-oriented spectrum and $X \in \mathcal{S}m_S$. Then*

$$th(\mathbf{1}_X^{G,n}) = \Sigma_{\mathrm{T}}^n 1 \in A^{0,0}(X; \mathbf{1}_X^{G,n}).$$

PROOF. For $X = S$ the claim follows from the properties 3 and 4 of Definition 3.3. For a general X the claim follows from $X = S$, properties 1 and 2 of Definition 3.3 and the canonical isomorphism $\mathbf{1}_X^{G,n} \cong p^*\mathbf{1}_S^{G,n}$ for the projection $p\colon X \to S$. □

DEFINITION 3.6. Let $A \in \mathcal{SH}(S)$ be a commutative ring spectrum and $\mathcal{E}$ be a rank n vector G-bundle over $X \in \mathcal{S}m_S$. We say that $th \in A^{0,0}(X;\mathcal{E})$ is *locally G-standard* if there exists a Zariski open cover $X = \bigcup_{i=1}^m U_i$ and isomorphisms of vector G-bundles $\theta_i\colon \mathcal{E}|_{U_i} \xrightarrow{\simeq} \mathbf{1}_{U_i}^{G,n}$ such that $th|_{U_i} = \theta_i^A \Sigma_{\mathrm{T}}^n 1$. In the case of $G = \mathrm{SL}^c$ following the convention of Definition 2.3 we abuse the notation writing $\theta_i^A \Sigma_{\mathrm{T}}^n 1$ where we should write $\underline{\theta}_i^A \Sigma_{\mathrm{T}}^n 1$ with $\theta_i = (\underline{\theta}_i, \hat{\theta}_i)\colon \mathcal{E}|_{U_i} \xrightarrow{\simeq} \mathbf{1}_{U_i}^{\mathrm{SL}^c,n}$ being the isomorphisms.

LEMMA 3.7. *Let $A \in \mathcal{SH}(S)$ be a commutative ring spectrum, $\mathcal{E}$ be a rank n vector G-bundle over $X \in \mathcal{S}m_S$ and $th \in A^{0,0}(X;\mathcal{E})$ be locally G-standard. Then for every morphism of smooth S-schemes $p\colon Y \to X$ the element $p^A(th) \in A^{0,0}(Y;\mathcal{E})$ is locally G-standard.*

PROOF. Let $X = \bigcup_{i=1}^m U_i$ be a Zariski open cover and $\theta_i\colon \mathcal{E}|_{U_i} \xrightarrow{\simeq} \mathbf{1}_{U_i}^{G,n}$ be isomorphisms of vector G-bundles such that $th|_{U_i} = \theta_i^A \Sigma_{\mathrm{T}}^n 1$. Then for the open cover $Y = \bigcup_{i=1}^m p^{-1}(U_i)$ and isomorphisms $p^*\theta_i\colon p^*\mathcal{E}|_{p^{-1}(U_i)} \xrightarrow{\simeq} \mathbf{1}_{p^{-1}(U_i)}^{G,n}$ one has $th|_{p^{-1}(U_i)} = (p^*\theta_i)^A \Sigma_{\mathrm{T}}^n 1$. □

LEMMA 3.8. *Let $A \in \mathcal{SH}(S)$ be a commutative ring spectrum and $\mathcal{E}$ be a rank n vector G-bundle over $X \in \mathcal{S}m_S$. Suppose that $th \in A^{0,0}(X;\mathcal{E})$ is locally G-standard. Then*

$$- \cup th\colon A^{*,*}(-) \to A^{*,*}(-;\mathcal{E})$$

is an isomorphism of presheaves on the slice category $\mathcal{S}m_S/X$.

PROOF. In view of Lemma 3.7 it is sufficient to check that

$$- \cup th\colon A^{*,*}(X) \to A^{*,*}(X;\mathcal{E})$$

is an isomorphism. The proof follows via a standard Mayer-Vietoris argument.

We argue by induction on the minimal possible size of the Zariski open cover $X = \bigcup_{i=1}^m U_i$ satisfying $th|_{U_i} = \theta_i^A \Sigma_{\mathrm{T}}^n 1$ for some isomorphisms of vector G-bundles $\theta_i\colon \mathcal{E}|_{U_i} \xrightarrow{\simeq} \mathbf{1}_{U_i}^{G,n}$. First suppose that $m = 1$, i.e. there exists a trivialization $\theta\colon \mathcal{E} \xrightarrow{\simeq} \mathbf{1}_X^{G,n}$. Then $- \cup th$ is the composition of the isomorphisms

$$\theta^A \circ \Sigma_{\mathrm{T}}^n\colon A^{*,*}(X) \xrightarrow{\simeq} A^{*,*}(X; \mathbf{1}_X^{G,n}) \xrightarrow{\simeq} A^{*,*}(X;\mathcal{E}).$$

For the inductive step choose a Zariski open cover $X = \bigcup_{i=1}^m U_i$ and isomorphisms of vector G-bundles $\theta_i\colon \mathcal{E}|_{U_i} \xrightarrow{\simeq} \mathbf{1}_{U_i}^{G,n}$ such that $th|_{U_i} = \theta_i^A \Sigma_{\mathrm{T}}^n 1$. Put $V = \bigcup_{i=1}^{m-1} U_i$ and $W = U_m$. Consider the morphism of Mayer-Vietoris long exact sequences induced by $-\cup th$.

$$\begin{array}{ccccccccc}
\dots \longrightarrow & A^{*,*}(X) & \longrightarrow & A^{*,*}(V) \oplus A^{*,*}(W) & \longrightarrow & A^{*,*}(V\cap W) & \longrightarrow \dots \\
& \Big\downarrow{\scriptstyle \cup th} & & \Big\downarrow{\scriptstyle (\cup th|_V)\oplus(\cup th|_W)} & & \Big\downarrow{\scriptstyle \cup th|_{V\cap W}} & \\
\dots \longrightarrow & A^{*,*}(X;\mathcal{E}) & \longrightarrow & A^{*,*}(V;\mathcal{E}) \oplus A^{*,*}(W;\mathcal{E}) & \longrightarrow & A^{*,*}(V\cap W;\mathcal{E}) & \longrightarrow \dots
\end{array}$$

The elements $th|_V$, $th|_W$ and $th|_{V\cap W}$ are locally G-standard for the covers $V = \bigcup_{i=1}^{m-1} U_i$, $W = U_m$ and $V\cap W = (\bigcup_{i=1}^{m-1} U_i)\cap U_m$ respectively whence the second and the third vertical homomorphisms are isomorphisms by the inductive assumption. It follows from the five lemma that the first vertical homomorphism is an isomorphism as well. □

COROLLARY 3.9. *Let $A \in \mathcal{SH}(S)$ be a G-oriented spectrum. Then for a rank n vector G-bundle $\mathcal{E}$ over $X \in \mathcal{S}m_S$ the homomorphism*

$$-\cup th(\mathcal{E})\colon A^{*,*}(-) \to A^{*,*}(-;\mathcal{E})$$

is an isomorphism of presheaves on the slice category $\mathcal{S}m_S/X$.

PROOF. Applying Lemma 2.7 we may choose a covering $X = \bigcup_{i=1}^m U_i$ such that $\mathcal{E}|_{U_i}$ is trivial for every i. Lemma 3.5 combined with the properties 1 and 2 of Definition 3.3 yields that $th(\mathcal{E}) \in A^{0,0}(X;\mathcal{E})$ is locally G-standard for the chosen cover and for every choice of trivializations $\theta_i\colon \mathcal{E}|_{U_i} \xrightarrow{\simeq} \mathbf{1}_{U_i}^{G,n}$. The claim follows by Lemma 3.8. □

REMARK 3.10. A (non-normalized) *G-orientation* of a commutative ring spectrum $A \in \mathcal{SH}(S)$ is a normalized G-orientation in the sense of Definition 3.3 with the property 4 replaced with

(4′) For every vector G-bundle $\mathcal{E}$ over $X \in \mathcal{S}m_S$ the homomorphism

$$-\cup th(\mathcal{E})\colon A^{*,*}(X) \to A^{*,*}(X;\mathcal{E})$$

is an isomorphism.

Corollary 3.9 yields that a normalized G-orientation is a G-orientation in the above sense. On the other hand, a G-orientation with Thom classes $th(\mathcal{E})$ gives rise to a normalized G-orientation as follows.

$G = \mathrm{GL}, \mathrm{SL}$ or SL^c. Put $\alpha = \Sigma_{\mathrm{T}}^{-1} th(\mathbf{1}_S^{G,1}) \in A^{0,0}(S)$. Property (4′) yields that

$$A^{0,0}(S) \cup \alpha = A^{0,0}(S),$$

i.e. that α is an invertible element of $A^{0,0}(S)$. Put $\widetilde{th}(\mathcal{E}) = \alpha^{-n} \cup th(\mathcal{E})$ for a rank n vector G-bundle $\mathcal{E}$. This rule defines a normalized G-orientation.

$G = \mathrm{Sp}$. Put $\alpha = \Sigma_{\mathrm{T}}^{-2} th(\mathbf{1}_S^{G,2}) \in A^{0,0}(S)$. As above, property (4′) yields that α is invertible in $A^{0,0}(S)$. Put $\widetilde{th}(\mathcal{E}) = \alpha^{-n} \cup th(\mathcal{E})$ for a rank $2n$ symplectic bundle $\mathcal{E}$. This rule defines a normalized symplectic orientation.

4. SL^c Thom isomorphisms for SL-oriented cohomology

LEMMA 4.1 (cf. [**An16b**, Lemma 2]). *Let E be a vector bundle over $X \in \mathcal{S}m_S$ and L be a line bundle over X. Then there exists an isomorphism*

$$\rho\colon Th(E \oplus L) \xrightarrow{\simeq} Th(E \oplus L^\vee)$$

in the motivic unstable homotopy category $\mathcal{H}_\bullet(S)$.

PROOF. Let E^o, L^o and $(L^\vee)^o$ be the complements to the zero sections of the respective bundles. The Thom spaces $Th(E \oplus L)$ and $Th(E \oplus L^\vee)$ can be realized as the total cofibers of the following diagrams.

$$\begin{array}{ccc} E^o \times_X L^o & \longrightarrow & L^o \\ \downarrow & & \downarrow \\ E^o & \longrightarrow & X \end{array} \qquad \begin{array}{ccc} E^o \times_X (L^\vee)^o & \longrightarrow & (L^\vee)^o \\ \downarrow & & \downarrow \\ E^o & \longrightarrow & X \end{array}$$

Here all the morphisms are the projections. There exists a unique isomorphism of X-schemes $L^o \to (L^\vee)^o$ that on the fibers takes a vector v to the functional f such that $f(v) = 1$. This isomorphism gives rise to a morphism of the above diagrams inducing the isomorphism ρ on the total cofibers. □

REMARK 4.2. It is clear that the isomorphism $\rho = \rho_{X,E,L}$ constructed in Lemma 4.1 is functorial in X, i.e. that for $Y \in \mathcal{S}m_S$ and a regular morphism $p\colon Y \to X$ the following diagram commutes.

$$\begin{array}{ccc} Th(p^*E \oplus p^*L) & \xrightarrow{\rho_{Y,p^*E,p^*L}} & Th(p^*E \oplus p^*L^\vee) \\ {\scriptstyle p_{E\oplus L}}\downarrow & & \downarrow{\scriptstyle p_{E\oplus L^\vee}} \\ Th(E \oplus L) & \xrightarrow{\rho_{X,E,L}} & Th(E \oplus L^\vee) \end{array}$$

Here $p_{E\oplus L}$ and $p_{E\oplus L^\vee}$ are induced by the canonical morphisms of total spaces $p^*E \oplus p^*L \to E \oplus L$ and $p^*E \oplus p^*L^\vee \to E \oplus L^\vee$.

THEOREM 4.3. *Let $A \in \mathcal{SH}(S)$ be an SL-oriented spectrum and $\mathcal{E} = (E, L, \lambda)$ be a rank n vector SL^c-bundle over $X \in \mathcal{S}m_S$. Then there exists an isomorphism*

$$A^{*,*}(-) \xrightarrow{\simeq} A^{*,*}(-; E)$$

of presheaves on the slice category $\mathcal{S}m_S/X$.

PROOF. The isomorphism $A^{*,*}(X) \xrightarrow{\simeq} A^{*,*}(X; E)$ is given by the following chain of isomorphisms:

$$A^{*,*}(X) \xrightarrow{\phi_1} A^{*,*}(X; E \oplus L^\vee \oplus L^\vee) \xrightarrow{\psi} A^{*,*}(X; E \oplus L^\vee \oplus L) \xrightarrow{\phi_2} A^{*,*}(X; E).$$

Here

- $\phi_1 = - \cup th(\widetilde{\mathcal{E}})$ is given by the cup-product with the Thom class for the rank $n+2$ vector SL-bundle $\widetilde{\mathcal{E}} = (E \oplus L^\vee \oplus L^\vee, \widetilde{\lambda})$. The isomorphism $\widetilde{\lambda}\colon \det(E \oplus L^\vee \oplus L^\vee) \xrightarrow{\simeq} \mathbf{1}_Y$ is given by the composition of the isomorphisms $\det(E \oplus L^\vee \oplus L^\vee) \xrightarrow{\simeq} \det E \otimes (L^\vee)^{\otimes 2} \xrightarrow{\lambda\otimes \mathrm{id}} L^{\otimes 2} \otimes (L^\vee)^{\otimes 2} \xrightarrow{\simeq} \mathbf{1}_X$.
- $\psi = \rho^A$ is the pullback along the isomorphism

  $$\rho\colon Th(E \oplus L^\vee \oplus L) \xrightarrow{\simeq} Th(E \oplus L^\vee \oplus L^\vee)$$

  given by Lemma 4.1.

- $\phi_2 = \widetilde{\phi}_2^{-1}$ is the inverse to the isomorphism

$$\widetilde{\phi}_2\colon A^{*,*}(X;E) \xrightarrow{-\cup th(L^\vee \oplus L, \lambda_{ev})} A^{*,*}(X; E \oplus L^\vee \oplus L)$$

given by the cup-product with the Thom class for the rank 2 vector SL-bundle $(L^\vee \oplus L, \lambda_{ev})$ where $\lambda_{ev}\colon \det(L^\vee \oplus L) \xrightarrow{\simeq} L^\vee \otimes L \xrightarrow{ev} \mathbf{1}_X$ is induced by the evaluation homomorphism. Note that $\widetilde{\phi}_2$ is an isomorphism by [**An16b**, Lemma 3].

It is clear that the construction is functorial in X whence the claim. □

REMARK 4.4. Let $A \in \mathcal{SH}(S)$ be an SL-oriented spectrum. Theorem 4.3 allows one to define a Thom class for an SL^c-bundle $\mathcal{E}$ over $X \in \mathcal{S}m_S$ as the image of $1 \in A^{0,0}(X)$ under the isomorphism $A^{0,0}(X) \xrightarrow{\simeq} A^{0,0}(X;\mathcal{E})$. A natural question is whether this rule gives rise to an SL^c-orientation of A and if so, what are the relations between this SL^c-orientation and the original SL-orientation. Unfortunately, while functoriality and normalization (properties 1, 2 and 4 of Definition 3.3) are straightforward, multiplicativity (property 3 of Definition 3.3) seems to be more involved and at the moment it is not clear whether this property holds. We leave these questions to future investigations.

5. Orienting A when $\mathbf{A^{0,0}}(-)$ is a sheaf

LEMMA 5.1. *Let $A \in \mathcal{SH}(S)$ be a commutative ring spectrum and suppose that $A^{0,0}(-)$ is a sheaf in the Zariski topology. Then A admits at most one normalized G-orientation.*

PROOF. Suppose that A admits normalized G-orientations with Thom classes $th(\mathcal{E})$ and $\widetilde{th}(\mathcal{E})$. Corollary 3.9 yields that for a rank n vector G-bundle $\mathcal{E}$ over $X \in \mathcal{S}m_S$ the homomorphism

$$-\cup th(\mathcal{E})\colon A^{0,0}(-) \to A^{0,0}(-;\mathcal{E})$$

is an isomorphism of presheaves on the slice category $\mathcal{S}m_S/X$ whence $A^{0,0}(-;\mathcal{E})$ is a sheaf on the small Zariski site of X. Apply Lemma 2.7 and choose a cover $X = \bigcup_{i=1}^m U_i$ together with trivializations $\theta_i\colon \mathcal{E}|_{U_i} \xrightarrow{\simeq} \mathbf{1}_{U_i}^{G,n}$. Lemma 3.5 combined with the properties 1 and 2 of Definition 3.3 yields that

$$th(\mathcal{E})|_{U_i} = th(\mathcal{E}|_{U_i}) = \theta_i^A \Sigma_{\mathrm{T}}^n 1 = \widetilde{th}(\mathcal{E}|_{U_i}) = \widetilde{th}(\mathcal{E})|_{U_i}$$

for every i. Thus $th(\mathcal{E})$ and $\widetilde{th}(\mathcal{E})$ coincide locally. We have already shown that $A^{0,0}(-;\mathcal{E})$ is a sheaf whence $th(\mathcal{E}) = \widetilde{th}(\mathcal{E})$ □

LEMMA 5.2. *Let $\mathcal{E}$ be a rank n vector SL^c-bundle over $X \in \mathcal{S}m_S$, let $A \in \mathcal{SH}(S)$ be a commutative ring spectrum and suppose that $A^{0,0}(-)$ restricts to a sheaf on the small Zariski site of X. Then there exists at most one locally SL^c-standard element in $A^{0,0}(X;\mathcal{E})$.*

PROOF. Let $th, \widetilde{th} \in A^{0,0}(X;\mathcal{E})$ be locally SL^c-standard elements. Lemma 3.8 yields that

$$-\cup th\colon A^{0,0}(-) \to A^{0,0}(-;\mathcal{E})$$

is an isomorphism of presheaves on the slice category $\mathcal{S}m_S/X$. Then $A^{0,0}(-;\mathcal{E})$ restricts to a sheaf on the small Zariski site of X.

Choose the covers and trivializations from Definition 3.6 for th and $\widetilde{th}$. Intersecting the elements of the covers and restricting trivializations we may choose a Zariski open cover $X = \bigcup_{i=1}^m U_i$ and trivializations $\theta_i, \widetilde{\theta}_i \colon \mathcal{E}_{U_i} \xrightarrow{\simeq} \mathbf{1}_{U_i}^{\mathrm{SL}^c,n}$ such that

$$th|_{U_i} = \theta_i^A \Sigma_{\mathrm{T}}^n 1, \quad \widetilde{th}|_{U_i} = \widetilde{\theta}_i^A \Sigma_{\mathrm{T}}^n 1.$$

The automorphisms

$$\widetilde{\theta}_i \circ \theta_i^{-1} \colon \mathbf{1}_{U_i}^{\mathrm{SL}^c,n} \xrightarrow{\simeq} \mathbf{1}_{U_i}^{\mathrm{SL}^c,n}$$

are given by

$$(g_i, \lambda_i) \in \mathrm{SL}_\mathrm{n}^\mathrm{c}(R_i) \subseteq \mathrm{GL}_n(R_i) \times R_i^*, \qquad R_i = \Gamma(U_i, \mathcal{O}_{U_i})$$

with $\det g_i = \lambda_i^2$. Put $g_i' = g_i d_i$ for the diagonal matrix $d_i = \mathrm{diag}(\lambda_i^{-2}, 1, \ldots 1)$. Then $\det g_i' = 1$. Every matrix of determinant 1 over a local ring is an elementary matrix (i.e. a product of transvections) whence for every i there exists a Zariski open covering $U_i = \bigcup_{j=1}^{m_i} U_{ij}$ such that $g_i'|_{U_{ij}}$ is an elementary matrix for every j. It is well known that the action of an elementary matrix on the Thom space of a trivial vector bundle is homotopy equivalent to the identity morphism (see, for example, [**An16b**, Lemma 1]). Moreover, multiplication by a perfect square on the Thom space of a trivial line bundle is also homotopy equivalent to the identity morphism, see, for example, [**An16b**, Lemma 5] (the standing assumption of the loc. cit. that the base scheme S is the spectrum of a field is not used in the proof of the lemma). Hence

$$\widetilde{th}|_{U_{ij}} = (\widetilde{\theta}_i^A \Sigma_{\mathrm{T}}^n 1)|_{U_{ij}} = (\widetilde{\theta}_i|_{U_{ij}})^A (\Sigma_{\mathrm{T}}^n 1) = (\theta_i|_{U_{ij}})^A \circ (\widetilde{\theta}_i \circ \theta_i^{-1})|_{U_{ij}}^A (\Sigma_{\mathrm{T}}^n 1) =$$
$$= (\theta_i|_{U_{ij}})^A \circ (d_i|_{U_{ij}})^A \circ (g_i'|_{U_{ij}})^A (\Sigma_{\mathrm{T}}^n 1) = (\theta_i|_{U_{ij}})^A (\Sigma_{\mathrm{T}}^n 1) = (\theta_i^A \Sigma_{\mathrm{T}}^n 1)|_{U_{ij}} = th|_{U_{ij}}.$$

Then th and $\widetilde{th}$ coincide locally and since we have already shown that $A^{0,0}(-;\mathcal{E})$ is a sheaf this yields $th = \widetilde{th}$. □

THEOREM 5.3. *Let $A \in \mathcal{SH}(S)$ be a commutative ring spectrum and suppose that $A^{0,0}(-)$ is a sheaf in the Zariski topology. Then A admits a unique normalized SL^c-orientation.*

PROOF. The proof goes in two steps: first we show that for every rank n vector SL^c-bundle $\mathcal{E}$ there exists a locally SL^c-standard element $th(\mathcal{E}) \in A^{0,0}(X;\mathcal{E})$ and then we apply the uniqueness of such elements given by Lemma 5.2 in order to show that they satisfy the properties of Definition 3.3.

Let $\mathcal{E}$ be a rank n vector SL^c-bundle over $X \in \mathcal{S}m_S$. Applying Lemma 2.7 choose a Zariski open cover $X = \bigcup_{i=1}^m U_i$ and isomorphisms of vector SL^c-bundles $\theta_i \colon \mathcal{E}|_{U_i} \xrightarrow{\simeq} \mathbf{1}_{U_i}^{\mathrm{SL}^c,n}$. For every i put

$$th_i = \theta_i^A \Sigma_{\mathrm{T}}^n 1 \in A^{0,0}(U_i;\mathcal{E}).$$

Denote $U_{\le j} = \bigcup_{i=1}^j U_i$. We will inductively construct locally SL^c-standard elements $th_{\le j} \in A^{0,0}(U_{\le j};\mathcal{E})$. Put $th_{\le 1} = th_1$. In order to construct $th_{\le j}$ consider the following fragment of the Mayer-Vietoris long exact sequence.

$$\ldots \to A^{0,0}(U_{\le j};\mathcal{E}) \to A^{0,0}(U_{\le j-1};\mathcal{E}) \oplus A^{0,0}(U_j;\mathcal{E}) \to A^{0,0}(U_{\le j-1} \cap U_j;\mathcal{E}) \to \ldots$$

Lemma 3.7 yields that both the elements

$$th_{\le j-1}|_{U_{\le j-1} \cap U_j},\ th_j|_{U_{\le j-1} \cap U_j} \in A^{0,0}(U_{\le j-1} \cap U_j;\mathcal{E})$$

are locally $\mathrm{SL^c}$-standard whence Lemma 5.2 yields

$$th_{\leq j-1}|_{U_{\leq j-1}\cap U_j} = th_j|_{U_{\leq j-1}\cap U_j}.$$

It follows from the above Mayer-Vietoris long exact sequence that there exists some

$$th_{\leq j} \in A^{0,0}(U_{\leq j};\mathcal{E})$$

such that $th_{\leq j}|_{U_{\leq j-1}} = th_{\leq j-1}$ and $th_{\leq j}|_{U_j} = th_j$. This element is clearly locally $\mathrm{SL^c}$-standard for the open cover $U_{\leq j} = \bigcup_{i=1}^{j} U_i$ and isomorphisms of vector $\mathrm{SL^c}$-bundles $\theta_i\colon \mathcal{E}|_{U_i} \xrightarrow{\simeq} \mathbf{1}_{U_i}^{\mathrm{SL^c},n}$. Put

$$th(\mathcal{E}) = th_{\leq m}.$$

Note that $th(\mathcal{E})$ does not depend on the choices made above, i.e. on the choice of an open cover $X = \bigcup_{i=1}^{m} U_i$ and isomorphisms θ_i, since $th(\mathcal{E})$ is locally $\mathrm{SL^c}$-standard by construction whence unique by Lemma 5.2.

Now we are going to check that the properties of Definition 3.3 hold for the constructed elements $th(\mathcal{E})$. Each time it is sufficient to check that the element at right-hand side of the corresponding equality is locally $\mathrm{SL^c}$-standard and then apply Lemma 5.2.

(1) For the property 1 choose a Zariski open cover $X = \bigcup_{i=1}^{m} U_i$ and isomorphisms of vector $\mathrm{SL^c}$-bundles $\theta_i\colon \mathcal{E}'|_{U_i} \xrightarrow{\simeq} \mathbf{1}_{U_i}^{\mathrm{SL^c},n}$ such that $th(\mathcal{E}')|_{U_i} = \theta_i^A\Sigma_{\mathrm{T}}^n 1$. Then $\theta^A th(\mathcal{E}')_{U_i} = (\theta_i\circ\theta|_{U_i})^A\Sigma_{\mathrm{T}}^n 1$ whence $\theta^A th(\mathcal{E}')$ is locally $\mathrm{SL^c}$-standard with the cover $X = \bigcup_{i=1}^{m} U_i$ and trivializations $\theta_i\circ\theta|_{U_i}$.

(2) For the property 2 note that the element $f^A th(\mathcal{E})$ is locally $\mathrm{SL^c}$-standard by Lemma 3.7.

(3) For the property 3 intersecting the elements of open covers for $\mathcal{E}$ and $\mathcal{E}'$ and restricting isomorphisms we may choose a Zariski open cover $X = \bigcup_{i=1}^{m} U_i$ and isomorphisms of vector $\mathrm{SL^c}$-bundles

$$\theta_i\colon \mathcal{E}|_{U_i} \xrightarrow{\simeq} \mathbf{1}_{U_i}^{\mathrm{SL^c},n}, \quad \widetilde{\theta}_i\colon \mathcal{E}'|_{U_i} \xrightarrow{\simeq} \mathbf{1}_{U_i}^{\mathrm{SL^c},n'}$$

such that $th(\mathcal{E})|_{U_i} = \theta_i^A\Sigma_{\mathrm{T}}^n 1$, $th(\mathcal{E}')|_{U_i} = \widetilde{\theta}_i^A\Sigma_{\mathrm{T}}^{n'} 1$. Then

$$(q_1^A th(\mathcal{E}) \cup q_2^A th(\mathcal{E}'))|_{U_i} = (\theta_i\oplus\widetilde{\theta}_i)^A(\Sigma_{\mathrm{T}}^{n+n'} 1)$$

whence $q_1^A th(\mathcal{E}) \cup q_2^A th(\mathcal{E}')$ is locally $\mathrm{SL^c}$-standard with the open cover $X = \bigcup_{i=1}^{m} U_i$ and trivializations $\theta_i\oplus\widetilde{\theta}_i$.

(4) For the property 4 note that $\Sigma_{\mathrm{T}}^n 1$ is tautologically locally $\mathrm{SL^c}$-standard.

The uniqueness of the normalized $\mathrm{SL^c}$-orientation follows from Lemma 5.1. □

COROLLARY 5.4. *Let $A \in \mathcal{SH}(S)$ be a commutative ring spectrum and suppose that $A^{0,0}(-)$ is a sheaf in the Zariski topology. Then A admits a unique normalized SL-orientation.*

PROOF. The existence follows from Lemma 3.4 and Theorem 5.3, the uniqueness follows from Lemma 5.1. □

REMARK 5.5. Let $S = \operatorname{Spec} k$ be the spectrum of a field. Theorem 5.3 and Corollary 5.4 provide a uniform approach to the $\mathrm{SL^c}$ and SL orientations of some well known motivic cohomology theories that are known to be SL^c-oriented by some other means. In particular, $A^{0,0}(-)$ is a sheaf in the following cases.

(1) A is the Eilenberg-MacLane spectrum for a homotopy module [**Mor04**, Section 5.2], i.e. A belongs to the heart of the homotopy t-structure on the category $\mathcal{SH}(k)$.
(2) $A = H\mathbb{Z}$ is the spectrum representing motivic cohomology. In this case $A^{0,0}(-) = H\mathbb{Z}^{0,0}(-)$ is well known to be the constant sheaf $\underline{\mathbb{Z}}$.
(3) $A = H_{MW}\mathbb{Z}$ is the spectrum representing Milnor-Witt motivic cohomology [**DF17**], k is infinite and perfect. Then $A^{0,0}(-) = H_{MW}\mathbb{Z}^{0,0}(-) \cong \underline{GW}$ is the unramified Grotendieck–Witt sheaf [**DF17**, Section 3.2.1].

6. Interlude on the Hopf element and the connecting homomorphism

DEFINITION 6.1 ([**An16a**, Definition 3.6]). Let $A \in \mathcal{SH}(S)$ be a commutative ring spectrum and $u \in \Gamma(X, \mathcal{O}_X^*)$ be an invertible regular function on $X \in \mathcal{S}m_S$. Put

$$\langle u \rangle = \langle u \rangle_A = \Sigma_{\mathrm{T}}^{-1} f_u^A \Sigma_{\mathrm{T}} 1 \in A^{0,0}(X)$$

for the morphism $f_u \colon X_+ \wedge \mathrm{T} \to X_+ \wedge \mathrm{T}$ given by $f_u(x,t) = (x, u(x)t)$.

LEMMA 6.2. *Let $u, v \in \Gamma(X, \mathcal{O}_X^*)$ be invertible regular functions on $X \in \mathcal{S}m_S$. Then for a commutative ring spectrum $A \in \mathcal{SH}(S)$ one has*

$$\langle v^2 u \rangle = \langle u \rangle \in A^{0,0}(X).$$

PROOF. The same proof as in [**An16b**, Lemma 5] yields that

$$\Sigma_{\mathrm{T}}^{\infty} f_u = \Sigma_{\mathrm{T}}^{\infty} f_{v^2 u} \colon \Sigma_{\mathrm{T}}^{\infty} X_+ \wedge \mathrm{T} \to \Sigma_{\mathrm{T}}^{\infty} X_+ \wedge \mathrm{T}$$

whence the claim. Note that the assumption $S = \operatorname{Spec} k$ of [**An16b**] is not needed for the proof, one applies literally the same reasoning over a general base. □

DEFINITION 6.3. The *Hopf map* is the morphism $H \colon \mathbb{A}^2 - (0,0) \to \mathbb{P}^1$ given by $H(x,y) = [x : y]$. Recall that there are canonical isomorphisms

$$(\mathbb{A}^2 - (0,0), (1,1)) \cong \mathrm{S}^{3,2}, \quad (\mathbb{P}^1, [1:1]) \cong \mathrm{S}^{2,1}$$

in the unstable homotopy category [**MV99**, Lemma 2.15, Example 2.20] whence the Hopf map gives rise to the *Hopf element* $\eta \in \pi^{-1,-1}(S)$. See [**An16a**, Definition 3.5] for the precise formula defining η out of H that we are going to use in the current paper. For a commutative ring spectrum $A \in \mathcal{SH}(S)$ we abuse the notation and denote by the same letter the associated Hopf element $\eta \in A^{-1,-1}(S)$.

LEMMA 6.4. *Let $X \in \mathcal{S}m_S$ and $t \in \Gamma(X \times \mathbb{G}_m, \mathcal{O}_{X \times \mathbb{G}_m}^*)$ be the invertible function on $X \times \mathbb{G}_m$ given by the coordinate function on $\mathbb{G}_m$. Then for a commutative ring spectrum $A \in \mathcal{SH}(S)$ one has*

$$\partial \langle -t \rangle = \Sigma_{\mathrm{T}} \eta$$

where ∂ is the connecting homomorphism in the localization long exact sequence

$$\cdots \to A^{*,*}(X_+ \wedge \mathrm{T}) \to A^{*,*}(X \times \mathbb{A}^1) \to A^{*,*}(X \times \mathbb{G}_m) \xrightarrow{\partial} A^{*+1,*}(X_+ \wedge \mathrm{T}) \to \cdots$$

PROOF. Lemma 6.2 yields that $\langle -t \rangle = \langle -t^{-1} \rangle$ Then the claim follows from [**An16a**, Theorem 3.8]. Note that the assumption $S = \operatorname{Spec} k$ of [**An16a**] is not needed for the proof, one applies literally the same reasoning over a general base. □

REMARK 6.5. One should not be too surprised to see $\langle -t\rangle$ in the statement of the above lemma since the computation depends on the particular choice of isomorphisms $(\mathbb{A}^2-(0,0),(1,1))\cong \mathrm{S}^{3,2}$ and $(\mathbb{P}^1,[1:1])\cong \mathrm{S}^{2,1}$ defining η. In the current paper we adopted the choices made in [**An16a**, Definition 3.5], if one defines $\widetilde{\eta}$ via the same isomorphism $(\mathbb{A}^2-\{0,0\},(1,1))\cong \mathrm{S}^{3,2}$ and the same isomorphism $(\mathbb{P}^1,[1:1])\cong \mathrm{S}^{2,1}$ precomposed with an automorphism $(\mathbb{P}^1,[1:1])\to(\mathbb{P}^1,[1:1])$ given by $[x:y]\mapsto[y:x]$ then one can show that $\partial\langle t\rangle=\Sigma_{\mathrm{T}}\widetilde{\eta}$.

7. Characteristic classes and Hopf element

DEFINITION 7.1. Let $A\in\mathcal{SH}(S)$ be a G-oriented spectrum and $\mathcal{E}$ be a rank n vector G-bundle over $X\in\mathcal{S}m_S$. Put

$$e(\mathcal{E})=z^A th(\mathcal{E})\in A^{2n,n}(X)$$

for the pullback

$$z^A\colon A^{0,0}(X;\mathcal{E})=A^{2n,n}(Th(\mathcal{E}))\to A^{2n,n}(X)$$

along the morphism $z\colon X\to Th(\mathcal{E})$ induced by the zero section of the bundle $\mathcal{E}$. We refer to $e(\mathcal{E})$ as the *Euler class* of $\mathcal{E}$.

LEMMA 7.2. *Let $A\in\mathcal{SH}(S)$ be a G-oriented spectrum. Then*

1. *$e(\mathcal{E})=e(\mathcal{E}')$ for isomorphic vector G-bundles $\mathcal{E}$ and $\mathcal{E}'$.*
2. *$f^Ae(\mathcal{E})=e(f^*\mathcal{E})$ for a morphism of smooth S-schemes $f\colon Y\to X$ and a vector G-bundle $\mathcal{E}$ over X.*
3. *$e(\mathcal{E}\oplus\mathcal{E}')=e(\mathcal{E})\cup e(\mathcal{E}')$ for vector G-bundles $\mathcal{E}$ and $\mathcal{E}'$ over $X\in\mathcal{S}m_S$.*

PROOF. Straightforward from the properties stated in Definition 3.3. □

LEMMA 7.3. *Let $A\in\mathcal{SH}(S)$ be an SL-oriented spectrum and (E,λ) be a rank n vector SL-bundle over $X\in\mathcal{S}m_S$. Then for $u\in\Gamma(X,\mathcal{O}_X^*)$ one has*

$$th(E,u\cdot\lambda)=\langle u\rangle\cup th(E,\lambda)\in A^{0,0}(X;E).$$

PROOF. Property 3 of Definition 3.3 yields

$$th(E,u\cdot\lambda)\cup th(\mathbf{1}_X,\mathrm{id})=th(E\oplus\mathbf{1}_X,u\cdot\lambda\otimes\mathrm{id})=th(E,\lambda)\cup th(\mathbf{1}_X,u\cdot\mathrm{id}).$$

It follows from Lemma 3.5 that $th(\mathbf{1}_X,\mathrm{id})=\Sigma_{\mathrm{T}}1$ whence

$$th(E,u\cdot\lambda)=th(E,\lambda)\cup\Sigma_{\mathrm{T}}^{-1}th(\mathbf{1}_X,u\cdot\mathrm{id})=th(E,\lambda)\cup\langle u\rangle.$$ □

THEOREM 7.4 (cf. [**Lev17**, Proposition 7.2]). *Let $A\in\mathcal{SH}(S)$ be an SL-oriented spectrum and $\mathcal{E}=(E,\lambda)$ be a vector SL-bundle over $X\in\mathcal{S}m_S$. Suppose that there exists an isomorphism of vector bundles $E\cong E_1\oplus E_2$ with E_1 being of odd rank. Then $\eta\cup e(\mathcal{E})=0$.*

PROOF. Choose an isomorphism $\theta\colon E_1\oplus E_2\xrightarrow{\simeq}E$. Without loss of generality we may assume $E=E_1\oplus E_2$, so we omit θ from notation. Let $\pi\colon X\times\mathbb{G}_m\to X$ be the projection and denote $t\in\Gamma(X\times\mathbb{G}_m,\mathcal{O}^*_{X\times\mathbb{G}_m})$ the invertible function on $X\times\mathbb{G}_m$ given by the coordinate function on $\mathbb{G}_m$. Consider the automorphism $\rho\colon\pi^*E_1\to\pi^*E_1$ induced by the multiplication by $-t$, i.e. $\rho(v)=-tv$ on the sections. Then ρ gives rise to an isomorphism of vector SL-bundles

$$(\rho\oplus\mathrm{id})\colon(\pi^*(E_1\oplus E_2),-t^{2n+1}\cdot\pi^*\lambda)\xrightarrow{\simeq}(\pi^*(E_1\oplus E_2),\pi^*\lambda)$$

with $2n+1 = \operatorname{rank} E_1$. Property 1 of Definition 3.3 yields

$$th(\pi^*(E_1 \oplus E_2), -t^{2n+1} \cdot \pi^*\lambda) = (\rho \oplus \mathrm{id})^A th(\pi^*(E_1 \oplus E_2), \pi^*\lambda).$$

Then it follows from Lemmas 7.3 and 6.2 that

$$(\rho \oplus \mathrm{id})^A th(\pi^*(E_1 \oplus E_2), \pi^*\lambda) = \langle -t^{2n+1} \rangle \cup th(\pi^*(E_1 \oplus E_2), \pi^*\lambda) = \\ = \langle -t \rangle \cup th(\pi^*(E_1 \oplus E_2), \pi^*\lambda).$$

For the morphism $z\colon X \times \mathbb{G}_m \to Th(\pi^*(E_1 \oplus E_2))$ induced by the zero section we have $(\rho \oplus \mathrm{id}) \circ z = z$ whence

$$\pi^A e(E_1 \oplus E_2, \lambda) = e(\pi^*(E_1 \oplus E_2), \pi^*\lambda) = z^A th(\pi^*(E_1 \oplus E_2), \pi^*\lambda) = \\ = z^A (\rho \oplus \mathrm{id})^A th(\pi^*(E_1 \oplus E_2), \pi^*\lambda) = z^A(\langle -t \rangle \cup th(\pi^*(E_1 \oplus E_2), \pi^*\lambda)) = \\ = \langle -t \rangle \cup z^A(th(\pi^*(E_1 \oplus E_2), \pi^*\lambda)) = \langle -t \rangle \cup \pi^A e(E_1 \oplus E_2, \lambda).$$

Let ∂ be the connecting homomorphism in the localization long exact sequence

$$\cdots \to A^{*,*}(X_+ \wedge \mathrm{T}) \to A^{*,*}(X \times \mathbb{A}^1) \to A^{*,*}(X \times \mathbb{G}_m) \xrightarrow{\partial} A^{*+1,*}(X_+ \wedge \mathrm{T}) \to \cdots$$

Then for every $\alpha \in A^{*,*}(X)$ one has $\partial(\pi^A \alpha) = 0$. Lemma 6.4 combined with the above yields

$$0 = \partial(\pi^A e(E_1 \oplus E_2, \lambda)) = \partial(\langle -t \rangle \cup \pi^A e(E_1 \oplus E_2, \lambda)) = \\ = \partial(\langle -t \rangle) \cup e(E_1 \oplus E_2, \lambda) = \eta \cup e(E_1 \oplus E_2, \lambda). \quad \square$$

REMARK 7.5. Theorem 7.4 improves [**An15**, Corollary 2] where it was proved that for a vector SL-bundle of odd rank one has $e(\mathcal{E}) = 0$ after inverting η.

DEFINITION 7.6 ([**PW10a**, Definition 14.1]). Let $A \in \mathcal{SH}(S)$ be a commutative ring spectrum. A theory of *Borel classes* on A is a rule which assigns to every symplectic bundle $\mathcal{E}$ over $X \in \mathcal{S}m_S$ a sequence of elements

$$b_i(\mathcal{E}) = b_i^A(\mathcal{E}) \in A^{4i,2i}(X), \quad i \geq 1,$$

with the following properties:

(1) For isomorphic symplectic bundles $\mathcal{E} \cong \mathcal{E}'$ one has $b_i(\mathcal{E}) = b_i(\mathcal{E}')$ for all i.
(2) For a morphism of smooth S-schemes $f\colon Y \to X$ and a symplectic bundle $\mathcal{E}$ over X one has $f^A b_i(\mathcal{E}) = b_i(f^*\mathcal{E})$ for all i.
(3) For $X \in \mathcal{S}m_S$ the homomorphism

$$A^{*,*}(X) \oplus A^{*-4,*-2}(X) \to A^{*,*}(\mathrm{HP}^1 \times_S X)$$

given by $a + a' \mapsto \pi_2^A(a) + \pi_2^A(a') \cup \pi_1^A(b_1(\mathcal{T}))$ is an isomorphism. Here
- $\mathrm{HP}^1 = \mathrm{Sp}_4/(\mathrm{Sp}_2 \times \mathrm{Sp}_2)$ is the quaternionic projective line that is the variety of nondegenerate symplectic planes in the 4-dimensional symplectic space $\mathcal{H}(\mathbf{1}_S^{\oplus 2})$.
- $\mathcal{T}$ is the tautological rank 2 symplectic bundle over HP^1.
- $\pi_1\colon \mathrm{HP}^1 \times_S X \to \mathrm{HP}^1$, $\pi_2\colon \mathrm{HP}^1 \times_S X \to X$ are the projections.

(4) For the trivialized rank 2 symplectic bundle $\mathcal{H}(\mathbf{1}_S) = \mathbf{1}_S^{\mathrm{Sp},2}$ over S one has $b_1(\mathcal{H}(\mathbf{1}_S)) = 0 \in A^{4,2}(S)$.
(5) For a rank $2n$ symplectic bundle $\mathcal{E}$ one has $b_i(\mathcal{E}) = 0$ for $i > n$.
(6) For symplectic bundles $\mathcal{E}, \mathcal{E}'$ over X one has $b_t(\mathcal{E})b_t(\mathcal{E}') = b_t(\mathcal{E} \oplus \mathcal{E}')$, where

$$b_t(\mathcal{E}) = 1 + b_1(\mathcal{E})t + b_2(\mathcal{E})t^2 + \cdots \in A^{*,*}(X)[t].$$

We refer to $b_i(E)$ as *Borel classes* of E and $b_t(E)$ is the *total Borel class.*

It follows from [**PW10a**, Theorem 14.4] that a symplectically oriented spectrum $A \in \mathcal{SH}(S)$ admits a unique theory of Borel classes such that $b_n(\mathcal{E}) = e(\mathcal{E})$ for every rank $2n$ symplectic bundle $\mathcal{E}$ over $X \in \mathcal{S}m_S$. Moreover, Lemma 3.4 yields that an SL-oriented spectrum is symplectically oriented in a canonical way. Thus every SL-oriented spectrum admits a canonical theory of Borel classes such that for every rank $2n$ symplectic bundle $\mathcal{E}$ one has

$$b_n(\mathcal{E}) = e(\Phi(\mathcal{E}))$$

with $\Phi(\mathcal{E})$ being the associated rank $2n$ vector SL-bundle (see Definition 2.2).

DEFINITION 7.7 (cf. [**An15**, Definition 7] and [**HW17**, Definition 5.6]). Let $A \in \mathcal{SH}(S)$ be a commutative ring spectrum with a chosen Borel classes theory (e.g. a symplectically or SL-oriented spectrum). For a rank n vector bundle E over $X \in \mathcal{S}m_S$ put

$$p_i(E) = (-1)^i b_{2i}(\mathcal{H}(E)) \in A^{8i,4i}(X), \quad i = 1, 2, \ldots,$$

for the associated rank $2n$ hyperbolic symplectic bundle $\mathcal{H}(E)$. We refer to $p_i(E)$ as *Pontryagin classes* of E and

$$p_t(E) = 1 + p_1(E)t + p_2(E)t^2 + \cdots \in A^{*,*}(X)[t]$$

is the *total Pontryagin class.*

LEMMA 7.8. *Let $A \in \mathcal{SH}(S)$ be an SL-oriented spectrum and E be a rank $2n+1$ vector bundle over $X \in \mathcal{S}m_S$. Then $\eta \cup b_{2n+1}(\mathcal{H}(E)) = 0$.*

PROOF. Let $\Psi(\mathcal{H}(E)) = (E \oplus E^\vee, \lambda)$ be the rank $4n+2$ vector SL-bundle associated with $\mathcal{H}(E)$ (see Definition 2.2). Then one has

$$b_{2n+1}(\mathcal{H}(E)) = e(E \oplus E^\vee, \lambda).$$

It follows from Theorem 7.4 that $\eta \cup e(E \oplus E^\vee, \lambda) = 0$. □

COROLLARY 7.9. *Suppose that $S = \operatorname{Spec} k$ for a field k of characteristic different from 2. Let $A \in \mathcal{SH}(k)$ be an SL-oriented spectrum. Then the following holds after inverting η in the coefficients, i.e. in $A^{*,*}(X)[\eta^{-1}]$.*

(1) *$b_{2i+1}(\mathcal{H}(E)) = 0, i \geq 0$, for a vector bundle E over $X \in \mathcal{S}m_k$.*
(2) *$p_t(E \oplus E') = p_t(E)p_t(E')$ for vector bundles E, E' over $X \in \mathcal{S}m_k$.*
(3) *$p_i(E) = 0$ for a rank n vector bundle E over $X \in \mathcal{S}m_k$ and $i > \frac{n}{2}$.*

PROOF. We work with η-inverted coefficients, i.e. in $A^{*,*}(X)[\eta^{-1}]$.

(1) It follows from Lemma 7.8 that

$$b_t(\mathcal{H}((\det E)^\vee)) = 1 + b_1(\mathcal{H}((\det E)^\vee))t = 1.$$

Then the multiplicativity property of the total Borel class yields

$$b_t(\mathcal{H}(E \oplus (\det E)^\vee)) = b_t(\mathcal{H}(E))b_t(\mathcal{H}((\det E)^\vee)) = b_t(\mathcal{H}(E)).$$

The equality $b_{2i+1}(\mathcal{H}(E \oplus (\det E)^\vee)) = 0$ follows from [**An15**, Corollary 3] whence the claim.

(2) It follows from the above that

$$p_{-t^2}(E) = b_t(\mathcal{H}(E)),\ p_{-t^2}(E') = b_t(\mathcal{H}(E')),\ p_{-t^2}(E \oplus E') = b_t(\mathcal{H}(E \oplus E')).$$

The claim follows from the multiplicativity property of the total Borel class.

(3) We have $p_t((\det E)^\vee) = 1$. It follows from the above that

$$p_t(E) = p_t(E)p_t((\det E)^\vee) = p_t(E \oplus (\det E)^\vee).$$

[**An15**, Corollary 3] yields $p_i(E \oplus (\det E)^\vee) = 0$ for $i > \frac{n+1}{2}$ whence $p_i(E) = 0$ for $i > \frac{n+1}{2}$. The only remaining case is $n = 2m + 1$, $i = m + 1$. In this case we have

$$p_{m+1}(E) = p_{m+1}(E \oplus (\det E)^\vee) = e(E \oplus (\det E)^\vee, \lambda)^2$$

where $\lambda\colon \det(E \oplus (\det E)^\vee) \xrightarrow{\simeq} \mathbf{1}_X$ is the canonical isomorphism and the last equality is given by [**An15**, Corollary 3]. Theorem 7.4 yields $e(E \oplus (\det E)^\vee, \lambda) = 0$ whence the claim. □

REMARK 7.10. The assumption $S = \operatorname{Spec} k$ with k being a field of characteristic different from 2 arises from the same assumption of [**An15**]. It seems that this assumption is redundant and all the results of the loc. cit. as well as the above corollary should hold over a general quasi-compact quasi-separated scheme. Moreover, almost all the reasoning given in [**An15**] works over a general base.

References

[An15] Alexey Ananyevskiy, *The special linear version of the projective bundle theorem*, Compos. Math. **151** (2015), no. 3, 461–501, DOI 10.1112/S0010437X14007702. MR3320569

[An16a] Alexey Ananyevskiy, *On the relation of special linear algebraic cobordism to Witt groups*, Homology Homotopy Appl. **18** (2016), no. 1, 204–230, DOI 10.4310/HHA.2016.v18.n1.a11. MR3491850

[An16b] Alexey Ananyevskiy, *On the push-forwards for motivic cohomology theories with invertible stable Hopf element*, Manuscripta Math. **150** (2016), no. 1-2, 21–44, DOI 10.1007/s00229-015-0799-6. MR3483167

[An17] Alexey Ananyevskiy, *Stable operations and cooperations in derived Witt theory with rational coefficients*, Ann. K-Theory **2** (2017), no. 4, 517–560, DOI 10.2140/akt.2017.2.517. MR3681106

[DF17] F. Déglise and J. Fasel, *The Milnor-Witt motivic ring spectrum and its associated theories,* arXiv:1708.06102

[HW17] J. Hornbostel and M. Wendt, *Chow-Witt rings of classifying spaces for symplectic and special linear groups,* arXiv:1703.05362v3

[Lev17] M. Levine, *Toward an enumerative geometry with quadratic forms,* arXiv:1703.03049v3

[LM07] M. Levine and F. Morel, *Algebraic cobordism*, Springer Monographs in Mathematics, Springer, Berlin, 2007. MR2286826

[MH73] John Milnor and Dale Husemoller, *Symmetric bilinear forms*, Springer-Verlag, New York-Heidelberg, 1973. Ergebnisse der Mathematik und ihrer Grenzgebiete, Band 73. MR0506372

[Mor04] Fabien Morel, *An introduction to $\mathbb{A}^1$-homotopy theory*, Contemporary developments in algebraic K-theory, ICTP Lect. Notes, XV, Abdus Salam Int. Cent. Theoret. Phys., Trieste, 2004, pp. 357–441. MR2175638

[Mor12] Fabien Morel, *$\mathbb{A}^1$-algebraic topology over a field*, Lecture Notes in Mathematics, vol. 2052, Springer, Heidelberg, 2012. MR2934577

[MV99] Fabien Morel and Vladimir Voevodsky, *$\mathbf{A}^1$-homotopy theory of schemes*, Inst. Hautes Études Sci. Publ. Math. **90** (1999), 45–143 (2001). MR1813224

[Pan03] I. Panin, *Oriented cohomology theories of algebraic varieties*, K-Theory **30** (2003), no. 3, 265–314, DOI 10.1023/B:KTHE.0000019788.33790.cb. Special issue in honor of Hyman Bass on his seventieth birthday. Part III. MR2064242

[Pan04] I. Panin, *Riemann-Roch theorems for oriented cohomology*, Axiomatic, enriched and motivic homotopy theory, NATO Sci. Ser. II Math. Phys. Chem., vol. 131, Kluwer Acad. Publ., Dordrecht, 2004, pp. 261–333, DOI 10.1007/978-94-007-0948-5_8. MR2061857

[PW10a] I. Panin and C. Walter, *Quaternionic Grassmannians and Borel classes in algebraic geometry,* arXiv:1011.0649

[PW10b] I. Panin and C. Walter, *On the algebraic cobordism spectra MSL and MSp*, arXiv:1011.0651

[PW10c] I. Panin and C. Walter, *On the motivic commutative ring spectrum* **BO**, Algebra i Analiz **30** (2018), no. 6, 43–96. MR3882540

[Sm07a] A. L. Smirnov, *Orientations and transfers in the cohomology of algebraic varieties* (Russian, with Russian summary), Algebra i Analiz **18** (2006), no. 2, 167–224, DOI 10.1090/S1061-0022-07-00952-1; English transl., St. Petersburg Math. J. **18** (2007), no. 2, 305–346. MR2244939

[Sm07b] A. L. Smirnov, *The Riemann-Roch theorem for operations in the cohomology of algebraic varieties* (Russian, with Russian summary), Algebra i Analiz **18** (2006), no. 5, 210–236, DOI 10.1090/S1061-0022-07-00976-4; English transl., St. Petersburg Math. J. **18** (2007), no. 5, 837–856. MR2301046

[Voe98] Vladimir Voevodsky, $\mathbf{A}^1$*-homotopy theory*, Proceedings of the International Congress of Mathematicians, Vol. I (Berlin, 1998), Doc. Math. **Extra Vol. I** (1998), 579–604. MR1648048

St. Petersburg Department, Steklov Math. Institute, Fontanka 27, St. Petersburg 191023 Russia, and Chebyshev Laboratory, St. Petersburg State University, 14th Line V.O., 29B, St. Petersburg 199178 Russia

Email address: alseang@gmail.com

Contemporary Mathematics
Volume **745**, 2020
https://doi.org/10.1090/conm/745/15021

The homotopy Leray spectral sequence

Aravind Asok, Frédéric Déglise, and Jan Nagel

ABSTRACT. In this work, we build a spectral sequence in motivic homotopy that is analogous to both the Serre spectral sequence in algebraic topology and the Leray spectral sequence in algebraic geometry. Here, we focus on laying the foundations necessary to build the spectral sequence and give a convenient description of its E_2-page. Our description of the E_2-page is in terms of homology of the local system of fibers, which is given using a theory similar to Rost's cycle modules. We close by providing some sample applications of the spectral sequence and some hints at future work.

CONTENTS

1. Introduction

The goal of this paper is to study an algebro-geometric version of the Leray–Serre spectral sequence for generalized cohomology theories. To explain the setup, suppose one is given a topological space X, a sheaf $\mathcal{F}$ of abelian groups on X and a filtration

$$X_\bullet = X_0 \subset \ldots \subset X_p \subset X_{p+1} \subset \ldots \subset X_n = X$$

by closed subsets. One naturally associates an exact couple with the preceding data and obtains a spectral sequence of the form:

$$E_1^{p,q}(X_\bullet, \mathcal{F}) \quad = \quad H^{p+q}(X_p \setminus X_{p-1}, \mathcal{F}) \Rightarrow H^{p+q}(X, \mathcal{F}).$$

If $X_\bullet$ is cellular with respect to $\mathcal{F}$, i.e., if $H^i(X_p \setminus X_{p-1}, \mathcal{F}) = 0$ for all $i \neq p$, then there are isomorphisms of the form $H^k(X, \mathcal{F}) \cong E_2^{k,0}(X_\bullet, \mathcal{F})$. If $f : X \to B$ is a continuous map of topological spaces and $B_\bullet$ is a filtration of B that is cellular

2010 *Mathematics Subject Classification.* 14F42, 55F20.

with respect to the direct image sheaves $R^q f_* \mathcal{F}$ for all q, then the E_2–term of Leray spectral sequence

$$E_2^{p,q} = H^p(B, R^q f_* \mathcal{F}) \Rightarrow H^{p+q}(X, \mathcal{F})$$

is isomorphic to $E_2^{p,q}(X_\bullet, \mathcal{F})$, where $X_\bullet$ is the inverse image of $B_\bullet$.

On the other hand, suppose

$$F \longrightarrow X \xrightarrow{f} B$$

is a Serre fibration of topological spaces, where B has the homotopy type of a (connected) finite CW complex, and E is a (generalized) cohomology theory in the sense of classical stable homotopy theory. One may consider an associated Atiyah–Hirzebruch spectral sequence (see, e.g., [**DK01**, §9.2-9.5]): the E_2-page of this spectral sequence is given in terms of the ordinary (co)homology of B with coefficients in local systems attached to the E-(co)homology of F and converges to the E-(co)homology of X. When E is ordinary cohomology and $\mathscr{F}$ is the constant sheaf $\mathbf{Z}_X$, the two spectral sequences coincide. Indeed, Serre showed that the direct image sheaves $R^q f_* \mathbf{Z}_X$ are local systems on B depending only on the cohomology of the fiber F and the action of the fundamental group of B on these cohomology groups.

To explain our algebro–geometric analog, recall that Arapura showed the Leray spectral sequence is "motivic" [**Ara05**]. In more detail, suppose $f : X \to B$ is a projective morphism of complex quasi–projective algebraic varieties. After reducing to a situation where B is affine (via the Jouanolou trick), work of Beilinson and Nori [**Nor02**] shows that any constructible sheaf on B can be made cellular with respect to a filtration by closed algebraic subsets. In this situation, Arapura compares the Leray spectral sequence

$$E_2^{a,b}(f) = H^a(B, R^q f_* \mathbf{Z}_X) \Rightarrow H^{a+b}(X, \mathbf{Z}_X)$$

and the spectral sequence associated with the skeletal filtration and uses this to show that the Leray spectral sequence essentially reflects suitably functorial algebro-geometric structure present on cohomology groups (e.g., it lifts to Nori's category of mixed motives). A similar result holds for the perverse Leray spectral sequence

$$E_2^{a,b} = H^a(B, {}^pR^b f_* \mathbf{Z}_X) \Rightarrow H^{a+b}(X, \mathbf{Z}_X)$$

that is obtained by replacing the classical truncation functor τ by the perverse truncation ${}^p\tau$ [**dCM10**].

In this paper, we consider a variant of this setup. Let us now consider a simple case to highlight our considerations. Suppose k is a field, B is a finite dimensional irreducible smooth k-variety, and $f : X \to B$ is a smooth morphism of k-varieties. As usual, f may be thought of as an étale locally trivial fibration in algebraic geometry. We write $B^{(n)}$ for the set of codimension n points of B. If B has dimension d, we view the collection $B^{(i)}$ as i ranges from 0 through d as an algebro-geometric analog of the skeletal filtration of B. In this situation, the scheme-theoretic fiber X_b for $b \in B^{(i)}$ is then a smooth variety over b.

There is a comparison theorem between the Leray spectral sequence and the spectral sequence associated to the Bloch-Ogus complex. If $H^*(X, A)$ is a suitable cohomology theory defined using a Grothendieck topology that is finer than the Zariski topology (e.g., étale cohomology, or Betti cohomology over $\mathbf{C}$), one can consider the Leray spectral sequence associated with the morphism of sites $\pi\colon X_{\text{fine}} \to X_{Zar}$. The higher direct image sheaves $R^q\pi_*A$ are the Zariski sheaves $\mathcal{H}^q$ associated with the presheaves $U \mapsto H^q(\pi^{-1}(U), A)$. Bloch and Ogus show that the Leray spectral sequence $E_2^{p,q} = H^p(X, \mathcal{H}^q) \Rightarrow H^{p+q}(X, A)$ can be identified from the E_2–term onward with the spectral sequence

$$E_1^{p,q} = \oplus_{x \in X^{(p)}} H^{q-p}(k(x)) \Rightarrow N^\bullet H^{p+q}(X)$$

that gives the coniveau filtration on cohomology. (The argument uses Deligne's technique of "décalage", see [**Par96**].) Furthermore, there are variants of the previous constructions where one replaces cohomology by Borel–Moore homology and works with the niveau (rather than coniveau) filtration. More generally, one can work with the bivariant theory

$$H^i(X \xrightarrow{f} Y) = \mathrm{Hom}(Rf_!\mathbf{Q}_X, \mathbf{Q}_Y[i])$$

that simultaneously generalizes both Borel–Moore homology and cohomology.

Building on this analysis, by analyzing formal properties of Gersten complexes, Rost developed a notion of "local coefficient system" on a k-scheme B, which he called a cycle module. Given a morphism $f : X \to B$ and a cycle module M on X, Rost defined Chow groups with coefficients in M, $A_*(X, M)$. For example, there is a cycle module built out of Milnor K-theory for which the associated Chow groups with coefficients coincide with usual Chow groups. Finally, Rost constructed a spectral sequence of the form:

$$E^2_{p,q} = A_p(B, A_q(X, M)) \Rightarrow A_{p+q}(X, M);$$

here $A_q(X, M)$ is a cycle module on B obtained by taking homology of the fibers.

Our approach in motivic homotopy theory essentially mixes all the ideas above and combines the classical Leray spectral sequence of a fibration with Rost's spectral sequence. The jumping off point is the work of the second author exploring the close relationship between the notion of cycle module in the sense of Rost and the heart of a t-structure on the motivic stable homotopy category. We use a relative version of Morel's homotopy t-structure, the so-called *perverse homotopy t-structure*, defined initially by Ayoub [**Ayo07**] and developed further in [**BD17**]. The approach to constructing this t-structure in [**BD17**] is rather flexible and works in great generality: the basic definition requires only the existence of a dimension function δ on the base scheme; for this reason, the t-structure was called the *δ-homotopy t-structure* in *op. cit.*.

The heart of the δ-homotopy t-structure consists of δ-homotopy modules, or simply *homotopy modules.* These objects are closely related to cycle modules and their homology (with respect to the truncation functors in the δ-homotopy t-structure) can be computed by a Gersten-type complex that is formally extremely similar to that written down by Rost (see Definition 3.2.4 for more details). The approach we take via t-structures has technical advantages: functoriality properties and multiplicative structure are easy to obtain (see Proposition 3.2.7 and Proposition 3.3.2 for more details). With these tools in hand, we state a version of our main result here (the notation and terminology in the statement that we have not yet mentioned may be found in the notations and conventions section below).

THEOREM 1 (see Theorem 4.1.2). *Fix a base scheme S with a dimension function δ, and suppose $f : X \to B$ is a separated morphism of S-schemes having finite* *type. Let $\mathbb{E}$ be a motivic spectrum (or a mixed motive) over S, and put: $\mathbb{E}^!_X := f^!\mathbb{E}$. Then there exists a convergent spectral sequence of the form:*

$$E^2_{p,q}(f,\mathbb{E}) = A^\delta_p(B, H^\delta_q(f_*\mathbb{E}^!_X)) \Rightarrow \mathbb{E}_{p+q}(X/S),$$

abutting to the $\mathbb{E}$-homology of X relative to S, or rather, the bivariant theory of X/S with coefficients in $\mathbb{E}$ (see below our Notations and conventions). In fact, $E^2_{p,q}$

i) *vanishes outside a range of columns bounded by the maximum and minimum of the dimension function δ on B; and*
ii) *is described as the homology of a Gersten type complex with coefficients in the homotopy module $H^\delta_q(f_*\mathbb{E}^!_X)$).*

The spectral sequence above is obtained by filtering $f_*(\mathbb{E}^!_X)$ by truncations with respect to the δ-homotopy t-structure. It is closely related to the spectral sequence one would obtain by filtering $\mathbb{E}^*(-/S)$ via the pullback of the (δ-)niveau filtration on B along f (for the dimension function δ). The following result gives a precise formulation.

PROPOSITION 2 (See Proposition 4.1.9). *Let ${}^\delta F^f_p\mathbb{E}_n(X/S)$, $p \in \mathbf{Z}$, be the abutting filtration of the homotopy Leray spectral sequence. Define the δ-niveau filtration on $\mathbb{E}_*(X/S)$ relative to f by the following formula:*

$${}^\delta N^f_p\mathbb{E}_*(X/S) = \bigcup_{i:Z\to X,\delta(Z)\leq p} \mathrm{Im}\big(i_* : \mathbb{E}_*(X \times_B Z/S) \to \mathbb{E}_*(X/S)\big).$$

For any pair of integer $(p,n) \in \mathbf{Z}^2$, the following relation holds:

$${}^\delta F^f_p\mathbb{E}_n(X/S) = {}^\delta N^f_{n-p}\mathbb{E}_n(X/S).$$

This proposition may be viewed as an analogue of Washnitzer's conjecture (that the coniveau filtration on de Rham cohomology coincides with the filtration arising from the second hypercohomology spectral sequence), which was established by Bloch and Ogus [**BO74**].

There is also a cohomological form of the homotopy Leray spectral sequence, which is better suited to analysis of product structures.

THEOREM 3 (See Theorem 4.2.5). *Fix a base scheme S with a dimension function δ, and suppose $f : X \to B$ is a separated morphism of S-schemes having finite type. If $\mathbb{E}$ is a motivic ring spectrum over X, then there is a convergent spectral sequence of the form*

$$E_2^{p,q}(f,\mathbb{E}) = A^p_\delta(B, H^q_\delta(f_*\mathbb{E}_X)) \Rightarrow H^{p+q}(X,\mathbb{E}),$$

where we write $H^*(X,\mathbb{E})$ for the groups usually denoted $\mathbb{E}^*(X)$ to emphasize the link with Leray spectral sequence. As before the E_2-page is a) concentrated in a range of columns bounded by the maximum and minimum of the dimension function δ on B, and b) described in terms of (co)homology of the Gersten-style complex associated with homotopy module $H^q_\delta(f_*\mathbb{E}_X)$. Moreover,

i) *each page E_r has the structure of a differential graded algebra, and*
ii) *the induced product structure on the abutment coincides with the $\mathbb{E}$-cohomology product structure.*

While the reader may consult the beginning of each section for further discussion of its contents, we close the introduction with a brief overview of the paper.

Section 2 is devoted to reviewing the basic properties of this homotopy t–structure. Section 3 then studies homotopy modules, Gersten complexes and the associated niveau spectral sequences. In particular, we describe multiplicative structure on the niveau spectral sequences when considering cohomology theories equipped with suitable multiplicative structure. Section 4 is the theoretical heart of the paper. We build the homotopy Leray spectral sequence (see, e.g., Theorem 4.1.2) and study an analog of "locally constant sheaves" in our context. Finally, Section 5 contains some sample applications of these spectral sequences. In particular, we analyze fibrations with $\mathbf{A}^1$-contractible fibers, construct Gysin and Wang sequences and, in the motivic case, prove a degeneration result for the homotopy Leray spectral sequence for relative cellular spaces.

Acknowledgements. The authors want to thank Adrien Dubouloz for fruitful discussions about fibrations, and, especially, Fabien Morel for providing the initial impetus to pursue this work. Aravind Asok was partially supported by National Science Foundation Awards DMS-1254892 and DMS-1802060. F. Déglise and J. Nagel received support from the French "Investissements d'Avenir" program, project ISITE-BFC (contract ANR-lS-IDEX-OOOB).

Notations and conventions

Geometry. All schemes in this paper will be noetherian, finite dimensional and assumed to come equipped with a dimension function, usually denoted δ. While we fix such a dimension function throughout, and while it may even appear explicitly in various notions we use, we emphasize that the choice is inessential in the sense that the most of the relevant notions do not depend on the fixed choice of dimension function up to a suitable notion of canonical equivalence; see Remark 2.1.6 for a more precise statement.[1] Anyway the dimension function is used to fix conventions regarding degrees of (co)homology as in 2.1.10.

Fix a base scheme S. An S-scheme X will be said to have essentially finite type or, equivalently, $X \to S$ has essentially finite type, if X can be written as a (co)filtered limit a S-schemes of finite type with affine étale transition morphisms. By a point of a scheme X, we will mean a map $x : \mathrm{Spec}(K) \to X$ where i) K is a field and ii) if k is the residue field of the image of x in X, then the extension K/k is finitely generated; equivalently, the morphism $x : \mathrm{Spec}(K) \to X$ has essentially finite type. We will simply write such a point by $x \in X(K)$.

The dimension function δ on S, is extended to a dimension function on X as follows:

$$\begin{aligned} \forall x/s \in X/S, \delta(x) &:= \delta(s) + \mathrm{trd}(\kappa(x)/\kappa(s)). \\ \delta(X) = \delta_+(X) &:= \max_{x\in X}(\delta(x)). \\ \delta_-(X) &:= \min_{x\in X}(\delta(x)). \end{aligned}$$

[1]There is an exception to this rule. this is when one considers the underlying monoidal structure and its interaction with the δ-homotopy t-structure. It will appear once in our applications, when we consider products on the homotopy Leray spectral sequence. We refer the reader to Sections 3.3 and 4.2.

Two examples to keep in mind include:

(S1) if S is the spectrum of a field k, $\delta = 0$, then $\delta(x) = \mathrm{trd}(\kappa(x)/k)$;

(S2) if S is an excellent regular scheme of dimension less or equal than 3 and δ is the Krull dimension.

In both cases, if X/S has finite type, then the integer $\delta(X)$ coincides with the Krull dimension of X.

Finally, the following formulas will be used in the paper:

(D1) if S is regular connected, $\delta = d - \mathrm{codim}_S$ where $d = \delta(S)$;

(D2) if the morphism $f : X \to S$ has essentially finite type, and is lci with cotangent complex L_f, for any point $x \in X$, $\delta(x) = \mathrm{rk}\big(L_{f,x}\big) + \delta(s)$. Moreover, if X and S are irreducible, and $d = \dim(f)$ is the rank of L_f, one has:

$$\delta(X) = d + \delta(S).$$

Motivic stable homotopy. We fix a motivic triangulated category $\mathscr{T}$ in the sense of [**CD09**, Def. 2.4.45], equipped with a motivic adjunction:

$$\mathrm{SH} \leftrightarrows \mathscr{T}.$$

In brief, $\mathscr{T}$ consists of the following data: for any scheme S, a triangulated closed symmetric monoidal category $\mathscr{T}(S)$; for any morphism of schemes f and any separated morphism of finite type p, pairs of adjoint functors (f^*, f_*), $(p_!, p^!)$ satisfying the so-called Grothendieck six functor formalism (see [**CD09**, Th. 2.4.50] for a precise statement). Following the terminology from stable homotopy theory objects of $\mathscr{T}(S)$ will be called *$\mathscr{T}$-spectra over S*.

We write $\mathbb{1}_S$ for the monoidal unit in $\mathscr{T}(S)$, and $\mathbb{1}_S(1)$ for the Tate twist. In the sequel, various combinations of Tate twists and shifts naturally arise, and we introduce a separate notation for these twists:

$$\mathbb{1}_S\{1\} := \mathbb{1}_S(1)[1], \ \mathbb{1}_S\langle 1\rangle := \mathbb{1}_S(1)[2].$$

Given an object $\mathbb{E}$ in $\mathscr{T}(S)$, and a morphism $f : X \to S$ (resp. a separated morphism having finite type) we may define $\mathbb{E}$-cohomology and bivariant $\mathbb{E}$-theory by means of the formulas:

- (*Cohomology*) $\mathbb{E}^{n,i}(X) = \mathrm{Hom}_{\mathscr{T}(X)}(\mathbb{1}_X, f^*\mathbb{E}(i)[n])$.
- (*Bivariant theory* $\mathbb{E}_{n,i}(X/S) = \mathrm{Hom}_{\mathscr{T}(X)}\big(\mathbb{1}_X(i)[n], f^!\mathbb{E}\big)$.

This notation is a standard in motivic homotopy theory. When dealing with spectral sequences, and homotopy modules, it will both be useful and meaningful to use the following notations:

- $H^n(X, \mathbb{E}) = \mathbb{E}^{n,0}(X)$.
- $H_n(X/S, \mathbb{E}) = \mathbb{E}_{n,0}(X/S)$.

We will use the later notations exclusively in Sections 4 and 5.

The definition of the bivariant theory attached to $\mathbb{E}$ will be extended to the situation where X has essentially finite type over S (see Paragraph 2.1.1). The indexing for cohomology used above may also be expressed in terms of the other conventions for twists mentioned above, with notation changed accordingly: one has equalities of the form

$$\mathbb{E}^{n,i}(X) = \mathbb{E}^{n-i,\{i\}}(X) = \mathbb{E}^{n-2i,\langle i\rangle}(X);$$

similar notation will be used for the associated bivariant theory. When $\mathbb{E} = \mathbb{1}_S$, the various cohomology groups will be referred to as $\mathscr{T}$-cohomology and bivariant

$\mathscr{T}$-theory respectively, and we write $H^{n,i}(X,\mathscr{T})$ (resp. $H^{BM}_{n,i}(X/S,\mathscr{T})$) for these groups.

Write $\underline{\mathrm{K}}(S)$ for the category of virtual vector bundles over S, associated with the category of vector bundles over S with morphisms the isomorphisms of vector bundles. The Thom space construction may be viewed as a functor

$$\mathrm{Th}_S : \underline{\mathrm{K}}(S) \to \mathscr{T}(S)$$

that sends sums to tensor products. Following [**DJK18**], we may twist cohomology and bivariant theories by pairs $(n,v) \in \mathbf{Z} \times \underline{\mathrm{K}}(X)$; we use the following notation for these twists:

$$\mathbb{E}^n(X,v) = \mathrm{Hom}_{\mathscr{T}(X)}(\mathbb{1}_X, f^*\mathbb{E} \otimes Th_X(v)[n])$$
$$\mathbb{E}_n(X/S,v) = \mathrm{Hom}_{\mathscr{T}(X)}\big(\mathrm{Th}_X(v)[n], f^!\mathbb{E}\big).$$

Given an integer $i \in \mathbf{Z}$, we denote by $\langle i \rangle$ the unique free virtual vector bundle of rank i – the underlying scheme is implicit – so that this notation is compatible with our conventions on twists. If K is a perfect complex over X, we denote by $\langle K \rangle$ the associated virtual bundle (see [**DJK18**, 2.1.5]).

In order to be able to apply the construction of the δ-homotopy t-structure of [**BD17**], we will require that $\mathscr{T}$ satisfies the following assumptions:

(T1) $\mathscr{T}$ is generated by Tate twists of smooth schemes: more precisely, for any scheme S, $\mathscr{T}(S)$ is generated as a triangulated category by objects of the form $M_S(X)(i)$ with X/S smooth and $i \in \mathbf{Z}$.[2]

(T2) $\mathscr{T}$ is continuous with respect to Tate twists: see [**CD09**, 4.3.2].[3]

(T3) $\mathscr{T}$ is *homotopically compatible*: see [**BD17**, 3.2.12].[4]

(T4) A suitable form of resolution of singularities holds: property (Resol) of [**BD17**, 2.4.1].[5]

Examples. The abstract setting being given, we now give our list of concrete frameworks, that will be used in all of our examples. We first fix some absolute base S_0 and restrict our schemes to S_0-schemes essentially of finite type.

(1) **Motivic case**. Let R be a ring of coefficients, and assume one of the following situation:
- S_0 is the spectrum of a field k whose characteristic exponent is invertible in R and $\mathscr{T} = \mathrm{DM}(-,R)$ is Voevodsky's cdh-local category of triangulated mixed motives, introduced in [**CD15**].

---

[2] All objects in $\mathscr{T}(S)$ are obtained by taking extensions of arbitrary coproducts. Equivalently, an object K of $\mathscr{T}(S)$ is zero if and only if:

$$\forall X/S \text{ smooth}, (n,i) \in \mathbf{Z}^2, \mathrm{Hom}_{\mathscr{T}(S)}\big(M_S(X)(i)[n], K\big) = 0.$$

[3] Recall that this expresses the compatibility of $\mathscr{T}$ with projective limits.

[4] This condition is automatically verified (see [**BD17**, 3.2.13]) if $\mathscr{T}$ satisfies absolute purity and the following vanishing statement holds:

$$\forall \text{fields } E, \forall n > m, H^{n,m}(\mathrm{Spec}(E),\mathscr{T}) = 0.$$

[5] Under two geometric assumptions, this means that:
- In case (S1), we will require that, for p the characteristic exponent of k, $\mathscr{T}$ is $\mathbf{Z}[1/p]$linear.
- In case (S2), we require that $\mathscr{T}$ is $\mathbf{Q}$-linear.

- S_0 is a Dedekind ring (or more generally an excellent scheme of dimension greater or equal than 3), R is a $\mathbf{Q}$-algebra and $\mathscr{T} = \mathrm{DM}(-, R)$ is one the many models of the triangulated category of R-motives introduced in [**CD09**].

(2) **Homotopical case**. The absolute base S_0 is the spectrum of a field with characteristic exponent p, and $\mathscr{T} = \mathrm{SH}[p^{-1}]$ is the $\mathbf{Z}[p^{-1}]$-linearization of the Morel-Voevodsky stable homotopy category of $\mathbb{P}^1$-spectra. Particular motivic ring spectra of interest to us will include:
   - $S^0_S = \mathbb{1}_S$ the sphere spectrum.
   - $\mathbf{H}R_S$ the motivic Eilenberg-MacLane ring spectrum with coefficients in R.[6]
   - $\mathbf{H}\tilde{R}_S$ the Milnor–Witt motivic ring spectrum with coefficients in R. See [**DF17**].

To fix ideas, the reader may take the dimension function δ_0 on S_0 satisfying the conditions (S1) or (S2) fixed above, and for any S_0-scheme essentially finite type, take the induced dimension function (as described above).

2. Homotopy t-structure and duality

The goal of this section is to review a variant of the homotopy t-structure analyzed in [**BD17**] that makes sense over rather general base schemes. In more detail, Section 2.1 reviews the necessary theory from [**BD17**] and fixes additional notations and conventions to be used in the sequel. We highlight here Theorem 2.1.4 which summarizes the required existence result and Point 2.1.8 which describes *homological* conventions for t-structures that will be in force throughout the paper. On the other hand, Section 2.2 is somewhat orthogonal to the above considerations and is concerned with recalling some facts about purity and duality that will be used in the remainder of the paper.

2.1. Recollections on the homotopy t-structure.

2.1.1. For any $\mathscr{T}$-spectrum $\mathbb{E}$ over a scheme S, bivariant $\mathbb{E}$-theory extends canonically from the category of separated S-schemes of finite type to that of separated S-schemes essentially of finite type. Indeed, suppose X is a separated S-scheme essentially of finite type. By assumption, there exists a pro-scheme $(X_\lambda)_\lambda$ with affine and étale transition morphisms, such that each X_λ is separated and has finite type over S, and with limit X. We set:

$$\mathbb{E}_{n,i}(X/S) := \varinjlim_{\lambda} \mathbb{E}_{n,i}(X_\lambda/S).$$

This definition is independent of the pro-scheme presenting X as a limit [**GD67**, §8.2]. The resulting definition presents a canonical extension of the original functor by the continuity assumption on $\mathscr{T}$ (i.e., property (T2) from our conventions). Moreover, when X has finite type over S the new definition agrees with the old definition by using again the continuity assumption on $\mathscr{T}$.

[6] Recall this ring spectrum is obtained as the image of the constant motive under the canonical map: $K \circ \gamma_* : \mathrm{DM}(S, \mathbf{Z}[p^{-1}]) \to SH(S)[p^{-1}]$, by forgetting the transfer and then taking the Nisnevich Eilenberg-MacLane functor. See for example [**CD09**].

PROPOSITION 2.1.2. *If $\mathbb{E}$ is a $\mathscr{T}$-spectrum over S, then the following conditions are equivalent.*

(i) *For any separated scheme X/S of finite type, one has:*

$$\begin{aligned}\mathbb{E}_{n,i}(X/S) = 0 \text{ when } n-i < \delta_-(X),\\ \text{respectively } n-i > \delta_+(X).\end{aligned}$$

(ii) *For any point $x \in S(K)$ (see our Notations and conventions), one has:*

$$\begin{aligned}\mathbb{E}_{n,i}(x) = 0 \text{ when } n-i < \delta(x),\\ \text{respectively } n-i > \delta(x).\end{aligned}$$

PROOF. The equivalence of (i) and (ii) follows by combining Theorem 3.3.1 and Corollary 3.3.5 of [**BD17**]. Alternatively, it is a straightforward consequence of the existence and convergence of the δ-niveau spectral sequence ([**BD17**, Def. 3.1.5] or Paragraph 3.2.2 in this paper). □

REMARK 2.1.3.

(1) In [**BD17**], the extension $\mathbb{E}_{**}$ to separated schemes essentially of finite type was denoted by $\hat{\mathbb{E}}_{**}$. Since, according to paragraph 2.1.1, this extension is unique and well-defined – using in particular the assumption (T2) for coherence, we will not follow this notational convention here (and we caution the reader that the decoration $\hat{\mathbb{E}}$ is used with a different meaning in this paper.)
(2) If we use the δ-niveau spectral sequence, then the proof of the previous proposition does not use the assumptions (T1), (T3) and (T4). Therefore, the preceding proposition is true without assuming these conditions.

We can now state the main theorem of [**BD17**] (see *loc. cit.* Th. 3.3.1 and Cor. 3.3.5).

THEOREM 2.1.4. *Given any scheme S, there exists a t-structure on $\mathscr{T}(S)$ whose homologically non-negative (resp. non-positive) objects are the $\mathscr{T}$-spectra $\mathbb{E}$ over S satisfying the equivalent conditions (i) and (ii) of the above proposition. This t-structure is, moreover, non-degenerate and satisfies gluing in the sense of* [**BBD82**, 1.4.10] *(see also Remark 2.1.12).*

DEFINITION 2.1.5. Given any scheme S, the t-structure on $\mathscr{T}(S)$ of the above theorem will be called the *δ-homotopy t-structure*. Objects of the heart of this t-structure, denoted by $\mathscr{T}(S)^\heartsuit$, will be called *$\delta$-homotopy modules*, or simply *homotopy modules* when this does not lead to confusion.

REMARK 2.1.6. The t-category $\mathscr{T}(S)$ is "independent" of the choice of δ in a sense we now explain. Given another choice δ', we know that the function $\delta' - \delta$ is constant on connected components of S. As $\mathscr{T}(S \sqcup S') = \mathscr{T}(S) \oplus \mathscr{T}(S')$, we may obviously assume S is connected so that $\delta' = \delta + n$. It follows from the above definition that

$$\phi_{\delta,\delta'} : (\mathscr{T}(S), t_\delta) \to (\mathscr{T}(S), t_{\delta'}), \mathbb{E} \mapsto \mathbb{E}[n]$$

is an equivalence of t-categories. It may be useful to remember the formula:

$$\tau^{\delta+n}_{\geq p} = \tau^{\delta}_{\geq p+n} \tag{2.1.6.a}$$

(see the conventions of Par. 2.1.8 for these truncation functors).

EXAMPLE 2.1.7.

(1) Motivic case: Let k be a perfect field and assume δ is the obvious dimension function on k. In the motivic case, the δ-homotopy t-structure on $\mathrm{DM}(k,R)$ coincides with the stable version of Voevodsky's homotopy t-structure introduced in [**Dég11**, Sec. 5.2]. See [**BD17**, Ex. 2.3.5] for details. In particular, a δ-homotopy module $\mathbb{E}$ over k is just a $\mathbf{Z}$-graded homotopy invariant sheaf with transfers equipped with an isomorphism:
$$(\mathbb{E}_{n+1})_{-1} \simeq \mathbb{E}_n.$$
We refer the reader to [**Dég11**, 1.17] for more details.

(2) Homotopical case: In the homotopical case, given a field k with the obvious dimension function, the δ-homotopy t-structure on $\mathrm{SH}(k)$ coincides with Morel's homotopy t-structure (see [**Mor03**, Sec. 5.2]). We refer the reader again to [**BD17**, Ex. 2.3.5] for more details. In this case, a δ-homotopy module $\mathbb{E}$ over k is a $\mathbf{Z}$-graded strictly $\mathbf{A}^1$-invariant Nisnevich sheaf over smooth k-schemes with a given isomorphism:
$$(\mathbb{E}_{n+1})_{-1} \simeq \mathbb{E}_n.$$
In other word, this is a homotopy module in the sense of Morel. Recall also that a spectrum $\mathbb{E}$ is called *orientable* if it admits the structure of a module over the ring spectrum **MGL**. Orientability for homotopy modules turns out to be equivalent to requiring that the Hopf map η (an element in the graded endomorphisms of the motivic sphere spectrum) acts trivially. In fact, orientability is also equivalent to requiring that $\mathbb{E}$ admits transfers, in which case these transfers are unique (see [**Dég13**, 4.1.5, 4.1.7] for further details).

(3) Over a general base S, in both the homotopical and motivic cases, the δ-homotopy t-structure can be compared with the *perverse homotopy t-structure* defined by Ayoub in [**Ayo07**, §2.2.4]. This comparison requires an appropriate choice of δ, and we refer the reader to [**BD17**, 2.3.11] for details.

2.1.8. *Homological conventions.*– For the most part, we adopt homological conventions, as they are better suited to issues that arise involving singularities. We will write $\mathbb{E} \geq_\delta n$ (resp. $\mathbb{E} \leq_\delta n$) to say that $\mathbb{E}$ is concentrated in homological degree above $n-1$ (resp. below $n+1$) and denote by $\tau^\delta_{\geq n}$ (resp. $\tau^\delta_{\geq n}$) the corresponding homological truncation functor. Using homological conventions, the truncation triangles read:
$$\tau^\delta_{\geq 0}(\mathbb{E}) \to \mathbb{E} \to \tau^\delta_{<0}(\mathbb{E}) \xrightarrow{+1}$$
and $\mathrm{Hom}(\mathbb{E},\mathbb{F}) = 0$ if $\mathbb{E} \geq_\delta 0$ and $\mathbb{F} <_\delta 0$.

We denote by H^δ_n the n-th homology functor with respect to the t_δ-homotopy t-structure. Finally, we summarize conventions with respect to suspensions:
$$H^\delta_n(\mathbb{E}[i]) = H^\delta_{n-i}(\mathbb{E}),\ \tau^\delta_{\geq n}(\mathbb{E}[i]) = \tau^\delta_{\geq n-i}(\mathbb{E})[i],\ \tau^\delta_{\leq n}(\mathbb{E}[i]) = \tau^\delta_{\leq n-i}(\mathbb{E})[i].$$

REMARK 2.1.9. We may pass from homological to cohomological indexing by writing indices as superscripts and reversing signs. In formulas:
$$H^n_\delta(\mathbb{E}) = H^\delta_{-n}(\mathbb{E}),\ \tau^{\leq n}_\delta = \tau^\delta_{\geq -n},\ \tau^{>n}_\delta = \tau^\delta_{<-n}.$$

2.1.10. For the record, we now give a number of equivalent characterizations of positivity or negativity with respect to the δ-homotopy t-structure – this is simply a consequence of Proposition 2.1.2, using the above conventions and the construction in Paragraph 2.1.1.

Given a $\mathscr{T}$-spectrum $\mathbb{E}$ over S and an integer $m \in \mathbf{Z}$, the following conditions are equivalent:

(i) $\mathbb{E} \geq_\delta m$ (resp. $\mathbb{E} \leq_\delta m$)

(ii) For any separated scheme X/S having finite type, one has:

$$\mathbb{E}_{n,i}(X/S) = 0 \text{ when } n - i < m + \delta_-(X), \\ \text{respectively } n - i > m + \delta_+(X).$$

(ii') For any separated scheme X/S being essentially of finite type, one has:

$$\mathbb{E}_{n,i}(X/S) = 0 \text{ when } n - i < m + \delta_-(X), \\ \text{respectively } n - i > m + \delta_+(X).$$

(iii) For any point $x \in S(K)$ (see again our Notations and conventions), one has:

$$\mathbb{E}_{n,i}(x) = 0 \text{ when } n - i < m + \delta(x), \\ \text{respectively } n - i > m + \delta(x).$$

2.1.11. *t-exactness of the six operations.*– Given a functor F between triangulated categories equipped with t-structures, one says that F is left (resp. right) t-exact if it respects homologically negative (resp. positive) objects; such a functor F is t-exact if it is both left and right t-exact. One says that F has homological amplitude $[a, b]$ if for any object $\mathbb{E}$:

- $\mathbb{E} \geq 0 \Rightarrow F(\mathbb{E}) \geq a$.
- $\mathbb{E} \leq 0 \Rightarrow F(\mathbb{E}) \leq b$.

Let $f : X \to S$ be a morphism essentially of finite type, and d the maximum dimension of its fibers. We consider the δ^f-homotopy t-structure on $\mathscr{T}(X)$, where δ^f is the dimension function on X induced by that of S with respect to the morphism f (see Notations and conventions page 25). Then one has the following results:

- $f^*[d]$ is right t_δ-exact.
- If f is smooth, $f^*[d]$ is t_δ-exact.
- The functor f_* has t_δ-amplitude $[0, d]$.

The first, second and third points are respectively proved in [**BD17**], 2.1.6(3), 2.1.12, 3.3.7.

If in addition f is separated of finite type, we get:

- $f_!$ is right t_δ-exact.
- $f^!$ is t_δ-exact.
- If $\delta \geq 0$ then $\otimes$ is right t_δ-exact.

These points are respectively proved in [**BD17**], 2.1.6(1), 3.3.7(4), 2.1.6(2).

Remark 2.1.12. As mentioned in Theorem 2.1.4, the δ-homotopy t-structure satisfies gluing. We recall here precisely what this means. Consider a closed immersion $i : Z \to S$ with complementary open immersion $j : U \to S$ and a $\mathscr{T}$-spectrum

$\mathbb{E}$ over S. The following conditions are equivalent:

- $\mathbb{E}$ is homologically non-t_δ-negative (resp. non-t_δ-positive).
- $j^*\mathbb{E}$ and $i^*\mathbb{E}$ are homologically non-t_δ-negative (resp. $j^*\mathbb{E}$ and $i^!\mathbb{E}$ are homologically non-t_δ-positive).

The equivalence of these conditions can be deduced from the localization property of $\mathscr{T}$ and the t-exactness stated above (see [**BD17**, Cor. 2.1.9]).

Let us now explicitly state the following consequence of the result on the tensor product.

PROPOSITION 2.1.13. *Assume the dimension function δ on S is non-negative. Let $\mathbb{E}$ be a $\mathscr{T}$-spectrum over S. Then for any pair $(p,q) \in \mathbf{Z}^2$, one has a canonical pairing:*

$$\bar{\phi} : \tau^\delta_{\geq p}(\mathbb{E}) \otimes \tau^\delta_{\geq q}(\mathbb{F}) \to \tau^\delta_{\geq p+q}(\mathbb{E} \otimes \mathbb{F})$$

which is bi-functorial in $\mathbb{E}$.

PROOF. According to the preceding paragraph, the assumption implies that the tensor product $\otimes$ in $\mathscr{T}(S)$ preserves non-negative objects. Consider the canonical map

$$\tau^\delta_{\geq p}(\mathbb{E}) \otimes \tau^\delta_{\geq q}(\mathbb{F}) \xrightarrow{\phi} \mathbb{E} \otimes \mathbb{F}.$$

According to the preceding assertion, the left hand-side is in homological degrees $\geq p+q$. Consider the distinguished triangle:

$$\tau^\delta_{\geq p+q}(\mathbb{E} \otimes \mathbb{F}) \xrightarrow{a} \mathbb{E} \otimes \mathbb{F} \xrightarrow{b} \tau^\delta_{<p+q}(\mathbb{E} \otimes \mathbb{F}) \xrightarrow{+1}$$

We deduce that the composition $b \circ \phi$ is zero. So ϕ uniquely factors through a giving us the desired map $\bar{\phi}$. The bifunctoriality of $\bar{\phi}$ follows from the uniqueness. □

EXAMPLE 2.1.14. Consider the assumptions of the preceding proposition.

The previous pairing is associative in an obvious sense - this follows from the uniqueness of the map $\bar{\phi}$. Hence the symmetric monoidal structure on $\mathscr{T}(X)$ induces a canonical symmetric monoidal structure on $\mathscr{T}(X)^\heartsuit$, using the formula, for δ-homotopy modules $\mathbb{E}$ and $\mathbb{F}$:

$$\mathbb{E} \otimes^H \mathbb{F} := \tau^\delta_{\leq 0}(\mathbb{E} \otimes \mathbb{F}).$$

Note in particular that the canonical functor:

$$\mathscr{T}(X)_{\geq_\delta 0} \to \mathscr{T}(X)^\heartsuit, \mathbb{E} \mapsto \tau^\delta_{\leq 0}(\mathbb{E}) = H^\delta_0(\mathbb{E})$$

is monoidal.

REMARK 2.1.15. Beware that the monoidal structure defined above a priori depends on $\delta \geq 0$. To be more precise, though the heart for two different choices of dimension functions $\delta \geq 0$, $\delta' \geq 0$ are equivalent as additive categories, according to Remark 2.1.6, this equivalence is not compatible in general with the tensor structures induced by δ and δ'.

Indeed, if k is a field and we are in the motivic (or the homotopical) case, if we take a dimension function δ on k such that $\delta(k) = n > 0$, then the induced tensor product on the heart is just the zero bi-functor ! The situation of a positive dimensional base is more complicated though, and it seems there are as many non-trivial tensor structures induced as in the above example as the dimension of the base.

2.2. Recollection on purity and duality.

2.2.1. Consider a $\mathscr{T}$-spectrum $\mathbb{E}$ over a scheme S.

Let us recall a construction from [**DJK18**]. Let $f : Y \to X$ be a quasi-projective lci morphism of S-schemes with cotangent complex L_f. Let $\mathbb{E}_X$ be the pullback of $\mathbb{E}$ along X/S. Then we associate to f a purity transformation (see [**DJK18**, 4.3.1]); evaluated at the object $\mathbb{E}_X$, it gives a canonical map:

$$\mathfrak{p}_f : f^*(\mathbb{E}_X) \otimes \mathrm{Th}(L_f) \to f^!(\mathbb{E}_X).$$

where $\mathrm{Th}(L_f)$ is the Thom space associated with the perfect complex L_f. The following definition extends classical considerations; in the motivic case, see [**DJK18**, 4.3.9].

DEFINITION 2.2.2. Consider the above notations. We will say that $\mathbb{E}$ is *absolutely pure* if for any quasi-projective morphism $f : Y \to X$ between regular schemes, the map $\mathfrak{p}_f$ is an isomorphism.

If S is regular, we will say that $\mathbb{E}$ is S*-pure* if for any quasi-projective morphism $f : X \to S$ with X regular, $\mathfrak{p}_f$ is an isomorphism.

EXAMPLE 2.2.3.

(1) Motivic case: the constant motive $\mathbb{1}_{S_0}$ is absolutely pure. Equivalently, $\mathbb{1}_S$ is S-pure for any regular scheme S.
(2) Homotopical case: Then any spectrum $\mathbb{E}$ over the base field k is absolutely pure (see [**DJK18**, 4.3.10(ii)]). Equivalently, any spectrum $\mathbb{E}$ over a regular base S is S-pure.

An almost immediate corollary of the S-purity assumption is the following duality statement.

PROPOSITION 2.2.4. *Let $f : X \to S$ be a morphism and $\mathbb{E}$ be a $\mathscr{T}$-spectrum over S. We assume one of the following hypothesis is fulfilled:*

- *f is essentially smooth.*
- *f is essentially quasi-projective, X and S are regular and $\mathbb{E}$ is S-pure.*

Then for any pair $(n, v) \in \mathbf{Z} \times \underline{\mathrm{K}}(X)$, the map $\mathfrak{p}_f$ gives an isomorphism:

$$\mathbb{E}^n(X, v) \xrightarrow{\sim} \mathbb{E}_{-n}(X/S, \langle L_f \rangle - v) \tag{2.2.4.a}$$

which is contravariantly natural in X with respect to étale maps.

PROOF. The case where f is of finite type is tautological, while the contravariance with respect to étale map follows from the compatibility of $\mathfrak{p}_f$ with étale pullbacks (apply [**DJK18**, 3.3.2(iii)] in the case where p is étale).

The general case is obtained using the previous one, together with the naturality with respect to étale maps, and the extension of bivariant theory described in 2.1.1. □

REMARK 2.2.5.

(1) This isomorphism can be described, at least when f is of finite type, as the cap-product by the fundamental class $\eta_f \in \mathbb{E}_0(X/S, \langle L_f \rangle)$ of f: see [**DJK18**, 4.3.9, 2.3.14]. In fact, the description of the above isomorphism in the general case immediately follows when one considers the obvious extension of the notion of fundamental classes to essentially quasi-projective lci morphisms.

(2) Recall that if either $\mathbb{E}$ is an object of $\mathrm{DM}(S,R)$ or if E is an **MGL**-module[7] over S then the Thom isomorphism implies we can canonically identify the twist by a virtual vector bundle with the Tate twist by its rank. In particular, the above isomorphism takes the following classical form:

$$\mathbb{E}^{n,i}(X) \xrightarrow{\sim} E_{2d-n,d-i}(X/S) \tag{2.2.5.a}$$

where d is the relative dimension of f.

3. Fiber homology and Gersten complexes

In this section, we investigate the heart of the homotopy t-structure discussed in the preceding section in greater detail. Section 3.1 is concerned with studying an approximation to the notation of homology associated with the truncation functors for the homotopy t-structure. Definition 3.1.2 introduces the notion of fiber δ-homology, which is esentially the restriction of δ-homology to points. This section concludes with Theorem 3.1.7, which introduces an "effective" variant of the t-stucture of Section 2.1, ameliorating some unboundedness issues that naturally arise because we work in a stable context. Section 3.2 recalls conventions for exact couples, and introduces Gersten complexes (see Definition 3.2.4), built out of fiber δ-homology, which essentially form the E_1-page of the niveau spectral sequence. We also give a detailed description of the differentials of the Gersten complexes and important formal properties of these complexes are summarized in Proposition 3.2.7. Finally, Section 3.3 is devoted to analyzing multiplicative structure in these complexes and the associated spectral sequences.

3.1. Fiber δ-homology.

3.1.1. Recall from the introduction that we have also considered $\mathbf{G}_m$-twists on bivariant theory. Indeed these twists are more natural with respect to the δ-homotopy t-structure, because the functor $-(1)[1]$ is t_δ-exact. Using this grading, one can reformulate Proposition 2.1.2 for a given $\mathscr{T}$-spectrum $\mathbb{E}$ over S as the equivalence of the following conditions:

(i) $\mathbb{E} \geq 0$ (resp. $\mathbb{E} \leq 0$).

(ii) For any separated scheme X/S essentially of finite type, one has:

$$\mathbb{E}_{n,\{*\}}(X/S) = 0 \text{ when } n < \delta_-(X) \ \big(\text{resp. } n > \delta_+(X)\big).$$

(iii) For any point $x \in S(K)$, one has:

$$\mathbb{E}_{n,\{*\}}(x) = 0 \text{ when } n < \delta(x) \ \big(\text{resp. } n > \delta(x)\big).$$

In view of this characterization of the δ-homotopy t-structure, we have adopted the following definition in [**BD17**].

DEFINITION 3.1.2. Let $\mathbb{E}$ be a $\mathscr{T}$-spectrum over S. We define the *fiber δ-homology* of $\mathbb{E}$ in degree $n \in \mathbf{Z}$ as the functor

$$\hat{H}^\delta_n(\mathbb{E}) : \mathrm{Pts}(S) \to \mathscr{A}b^{\mathbf{Z}}, x \mapsto \big(\mathbb{E}_{\delta(x)+n,\{\delta(x)-r\}}(x)\big)_{r\in\mathbf{Z}}$$

where $\mathrm{Pts}(S)$ is the discrete category of points of S. Dually, we define the *fiber δ-cohomology* of $\mathbb{E}$ as $\hat{H}^n_\delta(\mathbb{E}) = \hat{H}^\delta_{-n}(\mathbb{E})$ (the $\mathbf{G}_m$-grading does not change).

Given an object $\mathbb{F}$ of the δ-homotopy heart, we set $\hat{\mathbb{F}}^\delta_* := \hat{H}^\delta_0(\mathbb{F})$.

[7] in other words, an oriented spectrum; here module is to be understood in the sense of the monoidal category $\mathrm{SH}(S)$ *i.e.* in the weak homotopical sense.

Therefore $\mathbb{E} \geq 0$ (resp. $\mathbb{E} \leq 0$) if and only if $\hat{H}_n^\delta(\mathbb{E}) = 0$ for $n < 0$ (resp. $n > 0$).

REMARK 3.1.3. Fiber δ-homology is a good approximation of δ-homology. In fact, given an object $\mathbb{F}$ of the heart of the δ-homotopy t-structure on $\mathscr{T}(S)$, the functor $\hat{\mathbb{F}}_*^\delta$ is an approximation of a cycle module in the sense of Rost.

In the motivic case, this idea can be turned into an equivalence of categories between the δ-homotopy heart of $\mathrm{DM}(S, R)$ and the category of Rost R-linear cycle modules over S in the sense of [**Ros96**]. The details of such a theorem have not yet been written up, but see [**Dég14a**]. In the homotopical case, a generalization of Rost's theory is in the work (see [**Fel18**]).

In any case, we have already obtained in [**BD17**, 4.2.2] that, in the general case of an abstract triangulated motivic category $\mathscr{T}$ satisfying our general assumptions, the functor:

$$\mathscr{T}(S)^\heartsuit \to \mathrm{PSh}\big(\mathrm{Pts}(S), R - \mathrm{mod}^{\mathbf{Z}}\big), \mathbb{F} \mapsto \hat{\mathbb{F}}_*^\delta$$

is conservative, exact and commmutes with colimits. Similarly, the family of functors $(\hat{H}_n^\delta)_{n \in \mathbf{Z}}$ is conservative on the whole category $\mathscr{T}(S)$.

EXAMPLE 3.1.4. Let S be a regular connected scheme, and put $d = \delta(S)$. We fix a point $x : \mathrm{Spec}(K) \to S$.

(1) Abstractly, using any S-pure spectrum $\mathbb{E}$, one obtains, because of the duality isomorphism (2.2.4.a) and relation (D2) of dimension functions, a canonical isomorphism:

$$\hat{H}_n^\delta(\mathbb{E})_r(x) \simeq \mathbb{E}^{-n-r}\big(\mathrm{Spec}(K), \langle L_{x/S}\rangle - \langle \delta(x)\rangle + \langle r\rangle\big).$$

(2) Assume we are in the motivic case. One then obtains (using the previous computation) a canonical isomorphism:

$$\hat{H}^n_\delta(\mathbb{1}_S)_r(x) = H_M^{r-n-2d, r-d}(\mathrm{Spec}(K), R).$$

This follows because under our assumptions, motivic cohomology satisfies absolute purity, and is oriented.

Observe, for example, that:

$$\hat{H}_d^\delta(\mathbb{1}_S)_* = \hat{H}_\delta^{-d}(\mathbb{1}_S)_* = K_*^M|_S,$$

the restriction of the Milnor K-theory functor to the discrete category $\mathrm{Pts}(S)$.

Additionally, the only vanishing that we have is: $\hat{H}_n^\delta(\mathbb{1}_S) = 0$ if $n > d$. In other words, $\mathbb{1}_S$ is concentrated in t_δ-homological degrees $]-\infty, d]$. On the other hand, we know for a fact, at least when S is a complex scheme, that for all $n \leq d$, $\hat{H}_n^\delta(S) \neq 0$ (this is due to the existence and non-triviality of polylogarithm elements; see [**BD17**, Ex. 3.3.2]). So $\mathbb{1}_S$ is t_δ-unbounded below.

(3) Assume we are in the homotopical case. Then one obtains, using the computation of point (1), a non-canonical isomorphism:

$$\hat{H}_n^\delta(\mathbb{1}_S)_r(x) \simeq \pi_{n+d}^{\mathbf{A}^1}(S_K^0)_{r-d}[1/p].$$

This is because absolute purity is automatic in the homotopical case: see point (2) of Example 2.2.3.

In particular, Morel's computation of the zeroth stable $\mathbf{A}^1$-homotopy sheaf of the sphere gives an isomorphism, again non-canonical:

$$\hat{H}_d^\delta(\mathbb{1}_S)_r(x) \simeq K_r^{MW}(K)$$

where the right hand side is the r-th Milnor–Witt cohomology group of the field K.

Moreover, the zero sphere spectrum $\mathbb{1}_S$ is concentrated in homological degrees $]-\infty, d]$ for the δ-homotopy t-structure. Not much else can be said in this case since, when S is a complex scheme, the rational homology of the zero sphere spectrum agrees with the rational homology of the motivic spectrum, which is non-trivial. Thus, $\mathbb{1}_S$ is unbounded below for the δ-homotopy t-structure.

(4) Again in the homotopical case, we can consider the Milnor–Witt motivic ring spectrum $\mathbf{H}\tilde{R}_S$. Known computations of Milnor–Witt cohomology imply, as in the case of motivic cohomology, that $\mathbf{H}\tilde{R}_S$ is concentrated in homological degrees $]-\infty, d]$. Moreover, the unit of this ring spectrum induces a canonical isomorphism:

$$\hat{H}^\delta_d(\mathbb{1}_S)_* \xrightarrow{\sim} \hat{H}^\delta_d(\mathbf{H}\tilde{R}_S)_*.$$

REMARK 3.1.5. In the non-oriented context, it appears more natural to introduce another twist for a virtual vector bundle v over a scheme X, namely:

$$\mathbb{1}_X\{v\} := \mathrm{Th}_X(v)[-\operatorname{rk}(v)].$$

This notation is consistent with our notation $\mathbb{1}_X\{r\}$ for the $\mathbf{G}_m$-twist (where r corresponds to the trivial bundle of rank r).

Assuming $\delta(S) = 0$ to simplify, the isomorphism of point (1) can be rewritten as:

$$\hat{H}^\delta_n(\mathbb{E})_r(x) \simeq \mathbb{E}^{-n}\big(\operatorname{Spec}(K), \{L_{x/S}\} + \{r\}\big).$$

The unboundedness of the constant object, observed in Points (2), (3), (4) of the previous example, appears to be a consequence of working in the stable context – recall that stability with respect to Tate twists is among the axioms of triangulated motivic categories used in [**CD09**]. This potentially unpleasant feature can be corrected as follows.

DEFINITION 3.1.6. (see [**BD17**, Def. 2.2.1]) Let S be an arbitrary scheme. We define the triangulated category of δ-effective $\mathscr{T}$-spectra over S, denoted by $\mathscr{T}^{\delta-\mathrm{eff}}(S)$, as the full localizing subcategory of $\mathscr{T}(S)$ generated by objects of the form:

$$f_!(\mathbb{1}_X)(n)$$

where $f : X \to S$ is separated of finite type and $\delta(X) \geq n$.

Given a $\mathscr{T}$-spectrum $\mathbb{E}$ over S, we define its *effective fiber δ-homology* $H^{\delta\mathrm{eff}}_n(\mathbb{E})$ as the negatively graded functor obtained from $\hat{H}^\delta_n(\mathbb{E})$ by restricting the $\mathbf{G}_m$-grading to negative integers.

It follows from [**BD17**, 3.1.1] that one can define an effective version of the δ-homotopy t-structure.

THEOREM 3.1.7. *Consider the notations of the previous definition. For any scheme S, there exists a t-structure on $\mathscr{T}^{\delta\mathrm{eff}}(S)$ whose homologically non-negative (resp. non-positive) objects are those $\mathscr{T}$-spectra $\mathbb{E}$ over S such that $\hat{H}^{\delta\mathrm{eff}}_n(\mathbb{E}) = 0$ for $n < 0$ (resp. $n > 0$). This t-structure is non-degenerate and satisfies glueing (recall:* [**BBD82**, 1.4.10], *Remark* 2.1.12*).*

EXAMPLE 3.1.8. Let S be a regular connected scheme, and put $d = \delta(S)$.

(1) Assume we are in the motivic case. Then the computation done in Example 3.1.4(1) shows that $\hat{H}^\delta_n(\mathbb{1}_S)_{\leq 0} = 0$ if $n \neq d$. In other words, $\mathbb{1}_S[-d]$ is in the heart of the effective δ-homotopy t-structure. This is exactly what happens for the perverse t-structure.

(2) Assume we are in the homotopical case. The preceding example shows that the Eilenberg-MacLane motivic ring spectrum $\mathbf{H}R_S[-d]$ is in the heart of the effective δ-homotopy t-structure.

The computations of higher stable homotopy groups of spheres do not allow us at the moment to conclude anything about the sphere spectrum. However, the computations of Milnor-Witt motivic cohomology imply, as in the motivic case, that $\mathbf{H}_{MW}R_S[-d]$ is in the heart of the effective δ-homotopy t-structure.

3.1.9. The δ-effective categories $\mathscr{T}^{\delta\text{eff}}(S)$, with the t-structure just defined, satisfy good properties. We refer the reader to [**BD17**] for the following facts. First, by construction, one has a pair of adjoint functors:

$$s : \mathscr{T}^{\delta\text{eff}}(S) \leftrightarrows \mathscr{T}(S) : w$$

such that s is fully faithful and w is t_δ-exact.

- If $\delta \geq 0$, the subcategory $\mathscr{T}^{\delta\text{eff}}(S)$ of $\mathscr{T}(S)$ is stable under tensor products. So $\mathscr{T}^{\delta\text{eff}}(S)$ becomes a closed symmetric monoidal category with internal hom given by:

$$\underline{\mathrm{Hom}}_{\mathscr{T}^{\delta\text{eff}}}(S) = w \circ \underline{\mathrm{Hom}}_{\mathscr{T}(S)}(M, N).$$

  Moreover, the tensor product on $\mathscr{T}^{\delta\text{eff}}(S)$ is right t_δ-exact.
- For $f : X \to S$ essentially of finite type with $\dim(f) \leq d$, we get an adjunction of t-categories:

$$f^*(d)[2d] : \mathscr{T}^{\delta\text{eff}}(S) \leftrightarrows \mathscr{T}^{\delta\text{eff}}(X) : w \circ \big(f_*(-d)[-2d]\big).$$

  When f is smooth of pure dimension d, $f^*(d)[2d]$ is t_δ-exact.
- For $f : X \to S$ separated and essentially of finite type, we get an adjunction of t-categories:

$$f_! : \mathscr{T}^{\delta\text{eff}}(X) \leftrightarrows \mathscr{T}^{\delta\text{eff}}(S) : w \circ f^!.$$

  Moreover, $w \circ f^!$ is t_δ-exact.

In the last two points, the dimension function on X is that induced by the one fixed on S (see our notations and conventions).

3.2. Gersten complexes. We now build a spectral sequence using the theory of exact couples. Since conventions for exact couples vary in the literature, we now fix our conventions, which agree with those used in [**BD17**, 3.1.2].

DEFINITION 3.2.1. Let $\mathscr{A}$ be an abelian category. A *homological exact couple of degree $d > 0$ in $\mathscr{A}$* is the data of a pair of bigraded objects (D, E) of $\mathscr{A}$ together with a triangle of homogeneous maps (a, b, c)

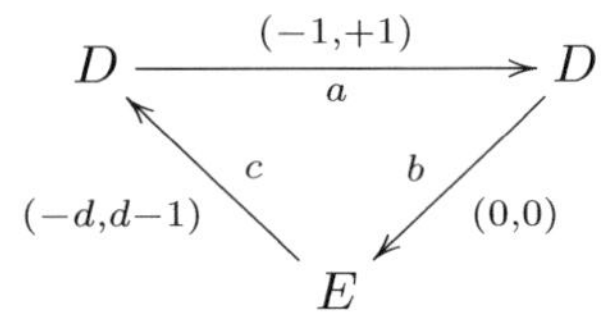

such that consecutive maps fit together to form an exact sequence.

As usual, one derives a (homologically indexed) spectral sequence from such an exact couple [**McC01**]: at page r, the r-th term is equal to the differential bigraded abelian group $(E, b \circ c)$.

3.2.2. We first recall δ-niveau spectral sequences, in the abstract setting and using dimension function following [**BD17**, 3.1.5].[8]

Let X be a separated S-scheme essentially of finite type. Recall from [**BD17**, 3.1.1] that a δ-flag of X is an increasing sequence of reduced closed subschemes $Z_* = (Z_p)_{p \in \mathbf{Z}}$ of X such that $\delta(Z_p) \leq p$ for all p. The set $\mathcal{F}(X)$ of δ-flags, ordered by term-wise inclusion, is cofiltered. Given such a δ-flag, the classical properties of the bivariant theory $\mathbb{E}_*(-/-)$ (see our Notations and conventions) imply that we have a long exact sequence:

$$\mathbb{E}_{p+q}(Z_{p-1}/S) \xrightarrow{a} \mathbb{E}_{p+q}(Z_p/S) \xrightarrow{b} \mathbb{E}_{p+q}(Z_p - Z_{p-1}/S) \xrightarrow{c} \mathbb{E}_{p+q-1}(Z_{p-1}/S)$$

where a (resp. b) is pushfoward (resp. pullback) along the obvious closed (resp. open) immersion, and c the boundary map. These long exact sequences are covariantly functorial with respect to inclusion of flags. Thus we can combine them into the following homological exact couple of degree 1:

$$\begin{aligned} {}^{\delta}D_{p,q} &= \varinjlim_{Z_* \in \mathcal{F}(X)} \big(\mathbb{E}_{p+q}(Z_p/S)\big), \\ {}^{\delta}E^1_{p,q} &= \varinjlim_{Z_* \in \mathcal{F}(X)} \big(\mathbb{E}_{p+q}(Z_p - Z_{p-1}/S)\big). \end{aligned}$$

Finally, observe that one can express the E_1-term as follows:

$${}^{\delta}E^1_{p,q} = \oplus_{x \in X_{(p)}} \mathbb{E}_{p+q}(x/S),$$

where

$$X_{(p)} = \{x \in X \mid \delta(x) = p\}.$$

Definition 3.2.3. Under the assumptions and notations above, the spectral sequence:

$${}^{\delta}E^1_{p,q} = \bigoplus_{x \in X_{(p)}} \mathbb{E}_{p+q}(x/S) \Rightarrow \mathbb{E}_{p+q}(X/S)$$

will be called the δ-niveau spectral sequence of X/S with coefficients in $\mathbb{E}$.

[8]The main idea is of course classical, introduced by Grothendieck, and thoroughly developed in [**BO74**]. The main novelty of [**BD17**] is to work over an arbitrary base scheme S (rather than a field) and appeal to abstract dimension functions. This use of dimension functions can also be found in the homological indexing for Chow groups in [**wc15**]. See [**BD17**, 3.1.7] for more on this point.

The spectral sequence converges to the following filtration, called the δ-niveau filtration:

$$N_p\mathbb{E}_*(X/S) = \bigcup_{i:Z\to X,\delta(Z)=p} \operatorname{Im}\big(i_* : \mathbb{E}_*(Z/S) \to \mathbb{E}_*(X/S)\big),$$

where i runs over the closed immersions whose source has δ-dimension p.

It is standard to consider this spectral sequence not just for $\mathbb{E}$ itself, but also for all Tate twists. In fact, if we apply this definition to the graded spectrum $\mathbb{E}(n)$ for an integer $n \in \mathbf{Z}$, we get the following form:

$${}^{\delta}E^1_{p,q} = \bigoplus_{x\in X_{(p)}} \mathbb{E}_{p+q,n}(x/S) \Rightarrow \mathbb{E}_{p+q,n}(X/S).$$

In view of Rost's theory of cycle modules, it will be useful to introduce the following definition.

DEFINITION 3.2.4. Under the assumptions of the preceding definition, we define the *Gersten (δ-homological) complex* of X/S with coefficients in $\mathbb{E}$, denoted by $C^\delta_*(X,\mathbb{E})$, as the complex of abelian groups located at the line $q=0$ of the E_1-term of the δ-niveau spectral sequence. The p-th homology of this complex will be called the *Gersten δ-homology of* $\mathbb{E}$ and be denoted by:

$$A^\delta_p(X,\mathbb{E}) = H_p\big(C^\delta_*(X,\mathbb{E})\big).$$

Note that the Gersten complex is concentrated in degrees $[\delta_-(X),\delta_+(X)]$. Moreover, using Definition 3.1.2 for fiber δ-homology, the p-th term of this complex takes the following form:

$$C^\delta_p(X,\mathbb{E}) = \bigoplus_{x\in X_{(p)}} H^\delta_0\mathbb{E}_p(x/S).$$

REMARK 3.2.5. The complex defined above is closely linked with Rost's theory of cycle modules. We have used a simplification here. Indeed recall that cycle modules, as well as their associated complexes, are $\mathbf{Z}$-graded. We can recover this $\mathbf{Z}$-grading by applying the above definition to the $\mathbf{Z}$-graded spectrum:

$$\mathbb{E}\{*\} = \big(\mathbb{E}(n)[n]\big)_{n\in\mathbf{Z}}.$$

It is also standard in stable $\mathbb{A}^1$-homotopy theory to consider this grading, called the *$\mathbf{G}_m$-grading.*

The complex $C^\delta_*(X,\mathbb{E}\{*\})$ is $\mathbf{Z}$-graded, and has the following form:

$$C^\delta_*(X,\mathbb{E}\{*\}) = \bigoplus_{x\in X_{(p)}} H^\delta_0\mathbb{E}_{p-*}(x/S).$$

The main reason to use this grading is that the differentials of the Gersten complex are then homogeneous of degree -1.

3.2.6. Consider the notations of the above definition. The differentials of the Gersten complex associated with $\mathbb{E}$ and X/S take the following form:

$$C^\delta_p(X,\mathbb{E}) = \bigoplus_{x\in X_{(p)}} \mathbb{E}_p(x/S) \xrightarrow{d_p} \bigoplus_{s\in X_{(p-1)}} \mathbb{E}_p(s/S) = C^\delta_{p-1}(X,\mathbb{E})$$

The differential d_p is obtained as the inductive limit of differentials d_p^Z associated with a δ-flag Z_* and is described as follows:

$$\begin{array}{ccccc}
\mathbb{E}_p(Z_p/S) \longrightarrow E_p(Z_p - Z_{p-1}/S) & \xrightarrow{c} & E_{p-1}(Z_{p-1}/S) & & \\
& & \| & \overset{d_p}{\searrow} & \\
E_{p-1}(Z_{p-2}/S) & \longrightarrow & E_{p-1}(Z_{p-1}/S) & \xrightarrow[b]{} & E_{p-1}(Z_{p-1} - Z_{p-3}/S)
\end{array}$$

where the rows arise as pieces of localization long exact sequences.

Using the functoriality of the localization long exact sequences with respect to pullbacks along open immersions, it is therefore possible to explicitly describe the differentials of this complex as follows. Consider a pair $(x, s) \in X_{(p)} \times X_{(p-1)}$ and denote by $(d_p)_y^x : \mathbb{E}_p(x/S)_{r-p} \to \mathbb{E}_{p-1}(y/S)_{r-p}$ the corresponding component of the above differential. Let us write $Z(x)$ the reduced closure of x in X. From the construction of the Gersten complex given above, and the functoriality of localization long exact sequence in bivariant theory with respect to pullbacks along open immersions and pushforward along closed immersions, one then deduces that:

- if $s \in Z(x)$: we let $Z(x)_{(s)}$ be the localization of $Z(x)$ at s; this is a 1-dimensional scheme so that $Z(x)_{(s)} = \{x, s\}$. Then $(d_p)_s^x$ is the boundary map of the localization long exact sequence associated with the closed immersion $i : \{s\} \to Z(x)_{(s)}$. Explicitly, it is the middle map in the following exact sequence:

$$\mathbb{E}_p\big(Z(x)_{(s)}/S\big) \xrightarrow{j^*} \mathbb{E}_p(x/S) \xrightarrow{(d_p)_s^x} \mathbb{E}_{p-1}(s/S) \xrightarrow{i_*} \mathbb{E}_{p-1}\big(Z(x)_{(s)}/S\big)$$

- Otherwise, $(d_p)_s^x = 0$.

The next result summarizes the formal properties of Gersten complexes.

PROPOSITION 3.2.7. *Consider the notations of the above definition.*

(1) *The complex $C_*^\delta(X, \mathbb{E})$ is covariantly functorial in $\mathbb{E}$. Given any integer p, the induced maps*

$$C_*^\delta(X, \mathbb{E}) \to C_*^\delta(X, \tau_{\leq p}\mathbb{E}),\ C_*^\delta(X, \tau_{\geq -p}\mathbb{E}) \to C_*^\delta(X, \mathbb{E})$$

are isomorphisms provided $p \geq 0$. In particular, one has a canonical isomorphism:

$$C_*^\delta(X, \mathbb{E}) \simeq C_*^\delta(X, H_0^\delta\mathbb{E}).$$

(2) *The complex $C_*^\delta(X, \mathbb{E})$ is functorial in X/S, covariantly with respect to proper maps and contravariantly with respect to étale maps.*

(3) *If the $\mathscr{T}$-spectrum $\mathbb{E}$ is homologically non-t_δ-negative (resp. non-t_δ-positive), there exists a canonical epimorphism (resp. monomorphism):*

$$\mathbb{E}_p(X/S) \to A_p^\delta(X, \mathbb{E})$$
$$\text{resp. } A_p^\delta(X, \mathbb{E}) \to \mathbb{E}_p(X/S).$$

These two maps are functorial in the S-scheme X, covariantly with respect to proper maps and contravariantly with respect to étale maps.

If $\mathbb{E}$ is in the δ-homotopy heart, or more generally is concentrated in one degree for the δ-homotopy t-structure, these two maps are inverse isomorphisms giving a functorial identification:

$$\mathbb{E}_p(X/S) \simeq A_p^\delta(X, \mathbb{E}).$$

PROOF. Point (1) follows from the δ-niveau spectral sequence and its obvious functoriality in $\mathbb{E}$ (Paragraph 3.2.2).

Point (2) follows from the classical functoriality of the (δ-)niveau spectral sequences. We refer the reader to [**Jin16**], proof of Proposition 3.11 for proper functoriality, and proof of Proposition 3.12 for étale contravariance. Note an important technical point here: we deal with the case of a general triangulated category $\mathscr{T}$, in contrast with the special case $\mathscr{T} = \mathrm{DM}$ which is oriented. In the case of proper functoriality, this does not come into play. The case of étale contravariance works as well as the tangent bundle of an étale map is trivial.

Point (3) is a consequence of the convergence of the δ-niveau spectral sequence and of the computation of its E_1-term. □

REMARK 3.2.8.

(1) According to the isomorphism of Point (3) above, one obtains using the main result of [**DJK18**] that, given a δ-homotopy module $\mathbb{E}$, the Gersten homology $A^\delta_*(X,\mathbb{E})$ is contravariant in X with respect to any smoothable lci morphism $f : Y \to X$. More precisely, one gets a canonical morphism:

$$A^\delta_*(X,\mathbb{E}) \simeq \mathbb{E}_p(X/S) \to \mathbb{E}_p(Y/S, \langle L_{Y/X}\rangle) \simeq A^\delta_*(Y, \mathbb{E}\langle L_{Y/X}\rangle)$$

where the last identification uses the fact $\mathbb{E}\langle L_{Y/X}\rangle$ is concentrated in one degree over any connected component of Y — which is the virtual rank of $L_{Y/X,\eta}$ where η is any generic point of the chosen connected component.

(2) As another illustration of the connexion of our Gersten complexes with Rost's theory of cycle complexes, when S is the spectrum of a perfect field k, δ is the obvious dimension function, and we are in the motivic case $\mathscr{T} = \mathrm{DM}(k,R)$. Then any object $\mathbb{E}$ in the heart of $\mathrm{DM}(k,R)$ is a homotopy module with transfers in the sense of [**Dég11**] which canonically corresponds to a cycle modules $\hat{\mathbb{E}}_* = \hat{\mathbb{E}}^\delta_*$ — this notation corresponds to the one of Definition 3.1.2. Moreover, according to [**Dég12**, 2.7(ii)], there exists a canonical isomorphism of complexes

$$C^\delta_*(X,\mathbb{E}) \simeq C_*(X, \hat{\mathbb{E}}_*)$$

where the right hand-side is Rost cycle complex associated with the cycle module $\hat{\mathbb{E}}_*$.

3.3. Products.

3.3.1. Consider a $\mathscr{T}$-spectrum $\mathbb{E}$ over S.

In order to describe products, we will use cohomological notations. We put:

$$C^p_\delta(S,\mathbb{E}) = C^\delta_{-p}(S,\mathbb{E}), \text{ resp. } A^p_\delta(S,\mathbb{E}) := A^\delta_{-p}(S,\mathbb{E})$$

and call it the *Gersten δ-cohomological complex* (resp. *δ-cohomology*) of S with coefficients in $\mathbb{E}$. Note in particular that, when $\mathbb{E}$ is in the heart, we get from Proposition 3.2.7 a canonical isomorphism:

$$A^p_\delta(S,\mathbb{E}) \simeq H^p(S,\mathbb{E}).$$

Consider now a morphism of $\mathscr{T}$-spectra over S:

$$\mu : \mathbb{E} \otimes \mathbb{F} \to \mathbb{G}.$$

We deduce as usual a morphism at the level of cohomologies:

$$\mathbb{E}^p(S) \otimes_{\mathbf{Z}} \mathbb{F}^q(S) \to \mathbb{G}^{p+q}(S)$$

by sending a pair of maps $(a : \mathbb{1}_S \to \mathbb{E}[p], b : \mathbb{1}_S \to \mathbb{F}[q])$ to the following map:

$$\mathbb{1}_S \xrightarrow{a \otimes b} \mathbb{E} \otimes \mathbb{F}[p+q] \xrightarrow{\mu} \mathbb{G}[p+q].$$

Using the preceding isomorphism, we deduce a canonical pairing:

$$A^p_\delta(S, \mathbb{E}) \otimes_{\mathbf{Z}} A^q_\delta(S, \mathbb{F}) \to A^{p+q}_\delta(S, \mathbb{G}).$$

We deduce the following results.

Proposition 3.3.2. *If $\mathbb{E}$ is a ring $\mathscr{T}$-spectrum over S, then $A^*_\delta(S, \mathbb{E})$ has a ring structure. If $\mathbb{F}$ is a $\mathscr{T}$-spectrum with equipped with an $\mathbb{E}$-module-structure, then $A^*(S, \mathbb{F})$ is equipped with the structure of a module over $A^*_\delta(S, \mathbb{E})$.*

Example 3.3.3. If $\mathbb{E}$ be a ring $\mathscr{T}$-spectrum over S, then according to Example 2.1.14, the graded δ-homotopy module $H^\delta_*(\mathbb{E})$ has the structure of a ring spectrum. According to the preceding proposition, $A^*_\delta(S, H^\delta_*(\mathbb{E}))$ has the structure of a bigraded ring.

Remark 3.3.4.

(1) Products are thus easy to obtain on our Gersten δ-homology. It is possible to follow Rost's approach in [**Ros96**] and to get a definition of the product on the level of complexes. Using classical techniques due to Levine, it is even possible to find a dg-algebra underlying our Gersten δ-homology (see [**Lev06, Lev08, BY18**]). An advantage of our approach is that it circumvents these technicalities.

(2) One can also extend the preceding considerations to the case of smooth S-schemes. Indeed, in that case, one has:

$$A^p_\delta(X, \mathbb{E}) = A^\delta_{-p}(X, \mathbb{E}) \simeq \mathbb{E}_{-p}(X/S) = H^p(X, \mathbb{E}\langle L_{X/S}\rangle)$$

according to Proposition 2.2.4. The preceding proposition obviously extends to that case. Note that the product obtained for smooth S-schemes is now compatible with étale pullbacks as defined in Proposition 3.2.7, and even with respect to smooth pullbacks.

4. The homotopy Leray spectral sequences

This section contains the main theoretical results of the paper and constructs the spectral sequences we mentioned in the introduction. We introduce two versions of this spectral sequence. Section 4.1 studies a homological version of the Leray spectral sequence; the main result is Theorem 4.1.2, which gives a description of the E_2-page of the relevant spectral sequence. Proposition 4.1.4 summarizes formal properties of the resulting spectral sequences. On the other hand, Section 4.2 contains a cohomological version of the spectral sequence (summarized in Theorem 4.2.5) and discusses compatibility with product structures. Finally, Section 4.3 contains a discussion of an analog of locally constant sheaves in our situation.

4.1. The homological version.

4.1.1. Let us fix a $\mathscr{T}$-spectrum $\mathbb{E}$ over S and consider the following geometric situation:

$$\begin{array}{ccc} X & \xrightarrow{f} & B \\ & \searrow^{\pi} & \\ & & S \end{array}$$

where f is any morphism of schemes, which plays the role of the fibration. We assume π is a separated morphism essentially of finite type and put $\mathbb{E}^!_X = \pi^!\mathbb{E}$. Given a $\mathscr{T}$-spectrum $\mathbb{E}$ over B, we put according to what was announced in the paragraph on Notations and conventions:

$$H_p(B, \mathbb{E}) = \mathrm{Hom}_{\mathscr{T}(B)}(\mathbb{1}_B[p], \mathbb{E}).$$

Then one can look at the tower of homological truncations of $f_*(\mathbb{E}^!_X)$ for the δ-homotopy t-structure, which is the analogue of the Postnikov tower in our situation:

$$\ldots \to \tau^\delta_{\geq q+1}(f_*\mathbb{E}^!_X) \to \tau^\delta_{\geq q}(f_*\mathbb{E}^!_X) \to \tau^\delta_{\geq q-1}(f_*\mathbb{E}^!_X) \to \ldots$$

It is standard to deduce from this filtration a spectral sequence. Let us be more precise. We first consider the canonical distinguished triangle:

$$\tau^\delta_{\geq q+1}(f_*\mathbb{E}^!_X) \to \tau^\delta_{\geq q}(f_*\mathbb{E}^!_X) \to \tau^\delta_{=q}(f_*\mathbb{E}^!_X) \to \tau^\delta_{\geq q+1}(f_*\mathbb{E}^!_X)[1]$$

where:

$$\tau^\delta_{=q}(f_*\mathbb{E}^!_X) = \tau^\delta_{\leq q}\tau^\delta_{\geq q}(f_*\mathbb{E}^!_X) = H^\delta_q(f_*\mathbb{E}^!_X)[q].$$

Applying the functor $H_{p+q}(B, -)$ to the preceding distinguished triangle, we get a long exact sequence:

$$H_{p+q}\left(B, \tau^\delta_{\geq q+1}(f_*\mathbb{E}^!_X)\right) \xrightarrow{a} H_{p+q}\left(B, \tau^\delta_{\geq q}(f_*\mathbb{E}^!_X)\right)$$
$$\xrightarrow{b} H_p\left(B, H^\delta_q(f_*\mathbb{E}^!_X)\right) \xrightarrow{c} H_{p+q-1}\left(B, \tau^\delta_{\geq q+1}(f_*\mathbb{E}^!_X)\right).$$

Using the conventions of definition 3.2.1, we then get a homological exact couple of degree 2 such that:

$$D_{p,q} = H_{p+q}\left(B, \tau^\delta_{\geq q}(f_*\mathbb{E}^!_X))\right),$$
$$E_{p,q} = H_p\left(B, H^\delta_q(f_*\mathbb{E}^!_X)\right).$$

We deduce from that exact couple our main construction.

THEOREM 4.1.2. *Fix assumptions and notations as in* 4.1.1. *The exact couple defined above gives a convergent spectral sequence of the form:*

$$E^2_{p,q}(f, \mathbb{E}) = H_p\left(B, H^\delta_q(f_*\mathbb{E}^!_X)\right) \simeq A^\delta_p(B, H^\delta_q(f_*\mathbb{E}^!_X))) \Rightarrow \mathbb{E}_{p+q}(X/S).$$

The $\mathbb{E}_2$*-term is concentrated in the range* $p \in [\delta_-(X), \delta_+(X)]$ *and is the homology in degree* p *of the* δ*-homological Gersten complex* $C^\delta_*(B, H^\delta_q(f_*\mathbb{E})))$ *(see Definition* 3.2.4*) which takes the form:*

$$\ldots \to \bigoplus_{x \in B_{(p)}} \mathbb{E}_{p+q}(X_x/S) \xrightarrow{d^G_p} \bigoplus_{s \in B_{(p-1)}} \mathbb{E}_{p+q-1}(X_s/S) \to \ldots$$

where X_x *is the fiber of* f *above the point* $x \in B$.

Note, in particular, that the differentials $d^r_{p,q}$ are trivial for $r > \big(\delta_+(X) - \delta_-(X)\big)$.

PROOF. The spectral sequence follows from the theory of (derived) exact couples. One can compute the E_2-term as in the above statement by using Proposition 3.2.7:

$$H_p\left(B, H^\delta_q(f_*\mathbb{E}^!_X)\right) \simeq A^\delta_p\left(B, H^\delta_q(f_*\mathbb{E}^!_X)\right).$$

Going back to the definition, this is the p-th homology of the following Gersten δ-homological complex:

$$\ldots \to \bigoplus_{x \in B_{(p)}} \mathbb{E}_{p+q}(x/B, f_*\mathbb{E}^!_X) \xrightarrow{d^G_p} \bigoplus_{s \in B_{(p-1)}} \mathbb{E}_{p+q-1}(s/B, f_*\mathbb{E}^!_X) \to \ldots$$

Given any point $x \in B^{(p)}$ one considers the cartesian square:

$$\begin{array}{ccc} X_x & \xrightarrow{i_x} & X \\ {\scriptstyle f_x}\downarrow & & \downarrow{\scriptstyle f} \\ \operatorname{Spec}(\kappa_x) & \xrightarrow{x} & B \end{array}$$

The computation:

$$\begin{aligned} \mathbb{E}_*(x/B, f_*\mathbb{E}^!_X) &= [x_! \mathbb{1}_B[*], f_*\mathbb{E}^!_X] = [\mathbb{1}_x[*], x^! f_*\mathbb{E}^!_X] \\ &= [\mathbb{1}_x[*], f_{x*} i^!_x \mathbb{E}^!_X] = [\mathbb{1}_x[*], i^!_x \pi^! \mathbb{E}] = \mathbb{E}_*(X_x/S) \end{aligned}$$

yields the E_2-term in the form of the statement and implies that the spectral sequence converges. □

Definition 4.1.3. The above spectral sequence will be called the *(homological) homotopy Leray spectral sequence* associated with f and with coefficients in $\mathbb{E}$.

The spectral sequence is obviously functorial in the $\mathscr{T}$-spectrum $\mathbb{E}$ over S. If we want to analyze the behavior with respect to Tate twists, we can look at the spectral sequence with coefficients in $\mathbb{E}(n)$, which takes the following form:

$$E^2_{p,q} = A^\delta_p(B, H^\delta_q(f_*\mathbb{E}^!_X)(n))) \Rightarrow \mathbb{E}_{p+q,n}(X/S).$$

Similarly, twists by more general Thom spaces man be considered. The following result summarizes the formal properties of the spectral sequence.

Proposition 4.1.4. *We consider the following diagram of schemes:*

$$\begin{array}{ccccc} & & Y & & \\ & \swarrow^{\pi'} & \downarrow{\scriptstyle p} & \searrow^{f'} & \\ S & & & & B \\ & \nwarrow_{\pi} & X & \nearrow_{f} & \end{array}$$

and a $\mathscr{T}$-spectrum $\mathbb{E}$ over S. Set $\mathbb{E}^!_X = \pi^!\mathbb{E}$ and $\mathbb{E}^!_Y = \pi'^!(\mathbb{E})$. Given any point $x \in B$, let $p_x : Y_x \to X_x$ be the induced morphism on the fibers over a point $x \in B$.

- *If p is proper, then there is an adjunction map:*

$$\phi_p : f'_*\mathbb{E}^!_Y = f'_*\pi'^!(\mathbb{E}) = f_*p_!p^!\pi^!(\mathbb{E}) \xrightarrow{ad(p_!,p^!)} f_*\pi^!(\mathbb{E}) = f_*(\mathbb{E}^!_X)$$

which induces a morphism of spectral sequences converging to the indicated map on the abutment:

$$\begin{array}{ccc} A^\delta_p(B, H^\delta_q(f'_*\mathbb{E}^!_Y))) & \Rightarrow & \mathbb{E}_{p+q}(Y/S) \\ {\scriptstyle \phi_{p*}}\downarrow & & \downarrow{\scriptstyle p_*} \\ A^\delta_p(B, H^\delta_q(f_*\mathbb{E}^!_X))) & \Rightarrow & \mathbb{E}_{p+q}(X/S). \end{array}$$

Moreover, the map ϕ_{p*} *is induced by the following morphism of Gersten complexes:*

$$\begin{array}{ccccccc} \dots \longrightarrow & \bigoplus_{x \in B_{(p)}} \mathbb{E}_{p+q}(Y_x/S) & \xrightarrow{d^G_{p,Y}} & \bigoplus_{s \in B_{(p-1)}} \mathbb{E}_{p+q-1}(Y_s/S) & \longrightarrow \dots \\ & \downarrow{\scriptstyle \sum_x p_{x*}} & & \downarrow{\scriptstyle \sum_s p_{s*}} & \\ \dots \longrightarrow & \bigoplus_{x \in B_{(p)}} \mathbb{E}_{p+q}(X_x/S) & \xrightarrow{d^G_{p,X}} & \bigoplus_{s \in B_{(p-1)}} \mathbb{E}_{p+q-1}(X_s/S) & \longrightarrow \dots \end{array}$$

- *If p is étale, then there is an adjunction map:*

$$\psi_p : f_*(\mathbb{E}^!_X) = f_*\pi^!(\mathbb{E}) \xrightarrow{ad(p^*,p_*)} f_*p_*p^*\pi^!(E) = f'_*\pi'^!(\mathbb{E}) = f'_*\mathbb{E}^!_Y$$

which induces a morphism of spectral sequences, converging to the indicated map on the abutment:

$$\begin{array}{ccc} A^\delta_p(B, H^\delta_q(f_*\mathbb{E}^!_X))) & \Rightarrow & \mathbb{E}_{p+q}(X/S) \\ \downarrow{\scriptstyle \psi_{p*}} & & \downarrow{\scriptstyle p^*} \\ A^\delta_p(B, H^\delta_q(f'_*\mathbb{E}^!_Y))) & \Rightarrow & \mathbb{E}_{p+q}(Y/S). \end{array}$$

Moreover, the map ϕ_{p*} *is induced by the following morphism of Gersten complexes:*

$$\begin{array}{ccccccc} \dots \longrightarrow & \bigoplus_{x \in B_{(p)}} \mathbb{E}_{p+q}(X_x/S) & \xrightarrow{d^G_{p,X}} & \bigoplus_{s \in B_{(p-1)}} \mathbb{E}_{p+q-1}(X_s/S) & \longrightarrow \dots \\ & \downarrow{\scriptstyle \sum_x p^*_x} & & \downarrow{\scriptstyle \sum_s p^*_s} & \\ \dots \longrightarrow & \bigoplus_{x \in B_{(p)}} \mathbb{E}_{p+q}(Y_x/S) & \xrightarrow{d^G_{p,Y}} & \bigoplus_{s \in B_{(p-1)}} \mathbb{E}_{p+q-1}(Y_s/S) & \longrightarrow \dots \end{array}$$

PROOF. Each point is obtained by using the functoriality of the homotopy Leray spectral sequence with respect to the $\mathscr{T}$-spectrum $\mathbb{E}$. The computation of the map on the abutment follows from the definition of the functoriality of the bivariant theory. The map on the E_2-terms follows from point (2) of Proposition 3.2.7. □

REMARK 4.1.5. Using fundamental classes as defined in [**DJK18**], and the induced functoriality on bivariant theory, one can extend the contravariant étale functoriality to smoothable lci morphisms $f : Y \to X$, up to considering twists by the Thom space of the cotangent complex of f. We leave the formulation to the reader.

It is possible to describe the filtration on the abutment of the homotopy Leray spectral sequence in geometric terms. The following definition is the obvious generalization of the classical definition of Grothendieck (see [**BO74**]).

DEFINITION 4.1.6. Consider the setting of Paragraph 4.1.1. We define the δ-niveau filtration on $\mathbb{E}_*(X/S)$ relative to f as:

$${}^\delta N^f_p \mathbb{E}_*(X/S) = \bigcup_{i:Z \to X, \delta(Z) \leq p} \operatorname{Im}\big(i_* : \mathbb{E}_*(X \times_B Z/S) \to \mathbb{E}_*(X/S)\big).$$

where i runs over the closed immersions with target X.

REMARK 4.1.7. One can also describe this filtration by the following formula:

$$^{\delta}N_p^f\mathbb{E}_*(X/S) = \bigcup_{\pi:Y\to X,\delta(Y)\leq p} \operatorname{Im}\big(\pi_* : \mathbb{E}_*(X\times_B Y/S) \to \mathbb{E}_*(X/S)\big).$$

where $\pi : Y \to X$ runs over the proper morphism with target X. Indeed, such a morphism always factors as

$$Y \xrightarrow{\bar{\pi}} \pi(Y) \xrightarrow{i} X$$

where $\pi(Y)$ denotes the image of Y, with its canonical structure of a closed subscheme of X. As the map $\bar{\pi}$ is surjective, one has $\delta(\pi(Y)) \leq p$. This concludes.

4.1.8. Before stating the computation of the filtration on the δ-homotopy spectral sequence, we will introduce notations in order to simplify the proof. We consider again the assumptions and notations of Paragraph 4.1.1.

First, note that by definition, the filtration induced by the δ-homotopy spectral sequence on $\mathbb{E}_*(X/S)$ is defined as:

$$^{\delta}F_q^f\mathbb{E}_*(X/S) = \operatorname{Im}\big(H_*(B, \tau_{\geq q}f_*\mathbb{E}_X^!) \to \mathbb{E}_*(X/S)\big).$$

Second, we can define the δ-niveau filtration relative to f at the level of schemes, by considering the following ind-schemes:

$$B_{\leq p} = \text{"}\varinjlim_{Z_*\in\mathcal{F}(B)}\text{"}\ Z_p.$$

Therefore we get a closed ind-immersion $B_{\leq p} \xrightarrow{i_p} B$. We can define a kind of complementary immersion by considering the following pro-objects:

$$B_{>p} = \text{"}\varprojlim_{Z_*\in\mathcal{F}(B)}\text{"}\ (B - Z_p)$$

together with the pro-open immersion $B_{>p} \xrightarrow{j_p} B$.

Using these notations, we can consider the localization long exact sequence

$$\mathbb{E}_*(X\times_B B_{\leq p}/S) \xrightarrow{i_{p*}} \mathbb{E}_*(X/S) \xrightarrow{j_p^*} \mathbb{E}_*(X\times_B B_{>p}/S) \xrightarrow{\partial_p} \mathbb{E}_*(X\times_B B_{\leq p}/S)$$

where the third (resp. first, fourth) member(s) is the obvious colimit, using the contravariance of $\mathbb{E}_*(-/S)$ with respect to open immersion (resp. contravariance with respect to closed immersions). This long exact sequence is nothing else than the filtered colimit of the localisation sequences with respect to the closed immersions $X \times_B Z_p \to X$ for δ-flags Z_* of B.

With these notations, the δ-niveau filtration relative to f simply equals the image of i_{p*}.

PROPOSITION 4.1.9. *Consider the above assumptions and notation. Then for any pair of integer $(p, n) \in \mathbf{Z}^2$, one has the following relation:*

$$^{\delta}F_p^f\mathbb{E}_n(X/S) = {}^{\delta}N_{n-p}^f\mathbb{E}_n(X/S),$$

where the left hand-side is the filtration on the abutment of the homotopy Leray spectral sequence associated with f and the right-hand side is the δ-niveau filtration relative to f.

PROOF. Let us put $\mathbb{E}' = f_*\mathbb{E}^!_X = f_*\pi^!\mathbb{E}$. We want to compare the following filtrations:

$$\begin{aligned} {}^\delta F^f_q\mathbb{E}_*(X/S) &= \mathrm{Im}\big(H_*(B, \tau_{\geq q}\mathbb{E}') \to H_*(B, \mathbb{E}')\big), \\ {}^\delta N^f_p\mathbb{E}_*(X/S) &= \mathrm{Im}\big(H_*(B_{\leq p}, \mathbb{E}') \to H_*(B, \mathbb{E}')\big). \end{aligned}$$

So, reasoning with $\mathbb{E}'$ instead of $\mathbb{E}$, we reduce to the case $X = B = S$, $f = Id_X$, $\pi = Id_X$.

We will start with the following lemma:

LEMMA 4.1.10. *Let $\mathbb{E}$ be a $\mathscr{T}$-spectrum over a scheme X. Then one has the following vanishing:*

$$\begin{aligned} H_n\big(X_{\leq p}, \tau^\delta_{<q}\mathbb{E}\big) &= 0 \text{ if } n \geq p + q, \\ H_n\big(X_{>p}, \tau^\delta_{\geq q}\mathbb{E}\big) &= 0 \text{ if } n \leq p + q. \end{aligned}$$

This follows from the case $X = S$, $m = q$, $i = 0$ of the equivalent conditions of Paragraph 2.1.10.

We can now consider the following commutative diagram, whose rows and columns are exact sequences:

$$\begin{array}{ccccccc} & & H_{n+1}\big(X_{\leq p}, \tau^\delta_{<q}\mathbb{E}\big) & & & & \\ & & \downarrow & & & & \\ H_{n+1}\big(X_{>p}, \tau^\delta_{\geq q}\mathbb{E}\big) & \longrightarrow & H_n\big(X_{\leq p}, \tau^\delta_{\geq q}\mathbb{E}\big) & \xrightarrow{b} & H_n\big(X, \tau^\delta_{\geq q}\mathbb{E}\big) & \longrightarrow & H_n\big(X_{>p}, \tau^\delta_{\geq q}\mathbb{E}\big) \\ & & \downarrow{\scriptstyle a} & & \downarrow{\scriptstyle c} & & \\ & & H_n\big(X_{\leq p}, \mathbb{E}\big) & \xrightarrow[d]{} & H_n\big(X, \mathbb{E}\big) & & \\ & & \downarrow & & & & \\ & & H_n\big(X_{\leq p}\tau^\delta_{<q}, \mathbb{E}\big) & & & & \end{array}$$

According to the preceding lemma, one gets that:

- a is an isomorphism if $n \geq p + q$.
- b is an isomorphism if $n \leq p + q - 1$, and an epimorphism if $n = p + q$.

Therefore, we obtain that $\mathrm{Im}(c) = \mathrm{Im}(d)$ if $n = p + q$. This concludes. □

EXAMPLE 4.1.11. The preceding proposition recovers the conjecture of Washnitzer proved by Bloch and Ogus in [**BO74**, 6.9]. This is obtained as follows:

- S is the spectrum of a perfect field k, X is smooth over k;
- $\mathbb{E} = \mathbb{E}_{dR}$ is the (ring) spectrum representing De Rham cohomology as in [**CD12**, §3.1] while $\mathscr{T} = \mathrm{DM}(-, \mathbf{Q})$.

The fact the truncation of $\mathbb{E}_{dR}$ for the homotopy t-structure agrees with the truncation of the De Rham complex follows from the construction of $\mathbb{E}_{dR}$ ([**CD12**, 3.1.5]).

REMARK 4.1.12. The proof of Bloch and Ogus is less precise and more theoretical. It consists in proving that the niveau spectral sequence for De Rham cohomology agrees with the hypercohomology spectral sequence associated with the De Rham complex.

We do not need such a comparison to prove our result, and our proof is more direct. But however, let us indicate that there is also an underlying comparison of spectral sequences. In fact, it is possible to prove that the homotopy Leray spectral sequence defined above agrees from E_2-on with the δ-niveau spectral sequence 3.2.3 of X/S with coefficients in the spectrum $f_*(\mathbb{E})$. The case where $X = B$, $f = Id_B$, S is the spectrum of a perfect field was proved in [**Bon10**] and [**Dég14b**]. The general case will be treated in future work.

4.2. The cohomological version.

4.2.1. Let us consider again the assumptions of Paragraph 4.1.1, but with $X = S$. So we fix a morphism of schemes:

$$f : X \to B$$

and a $\mathscr{T}$-spectrum $\mathbb{E}$ over X.

To get products on the δ-homotopy spectral sequence of $(f, \mathbb{E})$, we will use a classical construction of Douady (see [**Dou59**, Section II]) which uses the theory of *spectral diagrams* from [**CE99**, XV.7] rather than that of exact couples.

Here is how one gets such a spectral diagram underlying the spectral sequence (4.2.8.a). We first define the following tautological functors in a $\mathscr{T}$-spectrum $\mathbb{F}$ over B (in the end $\mathbb{F}$ plays the role of $f_*\mathbb{E}$):

$$\tau^{\delta}_{\geq -\infty}(\mathbb{F}) = \mathbb{F},\ \tau^{\delta}_{\geq +\infty}(\mathbb{F}) = 0,$$
$$\tau^{\delta}_{<-\infty}(\mathbb{F}) = 0,\ \tau^{\delta}_{<+\infty}(\mathbb{F}) = \mathbb{F}.$$

Consider the following poset:

$$\mathcal{P} = \{(p, q) \mid p, q \in \mathbf{Z} \cup \{\pm\infty\}, p \leq q\}.$$

such that $(p, q) \leq (p', q')$ when $p \leq p'$ and $q \leq q'$. Then for any $(p, q) \in \mathcal{P}$, we put:

$$\tau^{\delta}_{[p,q[}(\mathbb{F}) := \tau^{\delta}_{\geq p}\tau^{\delta}_{<q}(\mathbb{F}).$$

This defines a contravariant functor

$$\tau^{\delta}(\mathbb{F}) : \mathcal{P} \to \mathscr{T}(X), (p, q) \mapsto \tau^{\delta}_{[p,q[}(\mathbb{F}).$$

Using the truncation triangles associated with the δ-homotopy t-structure, we obtain distinguished triangles for $p \leq q \leq r$:

$$\tau^{\delta}_{[q,r[}(\mathbb{F}) \to \tau^{\delta}_{[p,r[}(\mathbb{F}) \to \tau^{\delta}_{[p,q[}(\mathbb{F}) \xrightarrow{\partial} \tau^{\delta}_{[q,r[}(\mathbb{F})[1]$$

where the first two maps are given by the functoriality of τ^{δ}. These formulas imply that the contravariant functor:

$$\mathcal{P} \mapsto \mathscr{A}b, (p, q) \mapsto H_*\big(B, \tau^{\delta}_{[p,q[}(f_*\mathbb{E})\big)$$

defines a $\mathbf{Z}$-graded spectral diagram in the sense of *loc. cit.*.[9] Again, this construction is obviously functorial in $\mathbb{E}$.

Note that for $\mathbb{F} = f_*(\mathbb{E})$, the exact couple of Paragraph 4.1.1 is contained in the preceding spectral diagram — it corresponds to pairs $(p, p + 1)$. Therefore, the spectral sequence associated with the preceding spectral diagram in [**CE99**] coincides with the δ-homotopy Leray spectral sequence of $(f, \mathbb{E})$. In particular, we get the following formula:

$$E^r_{p,q}(f, \mathbb{E}) = \mathrm{Im}\Big(H_{p+q}\big(B, \tau^{\delta}_{[p-r,p[}(f_*\mathbb{E})\big) \to H_{p+q}\big(B, \tau^{\delta}_{[p-1,p+r-1[}(f_*\mathbb{E})\big)\Big).$$

[9] Property (SP.5) of *loc. cit.* is not immediate. It follows from the convergence of the homotopy Leray spectral sequence — see Theorem 4.1.2.

REMARK 4.2.2. After the seminal work of Cartan and Eilenberg, spectral diagrams have appeared in other forms in the literature. In the triangulated context, one can refer, mainly for historical purposes, to [**Ver96**]. The ∞-categorical context is much more recent, but also much more powerful and satisfactory. The notion of spectral diagram in the ∞-categorical context is introduced by Lurie in [**Lur18**, Section 1.2.2], under the name of $\mathbf{Z}$-complex. The advantage of using ∞-category with pushouts (eg: stable ∞-categories) is that a $\mathbf{Z}$-complex is essentially equivalent to a tower of objects: see [**Lur18**, Lemma 1.2.2.4]. The fact that we can stay in the old-fashioned world of triangulated categories stems from the good behavior of t-structures.

To obtain products on the homotopy Leray spectral sequence following Douady, we need to refine Proposition 2.1.13 as follows.

PROPOSITION 4.2.3. *We consider the preceding notations and assume $\delta \geq 0$. Suppose we are given a morphism of $\mathscr{T}$-spectra over a scheme B:*

$$\mu : \mathbb{F} \otimes \mathbb{F}' \to \mathbb{F}''.$$

Then for any triple of integers (p, q, r) such that $r \geq 0$, there exists a canonical map:

$$\tau^{\delta}_{[p,p+r[}(\mathbb{F}) \otimes \tau^{\delta}_{[q,q+r[}(\mathbb{F}') \xrightarrow{\varphi_{p,q,r}} \tau^{\delta}_{[p+q,p+q+r[}(\mathbb{F}'').$$

Moreover, this pairing satisfies the formulas (SPP 1) and (SPP 2) of [**Dou59**, II.A, Définition, p. 19-06].[10]

PROOF. It is sufficient to treat the case where $\mathbb{F}'' = \mathbb{F} \otimes \mathbb{F}'$ and μ is the identity. As by assumption $\delta \geq 0$, the tensor product respects homologically positive objects (see the end of Paragraph 2.1.11).

According to Proposition 2.1.13, we get a canonical pairing:

$$\tau^{\delta}_{\geq p}(\mathbb{E}) \otimes \tau^{\delta}_{\geq q}(\mathbb{F}) \to \tau^{\delta}_{\geq p+q}(\mathbb{E} \otimes \mathbb{F}).$$

Let us consider the following diagram:

$$\begin{array}{ccc}
\tau^{\delta}_{\geq p}(\mathbb{E}) \otimes \tau^{\delta}_{\geq q+r}(\mathbb{F}) & \longrightarrow & \tau^{\delta}_{\geq p+q+r}(\mathbb{E} \otimes \mathbb{F}) \\
\downarrow & & \downarrow \\
\tau^{\delta}_{\geq p}(\mathbb{E}) \otimes \tau^{\delta}_{\geq q}(\mathbb{F}) & \longrightarrow & \tau^{\delta}_{\geq p+q}(\mathbb{E} \otimes \mathbb{F}) \\
\downarrow & & \downarrow \\
\tau^{\delta}_{\geq p}(\mathbb{F}) \otimes \tau^{\delta}_{[q,q+r[}(\mathbb{F}') & \dashrightarrow & \tau^{\delta}_{[p+q,p+q+r[}(\mathbb{F} \otimes \mathbb{F}')
\end{array}$$

where the solid arrows form a commutative diagram. The two columns of this diagram come from distinguished triangles. First this implies that the slanted dotted arrow is zero. Second, it implies that there exists a unique dotted horizontal arrow making the bottom square commutative.

---

[10] (SPP1) is the functoriality of this pairing with respect to $(p, p + r)$ (resp. $(q, q + r)$) considered as an object of $\mathcal{P}$. (SPP2) is the Leibniz rule for the boundary map of type ∂.

Then we consider the following diagram:

$$\begin{array}{ccc}
\tau^\delta_{\geq p+r}(\mathbb{E}) \otimes \tau^\delta_{[q,q+r[}(\mathbb{F}) & & \\
\downarrow & \overset{(1)}{\searrow} & \\
\tau^\delta_{\geq p}(\mathbb{E}) \otimes \tau^\delta_{[q,q+r[}(\mathbb{F}) & \longrightarrow & \tau^\delta_{[p+q,p+q+r[}(\mathbb{E} \otimes \mathbb{F}) \\
\downarrow & \underset{(2)}{\nearrow} & \\
\tau^\delta_{[p,p+r[}(\mathbb{F}) \otimes \tau^\delta_{[q,q+r[}(\mathbb{F}') & &
\end{array}$$

As the tensor product is right t_δ-exact, the object $\tau^\delta_{\geq p+r}(\mathbb{E}) \otimes \tau^\delta_{[q,q+r[}(\mathbb{F})$ is in homological degree $\geq p+q+r$. By definition, the object $\tau^\delta_{[p+q,p+q+r[}(\mathbb{E} \otimes \mathbb{F})$ is in homological degree $< p+q+r$. So the map labeled (1) must be 0. Therefore, the map (2) must exist and gives the existence of the pairing $\varphi_{p,q,r}$. The uniqueness of the construction then guarantees Douady's coherence properties (SPP1) and (SPP2). □

4.2.4. Granted Proposition 4.2.3, one may now apply the construction of [**Dou59**, Th. II]. Going back to the setting of Paragraph 4.2.1, and to our morphism $f : X \to B$, we consider a pairing of $\mathscr{T}$-spectra over X:

$$\mu : \mathbb{E} \otimes \mathbb{E}' \to \mathbb{E}''.$$

As f_* is weakly monoidal (left adjoint of a monoidal functor), we get a pairing:

$$\mu : f_*(\mathbb{E}) \otimes f_*(\mathbb{E}') \to f_*(\mathbb{E}'').$$

Applying the preceding proposition and Douady's construction, we get a pairing of spectral sequences:

$$E_r^{p,q}(f, \mathbb{E}) \otimes_{\mathbf{Z}} E_r^{s,t}(f, \mathbb{E}') \to E_r^{p+s,q+t}(f, \mathbb{E}'')$$

such that the differentials d_r satisfy the usual Leibniz rule.

Following standard usage, when considering products, we renumber the δ-homological spectral sequence cohomologically, and use our cohomological conventions (see in particular Paragraph 3.3.1). We then obtain the following result.

THEOREM 4.2.5. *Suppose $f : X \to B$ is a morphism of schemes, and let $\mathbb{E}$ be a $\mathscr{T}$-spectrum over X. The constructions of Paragraph* 4.2.1 *and* 4.2.4 *yield a convergent spectral sequence of the form:*

$$E_2^{p,q}(f, \mathbb{E}) = A^p_\delta(B, H^q_\delta(f_*\mathbb{E})) \Rightarrow H^{p+q}(X, \mathbb{E}).$$

If $\mathbb{E}$ admits a ring structure, then the spectral sequence is equipped with a multiplicative structure; the product on the E_2-term is induced by the construction of Example 3.3.3.

Before proceeding to a computation of the above E_2-term, let us introduce the following general construction within the six functors formalism satisfied by $\mathscr{T}$:

PROPOSITION 4.2.6. *Let $f : X \to S$ be a separated morphism essentially of finite type.*

Then there exists a pair of adjoint functors:

$$f_! : \mathscr{T}(X) \to \mathscr{T}(S) : f^!$$

such that for any factorisation $X \xrightarrow{u} X_0 \xrightarrow{\bar{f}} S$ *where* u *is pro-étale and* $\bar{f}$ *is of finite type, one has:* $f^! = u^* \bar{f}^!$.

PROOF. We write X as a projective limit of a pro-S-scheme $(X_i)_{i \in I}$ whose transition morphisms are étale affine of finite type and X_i/S is separated of finite type. If we choose an index $i \in I$, considering the induced factorisation

$$X \xrightarrow{u_i} X_i \xrightarrow{f_i} S$$

we put $f_! = f_{i!} u_{i\sharp}$, and $f^! = u_i^* f_i^!$. This definition does not depend on the chosen index i. First assume we have two indexes $i, j \in I$ with a map $j \to i$. Then we get a commutative diagram:

$$\begin{array}{ccccc} X & \xrightarrow{u_j} & X_j & \xrightarrow{f_j} & X \\ \| & & \downarrow{\scriptstyle \varphi_{ij}} & & \| \\ X & \xrightarrow{u_i} & X_i & \xrightarrow{f_i} & X \end{array}$$

and we get:

$$f^! = u_i^* f_i^! = u_j^* \varphi_{ij}^* f_i^! = u_j^* \varphi_{ij}^! f_i^! = u_j^* f_j^!.$$

In general, given two indices $i, j \in I$, there exists an third index $k \in I$ and maps $k \to i$, $k \to j$ which gives the canonical identification.

The definition does not depend on the pro-object chosen to present X/S. Indeed, any two such pro-objects are isomorphic. Finally, any factorisation as in the statement of the proposition can be taken as an element of a pro-object presenting X/S. □

REMARK 4.2.7.

(1) A more rigorous approach for the preceding proof will use a limit argument as in Deligne's construction of the functor $f_!$ when f is separated of finite type. Details are left to the reader.
(2) The procedure described in the previous proposition extends a classical trick used in [**BBD82**, 2.2.12].
(3) The extension of the pair of adjoint functors $(f_!, f^!)$ to the case where $f : X \to S$ is essentially of finite type immediately gives an extension of the definition of bivariant theories to schemes X/S essentially of finite type. It is straightforward to check this extension, under the continuity assumption (T2) agrees with that defined in Paragraph 2.1.1.

4.2.8. We endeavor to describe the E_2-term of the cohomological form of the homotopy Leray spectral sequence associated with a morphism $f : X \to B$ and a $\mathscr{T}$-spectrum $\mathbb{E}$ over X. Recall from Theorem 4.1.2 (with $X = S$ and with cohomological conventions) that $E_2^{p,q}(f, \mathbb{E})$ is the p-cohomology of the Gersten δ-cohomological complex $C_\delta^*(B, H_\delta^q f_* \mathbb{E})$. Moreover, for $p \in \mathbf{Z}$, one has:

(4.2.8.a) $$C_\delta^p(B, H_\delta^q f_* \mathbb{E}) = \bigoplus_{x \in B^{(p)}} H^{p+q}(X_x/X, \mathbb{E}),$$

where we have used the extension of the bivariant theory defined by $\mathbb{E}$ as explained in the preceding remark.

PROPOSITION 4.2.9. *Notations and assumptions as in 4.2.8, the following statements hold.*

(1) *If f is smooth and $\mathbb{E} = f^*\mathbb{E}_0$ for a $\mathscr{T}$-spectrum $\mathbb{E}_0$ over B, then, for any $p \in \mathbf{Z}$, one has:*

$$C^p_\delta(B, H^q_\delta f_*\mathbb{E}) = \bigoplus_{x \in B^{(p)}} H^{p+q}\big(X_x, \mathbb{E}_x\big)$$

where $\mathbb{E}_x = x^!\mathbb{E}$. More generally, given any point $x : \operatorname{Spec}(K) \to B$ such that $\delta(x) = -p$, and any integer $r \in \mathbf{Z}$, one has (recall Definition 3.1.2 and Paragraph 3.3.1):

$$\hat{H}^q_\delta f_*\mathbb{E}_r(x) = H^{p+q}\big(X_x, \mathbb{E}_x\{p+r\}\big),$$

again with $X_x = x^{-1}X$ and $E_x = x^!\mathbb{E}$.

(2) *If X is regular, the fibers of f are regular and $\mathbb{E}$ is X-pure (Definition 2.2.2), then for any $p \in \mathbf{Z}$, one has:*

$$C^p_\delta(B, H^q_\delta f_*\mathbb{E}) = \bigoplus_{x \in B^{(p)}} H^{p+q}\big(X_x, \mathbb{E}\langle -N(X_x/X)\rangle\big)$$

where $N(X_x/X)$ is the normal bundle of the regular closed immersion $i_x : X_x \to X$ and we set $\mathbb{E}\langle -N(X_x/X)\rangle = i_x^!\mathbb{E} \otimes \operatorname{Th}(-N(X_x/X))$.

PROOF. Consider the first point. We directly prove the assertion concerning a general point $x : \operatorname{Spec} K \to B$, as it implies the remaining assertion.[11] As in the proof of Theorem 4.1.2, consider the following cartesian diagram:

$$\begin{array}{ccc} X_x & \xrightarrow{i_x} & X \\ {\scriptstyle f_x}\downarrow & & \downarrow{\scriptstyle f} \\ \operatorname{Spec}(K) & \xrightarrow{x} & B \end{array}$$

Since f is smooth, there is a canonical isomorphism of functors: $i_x^! f^* = f_x^* x^!$ (using the notation introduced just before the statement of the proposition for the shriek functors). The following computation then concludes the proof:

$$\begin{aligned} H^{p+q}(X_x/X, \mathbb{E}) &= \big[\mathbb{1}_{X_x}, i_x^! f^*(\mathbb{E})[p+q]\big] = \big[\mathbb{1}_{X_x}, f_x^* x^!(\mathbb{E})[p+q]\big] \\ &= \big[f_{x\sharp}(\mathbb{1}_x), \mathbb{E}_x[p+q]\big] = H^{p+q}\big(X_x, \mathbb{E}_x\big). \end{aligned}$$

The second point is a direct consequence of the form (4.2.8.a) of the Gersten complex and of Proposition 2.2.4. $\square$

Again, this spectral sequence has good functoriality properties (e.g., in the (ring) $\mathscr{T}$-spectrum E). The next result summarizes other functoriality properties.

PROPOSITION 4.2.10. *Consider a commutative diagram:*

$$\begin{array}{ccc} Y & \xrightarrow{q} & X \\ & {\scriptstyle g}\searrow \quad \swarrow{\scriptstyle f} & \\ & B & \end{array}$$

[11] To be clear, recall:
$$H^{p+q}(X_x/X, \mathbb{E}) = (\hat{H}^q_\delta f_*\mathbb{E})_{-p}(x).$$

and a ring $\mathscr{T}$-spectrum $\mathbb{E}$. Let us put $\mathbb{E}_Y = q^*\mathbb{E}$. There exists a morphism of converging spectral sequences:

$$\begin{array}{ccc} E_2^{p,q}(f,\mathbb{E}) = A^p_\delta(B, H^q_\delta(f_*\mathbb{E})) & \Longrightarrow & H^{p+q}(X,\mathbb{E}) \\ {\scriptstyle p^*}\downarrow & & \downarrow{\scriptstyle q^*} \\ E_2^{p,q}(f',\mathbb{E}') = A^p_\delta(B, H^q_\delta(g_*\mathbb{E}')) & \Rightarrow & H^{p+q}(Y,\mathbb{E}_Y) \end{array}$$

where q^* is the usual pullback on cohomology.

The morphism of spectral sequences is simply obtained using the functoriality in $\mathbb{E}$ with respect to the following map:

$$\mathbb{E} \to q_*q^*(\mathbb{E}) = q_*\mathbb{E}_Y.$$

4.3. Remarkable properties of homotopy modules.

4.3.1. *δ-effectivity.*– Consider the situation of Paragraph 4.1.1: $\mathbb{E}$ is a $\mathscr{T}$-spectrum over S and one looks at morphisms:

$$\begin{array}{ccc} X & \xrightarrow{f} & B \\ & {\scriptstyle \pi}\searrow & \\ & & S \end{array}$$

where π is separated essentially of finite type, and put $\mathbb{E}^!_X = \pi^!\mathbb{E}$. We assume in addition:

(1) $\mathbb{E}^!_X$ and $\mathbb{1}_B$ are δ-effective (Definition 3.1.6).
(2) f is proper.

Then the first and second assumptions imply $f_*(\mathbb{E}^!_X) = f_!(\mathbb{E}^!_X)$ is δ-effective (see Paragraph 3.1.9). Moreover, one can compute the E_2-term of the homotopy Leray spectral sequence as follows:

$$\begin{aligned} \mathrm{Hom}_{\mathscr{T}(B)}(\mathbb{1}_B[p], H^\delta_q f_*\mathbb{E}) &= \mathrm{Hom}_{\mathscr{T}(B)}(s\mathbb{1}_B[p], H^\delta_q sf_*\mathbb{E}) \\ &= \mathrm{Hom}_{\mathscr{T}^{\delta\text{eff}}(B)}(\mathbb{1}_B[p], wH^\delta_q sf_*\mathbb{E}) \\ &= \mathrm{Hom}_{\mathscr{T}^{\delta\text{eff}}(B)}(\mathbb{1}_B[p], H^{\delta\text{eff}}_q wsf_*\mathbb{E}) \\ &= \mathrm{Hom}_{\mathscr{T}^{\delta\text{eff}}(B)}(\mathbb{1}_B[p], H^{\delta\text{eff}}_q f_*\mathbb{E}). \end{aligned}$$

The first identification uses the assumption (1), the second the adjunction (s,w), the third the fact w is t_δ-exact and the last one the fact s is fully faithful.

We have obtained the following remarkable result.

PROPOSITION 4.3.2. *With the assumptions of 4.3.1 in place, the homotopy Leray spectral sequence takes the following form:*

$$E^2_{p,q} = A^\delta_p(B, H^{\delta\text{eff}}_q f_*\mathbb{E}^!_X) \Rightarrow \mathbb{E}_{p+q}(X/S).$$

If $X = S$, then we can also consider the cohomological form of the δ-homotopy Leray spectral sequence:

$$E_2^{p,q} = A^p_\delta(B, H^q_{\delta\text{eff}} f_*\mathbb{E}) \Rightarrow \mathbb{E}^{p+q}(X).$$

EXAMPLE 4.3.3. The interest of the preceding proposition is that it is easier to get bounded objects with respect to the δ-effective category: see Example 3.1.8.

Let us consider either the homotopical or motivic case. We consider a proper morphism $f : X \to B$ of schemes essentially of finite type over k such that X is

k-smooth. We let our dimension functions be induced by that computed relative to k. Then $\mathbb{1}_X(n)$ for $n \geq 0$ and $\mathbb{1}_B$ are both δ-positive.

If we assume X is of pure dimension d, so that $\delta(X) = d$ according to the preceding choice, then the constant object $\mathbb{1}_X$ is concentrated in homological degree d. Therefore the preceding spectral sequence, in its homological form, is concentrated in degrees $q \in [d, d + \dim(f)]$.

The following definition is an analog of the notion of local system in classical topology.

DEFINITION 4.3.4. Suppose $p : B \to S$ is a separated morphism of finite type and $\mathbb{E}$ is a $\mathscr{T}$-spectrum over B. We will say that $\mathbb{E}$ is S-*simple* if there exists a $\mathscr{T}$-spectrum $\mathbb{E}_0$ over S and an isomorphism $\mathbb{E} \simeq p^!\mathbb{E}_0$. We will say that $\mathscr{T}$ is *locally simple over* S (or locally S-simple) if there exists a Nisnevich cover $\pi : W \to B$ such that $\pi^*\mathbb{E}$ is S-simple (with respect to the projection $p \circ \pi$).

Note that the terminology S-simple is analogous to the classical terminology of "simple local system". This corresponds to the case of trivial monodromy.

REMARK 4.3.5. This notion will come into play mainly when S is the spectrum of a base field k, or, in the motivic case, of a base Dedekind ring A. For us, the interest comes, as in topology, in the study of the homotopy Leray spectral sequence (Definition 4.1.3). Usually, we will start with a k-simple $\mathscr{T}$-spectrum $\mathbb{E}$ over S – thus $\mathbb{E}_X$ is k-simple. If we know that the homotopy module $H^\delta_q f_*\mathbb{E}$ over B is k-simple, then the E_2-term depends only on the homotopy type (resp. motive) of B over k. We will give examples in Section 5.

REMARK 4.3.6.

(1) Obviously, locally S-simple $\mathscr{T}$-spectra over B are stable under suspensions, twists and even tensor products by Thom spaces of virtual bundles over B. The same is true for S-simple $\mathscr{T}$-spectra except that one can only twist them by virtual bundles over B that come from S. Neither of these notions is stable under extensions or even direct factors in general.
(2) Let $f : B \to S$ be smooth and consider the motivic case. Then $f^! = f^*(d)[2d]$ where d is the relative dimension of f. One deduces that S-simple motives over B are stable under tensor products. The same remark applies to oriented spectra over B, but not to arbitrary spectra.
(3) Note that if k is a field, k-simple over a scheme B implies B-pure (Definition 2.2.2).
(4) One could say that a homotopy module over B is S-constructible if it is obtained by a finite number of extensions and direct factors of locally S-simple homotopy modules, within the abelian category of homotopy modules over B.

    This notion is not so well-behaved, compared to its model for torsion étale sheaves, as it lacks some notion of finiteness. It would be desirable to have some good finiteness condition on S-simple homotopy modules. But even when S is a base field, it is not obvious to find such a finiteness condition; see [**Dég11**, Rem. 6.7] for further discussion.

Note the following fact.

LEMMA 4.3.7. *Consider a separated morphism $f : B \to S$ of finite type, and a $\mathscr{T}$-spectrum $\mathbb{E}$ over B. Let us fix unrelated dimension functions δ on B and δ_0 on S.*

If $\mathbb{E}$ is S-simple, then for any $p \in \mathbf{Z}$, the homotopy module $H_q^\delta(\mathbb{E})$ over B is S-simple.

PROOF. We just need to be precise about dimension functions. By additivity, we can assume that B is connected. Then the dimension function δ_0^f on B, induced by δ_0, satisfies the relation $\delta_0^f = \delta + n$ for a fixed integer $n \in \mathbf{Z}$ (see Remark 2.1.6). In particular, $H_q^{\delta_0}(\mathbb{E}) = H_{q+n}^{\delta}(\mathbb{E})$ according to *loc. cit.*

By assumption, $\mathbb{E} = f^!\mathbb{E}_0$. it remains to apply the fact $f^!$ is t_{δ_0}-exact to conclude:

$$H_q^\delta(\mathbb{E}) = H_{q-n}^{\delta_0}(\mathbb{E}) \simeq H_{q-n}^{\delta_0}(f^!\mathbb{E}_0) = f^! H_{q-n}^{\delta_0}(\mathbb{E}_0).$$

□

EXAMPLE 4.3.8. In the motivic case, if $f : X \to S$ is smooth, then the constant object $\mathbb{1}_X$ is S-simple. In contrast, if f has sufficiently complicated singularities, then $\mathbb{1}_X$ may fail to be S-simple.

Similarly, the classical oriented ring spectra $\mathbf{H}R_X$, $\mathbf{KGL}_X$ (K-theory), $\mathbf{MGL}_X$ (algebraic cobordism), $\hat{\mathbf{H}}\mathbf{Q}_{\ell,X}$ (representing continuous ℓ-adic cohomology) are all S-simple.

On the contrary, spectra representing non-orientable theories such as $\mathbf{H}\tilde{R}_X$, KQ_X (hermitian K-theory), or the sphere spectrum $\mathbf{S}_X^0$, are not S-simple except when the tangent bundle of f is trivial (or is the pullback of a vector bundle over S). In any case, they all are locally S-simple.

Note finally that when $f : X \to S$ is arbitrary separated of finite type, the main result of [**Jin18**], Theorem 1.3, tells us that the spectrum $\mathbf{GGL}_X$ representing algebraic G-theory in $\mathrm{SH}(X)$ is S-simple.

As expected, here is the generic case where the homotopy modules appearing in the E_2-term of the homotopy Leray spectral sequence are simple.

PROPOSITION 4.3.9. *Let B and F be S-schemes separated essentially of finite type. We consider the trivial fibration $f : X = F \times_S B \to B$. Then for any S-simple $\mathscr{T}$-spectrum $\mathbb{E}$ over X, the $\mathscr{T}$-spectrum $f_*\mathbb{E}$ is S-simple.*

PROOF. This is a trivial exercise on the six functors formalism. We consider the cartesian square:

$$\begin{array}{ccc} X & \xrightarrow{f} & B \\ {\scriptstyle q}\downarrow & \Delta & \downarrow{\scriptstyle p} \\ F & \xrightarrow{f_0} & S. \end{array}$$

Then we get an associated exchange isomorphism: $Ex(\Delta_*^!) : p^! f_{0*} \xrightarrow{\sim} f_* q^!$. By assumption, there exists a $\mathscr{T}$-spectrum $\mathbb{E}_0$ over S such that $\mathbb{E} = h^!\mathbb{E}_0$, $h = pf = f_0 q$. Thus we can do the computation:

$$f_*\mathbb{E} = f_* h^! \mathbb{E}_0 = f_* q^! f_0^! \mathbb{E}_0 \xrightarrow{Ex(\Delta_*^!)^{-1}} p^! f_{0*} f_0^! \mathbb{E}_0.$$

This concludes the proof. □

COROLLARY 4.3.10. *Let $f : X \to B$ be Nisnevich locally principle fibration of S-schemes. Then for any locally S-simple $\mathscr{T}$-spectrum $\mathbb{E}$ over B, the $\mathscr{T}$-spectrum $f_*\mathbb{E}$ is locally S-simple.*

5. Applications

In this section, we study the homotopy Leray spectral sequence in various simple cases and include some applications. Section 5.1 studies morphism with (stably) $\mathbf{A}^1$-contractible fibers, and then morphisms with fibers that are "$\mathbf{A}^1$-homology spheres". Section 5.2 studies "fibrations" with either base or fiber that are motivic spheres. Finally, Section 5.3 is concerned with some applications of the spectral sequence to relative cellular spaces.

5.1. Morphisms with $\mathbf{A}^1$-contractible fibers. We first analyze the case where the fibers are $\mathbf{A}^1$-homotopically "as simple as possible", i.e., $\mathbf{A}^1$-contractible.

PROPOSITION 5.1.1. *Suppose $f : X \to B$ be a smooth morphism with $\mathbf{A}^1$-contractible fibers. Let M be a spectrum (resp. a motive) that is a homotopy module over B and set $M_X := f^*M$. Then,*

$$H^q_\delta f_* M_X = \begin{cases} M & q = 0, \\ 0 & q \neq 0. \end{cases}$$

In particular, the homotopy Leray spectral sequence is concentrated on the line $q = 0$, and thus degenerates; pullback along f yields identifications (in cohomological notation) of the form:

$$H^p(X, M_X) = H^p(B, H^0_\delta f_*(M_X)) = H^p(B, M).$$

Note in particular that f_* respects S-simple and locally S-constant objects.

PROOF. Consider the map

$$\alpha : H^q_\delta(M) \to H^q_\delta(f_* M_X)$$

induced by the adjunction map $M \to f_*f^*(M)$. We need only to prove it is an isomorphism on fiber homology (Remark 3.1.3). We compute the above map, evatuated at a point $x : \operatorname{Spec} K \to B$ and in $\mathbf{G}_m$-degree $r \in \mathbf{Z}$:

$$H^q_\delta(M)_r(x) \to H^q_\delta(f_* M_X)_r(x).$$

We use the computation of Proposition 4.2.9(1), which we can apply as f is smooth. So if we put $\delta(x) = -p$ and $M_x = x^! M$ (using the notation of Proposition 4.2.6), one obtains:

$$H^q_\delta(M_X)_r(x) = H^{p+q}(x, M_x\{p+r\}),$$
$$H^q_\delta(f_* M_X)_r(x) = H^{p+q}(X_x, M_x\{p+r\}).$$

Moreover, the canonical map α is isomorphic to the pullback map:

$$f_x^* : H^{q+r-2p, r-p}(x, M_x) \to H^{q+r-2p, r-p}(X_x, M_x),$$

which is itself an isomorphism as f_x is assumed to be an $\mathbf{A}^1$-weak equivalence. Note also that, because the spectral sequence is functorial with respect to pullbacks (Proposition 4.2.10), we know that the identification of the statement arises from the pullback map along f. The remaining assertions are straightforward. □

EXAMPLE 5.1.2. There exist many examples of morphisms satisfying the hypotheses of Proposition 5.1.1. For concreteness, assume k is a field having characteristic 0. In [**AD07**, Theorem 1.3], it is shown that there exist connected k-schemes B of arbitrary dimension and smooth morphisms $f : X \to B$ of relative dimension ≥ 6 whose fibers are $\mathbb{A}^1$-contractible and such that fibers over distinct k-points of B are pairwise non-isomorphic. Because of the last point, such morphisms f are *not* Zariski locally trivial. These results were improved in [**DF18, DPØ18**], where it was shown that one could build f as above that are smooth of relative dimension ≥ 3. On the other hand, it is expected that the only $\mathbf{A}^1$-contractible smooth k-scheme of dimension 2 is $\mathbf{A}^2$, and a long-standing conjecture of Dolgachev–Weisfeiler [**VD74**, 3.8.5] states that every flat morphism of (say) smooth schemes with all fibers isomorphic to affine space is Zariski locally trivial.

Suppose f is a smooth morphism f with $\mathbf{A}^1$-contractible fibers. It is not clear to the authors whether such an f is *unstably* an $\mathbf{A}^1$-weak equivalence without imposing further hypotheses (e.g., that f is Nisnevich locally trivial). Nevertheless, the following remark demonstrates that such f are *stable* $\mathbf{A}^1$-weak equivalences in a strong sense, which makes Proposition 5.1.1 somewhat unsurprising.

REMARK 5.1.3. Given a map $f : X \to B$ as in Proposition 5.1.1, one can directly show that the adjunction map:

$$Id \to f_* f^*$$

is an isomorphism of functors. Indeed, we can use the same argument as above and the fact the family of functors $x^! : \mathrm{SH}(B) \to \mathrm{SH}(\mathrm{Spec}(\kappa(x)))$, indexed by schematic points $x \in B$, is conservative.[12] In particular, one does not need to work over a base field; nor does one need to invert any integers.

As f is smooth, we also get a natural transformation $f_\sharp f^* \to Id$, which is an isomorphism according to the result of the preceding paragraph. In particular, the map

$$\Sigma^\infty X_+ = f_\sharp f^*(\mathbb{1}_B) \to \mathbb{1}_B = \Sigma^\infty B_+$$

is an isomorphism in $\mathrm{SH}(B)$. In fact, f is a *universal* stable $\mathbf{A}^1$-weak equivalence since f is a stable $\mathbf{A}^1$-weak equivalence and the property mentioned above remains true after base change.

5.2. Gysin and Wang sequences. The relative Atiyah–Hirzebruch spectral sequence takes a particularly simple form when the Serre fibration $F \to X \to B$ under consideration has either the property that B is a sphere, or F is a homology sphere and the associated local system on B is trivial. In those cases, the spectral sequences yields the so-called Wang or Gysin long exact sequences. The fact that the differentials in the cohomological form of the spectral sequence are derivations yields additional structure in these long exact sequences that is frequently useful in computations.

$\mathbb{A}^1$-*homology spheres.* In motivic homotopy theory, there are many smooth schemes over a base S that have the stable $\mathbf{A}^1$-homotopy type of a motivic sphere $\Sigma^i \mathbf{G}_m^{\wedge j}$. For example, Morel–Voevodsky showed that $\mathbf{A}^n \setminus 0_S$ has the $\mathbf{A}^1$-homotopy type of $\Sigma^{n-1}\mathbf{G}_m^{\wedge n}$. Likewise, the split smooth affine quadric Q_{2n-1} defined by the hypersurface $\sum_i x_i x_{n+i} = 1$ in $\mathbf{A}^{2n}_S$ is $\mathbf{A}^1$-weakly equivalent to $\mathbf{A}^n \setminus 0$ and

[12] One uses for example Proposition 4.2.6 to define $x^!$. The conservativity property is obtained using the continuity property of SH and the localization property.

[**ADF17**, Theorem 2] demonstrates that the smooth affine quadric Q_{2n} defined by the equation $\sum_i x_i x_{n+i} = x_{2n+1}(1-x_{2n+1})$ in $\mathbf{A}^{2n+1}_S$ is a model of $\Sigma^n \mathbf{G}_m^{\wedge n}$. On the other hand, it is known that $\Sigma^i \mathbf{G}_m^{\wedge j}$ has no smooth model if $i > j$ and conjecturally has no smooth model if $i < j - 1$. We now formulate a definition of "homology sphere" in $\mathbf{A}^1$-homotopy theory.

DEFINITION 5.2.1. We say that a (pointed) smooth S-scheme X is an $\mathbf{A}^1$-homology sphere if there exist integers $p, q, r \geq 0$ and an $\mathbf{A}^1$-weak equivalence $\Sigma^r X \sim \Sigma^p \mathbf{G}_m^{\wedge q}$.

The next proposition gives a construction of many $\mathbf{A}^1$-homology spheres, at least over a field.

PROPOSITION 5.2.2. *Fix a smooth base scheme S, and an $\mathbf{A}^1$-contractible smooth S-scheme X. Assume there exists a closed immersion of S-schemes $x : S \to X$ with trivial normal bundle $\nu_{x/X}$.*

(1) *A choice of trivialization of $\nu_{x/X}$ determines an $\mathbf{A}^1$-weak equivalence $\Sigma X \setminus x \cong \Sigma \mathbf{A}^d \setminus 0$, i.e., $X \setminus x$ is an $\mathbf{A}^1$-homology sphere.*
(2) *If $S = \operatorname{Spec}(k)$ for a perfect field k, X has dimension $d \geq 3$ and is $\mathbf{A}^1$-connected, then $X \setminus x$ is $\mathbf{A}^1$-simply connected as well.*

PROOF. For the first point, note that there is a cofiber sequence of the form

$$X \setminus x \longrightarrow X \longrightarrow X/(X \setminus x) \longrightarrow \Sigma X \setminus x \longrightarrow \cdots$$

Because the $\mathbf{A}^1$-local model structure is left proper, the fact that X is $\mathbf{A}^1$-contractible implies that the map $Th(\nu_{x/X}) \to \Sigma X \setminus x$ is an $\mathbb{A}^1$-weak equivalence. Under the assumptions on S, there is a homotopy purity isomorphism $X/(X \setminus x) \cong Th(\nu_{x/X})$ and a choice of trivialization of $\nu_{x/X}$ determines an $\mathbb{A}^1$-weak equivalence $Th(\nu_{x/X}) \xrightarrow{\sim} \Sigma^d \mathbf{G}_m^{\wedge d}$, which can be written $\Sigma \mathbf{A}^d \setminus 0$.

For the second point, since $X \setminus x$ is $\mathbf{A}^1$-connected, $X \setminus x$ has a non-empty set of k-points by the unstable 0-connectivity theorem [**MV99**, §2 Corollary 3.22]. Fix a base k-point in $X \setminus x$, and point X by its composite with the open immersion $X \setminus x \to X$. Finally, since $d \geq 3$, we may appeal to [**AD09**, Theorem 4.1] to conclude that the morphism $\pi_1^{\mathbf{A}^1}(X \setminus x) \to \pi_1^{\mathbf{A}^1}(X)$ is an isomorphism. Since X is $\mathbf{A}^1$-contractible, the latter sheaf is trivial and $X \setminus x$ is thus $\mathbf{A}^1$-simply connected. $\square$

REMARK 5.2.3. If k is a field, and X is furthermore affine, then $X \setminus x$ is isomorphic to $\mathbf{A}^d \setminus 0$ if and only if X is isomorphic to $\mathbf{A}^d$. Indeed, if there exists an isomorphism from $X \setminus x$ to $\mathbf{A}^d \setminus 0$, then normality of X allows one to extend this isomorphism to an isomorphism of X with $\mathbf{A}^d$; the other implication is immediate.

EXAMPLE 5.2.4. If X is the smooth affine threefold defined by $x + x^2 y + z^2 + t^3 = 0$, then the main result of [**DF18**] implies that X is $\mathbf{A}^1$-contractible, at least if k is an infinite field. In that case, for any extension L/k, B. Antieau observed that X is connected by chains of affine lines (see [**DPØ18**, Example 2.28] for a proof). If L is an infinite field, we may always assume our chains avoid a codimension ≥ 2 subset, and in particular it follows that $X \setminus x$ is connected by chains of affine lines; it follows that, $X \setminus x$ is $\mathbf{A}^1$-connected. Thus, if k is infinite and perfect, $X \setminus x$ satisfies the hypotheses of the proposition and yields an "exotic motivic sphere". In dimension $d \geq 4$, the examples in [**AD07**] or [**ADF17**] also satisfy the hypotheses of the theorem, at least over an infinite base field.

Gysin and Wang sequences for homotopy modules. Our goal now is to analyze the homotopy Leray spectral sequence for $f : X \to B$ a smooth morphism where B is an $\mathbf{A}^1$-homology sphere in the sense above; the outcome will be a version of the Gysin sequence. We begin by observing that from the appropriate cohomological standpoint, $\mathbf{A}^1$-homology spheres behave in a fashion analogous to spheres in classical homotopy theory, i.e., cohomology with "locally constant coefficients" is concentrated in precisely 2 degrees. To make this precise, assume we work over a field, and let δ be the usual dimension function relative to k. We consider homotopy modules M over k such that f_*M is *k-simple* in the sense of Definition 4.3.4.

LEMMA 5.2.5. *Assume k is a field, and X is a (pointed) smooth k-scheme that is an $\mathbf{A}^1$-homology sphere as in Definition* 5.2.1, *with $p, q \geq 1$. Let $f : X \to \operatorname{Spec}(k)$ be the canonical projection. For any homotopy module M over k, $M_X := f^*M$,*

$$H^i_\delta(f_*M_X) = \begin{cases} M & \text{if } i = 0 \\ M_{-q} & \text{if } i = p - r \\ 0 & \text{otherwise.} \end{cases}$$

PROOF. This result follows essentially immediately from [**AF14**, Lemma 4.5]. However, to keep the presentation self-contained, we sketch a proof in the spirit of this paper: one proceeds along the same lines as the proof of Proposition 5.1.1 and appeals to Proposition 4.2.9(1). Again we need only to compute fiber homology of $f_*(M_X)$. To this end, take a pair (x, n) where $x : \operatorname{Spec}(K) \to \operatorname{Spec}(k)$ is a induced by a field extension of finite type, and $n \in \mathbf{Z}$. To simplify the notation, let us assume that $\delta(x) = 0$; set $M_x = x^!M$.

We may then appeal to the computation of Proposition 4.2.9(1):

$$H^i_\delta(f_*M_X)_n(x) \cong H^i(X_x, M_x\{r\}) = H^i(K, M_x\{n\}) \oplus \tilde{H}^i(X_x, M_x\{n\}).$$

As X/k is pointed, X_x/K is also pointed and we obtain an identification:

$$H^i(X_x, M_x\{n\}) = H^i(K, M_x\{n\}) \oplus \tilde{H}^i(X_x, M_x\{n\})$$

where $\tilde{H}^*$ stands for the reduced cohomology of a pointed scheme. Furthermore there are isomorphisms of the form:

$$\tilde{H}^i(X_x, M_x\{n\}) \cong \tilde{H}^{i+r}(\Sigma^r X_x, M_x) \cong \tilde{H}^{i+r}(\Sigma^p \mathbf{G}_m^{\wedge q}, M_x) \cong \tilde{H}^{i+r-p}(\mathbf{G}_m^{\wedge q}, M_x).$$

Since M_x is a homotopy module over K, the last group vanishes if $i + r - p$ is not equal to 0 and is precisely $M_{-q}(\operatorname{Spec}(k))$ if $i = p - r$. The result then follows by unwinding the definitions. □

PROPOSITION 5.2.6. *Assume k is a field and $f : X \to B$ is a Zariski locally trivial smooth morphism of k-varieties with B connected and where the fibers F of f are $\mathbf{A}^1$-homology spheres (where $\Sigma^r F \sim \Sigma^p \mathbf{G}_m^{\wedge q}$ with $p, q \geq 1$). Assume furthermore that f trivializes on a Zariski open cover $\mathcal{U} = \{U_i\}_{i\in I}$ of B, and that we may fix $x \in \cap_{i\in I} U_i(k)$ (i.e., the intersection is non-empty). If M is a homotopy module and $H^j_\delta(f_*M_X)$ is k-simple for each $i \geq 0$, then*

$$H^j_\delta(f_*M_X) = \begin{cases} M_B & \text{if } j = 0 \\ (M_{-q})_B & \text{if } j = p - r\ , \\ 0 & \text{otherwise.} \end{cases}$$

and there is a long exact "Gysin" sequence of the form:

$$\cdots \longrightarrow H^i(B,M) \longrightarrow H^i(X,M) \longrightarrow H^{i+r-p}(B,M_{-q}) \xrightarrow{\partial} H^{i+1}(B,M) \longrightarrow \cdots.$$

PROOF. By Theorem 4.2.5, there is a spectral sequence with

$$E_2^{i,j} = A^i(B, H^j_\delta(f_*M_X)) \Longrightarrow H^{i+j}(X,M).$$

Granted the first statement, the spectral sequence is concentrated in two rows, and the existence of the resulting long exact sequence is immediate.

Thus, it remains to prove the first statement. If B is a field, the computation of $H^j_\delta(f_*M_X)$ is simply Lemma 5.2.5. The general case reduces to that one using the simplicity assumption and a Zariski patching argument. In more detail, by assumption, we may choose a k-point $x : \mathrm{Spec}(k) \to B$. By Lemma 5.2.5 there are induced morphisms $M \to x^*H^0_\delta(f_*M_X)$ and $M_{-q} \to x^*H^{p-r}_\delta(f_*M_X)$. Since $H^j_\delta(f_*M_X)$ is k-simple by assumption, pulling back these morphisms along the structure map $B \to \mathrm{Spec}(k)$ yields morphisms $M_B \to H^0_\delta(f_*M_X)$ and $(M_{-q})_B \to H^{p-r}_\delta(f_*M_X)$; moreover, these maps are compatible with restrictions to open sets $U \subset B$. The vanishing statement for $H^j_\delta(f_*M_X)$ will be established in an identical manner. Since f is Zariski locally trivial, upon fixing a trivialization over the open cover $\mathcal{U}$, appeal to Proposition 4.3.9 yields isomorphisms $M_{U_i} \xrightarrow{\sim} H^0_\delta(f_*M_X)|_{U_i}$ for any $i \in I$, and similarly isomorphisms of the form $(M_{-q})_{U_i} \xrightarrow{\sim} H^{p-r}_\delta(f_*M_X)|_{U_i}$. To check the maps of sheaves on B are isomorphisms, it suffices to check that the isomorphisms upon restriction to U_i just described are compatible on 2-fold intersections. However, since the all the sheaves in question are k-simple, this compatibility may be checked after pullback along x, where it is immediate. □

REMARK 5.2.7. Proposition 5.2.6 avoids the "degenerate" case where $f : X \to B$ is a $\mathbf{G}_m$-torsor. Such morphisms are, of course, always Zariski locally trivial, but behave like covering spaces in topology. Indeed, under the additional hypotheses in the statement, the sheaf $H^j_\delta(f_*M_X)$ is only non-vanishing when $j = 0$, in which case it is isomorphic to $M_B \oplus (M_{-1})_B$.

Proposition 5.2.6 admits a refinement when M is a homotopy module that admits a ring structure (e.g., K^M_* or K^{MW}_*). In such cases, $H^0(B,M)$ is a ring, and we may fix a generator ξ of $H^0(B, f_*M_{X,q}) \cong H^0(B,M)$ as a module over this ring. The class $\partial(\xi)$ then determines an element of $H^{p-r+1}(B,M_q)$ that we will refer to as the Euler class of f.

THEOREM 5.2.8. *Assume k is a field and $f : X \to B$ is a Zariski locally trivial smooth morphism of k-varieties with B connected and where the fibers F of f are $\mathbf{A}^1$-homology spheres (where $\Sigma^r F \sim \Sigma^p \mathbf{G}_m^{\wedge q}$ with $p, q \geq 1$). Assume furthermore that f trivializes on a Zariski open cover $\mathcal{U} = \{U_i\}_{i\in I}$ of B, and that we may fix $x \in \cap_{i\in I} U_i(k)$ (i.e., the intersection is non-empty). Suppose M is a homotopy module that admits a ring structure. If $H^j_\delta(f_*M_X)$ is k-simple for each $i \geq 0$, then a choice of generator ξ of $H^0(B, f_*M_{X,q}) \cong H^0(B,M)$ determines an Euler class $e(f) \in H^{p-r+1}(B,M_q)$, and the exact sequence of Proposition 5.2.6 takes the form:*

$$\cdots \longrightarrow H^i(B,M) \longrightarrow H^i(X,M) \longrightarrow H^{i+r-p}(B,M_{-q}) \xrightarrow{\cdot e(f)} H^{i+1}(B,M) \longrightarrow \cdots,$$

where the connecting homomorphism is given by product with $e(f)$ arising from the ring structure.

PROOF. Granted the identifications mentioned before the statement, the result follows from the existence of multiplicative structure in the homotopy Leray spectral sequence as discussed in Paragraph 4.2.4. □

PROPOSITION 5.2.9. *Assume k is a field, and $f : X \to B$ is a smooth morphism where B is an $\mathbf{A}^1$-homology sphere (where $\Sigma^r X \sim \Sigma^p \mathbf{G}_m^{\wedge q}$ with $p, q \geq 1$). If M is a homotopy module such that $H^j_\delta(f_* M)$ is k-simple for each $j \geq 0$, then there is a long exact "Wang" sequence of the form:*

$$\cdots \to H^i(X, M) \to H^i_\delta(f_* M)(k) \xrightarrow{\theta} H^{i+1+r-p}_\delta(f_* M)_{-p+r}(k) \to H^{i+1}(X, M) \to \cdots$$

where the map θ is a graded derivation.

PROOF. This result follows immediately by combining Theorem 4.2.5 and Lemma 5.2.5, which shows that the resulting spectral sequence is concentrated in two columns. □

5.3. Relative cellular spaces. The notion of "algebraic cell decomposition" and the related notion of "relative algebraic cell decomposition" has a long history. E.g., if a split torus acts on a smooth projective scheme X over a field, then Bialynicki–Birula [**ByB73**] showed that X may be decomposed as a disjoint union of smooth varieties that are total spaces of vector bundles over connected components of the fixed point loci (which are necessarily smooth). The cohomological consequences of the existence of such filtrations were observed almost immediately (e.g., one immediately computes Chow groups for smooth projective varieties equipped with a torus action with isolated fixed points).

Karpenko was one of the first to exploit the existence of such algebro-geometric cell decompositions to produce motivic decompositions of varieties ([**Kar00**]), in his case absolute Chow motives (over a base field). Such results have been developed in numerous directions, but we will mainly be concerned with the relative version studied in [**MNP13**, 8.4.2]. We begin by introducing our own version of "cellularity".

DEFINITION 5.3.1. Suppose $f : X \to B$ be a morphism of schemes. Say that f (or X/B) admits a flat (resp. lci) cellular structure if there exists a filtration:

$$\emptyset = X_{-1} \subset X_0 \subset \ldots \subset X_n \subset X_{n+1} = X$$

of closed subschemes of X such that for all α, the following conditions are satisfied:

(a) $f_\alpha = f|_{X_\alpha} : X_\alpha \to B$ is flat (resp. lci);
(b) setting $U_\alpha := X_\alpha - X_{\alpha-1}$, the restriction $p_\alpha = f|_{U\alpha} : U_\alpha \to S$ is a smooth morphism and stable $\mathbf{A}^1$-weak equivalence (eg: see Remark 5.1.3).

REMARK 5.3.2. In Karpenko's definition, the morphisms p_α are assumed to be vector bundles, which is sufficient for his purposes since he essentially uses the method of Bialynicki-Birula, where the geometric decomposition arises from the action of a split torus on a smooth projective variety. In contrast, our definition allows, e.g., cellular spaces where the "cells" are merely $\mathbf{A}^1$-contractible and we allow ourselves to consider varieties that are not necessarily projective. Examples show that this additional generality is natural. For example, the smooth affine quadrics Q_{2n} of dimension $2n$ (the hypersurface in affine space defined by the equation $\sum_i x_i y_i = z(1-z)$) admit a decomposition as an affine space ($x_1 = \cdots = x_n = z = 0$) of dimension n, and an open complement that is a strictly quasi-affine $\mathbf{A}^1$-contractible scheme [**ADF17**, Theorems 2.2.5 and 3.1.1] (in particular the complement is an $\mathbf{A}^1$-contractible variety that is not isomorphic to affine space). Likewise, the Panin–Walter model of quaternionic projective space HP^n admits a "cell decomposition" where the cells are strictly quasi-affine $\mathbf{A}^1$-contractible schemes [**PW10**, Theorem 3.1].

Since the morphism p_α is smooth, its relative dimension is a Zariski locally constant function on U_α that we denote by d_α. In other words, d_α simply consists of integers for each connected component of U_α. To state the next results, we will use the following notation:

$$\mathbb{1}_B(d_\alpha)[2d_\alpha] = \bigoplus_{x \in U_\alpha^{(0)}} \mathbb{1}_B(d_\alpha(x))[2d_\alpha(x)]. \tag{5.3.2.a}$$

5.3.3. Let us first recall the computation obtained in [**MNP13**, 8.4.3], for Chow motives. We work over a quasi-projective base B over a perfect field k. Taking into account the comparison result of Jin [**Jin16**, Th. 3.17], the category of relative Chow motives defined by Corti-Hanamura corresponds to the weight 0 part of the triangulated category of rational mixed motives $\mathrm{DM}(B, \mathbf{Q})$.

Given a projective morphism $f : X \to B$ such that X is smooth over k, one defines the Chow motive (*i.e.* cohomological motive) of X/B as:

$$h_B(X) = f_*(\mathbb{1}_X).$$

With this notation, [**MNP13**, Prop. 8.4.3] can be stated as follows.

PROPOSITION 5.3.4. *Assume $f : X \to B$ is a projective morphism of k-varieties, with X smooth and admitting a flat relative cellular structure. There is a canonical isomorphism (using notation* (5.3.2.a)*) of the form:*

$$h_B(X) = \bigoplus_\alpha \mathbb{1}_B(-d_\alpha)[-2d_\alpha].$$

Note [**MNP13**, 8.4.3] is stated for $k = \mathbb{C}$ but this assumption is not used in the proof, which works over an arbitrary base field k. We appeal to the assumption k perfect implicitly via the comparison result (i.e., [**Jin16**]).

5.3.5. We can actually extend the previous computation to $\mathbf{A}^1$-homotopy, if one restricts to *oriented spectra.* Let us adopt the following definition. We will say that the abstract triangulated motivic category $\mathscr{T}$ is *oriented* if:

- There exists a premotivic adjunction $\varphi^* : \mathrm{SH} \leftrightarrows \mathscr{T} : \varphi_*$;
- The ring spectrum $\varphi_*(\mathbb{1}_S)$ is oriented.

These assumptions in place, we will use the theory of Borel-Moore objects as described in [**BD17**, section 1.3], together with the theory of fundamental classes as developed in [**Dég18**] (see also [**DJK18**] for a more general account).

For the convenience of the reader, we recall the definitions and properties we use. Given any separated B-scheme of finite type $\pi : Y \to B$, we define its *Borel-Moore $\mathscr{T}$-motive*[13] by the formula:

$$\mathrm{M}^{\mathrm{BM}}(Y/B) := \pi_!(\mathbb{1}_Y).$$

[13] Another appropriate terminology would be cohomological $\mathscr{T}$-motive with compact support.

We will need the following properties:

(P1) According to the localization property of triangulated motivic categories, given any closed immersion $i : Z \to X$ of B-schemes with complementary open immersion $j : U \to X$, one obtains a distinguished triangle:

$$\mathrm{M}^{\mathrm{BM}}(U/B) \xrightarrow{j_*} \mathrm{M}^{\mathrm{BM}}(X/B) \xrightarrow{i^*} \mathrm{M}^{\mathrm{BM}}(Z/B) \to \mathrm{M}^{\mathrm{BM}}(U/B)[1].$$

(P2) Using the main construction of [**Dég18**, Th. 2.5.3], one associates to any quasi-projective morphism $f : Y \to X$ of relative dimension d, a morphism (also called a fundamental class in *loc. cit.*):

$$f_! : \mathrm{M}^{\mathrm{BM}}(Y/X) = f_!(\mathbb{1}_Y) \to \mathbb{1}_X(-d)[-2d].$$

(P3) If in addition to the assumptions of the previous point, one suppose that f is smooth, then the following diagram is commutative:

$$\begin{array}{ccc} f_!(\mathbb{1}_Y) & \xrightarrow{\quad f_! \quad} & \\ {\scriptstyle \mathfrak{p}_f}\downarrow\wr & & \mathbb{1}_X(-d)[-2d] \\ f_\sharp(\mathbb{1}_Y)(-d)[-2d] & \xrightarrow{f_*(-d)[-2d]} & \end{array}$$

where $\mathfrak{p}_f$ is the purity isomorphism associated with the smooth morphism f in the six functors formalism, $f_\sharp(\mathbb{1}_Y)$ is the $\mathscr{T}$-motive represented by Y/B, and f_* refers to the classical functoriality. Note in particular that f_* is the image of the canonical map $\Sigma^\infty Y_+ \to \Sigma^\infty B_+$ of spectra over B under the map $\varphi^* : \mathrm{SH}(B) \to \mathscr{T}(B)$.

(P4) If under the assumptions of point (2), X is separated of finite type over a scheme B, with structural morphism p, we obtain by applying $p_!$ a Gysin morphism:

$$f_! : \mathrm{M}^{\mathrm{BM}}(Y/B) \to \mathrm{M}^{\mathrm{BM}}(X/B)(-d)[-2d].$$

It follows from the compatibility of fundamental classes with composition [**Dég18**, 2.5.3] that these Gysin morphism are compatible with composition.

PROPOSITION 5.3.6. *We consider a morphism $f : X \to B$ with an lci cellular structure, and B an arbitrary scheme.*

Then, using notation (5.3.2.a), *there exists a canonical isomorphism in* **MGL**–*mod*(B)*:*

$$f_!(\mathbb{1}_X) = \bigoplus_\alpha \mathbb{1}_B(-d_\alpha)[-2d_\alpha]$$

PROOF. We look at the localization triangle of property (P1) associated with the closed immersion $i_\alpha : X_\alpha : X_{\alpha-1} \to X_\alpha$ with complementary open immersion $j_\alpha : U_\alpha \to X_\alpha$:

$$\mathrm{M}^{\mathrm{BM}}(U_\alpha/B) \xrightarrow{j_{\alpha *}} \mathrm{M}^{\mathrm{BM}}(X_\alpha/B) \xrightarrow{i_\alpha^*} \mathrm{M}^{\mathrm{BM}}(X_{\alpha-1}/B) \to \mathrm{M}^{\mathrm{BM}}(U_\alpha/B)[1].$$

Next we apply property (P4) to get a commutative diagram:

$$\begin{array}{ccc} \mathrm{M}^{\mathrm{BM}}(U_\alpha/B) & \xrightarrow{j_{\alpha *}} & \mathrm{M}^{\mathrm{BM}}(X_\alpha/B) \\ & {\scriptstyle p_{\alpha *}}\searrow \quad \swarrow{\scriptstyle f_{\alpha *}} & \\ & \mathrm{M}^{\mathrm{BM}}(B/B)(-d_\alpha)[-2d_\alpha] & \end{array}$$

Point (P3) applied to the smooth morphism p_α, which by assumption is a stable weak $\mathbf{A}^1$-equivalence, implies the map $p_{\alpha *}$ is an isomorphism. In particular, $j_{\alpha *}$ is a split monomorphism (with spliting $p_{\alpha *}^{-1} f_{\alpha *}$) and one gets an isomorphism $\mathrm{M}^{\mathrm{BM}}(U_\alpha/B) = \mathbb{1}_B(-d_\alpha)[-2d_\alpha]$.

This completes the proof. □

COROLLARY 5.3.7. *Assume that $f : X \to B$ is a projective morphism with lci cellular structure.*

(1) *Then one gets an isomorphism of the following* cohomological *$\mathscr{T}$-motives:*

$$f_*(\mathbb{1}_X) = \bigoplus_\alpha \mathbb{1}_B(-d_\alpha)[-2d_\alpha].$$

In particular, one gets an isomorphism of $\mathscr{T}$-cohomology:

$$H^{**}(X, \mathbb{1}_X) \simeq \bigoplus_\alpha H^{*-2d_\alpha, *-d_\alpha}(B, \mathbb{1}_B)$$

(2) *For any $\mathscr{T}$-spectrum $\mathbb{E}$ over B, there exists a canonical isomorphism:*

$$f_* f^*(\mathbb{E}) \simeq \bigoplus_\alpha \mathbb{1}_B(-d_\alpha)[-2d_\alpha].$$

Only the second point needs a proof. One simply uses the preceding proposition and the projection formula from the six functors formalism:

$$f_* f^*(\mathbb{E}) \simeq f_!(\mathbb{1}_X \otimes f^*(\mathbb{E})) \simeq f_!(\mathbb{1}_X) \otimes \mathbb{E}.$$

EXAMPLE 5.3.8.

(1) In the motivic case, one gets back the result of Proposition 5.3.4. In fact, with rational coefficients, we have shown that one can get rid of a base field, if one replaces the flatness by lci. (See also next Remark.)
(2) We can apply the preceding result to classical oriented ring spectra, such as **MGL** and **KGL**, by using the theory of modules over ring spectra (see [**CD09**, §7]). The corollary extends a previously known result in [**NZ06**], from characteristic 0 to the absolute case (without the need of a base field).

REMARK 5.3.9.

(1) The above corollary applies to strict **MGL**-modules M over B. That is the **MGL**-module structure on M must be defined at the model category level. This assumption is too strong in many situations. One can avoid it, assuming only that M is a spectrum in $\mathrm{SH}(B)$ with a module structure over $\mathbf{MGL}_B$. Indeed, it is possible to use the proof of Proposition 5.3.6 after replacing $f_!(\mathbb{1}_X) = f_! f^*(\mathbb{1}_X)$ with the spectrum $f_! f^*(M)$. One uses the theory of fundamental classes as developed in [**DJK18**] and the fact that one has Thom isomorphisms for M, using the $\mathbf{MGL}_B$-module structure.

We leave the details to the reader.

(2) The orientation assumptions in the preceding proposition and corollary is essential. Indeed, recall that the Chow-Witt ring does not satisfy the projective bundle formula though it is representable in SH by the Milnor-Witt homotopy module.

Nevertheless, one can extend the validity of the above result by considering weaker orientability conditions. For example, we can consider a symplectically oriented (ring) spectrum M (see [**PW18**]), provided that we give for each index α a symplectic structure on the tangent bundle of U_α/B.

(3) At least in the rational motivic case (which is equivalent to the rational orientable case), it is possible in principle to generalize both the flat and lci case. Indeed, it is visible in the proof that we only need a good theory of fundamental classes for the projections $X_\alpha \to B$. The recent work [**Jin18**] opens the way to define these fundamental classes for arbitrary *finite tor-dimension morphisms*, which contains both flat and lci cases. The main tool to do that is to extend Jin's work to a suitable representability theorem for relative K-theory (as defined in [**BGI71**, IV, 3.3]).

The relative cellular space provides us with a situation where the homotopy Leray spectral sequence is particularly simple.

PROPOSITION 5.3.10. *Let $f : X \to B$ be a proper morphism with an lci relative cellular structure. We consider a homotopy module M over B satisfying one of the following conditions:*

- *Homotopical case.– M is an oriented homotopy module in* $\mathrm{SH}(B)$.
- *Motivic case.– M is a homotopy module in* $\mathrm{DM}(B, R)$.

We consider the homotopy module $M_X = f^!M$. Then there exists an isomorphism:

$$f_*(M_X) \simeq \sum_{i=0}^{d} H^i_\delta(f_*M_X)[-i]$$

where d is the maximum of the dimension of the fibers of f.

Moreover, the homotopy Leray spectral sequence:

$$E_2^{p,q} = A^p(B, H^q_\delta(f_*M_X)) \Rightarrow A^{p+q}(X, M_X)$$

degenerates at E_2, and the abutting filtration splits giving an isomorphism:

$$A^n(X, M_X) \simeq \bigoplus_{p=0}^{n} A^p(B, H^{n-p}_\delta(f_*M_X)).$$

References

[AD07] Aravind Asok and Brent Doran, *On unipotent quotients and some $\mathbb{A}^1$-contractible smooth schemes*, Int. Math. Res. Pap. IMRP **2** (2007), Art. ID rpm005, 51. MR2335246

[AD09] Aravind Asok and Brent Doran, *$\mathbb{A}^1$-homotopy groups, excision, and solvable quotients*, Adv. Math. **221** (2009), no. 4, 1144–1190, DOI 10.1016/j.aim.2009.02.007. MR2518635

[ADF17] Aravind Asok, Brent Doran, and Jean Fasel, *Smooth models of motivic spheres and the clutching construction*, Int. Math. Res. Not. IMRN **6** (2017), 1890–1925, DOI 10.1093/imrn/rnw065. MR3658186

[AF14] Aravind Asok and Jean Fasel, *Algebraic vector bundles on spheres*, J. Topol. **7** (2014), no. 3, 894–926, DOI 10.1112/jtopol/jtt046. MR3252968

[Ara05] Donu Arapura, *The Leray spectral sequence is motivic*, Invent. Math. **160** (2005), no. 3, 567–589, DOI 10.1007/s00222-004-0416-x. MR2178703

[Ayo07] Joseph Ayoub, *Les six opérations de Grothendieck et le formalisme des cycles évanescents dans le monde motivique. II* (French, with English and French summaries), Astérisque **315** (2007), vi+364 pp. (2008). MR2438151

[BBD82] A. A. Beĭlinson, J. Bernstein, and P. Deligne, *Faisceaux pervers* (French), Analysis and topology on singular spaces, I (Luminy, 1981), Astérisque, vol. 100, Soc. Math. France, Paris, 1982, pp. 5–171. MR751966

[BD17] Mikhail Bondarko and Frédéric Déglise, *Dimensional homotopy t-structures in motivic homotopy theory*, Adv. Math. **311** (2017), 91–189, DOI 10.1016/j.aim.2017.02.003. MR3628213

[BGI71] P. Berthelot, A. Grothendieck, and L. Illusie, *Théorie des intersections et théorème de Riemann-Roch*, Lecture Notes in Mathematics, vol. 225, Springer-Verlag, 1971, Séminaire de Géométrie Algébrique du Bois–Marie 1966–67 (SGA 6). MR0354655

[BO74] Spencer Bloch and Arthur Ogus, *Gersten's conjecture and the homology of schemes*, Ann. Sci. École Norm. Sup. (4) **7** (1974), 181–201 (1975). MR0412191

[Bon10] Mikhail V. Bondarko, *Motivically functorial coniveau spectral sequences; direct summands of cohomology of function fields*, Doc. Math. **Extra vol.: Andrei A. Suslin sixtieth birthday** (2010), 33–117. MR2804250

[BY18] T. Bachmann and M. Yakerson, *Towards conservativity of $\mathbb{G}_m$-stabilization*, arXiv:1811.01541, November 2018.

[ByB73] A. Białynicki-Birula, *Some theorems on actions of algebraic groups*, Ann. of Math. (2) **98** (1973), 480–497, DOI 10.2307/1970915. MR0366940

[CD09] D.-C. Cisinski and F. Déglise, *Triangulated categories of mixed motives*, *Preprint*, available at `https://arxiv.org/abs/0912.2110`, 2009.

[CD12] Denis-Charles Cisinski and Frédéric Déglise, *Mixed Weil cohomologies*, Adv. Math. **230** (2012), no. 1, 55–130, DOI 10.1016/j.aim.2011.10.021. MR2900540

[CD15] D.-C. Cisinski and F. Déglise, *Integral mixed motives in equal characteristics*, Doc. Math. (2015), no. Extra volume: Alexander S. Merkurjev's sixtieth birthday, 145–194.

[CE99] Henri Cartan and Samuel Eilenberg, *Homological algebra*, Princeton Landmarks in Mathematics, Princeton University Press, Princeton, NJ, 1999. With an appendix by David A. Buchsbaum; Reprint of the 1956 original. MR1731415

[dCM10] Mark Andrea A. de Cataldo and Luca Migliorini, *The perverse filtration and the Lefschetz hyperplane theorem*, Ann. of Math. (2) **171** (2010), no. 3, 2089–2113, DOI 10.4007/annals.2010.171.2089. MR2680404

[Dég11] F. Déglise, *Modules homotopiques* (French, with English summary), Doc. Math. **16** (2011), 411–455. MR2823365

[Dég12] F. Déglise, *Coniveau filtration and motives*, Regulators, Contemporary Mathematics, vol. 571, 2012, pp. 51–76.

[Dég13] Frédéric Déglise, *Orientable homotopy modules*, Amer. J. Math. **135** (2013), no. 2, 519–560, DOI 10.1353/ajm.2013.0019. MR3038720

[Dég14a] Frédéric Déglise, *On the homotopy heart of mixed motives*, `http://deglise.perso.math.cnrs.fr/docs/2014/modhtpb.pdf`, 2014.

[Dég14b] Frédéric Déglise, *Suite spectrale du coniveau et t-structure homotopique* (French, with English and French summaries), Ann. Fac. Sci. Toulouse Math. (6) **23** (2014), no. 3, 591–609, DOI 10.5802/afst.1417. MR3266706

[Dég18] Frédéric Déglise, *Bivariant theories in motivic stable homotopy*, Doc. Math. **23** (2018), 997–1076. MR3874952

[DF17] F. Déglise and J. Fasel, *The milnor-witt motivic ring spectrum and its associated theories*, arXiv:1708.06102, 2017.

[DF18] Adrien Dubouloz and Jean Fasel, *Families of $\mathbb{A}^1$-contractible affine threefolds*, Algebr. Geom. **5** (2018), no. 1, 1–14. MR3734108

[DJK18] F. Déglise, F. Jin, and A. Khan, *Fundamental classes in motivic homotopy theory*, arXiv:1805.05920v2, 2018.

[DK01] James F. Davis and Paul Kirk, *Lecture notes in algebraic topology*, Graduate Studies in Mathematics, vol. 35, American Mathematical Society, Providence, RI, 2001. MR1841974

[Dou59] A. Douady, *La suite spectrale d'Adams: structure multiplicative, exp. 19*, Séminaire Henri Cartan, Ecole Normale Supérieure, vol. tome 11, n°2, 1958-1959, pp. 1–13.

[DPØ18] A. Dubouloz, S. Pauli, and P.-A. Østvær, *$\mathbb{A}^1$-contractibility of affine modifications*, *Preprint*, available at `https://arxiv.org/abs/1805.08959`, 2018.

[Fel18] N. Feld, *Milnor-witt cycle modules*, arXiv:1811.12163, 2018.

[GD67] A. Grothendieck, *Éléments de géométrie algébrique. IV. Étude locale des schémas et des morphismes de schémas. I* (French), Inst. Hautes Études Sci. Publ. Math. **20** (1964), 259. MR0173675

[Jin16] Jin Fangzhou, *Borel-Moore motivic homology and weight structure on mixed motives*, Math. Z. **283** (2016), no. 3-4, 1149–1183, DOI 10.1007/s00209-016-1636-7. MR3519998

[Jin18] Jin Fangzhou, *Algebraic g-theory in motivic homotopy categories*, arXiv:1806.03927, 2018.

[Kar00] N. A. Karpenko, *Cohomology of relative cellular spaces and of isotropic flag varieties* (Russian, with Russian summary), Algebra i Analiz **12** (2000), no. 1, 3–69; English transl., St. Petersburg Math. J. **12** (2001), no. 1, 1–50. MR1758562

[Lev06] Marc Levine, *Chow's moving lemma and the homotopy coniveau tower*, K-Theory **37** (2006), no. 1-2, 129–209, DOI 10.1007/s10977-006-0004-5. MR2274672

[Lev08] Marc Levine, *The homotopy coniveau tower*, J. Topol. **1** (2008), no. 1, 217–267, DOI 10.1112/jtopol/jtm004. MR2365658

[Lur18] J. Lurie, *Higher algebra*, `http://www.math.harvard.edu/~lurie/papers/HA.pdf`, Sep. 2018.

[McC01] John McCleary, *A user's guide to spectral sequences*, 2nd ed., Cambridge Studies in Advanced Mathematics, vol. 58, Cambridge University Press, Cambridge, 2001. MR1793722

[MNP13] Jacob P. Murre, Jan Nagel, and Chris A. M. Peters, *Lectures on the theory of pure motives*, University Lecture Series, vol. 61, American Mathematical Society, Providence, RI, 2013. MR3052734

[Mor03] Fabien Morel, *An introduction to $\mathbb{A}^1$-homotopy theory*, Contemporary developments in algebraic K-theory, ICTP Lect. Notes, XV, Abdus Salam Int. Cent. Theoret. Phys., Trieste, 2004, pp. 357–441. MR2175638

[MV99] Fabien Morel and Vladimir Voevodsky, *$\mathbf{A}^1$-homotopy theory of schemes*, Inst. Hautes Études Sci. Publ. Math. **90** (1999), 45–143 (2001). MR1813224

[Nor02] Madhav V. Nori, *Constructible sheaves*, Algebra, arithmetic and geometry, Part I, II (Mumbai, 2000), Tata Inst. Fund. Res. Stud. Math., vol. 16, Tata Inst. Fund. Res., Bombay, 2002, pp. 471–491. MR1940678

[NZ06] A. Nenashev and K. Zainoulline, *Oriented cohomology and motivic decompositions of relative cellular spaces*, J. Pure Appl. Algebra **205** (2006), no. 2, 323–340, DOI 10.1016/j.jpaa.2005.06.021. MR2203620

[Par96] K. H. Paranjape, *Some spectral sequences for filtered complexes and applications*, J. Algebra **186** (1996), no. 3, 793–806, DOI 10.1006/jabr.1996.0395. MR1424593

[PW10] I. Panin and C. Walter, *Quaternionic Grassmannians and Pontryagin classes in algebraic geometry*, *Preprint* available at `http://arxiv.org/abs/1011.0649`, 2010.

[PW18] I. Panin and C. Walter, *Quaternionic Grassmannians and Borel classes in algebraic geometry*, arXiv: 1011.0650v2, March 2018.

[Ros96] Markus Rost, *Chow groups with coefficients*, Doc. Math. **1** (1996), No. 16, 319–393. MR1418952

[VD74] B. Ju. Veĭsfeĭler and I. V. Dolgačev, *Unipotent group schemes over integral rings* (Russian), Izv. Akad. Nauk SSSR Ser. Mat. **38** (1974), 757–799. MR0376697

[Ver96] Jean-Louis Verdier, *Des catégories dérivées des catégories abéliennes* (French, with French summary), Astérisque **239** (1996), xii+253 pp. (1997). With a preface by Luc Illusie; Edited and with a note by Georges Maltsiniotis. MR1453167

[wc15] Collective work (198 collaborators), *The stacks project*, version 10/11/2018, 2015.

Department of Mathematics, University of Southern California, 3620 S. Vermont Ave., Los Angeles, CA 90089-2532, United States
Email address: `asok@usc.edu`

IMB - UMR5584, 9 avenue Alain Savary 21000 Dijon Cedex; France
Email address: `frederic.deglise@u-bourgogne.fr`

IMB - UMR5584, 9 avenue Alain Savary 21000 Dijon Cedex; France
Email address: `Johannes.Nagel@u-bourgogne.fr`

Contemporary Mathematics
Volume **745**, 2020
https://doi.org/10.1090/conm/745/15022

Examples of wild ramification in an enriched Riemann–Hurwitz formula

Candace Bethea, Jesse Leo Kass, and Kirsten Wickelgren

ABSTRACT. M. Levine proved an enrichment of the classical Riemann–Hurwitz formula to an equality in the Grothendieck–Witt group of quadratic forms. In its strongest form, Levine's theorem includes a technical hypothesis on ramification relevant in positive characteristic. We describe what happens when the hypotheses are weakened by showing an analogous Riemann–Hurwitz formula and describing an example suggested by S. Saito.

1. Introduction

In the recent preprint [**Lev17**], Marc Levine established an enriched version of the Riemann–Hurwitz formula that is valued in the Grothendieck–Witt group of nondegenerate symmetric bilinear forms. In positive characteristic, especially over an imperfect field, the strongest form of Levine's theorem includes technical hypotheses on the ramification. At the workshop *Motivic homotopy theory and refined enumerative geometry*, Shuji Saito asked whether a Riemann–Hurwitz formula should hold more generally. As an illustration of the situation he was interested in, he gave the example of the map $\mathbf{P}^1_{\mathbf{F}_p(t)} \to \mathbf{P}^1_{\mathbf{F}_p(t)}$ defined by $y \mapsto \frac{t-y^p}{y}$. In this article, we prove a theorem which strengthens Levine's result by establishing his result under weaker hypotheses so that it applies to Saito's function. We also illustrate the content of the theorem by computing the terms for Saito's example as well as for the Artin–Schreier cover $\mathbf{P}^1_{\mathbf{F}_p(t)} \to \mathbf{P}^1_{\mathbf{F}_p(t)}$ defined by $y \mapsto y^p - y$.

The problem of establishing an enriched Riemann–Hurwitz formula with weaker hypotheses in characteristic p is interesting because the classical Riemann–Hurwitz formula becomes more complicated when passing from characteristic 0 to characteristic $p > 0$. Over an algebraically closed field of characteristic 0, the formula, as described in Corollary 2.4 of [**Har77**, Chapter 4], states that a nonconstant map $f\colon Y \to X$ of curves satisfies

$$\chi(Y) = d \cdot \chi(X) - \sum (e(y) - 1). \tag{1.1}$$

Here d is the degree of f, $\chi(C) = 2 - 2g(C)$ is the topological Euler characteristic of a curve C, and $e(y)$ is the ramification index of f. Recall the ramification index is the normalized valuation $\nu_y(f^*(t))$ of the pullback of a uniformizer $t \in \mathcal{O}_{X,f(y)}$.

2010 *Mathematics Subject Classification.* Primary 14F42.

When k has positive characteristic (but is still assumed to be algebraically closed), Formula (1.1) becomes more complicated in two ways. First, we need to additionally require that f is separable (to avoid maps like $y \mapsto y^p$ which is ramified everywhere). Second, Equation (1.1) holds as stated when f is separable and the ramification indices are all coprime to p, i.e. when f is tamely ramified, but in general, the term $e(y)-1$ must be modified. Define the branch index $b(y)$ to be the length of the module of relative Kähler differentials, i.e. $b(y) := \text{length}(\Omega_{Y/X,y})$. We then have

$$\chi(Y) = d \cdot \chi(X) - \sum b(y), \tag{1.2}$$

and $b(y) \geq e(y) - 1$ with equality holding if and only if $e(y)$ is coprime to p. The branch index can alternatively be described in terms of the uniformizers. If $t \in \mathcal{O}_{X,f(y)}$ and $u \in \mathcal{O}_{Y,y}$ are uniformizers, then the branch index equals the valuation $v_y(dt/du)$ for dt/du the unique function satisfying $f^*(dt) = dt/du \cdot du$.

Formula (1.2) remains valid when k is nonalgebraically closed provided the branch index is defined by $b(y) := \text{length}(\Omega_{Y/X,y})$ [**Liu02**, Theorem 4.16, Remark 4.17]. The branch index does not, however, always equal $v_y(dt/du)$. Indeed, for Saito's example $y \mapsto \frac{t-y^p}{y}$, we have $dt/du = 0$. In this example, the residual extension $k(y)/k(x)$ is inseparable. When the residual extension is separable, an explicit expression for $b(y)$ is given by [**Ser79**, Chapter 3, Propositions 13 and 14; Chapter 4, Proposition 4] (where $b(y)$ appears as the valuation of the different).

Over the real numbers $k = \mathbf{R}$, the Riemann–Hurwitz formula admits a real-topological analogue. One implication of the hypotheses is that the manifold of real points $Y(\mathbf{R})$ is orientable. Once we fix an orientation, the topological degree $\deg^{\mathbf{R}}(f)$ of the map $f\colon Y(\mathbf{R}) \to X(\mathbf{R})$ on real points is well-defined and satisfies an analogue of the Riemann–Hurwitz formula, as was observed by Levine in [**Lev17**, Example 12.9]. Specifically, if $y \in Y$ is a point with residue field $\mathbf{R}$ and $t \in \mathcal{O}_{X,f(y)}$, $u \in \mathcal{O}_{Y,y}$ are uniformizers with t compatible with the orientation (so the function germ $t\colon X(\mathbf{R}) \to \mathbf{R}$ is orientation-preserving at $f(y)$), then define the real branch index $b^{\mathbf{R}}(y)$ to be the local degree of dt/du at y, so

$$b^{\mathbf{R}}(y) = \begin{cases} +1 & \text{if } dt/du \circ u^{-1} \text{ is increasing at } 0; \\ -1 & \text{if } dt/du \circ u^{-1} \text{ is decreasing at } 0; \\ 0 & \text{otherwise.} \end{cases}$$

With this definition, we have

$$\chi^{\mathbf{R}}(X) = d \cdot \chi^{\mathbf{R}}(Y) - \sum b^{\mathbf{R}}(y). \tag{1.3}$$

Here $\chi^{\mathbf{R}}(X)$ denotes the Euler characteristic of $X(\mathbf{R})$ of the real locus. These Euler characteristics vanish, so the formula simplifies to

$$0 = \sum b^{\mathbf{R}}(y).$$

This equation admits a particularly simple interpretation when $X = Y = \mathbf{P}^1_{\mathbf{R}}$ and f is defined by a monic polynomial $f = x^d + a_1 x^{d-1} + a_2 x^{d-2} + \cdots + a_d$. Considering f as a continuous function $f\colon \mathbf{R} \to \mathbf{R}$, a computation of $b(\infty)$ shows that Equation (1.3) takes the form

$$\#\text{local maxima of } f - \#\text{local minima of } f = \begin{cases} 0 & \text{if } d \text{ is odd;} \\ -1 & \text{if } d \text{ is even.} \end{cases}$$

This result is the real realization of Levine's enriched Riemann–Hurwitz formula. Over an arbitrary field, we replace the choice of an orientation of $X(\mathbf{R})$ with the choices of a line bundle M and an isomorphism $\alpha\colon M^{\otimes 2} \cong \mathrm{T}(X)$ of the square of M with the tangent bundle. Observe that, over $\mathbf{R}$, the pair (M, α) determines an orientation of $X(\mathbf{R})$, but not every real curve admits a pair (M, α). (Consider, for example, the Brauer–Severi curve $\{X^2 + Y^2 + Z^2 = 0\} \subset \mathbf{P}^2_{\mathbf{R}}$.)

Given $f\colon Y \to X$, Levine's strongest form of the enriched Riemann–Hurwitz formula holds under technical assumptions on the ramification that we recall below. Let t and u be uniformizers as before, but now require that t is compatible with (M, α) in the sense that, under the isomorphism on stalks $\alpha^{\vee}_{f(y)}\colon M^{-\otimes 2}_{f(y)} \cong \mathrm{T}^{\vee}_x(X)$, dt corresponds to a tensor of the form $s \otimes s$ (rather than $s \otimes s'$ for $s \neq s'$).

Write the pullback $f^*(t)$ of the uniformizing parameter t under the natural homomorphism $f^*\colon \mathcal{O}_{X,f(x)} \to \mathcal{O}_{Y,y}$ as $f^*(t) = a \cdot u^{e(y)}$ with $a \in \mathcal{O}^*_{Y,y}$. (Such an expression exists by the definition of $e(y)$.) Define the motivic branch index by

$$b^{\mathbf{A}^1}(y) = \langle a(y)e(y)\rangle \cdot \sum_{i=0}^{e-2} \langle (-1)^i \rangle \text{ in } \mathrm{GW}(k(y)).$$

Here $\mathrm{GW}(k(y))$ denotes the Grothendieck–Witt group of nondegenerate symmetric bilinear forms, $\langle u \rangle$ denotes the class of the rank 1 bilinear form with Gram matrix (u), and $a(y)$ denotes the image of $a \in \mathcal{O}^*_{Y,y}$ in the residue field.

With this notation, the hypothesis to [**Lev17**, Theorem 12.7] is that $\operatorname{char} k \neq 2$, f is separable, and every ramification point has the property that $k(y)$ is a separable extension of k and $e(y)$ is coprime to p. When these hypotheses are satisfied, the theorem states

$$\chi^{\mathbf{A}^1}(Y) = d \cdot \chi^{\mathbf{A}^1}(X) - \sum_{\{y \in Y\,:\, e(y) > 1\}} \mathrm{Tr}_{k(y)/k}(b^{\mathbf{A}^1}(y)) \tag{1.4}$$

Here the sum runs over all ramification points of f. The term $\chi^{\mathbf{A}^1}(C)$ is the Euler characteristic in $\mathbf{A}^1$-homotopy theory which equals $(1 - g(C)) \cdot (\langle +1 \rangle + \langle -1 \rangle)$ when C is a smooth curve. The expression $\mathrm{Tr}_{k(y)/k}(b^{\mathbf{A}^1}(y))$ is the class of the composition $\mathrm{Tr}_{k(y)/k} \circ \beta$ of the field trace map $\mathrm{Tr}_{k(y)/k}\colon k(y) \to k$ with a bilinear form β representing $b^{\mathbf{A}^1}(y)$.

Equation (1.4) is an enrichment of the earlier Riemann–Hurwitz formulas in the sense that those earlier statements can be deduced from it by comparing invariants. The fact that the left-hand and right-hand sides of (1.4) have the same rank is Equation (1.2). When $k = \mathbf{R}$, the fact that the signatures of the sides are equal is Equation (1.3).

Levine's hypotheses on the ramification points fail to hold in Saito's example and in the example of the Artin–Schreier cover. The Artin–Schreier cover fails to satisfy the hypotheses because $b(\infty) = p$. In Saito's example, the ramification point y defined by the ideal $(y^p - t)$ has the property that $e(y) = 1$ but the residue extension $k(y)/k(f(y))$ is inseparable. Levine deduces [**Lev17**, Theorem 12.7] from [**Lev17**, Corollary 10.9], and that corollary applies when f is wildly ramified, but it does not provide an explicit expression for $e^{\mathbf{A}^1}(y)$. As Levine remarks immediately after the corollary, the main result of [**KW19**] can be used to derive an explicit expression for these branch indices. In fact, we can reduce to the case where $\mathcal{O}_{Y,y}$ is a monogenic extension of $\mathcal{O}_{X,f(y)}$ by Proposition 4.3 below, and in this case the branch indices can be computed using the earlier work of Cazanave [**Caz12, Caz08**] (loc. cit. only treat the global $\mathbf{A}^1$-degree, but see [**KW18**] for the relation with the local $\mathbf{A}^1$-degree).

In this paper, we explain in more detail how to use the results of [**KW19**] to establish an enriched Riemann–Hurwitz formula with explicitly computable branch indices when f is allowed to have wild ramification. Rather than using the formalism developed in [**Lev17**], we establish an enriched Riemann–Hurwitz formula using the Euler class formalism in [**KW17**]. Under suitable hypotheses, the local index $\operatorname{ind}_y df$ is defined as the local $\mathbf{A}^1$-degree with respect to a coordinate system. This class is represented by an explicit bilinear form, and we recall a recipe for computing the form in Section 2.

The main result is

THEOREM 1.1. *Let k be any field. Let $f : Y \to X$ be a non-constant, separable map of smooth, proper, geometrically connected curves over k. We make the following assumptions:*

1. *X is oriented, meaning we have a line bundle M and a chosen isomorphism $M^{\otimes 2} \cong T^\vee X$.*
2. *$e(Y, \operatorname{Hom}(f^*T^\vee X, T^\vee Y), df) = e(Y, \operatorname{Hom}(f^*T^\vee X, T^\vee Y), \alpha df)$ for all α in k^*.*

Then there is an equality

$$\chi^{\mathbf{A}^1}(Y) = d \cdot \chi^{\mathbf{A}^1}(X) - \sum_{\{y: df(y)=0\}} \operatorname{ind}_y df. \tag{1.5}$$

This is proven in Section 4 below. The notation $e(Y, \operatorname{Hom}(f^*T^\vee X, T^\vee Y), df)$ is defined in Section 2. As we explain, when $f \colon Y \to X$ is described explicitly, the local indices of df can be effectively computed using the main results of [**Caz12, Caz08, KW19**].

REMARK 1.2. *We comment on the assumptions in Theorem 1.1.*

1. *Assumption 1 can always be achieved after base change to a finite extension M of k, because $T^\vee X$ has even degree and $\operatorname{Pic}^0(X)(\bar{k})$ is divisible.*
2. *There is work in progress to show that Assumption (2) always holds [**BW19**], as well as to show that the local indices agree with those of [**Lev17**].*

We demonstrate the theorem in Section 3 by explicitly working out the terms in Equation (1.5) for the Artin–Schreier cover and Saito's function $f(y) = (t - y^p)/y$. These examples are especially interesting in the context of work of Kato and Abbes–Saito [**Kat89, AS02, AS03, Sai12**] on the conductor of an extension of local fields with inseparable residue extension because classical work suggests that it would be interesting to explore the connection between that conductor and the term $b^{\mathbf{A}^1}(y)$. Recall that, when the residue field is separable, the classical index $b(y)$ equals the valuation of the different which is related to the conductor by Artin's Conductor-Discriminant formula [**Neu99**, Chapter VII,(11.9)]. Perhaps there is an enrichment of this formula that extends to the case where the residual extension is inseparable.

2. Notation

For a field k, let $\mathrm{GW}(k)$ denote the Grothendieck–Witt group of k, which is the group completion of the semi-ring under $\oplus$ and $\otimes$ of isomorphism classes of

k-valued nondegenerate symmetric bilinear forms on finite dimensional k-vector spaces. Since all such forms are stably diagonalizable, $\mathrm{GW}(k)$ is generated by 1-dimensional forms $\langle a\rangle$ with a in $k^*/(k^*)^2$, where $\langle a\rangle$ is the isomorphism class generated by the bilinear form

$$k \times k \to k,$$

$$(x,y) \mapsto axy.$$

The class of the hyperbolic form is denoted h and is given by $h = \langle 1\rangle + \langle -1\rangle$.

We recall some definitions from [**KW17**] that allow us to define an Euler number in $\mathrm{GW}(k)$. There are other definitions of such Euler numbers. Relevant references include [**GI80**], [**BM00**], [**Fas08**], [**Mor12**, Chapter 8.2], [**AF16**], [**DJK18**], and [**LR18**]. Please see, for example, the discussion in Section 1.1 of [**KW17**] entitled *Relation to other work.*

Let Y be a smooth k-scheme of dimension r. Given a point y of Y, *Nisnevich coordinates around* y are the data of a Zariski open neighborhood U of y and an étale map $\phi : U \to \mathbb{A}^r_k$ from U to affine r-space such that the induced map $k(\phi(y)) \to k(y)$ on residue fields is an isomorphism.

Let $\mathcal{V} \to Y$ be a vector bundle of rank r over Y. A relative orientation of $\mathcal{V}$ is a line bundle M and an isomorphism $\mathrm{Hom}(\det TY, \det \mathcal{V}) \cong M^{\otimes 2}$.

Let σ be a section of $\mathcal{V}$. Given Nisnevich local coordinates around an isolated zero y of σ, there is a local index (also called a *local degree*) $\mathrm{ind}_y\,\sigma$ in $\mathrm{GW}(k)$ of σ at y defined in [**KW17**, Definition 28]. Because our interest lies in being able to compute the local indices at points, especially points whose residue fields are inseparable extensions of k and when the order of vanishing of σ is divisible by the characteristic k, we recall the following computational recipe for $\mathrm{ind}_y\,\sigma$. One can also compute $\mathrm{ind}_y\,\sigma$ using the main result of [**Caz08**].

We choose a local trivialization of $\mathcal{V}$ near y which is compatible with the relative orientation. Under this trivialization, σ is identified with an element f of $\oplus_{i=1}^r \mathcal{O}_Y$. Let m_y denote the ideal corresponding to y. We can choose an element g of $\oplus_{i=1}^r m_y^N$ for N sufficiently large (relative to the order of vanishing of f) so that the function $f+g$ is in the image of $\phi^* : \oplus_{i=1}^r \mathcal{O}_{\mathbb{A}^r_k,\phi(y)} \to \oplus_{i=1}^r \mathcal{O}_{Y,y}$. Choose F in $\oplus_{i=1}^r \mathcal{O}_{\mathbb{A}^r_k,\phi(y)}$ such that $\phi^*(F) = f+g$. (Any choice of such a g and F will do.) Then F determines a function $F : W \to \mathbb{A}^r_k$ from an open subset $W \subset \mathbb{A}^r_k = \mathrm{Spec}\, k[y_1,\dots,y_r]$ to $\mathbb{A}^r_k$. Let $F = (F_1, F_2, \dots, F_r)$.

Since y is an isolated zero, $Q = k[y_1,\dots,y_r]_{\phi(y)}/\langle F_1,\dots,F_r\rangle$ is a finite dimensional k-vector space. Scheja–Storch [**SS75**, Section 3] construct the following bilinear form on Q, and the isomorphism class of this form is $\mathrm{ind}_y\,\sigma$. We can choose a_{ij} in $k[y_1,\dots,y_r] \otimes_k k[y_1,\dots,y_r]$ such that

$$F_j \otimes 1 - 1 \otimes F_j = \sum_i a_{ij}(y_i \otimes 1 - 1 \otimes y_i).$$

Let Δ denote the image of $\det(a_{ij})$ in $Q \otimes_k Q$. There is a canonical map $Q \otimes_k Q \to \mathrm{Hom}(\mathrm{Hom}(Q,k),Q)$ sending $b \otimes c$ to the linear map which sends μ to $\mu(b)c$. The image of Δ is an isomorphism $\Theta : \mathrm{Hom}(Q,k) \to Q$. Let $\eta = \Theta^{-1}(1)$. We obtain a bilinear form $Q \times Q \to k$ defined by

$$(b,c) \mapsto \eta(bc).$$

It follows from the main theorem of [**KW19**] that $\mathrm{ind}_y \sigma$ agrees with the local $\mathbb{A}^1$-degree of the associated function F at $\phi(y)$, at least when y is k-rational or F has a simple zero. We denote this latter element of $\mathrm{GW}(k)$ by $\deg_y^{\mathbb{A}^1} \sigma$.

Suppose now that additionally Y is proper, $\mathcal{V}$ is relatively oriented, and that σ is a section with only isolated zeros such that there are Nisnevich coordinates around every zero of σ. Then define the Euler number $e(Y, \mathcal{V}, \sigma)$ of $\mathcal{V}$ with respect to σ by

$$e(Y, \mathcal{V}, \sigma) = \sum_{x:\sigma(x)=0} \mathrm{ind}_x \sigma.$$

If σ and σ' are sections with only isolated zeros that can be connected by sections with only isolated zeros, potentially after base change by an odd degree field extension, then $e(Y, \mathcal{V}, \sigma) = e(Y, \mathcal{V}, \sigma')$ ([**KW17**, Corollary 36]).

3. Examples

Here we look at the Riemann–Hurwitz formula in two specific cases that illustrate some of the more delicate behavior of covers of curves in characteristic p. The first example is the example suggested by Shuji Saito. This example is a rational map $f\colon \mathbf{P}^1_{k(t)} \to \mathbf{P}^1_{k(t)}$ with a ramification point y such that $e(y) = 1$ but the residual extension $k(y)/k(f(y))$ is inseparable.

The second example is an Artin–Schreier cover $f\colon \mathbf{P}^1_k \to \mathbf{P}^1_k$. This cover has the property that it admits a ramification point y such that $k(y) = k(f(y))$ but the ramification index $e(y)$ equals the residual characteristic p.

We work over a field k of odd characteristic p. For a field extension L/k, let $\mathbf{P}^1_{L,y}$ denote 1-dimensional projective space over the field L with projective variables Y and W, i.e., $\mathbf{P}^1_{L,y} = \mathrm{Proj}\, L[W, Y]$ with affine coordinate $y = Y/W$, and let $\mathbf{P}^1_{L,x} = \mathrm{Proj}\, L[X, Z]$.

PROPOSITION 3.1. *The rational function $f : \mathbf{P}^1_{k(t),y} \to \mathbf{P}^1_{k(t),x}$ defined by $y \mapsto \frac{t-y^p}{y}$ has the property that there is a relative orientation such that the branch indices are*

$$\mathrm{ind}_{(y^p-t)}\, df = \left(\frac{p-1}{2}\right) \cdot h + \langle 1 \rangle$$
$$\mathrm{ind}_{\infty}\, df = \left(\frac{p-3}{2}\right) \cdot h + \langle -1 \rangle.$$

In particular, the enriched Riemann–Hurwitz formula (1.5) holds.

PROOF. We will show there is an equality

$$\sum_{\{y:df(y)=0\}} \mathrm{ind}_y^{\mathbb{A}^1} df = h\left(g(\mathbf{P}^1_{k(t),y}) - 1 + \deg f(1 - g(\mathbf{P}^1_{k(t),x}))\right)$$

in $\mathrm{GW}(k(t))$, where $\deg f$ refers to the degree of the extension of function fields. First observe that the right-hand side is

$$\begin{aligned} h\left(g(\mathbf{P}^1_{k(t),y}) - 1 + \deg f(1 - g(\mathbf{P}^1_{k(t),x}))\right) &= h(0 - 1 + p(1 - 0)) \\ &= h(p-1), \end{aligned}$$

so we need to show that $\sum_{\{y\colon df(y)=0\}} \text{ind}_y^{\mathbb{A}^1} df = h(p-1)$. Observe that

$$\{y\colon df(y)=0\} = \{y\colon f'(y)=0 \text{ or } h'(w)=0\},$$

where $w = \frac{1}{y}$ and $h = \frac{1}{f}$. Here we are using the affine coordinates $y = Y/W$ on $\operatorname{Spec} k(t)[y]$ and $w = W/Y$ on $\operatorname{Spec} k(t)[w]$. On $\operatorname{Spec} k(t)[y, y^{-1}]$, $x = f(y) = \frac{t-y^p}{y}$ and $f'(y) = \frac{y^p-t}{y^2} = 0$ only at the point defined by the ideal $(y^p - t)$. On $\operatorname{Spec} k(t)[w]$, $z = h(w) = \frac{w^{p-1}}{w^p t-1}$ and $h'(w) = \frac{(p-1)w^{p-2}}{w^p t-1} = 0$ at $w = \frac{1}{y} = 0$. We will compute $\text{ind}_{(y^p-t)}^{\mathbb{A}^1} df$ and $\text{ind}_{\infty}^{\mathbb{A}^1} df$ separately.

In order for local degrees to be well-defined, we must first construct a relative orientation for the line bundle $\text{Hom}(f^*T^\vee \mathbf{P}^1_{k(t),x}, T^\vee \mathbf{P}^1_{k(t),y}) \to \mathbf{P}^1_{k(t),y}$. We will use the definition of relative orientability given by Kass and Wickelgren in [**KW17**], which is related to the analogous definition in [**OT14**]. Given a line bundle E on a smooth curve C, a relative orientation of E is the datum of a line bundle M and an isomorphism $\text{Hom}(TC, E) \cong M^{\otimes 2}$. Observe that

$$\begin{aligned}
&\text{Hom}(T\mathbf{P}^1_{k(t),y}, \text{Hom}(f^*T^\vee \mathbf{P}^1_{k(t),x}, T^\vee \mathbf{P}^1_{k(t),y}))\\
&\quad\cong T^\vee \mathbf{P}^1_{k(t),y} \otimes \text{Hom}(f^*T^\vee \mathbf{P}^1_{k(t),x}, T^\vee \mathbf{P}^1_{k(t),y})\\
&\quad\cong T^\vee \mathbf{P}^1_{k(t),y} \otimes (f^*T^\vee \mathbf{P}^1_{k(t),x})^\vee \otimes T^\vee \mathbf{P}^1_{k(t),y}\\
&\quad\cong \mathcal{O}(-2) \otimes \mathcal{O}(2p) \otimes \mathcal{O}(-2)\\
&\quad\cong \mathcal{O}(p-2)^{\otimes 2},
\end{aligned}$$

so $\text{Hom}(f^*T^\vee \mathbf{P}^1_{k(t),x}, T^\vee \mathbf{P}^1_{k(t),y})$ is relatively orientable.

We will still make precise the explicit isomorphism $T\mathbf{P}^1_{k(t),y} \cong \mathcal{O}(1)^{\otimes 2}$ in order to compute local degrees using sections. Write $v := -w$ on $\text{Spec}(k(t)[v]) = \text{Spec}(k(t)[-w])$ and give $\mathbf{P}^1_{k(t),y}$ the affine coordinate $y = \frac{1}{v}$ so that $dy = \frac{1}{v^2} dv$. Define $T\mathbf{P}^1_{k(t),y} \to \mathcal{O}(1)^{\otimes 2}$ on affine patches by $v \mapsto \partial y$ on $\operatorname{Spec} k(t)[v]$ and $y \mapsto \partial v$ on $\operatorname{Spec} k(t)[y]$, where ∂y and ∂v are the respective duals of the sections dy and dv on $T^\vee \mathbf{P}^1_{k(t),y}$. Note then that $\partial y = v^2 \partial v$, as desired.

Let $U = f^{-1} \text{Spec}(k(t)[z]) \cap \text{Spec}(k(t)[v])$ for notational ease, and let ψ be the trivialization of $\text{Hom}(f^*T^\vee \mathbf{P}^1_{k(t),x}, T^\vee \mathbf{P}^1_{k(t),y})$ which associates $1 \in \mathcal{O}_{\mathbf{P}^1_{k(t),y}}$ to the function $\{dz \mapsto dv\}$. Since $z = h(w) = \frac{w^{p-1}}{w^p t-1}$, $dz = \frac{(1-p)v^{p-2}}{v^p t-1} dv$ and hence $df|_U = \frac{(1-p)v^{p-2}}{v^p t-1}\{dz \mapsto dv\}$ in $\text{Hom}(f^*T^\vee \mathbf{P}^1_{k(t),x}, T^\vee \mathbf{P}^1_{k(t),y})(U)$. Thus the global section df of $\text{Hom}(f^*T^\vee \mathbf{P}^1_{k(t),x}, T^\vee \mathbf{P}^1_{k(t),y})$ corresponds to the function $v \mapsto \frac{(1-p)v^{p-2}}{-v^p t-1}$ in the coordinate v on U. Therefore we need to compute $\text{ind}_{(y^p-t)}^{\mathbb{A}^1} df = \deg_{(y^p-t)} \frac{y^p-t}{y^2}$ and $\text{ind}_{\infty}^{\mathbb{A}^1} df = \deg_0^{\mathbb{A}^1} \frac{(1-p)v^{p-2}}{-v^p t-1}$.

The local degrees $\deg_{(y^p-t)} \frac{y^p-t}{y^2}$ and $\deg_0^{\mathbb{A}^1} \frac{(1-p)v^{p-2}}{-v^p t-1}$ can be computed using Cazanave's result on the naive homotopy class of a rational function from $\mathbf{P}^1_L$ to itself for any field L [**Caz12**]. Given a rational function $\frac{f_1}{f_2}\colon \mathbf{P}^1_L \to \mathbf{P}^1_L$, we can write $\frac{f_1(x)f_2(y)-f_1(y)f_2(x)}{x-y} =: \sum_{1\le i,j\le n} c_{ij} x^{i-1} y^{j-1}$. The Bézoutian of $\frac{f_1}{f_2}$, denoted $\text{Béz}(f_1, f_2)$, is defined to be the bilinear form with Gram matrix $[c_{ij}]_{1\le i,j\le n}$. Cazanave's main result is that $\text{Béz}(f_1, f_2)$ is a representative of the isomorphism class of $\deg^{\mathbb{A}^1} \frac{f_1}{f_2}$ in GW(L).

Cazanave's result allows us to compute the global degree, $\deg^{\mathbb{A}^1}(\frac{y^p-t}{y^2})$, which is equal to $\sum_{\{q:\, q\mapsto 0\}} \deg_q^{\mathbb{A}^1} \frac{y^p-t}{y^2}$ by [**KW19**]. In this particular case, the only zero of f is the closed point $\{(y^p-t)\}$, so a global degree computation using the Bézoutian also computes the local degree, $\deg^{\mathbb{A}^1}_{(y^p-t)} \frac{y^p-t}{y^2}$. If we write $f_1 = y^p - t$ and $f_2 = y^2$, then

$$\begin{aligned}\frac{f_1(x)f_2(y) - f_1(y)f_2(x)}{x-y} &= \frac{x^2y^2(x^{p-2} - y^{p-2}) + t(x^2-y^2)}{x-y}\\ &= x^2y^2(x^{p-3} + x^{p-4}y + \cdots + y^{p-3}) + t(x+y)\\ &= t(x+y) + x^{p-1}y^2 + x^{p-2}y^3 + \cdots + x^2y^{p-1}.\end{aligned}$$

Thus the Gram matrix of the Bézoutian of $\frac{y^p-t}{y^2}$, and hence a Gram matrix of $\mathrm{ind}^{\mathbb{A}^1}_{(y^p-t)} df$, is

$$\text{Béz}(y^p - t, y^2) = \begin{bmatrix} 0 & t & 0 & \dots & 0\\ t & 0 & 0 & \dots & 0\\ 0 & 0 & 0 & \dots & 1\\ \vdots & \vdots & \vdots & \dots & \vdots\\ 0 & 0 & 1 & \dots & 0\end{bmatrix}.$$

A diagonalization of this matrix is the Gram matrix of the diagonal nondegenerate symmetric bilinear form $\left(\frac{p-1}{2}\right)\cdot h + \langle 1\rangle$. Therefore we can conclude that $\mathrm{ind}^{\mathbb{A}^1}_{(y^p-t)} df = \left(\frac{p-1}{2}\right)\cdot h + \langle 1\rangle$ in $\mathrm{GW}(k(t))$.

Now we will compute $\deg_0^{\mathbb{A}^1} \frac{(1-p)v^{p-2}}{-v^pt-1} = \mathrm{ind}^{\mathbb{A}^1}_{\infty} df$. The global degree of $\frac{(1-p)v^{p-2}}{-v^pt-1}$ will be equal to the local degree at 0 plus the local degree at ∞, as $\{0,\infty\}$ is the fiber over 0. If we write $f_1 = (1-p)v^{p-2}$ and $f_2 = -v^pt - 1$, then

$$\begin{aligned}&\frac{f_1(x)f_2(y) - f_1(y)f_2(x)}{x-y}\\ &= \frac{(p-1)tx^{p-2}y^{p-2}(y^2-x^2) + (p-1)(y^{p-2} - x^{p-2})}{x-y}\\ &= (p-1)[tx^{p-2}y^{p-2}(y+x) + (y^{p-3} + y^{p-4}x + \cdots + yx^{p-4} + x^{p-3})]\\ &= (p-1)[tx^{p-2}y^{p-1} + tx^{p-1}y^{p-2} + y^{p-3} + y^{p-4}x + \cdots + yx^{p-4} + x^{p-3}].\end{aligned}$$

Thus the Gram matrix of the Bézoutian of $\frac{(1-p)v^{p-2}}{-v^pt-1}$ is

$$\begin{bmatrix} 0 & \dots & 0 & (p-1) & 0 & 0\\ 0 & \dots & (p-1) & 0 & 0 & 0\\ 0 & \dots & 0 & 0 & 0 & 0\\ \vdots & \ddots & 0 & 0 & \vdots & \vdots\\ (p-1) & \dots & 0 & 0 & 0 & 0\\ 0 & \dots & 0 & 0 & 0 & t(p-1)\\ 0 & \dots & 0 & 0 & t(p-1) & 0\end{bmatrix}.$$

The diagonalization of this Gram matrix of the diagonal bilinear form $\left(\frac{p-1}{2}\right)\cdot h + \langle (p-1)\rangle = \left(\frac{p-3}{2}\right)\cdot h + \langle -1\rangle$. Since $\deg_\infty^{\mathbb{A}^1} \frac{(1-p)v^{p-2}}{-v^pt-1} = \langle 1\rangle + \langle -1\rangle$, it follows that $\mathrm{ind}_\infty df = \deg_0^{\mathbb{A}^1} \frac{(1-p)v^{p-2}}{-v^pt-1} = \left(\frac{p-3}{2}\right)\cdot h + \langle -1\rangle$ in $\mathrm{GW}(k(t))$.

We conclude that

$$\begin{aligned}\sum_{\{y\colon df(y)=0\}} \operatorname{ind}_y^{\mathbb{A}^1} df &= \operatorname{ind}_{(y^p-t)}^{\mathbb{A}^1} df + \operatorname{ind}_\infty^{\mathbb{A}^1} df \\ &= \left(\frac{p-1}{2}\right) h + \langle 1\rangle + h\left(\frac{p-3}{2}\right) + \langle -1\rangle \\ &= (p-2)h + \langle 1\rangle + \langle -1\rangle \\ &= (p-1)h,\end{aligned}$$

as desired. □

PROPOSITION 3.2. *The rational function* $f : \mathbf{P}^1_{k(t),y} \to \mathbf{P}^1_{k(t),x}$ *defined by* $y \mapsto y^p - y$ *has the property that there is a relative orientation such that the branch indices are*

$$ind_\infty df = \left(\frac{p-1}{2}\right) \cdot h.$$

In particular, the enriched Riemann–Hurwitz formula (1.5) *holds.*

PROOF. Orient as in the previous proof. For the affine coordinates z and v as in that proof, we have that f is given by $v \mapsto v^p/(1-v^{p-1})$, so

$$df|U = \frac{-v^{2p-2}}{(1-v^{p-1})^2} \cdot dv.$$

We complete the proof by computing the local degree of $\frac{-v^{2p-2}}{(1-v^{p-1})^2}$ using [**Caz12**]. An algebra computation shows that the

$$\begin{aligned}&\frac{(-x^{2p-2})(1-y^{p-1})^2 - (-y^{2p-2})(1-x^{p-1})^2}{x-y} \\ &\quad = -(x^{2p-3} + x^{2p-4}y + \cdots + xy^{2p-4} + y^{2p-3}) \\ &\qquad\qquad + 2x^{p-1}y^{p-1}(x^{p-2} + x^{p-3}y + \cdots + xy^{p-3}y^{p-2}).\end{aligned}$$

We conclude that a Gram matrix for the local degree is

$$\begin{bmatrix} 0 & 0 & \dots & 0 & 0 & \dots & 0 & -1 \\ 0 & 0 & \dots & 0 & 0 & \dots & -1 & 0 \\ \vdots & \vdots & \ddots & \vdots & \vdots & \vdots & \vdots & \vdots \\ 0 & 0 & \dots & 0 & -1 & \dots & 0 & 2 \\ 0 & 0 & \dots & -1 & 0 & \dots & 2 & 0 \\ \vdots & \vdots & \ddots & \vdots & \vdots & \vdots & \vdots & \vdots \\ 0 & -1 & \dots & 0 & 2 & \dots & 0 & 0 \\ -1 & 0 & \dots & 2 & 0 & \dots & 0 & 0 \end{bmatrix}.$$

This Gram matrix is equivalent to the Gram matrix where the only nonzero entries are -1's along the antidiagonal, and this matrix represents $\left(\frac{p-1}{2}\right) \cdot h$. □

4. Main Theorem

We prove the enriched Riemann–Hurwitz formula, Theorem 1.1, over an arbitrary field k discussed in the introduction. Marc Levine has previously shown an enriched Riemann–Hurwitz formula [**Lev17**, Theorem 12.7]. The hypotheses and context of Theorem 1.1 differ from M. Levine's result, and our particular interest in the present context comes from the possibility of explicitly computing certain local indices, even in the presence of wild ramification and inseparable residue field extensions.

Let $f : Y \to X$ be a non-constant, separable map of smooth, proper, geometrically connected curves over k.

We will need Nisnevich coordinates around the closed points of Y (see Section 2 for the definition of Nisnevich coordinates). For k infinite, the existence of such coordinates is shown by [**Knu91**, Chapter 8, Proposition 3.2.1]. When $k \subseteq k(y)$ is separable, this is proven in [**KW17**, Lemma 18]. Combining these results gives the desired existence. We include a different proof here, directly extending [**KW17**, Lemma 18] in the case where the dimension of Y is 1, because it offers another perspective. Namely, the proof of [**KW17**, Lemma 18] holds under the weaker hypothesis that $k \subseteq k(y)$ is a simple extension of fields, meaning that $k(y)$ is obtained from k by adjoining a single element. We will show that $k \subseteq k(y)$ is always simple, using a modification of David Speyer's proof of the Primitive Element Theorem [**Spe10**].

LEMMA 4.1 (Speyer, Lemma 2, loc. cit). *Let $r(x)$ and $q(x)$ be polynomials with coefficients in a field with $q(0) \neq 0$. Then, for all but finitely many t, the polynomials $r(tx)$ and $q(x)$ have no common factor.*

Speyer's proof is cleverly designed to avoid field extensions; we include a prosaic one for convenience.

PROOF. In an algebraic closure, we may factor $r(x) = a(x-\alpha_1)(x-\alpha_2)\cdots(x-\alpha_n)$ and $q(x) = b(x-\beta_1)(x-\beta_2)\cdots(x-\beta_m)$. If $r(tx)$ and $q(x)$ have a common factor, then there is some α_i and β_j such that $tx - \alpha_i$ is a multiple of $x - \beta_j$, or equivalently $\alpha_i = t\beta_j$. As β_j is not zero, this eliminates only one value of t. □

LEMMA 4.2. *Let $k \subseteq M$ be a finite simple field extension. Let $M \subseteq E$ be a finite separable field extension. Then $k \subseteq E$ is a simple extension.*

PROOF. We may assume that k is an infinite field of characteristic $p > 0$, because the result is immediate when k is finite or characteristic 0.

By assumption, there is an element α of M such that $M = k[\alpha]$. Let f be the minimal polynomial of α over k.

Since $M \subseteq E$ is finite and separable, there exists β in E such that $E = M[\beta] = k[\alpha, \beta]$. Since β is separable over M, $E = M[\beta^{p^d}]$ for all d by [**Lan02**, Chapter V, Exercise 16, page 254]. There exists a d such that β^{p^d} is separable over k. Thus, by replacing β by β^{p^d}, we may assume that β is separable over k. Let g be the minimal polynomial of β over k.

Define $r(x)$ and $q(x)$ in $E[x]$ by $f(x) = (x-\alpha)r(x-\alpha)$ and $g(x) = (x-\beta)q(x-\beta)$. Since β is separable, we know that 0 is not a root of $q(x)$. Therefore by Lemma 4.1 and our assumption that k is infinite, we can choose t in k such that $r(tx)$ and $q(x)$ have no common factor. We claim that $E = k[\alpha - t\beta]$.

To see this, let $h(x) = f(tx + \alpha - t\beta)$. Note that $h(x)$ has coefficients in $k[\alpha - t\beta]$. Since the polynomial ring $k[\alpha - t\beta][x]$ is a principal ideal domain, the ideal $\langle h(x), g(x)\rangle$ is generated by a polynomial $s(x)$ in $k[\alpha - t\beta][x]$, which we may assume to be monic. (So $s(x)$ is the GCD, but we wish to emphasize the ambient ring.) The polynomial $s(x)$ also generates the ideal $\langle h(x), g(x)\rangle$ of $E[x]$ generated $h(x)$ and $g(x)$.

We compute $s(x)$ by computing the GCD of $h(x)$ and $g(x)$ in $E[x]$. We have $h(x) = f(tx+\alpha-t\beta) = t(x-\beta)r(t(x-\beta))$ and $g(x) = (x-\beta)q(x-\beta)$. By the choice to t, the polynomials $r(t(x-\beta))$ and $q(x-\beta)$ have no common factor. Therefore, the GCD of $h(x)$ and $g(x)$ is $x - \beta$. Greatest common divisors are well-defined up to a nonzero scalar. Since $x - \beta$ and $s(x)$ are both GCDs of $h(x)$ and $g(x)$ and are both monic polynomials, we have that $s(x) = x - \beta$. Thus β is in $k[\alpha - t\beta]$. Since t is in k, it follows that α is in $k[\alpha - t\beta]$. Thus $k[\alpha - t\beta] \supset k[\alpha, \beta] = E$, proving the claim. □

We combine the previous lemma to reprove the existence of the desired Nisnevich coordinates:

PROPOSITION 4.3. *Let Y be a smooth curve over k and let y be a closed point of Y. Then there exist Nisnevich coordinates around y.*

REMARK 4.4. *Proposition* 4.3 *follows from* [**Knu91**, Chapter 8, Proposition 3.2.1] *when k is infinite.*

PROOF. By the proof of [**KW17**, Lemma 18], Nisnevich coordinates exist around y when $k \subseteq k(y)$ is a finite, simple extension of fields. Since Y is smooth, there is an étale map ϕ from a Zariski open neighborhood of y to $\mathbb{A}^1_k$. It follows that $k \subseteq k(y)$ is of the form $k \subseteq k(\phi(y)) \subseteq k(y)$ where $k \subseteq k(\phi(y))$ is a simple extension and $k(\phi(y)) \subseteq k(y)$ is a separable extension. The proposition thus follows from Lemma 4.2. □

To prove Theorem 1.1, we prove that under the same hypotheses, there is an equality

$$\sum_{\{y:df(y)=0\}} \mathrm{ind}_y\, df = h(g(Y) - 1 + \deg f(1 - g(X)))$$

in $\mathrm{GW}(k)$ between the local indices (or degrees) of df at its zeros and the hyperbolic form $h = \langle -1\rangle + \langle 1\rangle$ multiplied by the integer $g(Y) - 1 + \deg f(1 - g(X))$, where $g(X)$ and $g(Y)$ denote the genera of X and Y, respectively, and $\deg f = [k(Y) : k(X)]$.

PROOF OF THEOREM 1.1. Since f is separable, the section df of

$$\mathcal{V} = \mathrm{Hom}(f^*T^\vee X, T^\vee Y)$$

is nonzero and therefore has only isolated zeros, because Y is a curve. By Proposition 4.3, there are Nisnevich coordinates around all the zeros of df.

We claim that $\mathcal{V}$ is relatively orientable. By Assumption (1), we may choose a line bundle M on X such that $M^{\otimes 2} \cong T^{\vee}X$. Then

$$\begin{aligned}\mathrm{Hom}(TY, \mathcal{V}) &\cong T^{\vee}Y \otimes \mathrm{Hom}(f^{\vee}T^{\vee}X, T^{\vee}Y)\\ &\cong T^{\vee}Y \otimes (f^{*}T^{\vee}X)^{\vee} \otimes T^{\vee}Y\\ &\cong (T^{\vee}Y)^{\otimes 2} \otimes ((f^{*}M)^{\vee})^{\otimes 2}\\ &\cong (T^{\vee}Y \otimes (f^{\vee}M)^{\vee})^{\otimes 2},\end{aligned}$$

and so $\mathrm{Hom}(TY, \mathcal{V})$ is a square. We may therefore choose a relative orientation, and we do this now. (It will not matter what the chosen relative orientation is in the present case.)

Therefore, $e(Y, \mathcal{V}, df)$ is defined, and by definition is equal to

$$e(Y, \mathcal{V}, df) = \sum_{y:df(y)=0} \mathrm{ind}_y\, df. \tag{4.1}$$

Since $\mathcal{V} = \mathrm{Hom}(f^{*}T^{\vee}X, T^{\vee}Y)$ is rank one, which is odd, $e(Y, \mathcal{V}, df)$ is a multiple of the hyperbolic element h by [**SW18**, Proposition 12] and Assumption 2. (In a different context and under different hypotheses [**SW18**, Proposition 12] is proven by M. Levine in [**Lev17**, Theorem 7.1]. M. Levine also credits J. Fasel.) Thus

$$e(Y, \mathcal{V}, df) = h(\deg \mathcal{V}/2). \tag{4.2}$$

Using the fact that $\mathrm{Hom}(f^{*}T^{\vee}X, T^{\vee}Y) = (f^{*}T^{\vee}X)^{\vee} \otimes T^{\vee}Y$, we have

$$\deg \mathrm{Hom}(f^{*}T^{\vee}X, T^{\vee}Y) = \deg(f^{*}T^{\vee}X)^{\vee} \otimes T^{\vee}Y \tag{4.3}$$
$$= \deg f^{*}TX + \deg T^{\vee}Y \tag{4.4}$$
$$= \deg f \deg TX + \deg T^{\vee}Y \tag{4.5}$$
$$= \deg f(2 - 2g(X)) + (2g(Y) - 2). \tag{4.6}$$

Combining Equations (4.1), (4.2), and (4.3) gives the desired result. □

4.1. Acknowledgements. We gratefully thank Shuji Saito for raising the question at the workshop *Motivic homotopy theory and refined enumerative geometry* in Essen, and for interesting discussions about it. We likewise thank Alexey Ananyevskiy and Ivan Panin for the reference to [**Knu91**, Chapter 8, Proposition 3.2.1] for the existence of Nisnevich coordinates.

Candace Bethea was partially supported by a SPARC Graduate Research Grant from the Office of the Vice President for Research at the University of South Carolina.

Jesse Kass was partially supported by the Simons Foundation under Award Number 429929.

Kirsten Wickelgren was partially supported by National Science Foundation Award DMS-1552730.

References

[AF16] A. Asok and J. Fasel, *Comparing Euler classes*, Q. J. Math. **67** (2016), no. 4, 603–635. MR3609848 ↑

[AS02] Ahmed Abbes and Takeshi Saito, *Ramification of local fields with imperfect residue fields*, Amer. J. Math. **124** (2002), no. 5, 879–920. MR1925338 ↑

[AS03] Ahmed Abbes and Takeshi Saito, *Ramification of local fields with imperfect residue fields. II*, Doc. Math. **Extra Vol.** (2003), 5–72. Kazuya Kato's fiftieth birthday. MR2046594 ↑

[BM00] Jean Barge and Fabien Morel, *Groupe de Chow des cycles orientés et classe d'Euler des fibrés vectoriels* (French, with English and French summaries), C. R. Acad. Sci. Paris Sér. I Math. **330** (2000), no. 4, 287–290, DOI 10.1016/S0764-4442(00)00158-0. MR1753295 ↑

[BW19] Tom Bachmann and Kirsten Wickelgren, *Three ways of looking at an A1-Euler class*, 2019. In preparation. ↑

[Caz08] Christophe Cazanave, *Classes d'homotopie de fractions rationnelles* (French, with English and French summaries), C. R. Math. Acad. Sci. Paris **346** (2008), no. 3-4, 129–133, DOI 10.1016/j.crma.2008.01.004. MR2393628 ↑

[Caz12] Christophe Cazanave, *Algebraic homotopy classes of rational functions* (English, with English and French summaries), Ann. Sci. Éc. Norm. Supér. (4) **45** (2012), no. 4, 511–534 (2013), DOI 10.24033/asens.2172. MR3059240 ↑

[DJK18] Frédéric Déglise, Fangzhou Jin, and Adheel Khan, *Fundamental classes in motivic homotopy theory*, 2018. *Preprint*, available at `https://arxiv.org/abs/1805.05920`. ↑

[Fas08] Jean Fasel, *Groupes de Chow-Witt* (French, with English and French summaries), Mém. Soc. Math. Fr. (N.S.) **113** (2008), viii+197, DOI 10.24033/msmf.425. MR2542148 ↑

[GI80] D. Ju. Grigor′ev and N. V. Ivanov, *On the Eisenbud-Levine formula over a perfect field* (Russian), Dokl. Akad. Nauk SSSR **252** (1980), no. 1, 24–27. MR572114 ↑

[Har77] Robin Hartshorne, *Algebraic geometry*, Springer-Verlag, New York-Heidelberg, 1977. Graduate Texts in Mathematics, No. 52. MR0463157 ↑

[Kat89] Kazuya Kato, *Swan conductors for characters of degree one in the imperfect residue field case*, Algebraic K-theory and algebraic number theory (Honolulu, HI, 1987), Contemp. Math., vol. 83, Amer. Math. Soc., Providence, RI, 1989, pp. 101–131, DOI 10.1090/conm/083/991978. MR991978 ↑

[Knu91] Max-Albert Knus, *Quadratic and Hermitian forms over rings*, Grundlehren der Mathematischen Wissenschaften [Fundamental Principles of Mathematical Sciences], vol. 294, Springer-Verlag, Berlin, 1991. With a foreword by I. Bertuccioni. MR1096299 ↑

[KW17] Jesse Kass and Kirsten Wickelgren, *An arithmetic count of the lines on a smooth cubic surface*, 2017. *Preprint*, available at arXiv:1708.01175. ↑

[KW18] Jesse Kass and Kirsten Wickelgren, *A classical proof that the algebraic homotopy class of a rational function is the residue pairing*, 2018. *Preprint*, available at arXiv:1602.08129v2. ↑

[KW19] Jesse Leo Kass and Kirsten Wickelgren, *The class of Eisenbud-Khimshiashvili-Levine is the local* $\mathbf{A}^1$*-Brouwer degree*, Duke Math. J. **168** (2019), no. 3, 429–469, DOI 10.1215/00127094-2018-0046. MR3909901 ↑

[Lan02] Serge Lang, *Algebra*, 3rd ed., Graduate Texts in Mathematics, vol. 211, Springer-Verlag, New York, 2002. MR1878556 ↑

[Lev17] Marc Levine, *Toward an enumerative geometry with quadratic forms*, 2017. *Preprint*, available at `https://arxiv.org/abs/1703.03049`. ↑

[Liu02] Qing Liu, *Algebraic geometry and arithmetic curves*, Oxford Graduate Texts in Mathematics, vol. 6, Oxford University Press, Oxford, 2002. Translated from the French by Reinie Erné; Oxford Science Publications. MR1917232 ↑

[LR18] Marc Levine and Arpon Raksit, *Motivic Gauss–Bonnet formulas*, 2018. *Preprint*, available at `https://arxiv.org/abs/1808.08385`. ↑

[Mor12] Fabien Morel, $\mathbb{A}^1$*-algebraic topology over a field*, Lecture Notes in Mathematics, vol. 2052, Springer, Heidelberg, 2012. MR2934577 ↑

[Neu99] Jürgen Neukirch, *Algebraic number theory*, Grundlehren der Mathematischen Wissenschaften [Fundamental Principles of Mathematical Sciences], vol. 322, Springer-Verlag, Berlin, 1999. Translated from the 1992 German original and with a note by Norbert Schappacher; With a foreword by G. Harder. MR1697859 ↑

[OT14] Christian Okonek and Andrei Teleman, *Intrinsic signs and lower bounds in real algebraic geometry*, J. Reine Angew. Math. **688** (2014), 219–241, DOI 10.1515/crelle-2012-0055. MR3176620 ↑

[Sai12] Takeshi Saito, *Ramification of local fields with imperfect residue fields III*, Math. Ann. **352** (2012), no. 3, 567–580, DOI 10.1007/s00208-011-0652-5. MR2885588 ↑

[Ser79] Jean-Pierre Serre, *Local fields*, Graduate Texts in Mathematics, vol. 67, Springer-Verlag, New York-Berlin, 1979. Translated from the French by Marvin Jay Greenberg. MR554237 ↑

[Spe10] David Speyer, *Math Overflow: primitive element theorem without building field extensions*, 2010. ↑

[SS75] Günter Scheja and Uwe Storch, *Über Spurfunktionen bei vollständigen Durchschnitten* (German), J. Reine Angew. Math. **278/279** (1975), 174–190. MR0393056 ↑

[SW18] Padmavathi Srinivasan and Kirsten Wickelgren, *An arithmetic count of the lines meeting four lines in* $\mathbb{P}^3$, 2018. *Preprint*, available at `https://arxiv.org/abs/1810.03503`. ↑

Contemporary Mathematics
Volume **745**, 2020
https://doi.org/10.1090/conm/745/15023

Lectures on Chow-Witt groups

Jean Fasel

ABSTRACT. We provide a toolkit to work with Chow-Witt groups, and more generally with the homology and cohomology of the Rost-Schmid complex associated to Milnor-Witt K-theory.

CONTENTS

Introduction

The purpose of these lectures (given in Essen in May 2018) is to give a quick introduction to Chow-Witt groups not involving prior knowledge in either Chow groups or the theory of symmetric bilinear forms. The goal is to stay as elementary as possible, avoiding altogether complicated proofs and focusing on explicit expressions allowing to perform computations. The lectures are based on the construction of the Rost-Schmid complex in [**23**] and on a primer on Milnor-Witt cohomology in preparation (jointly with Baptiste Calmès). The reader will eventually be able to refer to the latter for complete details, and in particular for exhaustive proofs of the results stated here. The program of the lectures is roughly the following. In the first part, Milnor-Witt K-theory of a field F is introduced, following the original presentation of [**23**, §3]. We state a few elementary results aimed at understanding the nature of this K-theory, mixing up Milnor K-theory and symmetric bilinear forms. Then, we pass to the fundamental construction of the residue homomorphism associated to a discrete valuation field. The subtle point of the construction is that this homomorphism doesn't depend only on the chosen valuation, but also on the choice of a uniformizing parameter. This fact parallels the situation of Witt groups, and there is a well known procedure to transform it in a homomorphism independent

2010 *Mathematics Subject Classification.* Primary: 14C17, 14C25, 14F43, 19-06, 19G99. Secondary: 11E04, 11E39, 11E70.

of the choice of the parameter. This requires the introduction of Milnor-Witt K-theory *twisted* by some extra information and leads to interesting but sometimes technically involved constructions. The major novelty here (with respect to [**23**] or [**12**]) is the notion of Milnor-Witt K-theory twisted by a *graded line bundle* following the original idea of [**10**]. Having the twisted residue homomorphisms at hand, we introduce in the second lecture the Rost-Schmid complexes, which are our main object of study. These complexes are defined for schemes essentially of finite type over a field k, but have some special properties when the scheme is essentially smooth. This leads to the notions of homological Rost-Schmid complex and cohomological Rost-Schmid complex, the latter being defined only for essentially smooth schemes. We then pass to the study of the functoriality properties of the complexes, the main result being that the homological complex is covariantly functorial for proper morphisms and contravariantly functorial for smooth morphisms, while the cohomological complex is contravariantly functorial for flat morphisms. In both cases, the (co-)homology of the complex is homotopy invariant allowing the definition of an Euler class associated to a vector bundle over a scheme X essentially of finite type over k. This Euler class is one of the main motivations for the introduction of Chow-Witt groups by Barge and Morel, but we only spend a few lines on these classes stating only the definition. More properties of these classes can be found in Levine's lectures in the same volume. In the third lecture, we construct a pull-back for morphisms between smooth schemes, and more generally for l.c.i. morphisms between finite type schemes. This pull-back is defined at the level of the (co-)homology and uses homotopy invariance and the deformation to the normal cone construction. This is the only point in these notes where we loose track of what happens at the level of (co-)cycles, mainly because of our use of homotopy invariance. This pull-back allows to define a ring structure at the level of the cohomology of the Rost-Schmid complex. We further study functoriality in this section, stating results on cdh and Nisnevich descent before passing to the base change and projection formulas. At this point, it must be said that our constructions are all based on the theory of cycle modules as studied in [**27**], consisting mostly of a refinement of these techniques as in [**23**].

To conclude this introduction, let us say a few words about the history of Chow-Witt groups. They were originally introduced by Barge-Morel ([**4**]) with the idea to define an Euler class able to tell when a rank d vector bundle E over a smooth affine scheme X actually splits a free bundle of rank one. This program was later fulfilled by Morel in [**23**]. The definition of the Chow-Witt groups was based on the one hand on the complex in Milnor K-theory and on the other hand on a Gersten-Witt complex defined in the PhD thesis of Schmid ([**30**]). Unfortunately, the proof that this complex was indeed a complex was missing at that time. Part of my PhD thesis ([**12**]) was to use another complex to define the Chow-Witt groups, namely the Gersten-Witt complex of Balmer-Walter ([**3**]). This complex involved the machinery of derived Witt groups as developed by Balmer, which allowed to give a definition of the Chow-Witt groups and obtain their basic functorial properties (all based on the corresponding functoriality of Witt groups). It was also the occasion to rename the *Chow groups of oriented cycles* of Barge-Morel to Chow-Witt groups. Indeed, the reference to an orientation seemed confusing at that time, simply because Chow-Witt groups are not oriented in the classical sense.

Acknowledgements. I wish to thank the editors, especially Federico Binda, for motiving me to write down these notes. I'm grateful to Peng Du for carefully reading preliminary versions of this text. This led to the elimination of countless typos and contributed to make the exposition better.

Conventions. We work over a field k, assumed to be perfect. The schemes are all k-schemes which are of finite type and separated. The fields we consider are all finitely generated field extensions of k.

1. Preliminaries

1.1. Milnor-Witt K-theory. Let F be a finitely generated field extension of k. Let $A(F)$ be the free (associative, unital) $\mathbb{Z}$-graded ring generated in degree 1 by symbols $[a]$, for $a \in F^\times$, and η in degree -1. The Milnor-Witt K-theory ring $\mathrm{K}_*^{\mathrm{MW}}(F)$ is the quotient of $A(F)$ generated by the following relations:

(1) $[a] \cdot [1-a] = 0$ for $a \neq 0, 1$.
(2) $[ab] = [a] + [b] + \eta \cdot [a] \cdot [b]$ for $a, b \in F^\times$.
(3) $\eta \cdot [a] = [a] \cdot \eta$ for $a \in F^\times$.
(4) $\eta \cdot (\eta \cdot [-1] + 2) = 0$.

In the sequel, we will simply denote by $[a_1, \ldots, a_n]$ the product of the $[a_i]$ (for $i = 1, \ldots, n$) and by $\eta[a]$ the product $\eta \cdot [a]$. By construction, the association

$$F \mapsto \mathrm{K}_*^{\mathrm{MW}}(F)$$

is functorial in F. Before stating elementary properties of Milnor-Witt K-theory, we first spend a few lines explaining its links with other objects in mathematics. First, note that the ideal generated by η is two-sided and that

$$\mathrm{K}_*^{\mathrm{MW}}(F)/(\eta) = \mathrm{K}_*^{\mathrm{M}}(F)$$

where the right-hand side is the Milnor K-theory ring as defined in [**20**]. The image of $[a_1, \ldots, a_n]$ is usually denoted by $\{a_1, \ldots, a_n\}$. Let now $\mathrm{GW}(F)$ be the Grothendieck-Witt ring of symmetric bilinear forms on F ([**21**]). For any $a \in F^\times$, we denote by $\langle a \rangle$ the class of the symmetric bilinear form on F defined by $(X, Y) \mapsto aXY$. The following result is fundamental ([**23**, Lemma 3.10]).

LEMMA 1.1. *The map*

$$\mathrm{GW}(F) \to \mathrm{K}_0^{\mathrm{MW}}(F)$$

given by $\langle a \rangle \mapsto 1 + \eta[a]$ passes to a well-defined isomorphism.

In view of this result, we will denote by $\langle a \rangle$ the symbol $1 + \eta[a]$ of degree 0. The second relation defining Milnor-Witt K-theory then becomes

$$[ab] = [a] + \langle a \rangle [b] = [b] + \langle b \rangle [a]$$

for any $a, b \in F^\times$. In particular, the fact that $\langle b^2 \rangle = \langle 1 \rangle$ yields $[ab^2] = [a] + [b^2]$.

Observe also that the symbol $\eta \cdot [-1] + 2$ appearing in the fourth relation becomes $\eta[-1] + 2 = \langle -1 \rangle + 1 = \langle -1, 1 \rangle$. This is the so-called hyperbolic form on F^2. Following the usual conventions, we set $h := \langle -1, 1 \rangle$ and we then see that the fourth relation becomes $\eta h = 0$.

For $i \geq 1$, we can consider the composite homomorphism

$$\mathrm{GW}(F) \to \mathrm{K}_0^{\mathrm{MW}}(F) \xrightarrow{\eta^i} \mathrm{K}_{-i}^{\mathrm{MW}}(F).$$

Using Lemma 1.1 above, it is not hard to see that this is surjective. By definition, we have $\eta h = 0$ and therefore we obtain a surjective homomorphism

$$\mathrm{W}(F) = \mathrm{GW}(F)/(h) \to \mathrm{K}_{-i}^{\mathrm{MW}}(F)$$

which is in fact an isomorphism ([**23**, Lemma 3.10]). The left-hand side is the Witt ring and is the basic object of study when dealing with symmetric bilinear forms on F. We have a homomorphism

$$r : \mathrm{GW}(F) \to \mathbb{Z}$$

which associate to a symmetric bilinear form its rank. The kernel of this homomorphism is the fundamental ideal $\mathrm{I}(F) \subset \mathrm{GW}(F)$. Since h is of rank 2, we may also consider the homomorphism $\mathrm{W}(F) \to \mathbb{Z}/2$ induced by r and observe that the kernel is also $\mathrm{I}(F)$. It is additively generated by the classes of $\langle -1, a\rangle$ in $\mathrm{W}(F)$ (or $\langle -1, a\rangle - h$ in $\mathrm{GW}(F)$). It follows that its powers $\mathrm{I}^n(F) \subset \mathrm{W}(F)$ (for $n \geq 1$) are generated by elements of the form

$$\langle\langle a_1, \ldots, a_n\rangle\rangle := \langle -1, a_1\rangle \cdot \ldots \cdot \langle -1, a_n\rangle$$

It is convenient to set $\mathrm{I}^n(F) = \mathrm{W}(F)$ for $n \leq 0$. We may also consider the quotient groups

$$\bar{\mathrm{I}}^n(F) := \mathrm{I}^n(F)/\mathrm{I}^{n+1}(F)$$

for any $n \in \mathbb{Z}$ (by definition, $\bar{\mathrm{I}}^n(F) = 0$ for $n \leq -1$).

LEMMA 1.2. *For any $n \in \mathbb{Z}$, there is a unique (surjective) homomorphism of $\mathrm{K}_0^{\mathrm{MW}}(F)$-modules*

$$j_n : \mathrm{K}_n^{\mathrm{MW}}(F) \to \mathrm{I}^n(F)$$

satisfying $[a_1, \ldots, a_n] \mapsto \langle\langle a_1, \ldots, a_n\rangle\rangle$.

PROOF. If $n \leq 0$, the definition of j_n is obvious. We then suppose that $n \geq 1$. We may use an alternative presentation of $\mathrm{K}_n^{\mathrm{MW}}(F)$ ([**23**, Definition 3.3, Lemma 2.4]). Its generators are of the form $\eta^m[a_1, \ldots, a_r]$ with $r - m = n$ and $r, m \in \mathbb{N}$, and the relations are as follows:

(1) $\eta^m[a_1, \ldots, a_r] = 0$ if $a_i + a_{i+1} = 1$ for some $i = 1, \ldots, r-1$.
(2) $\eta^m[a_1, \ldots, xy, \ldots, a_r] = \eta^m[a_1, \ldots, x, \ldots, a_r] + \eta^m[a_1, \ldots, y, \ldots, a_r] + \eta^{m+1}[a_1, \ldots, x, y, \ldots, a_r]$ for any $x, y \in F^\times$.
(3) $\eta^{m+2}[a_1, \ldots, a_{i-1}, -1, a_{i+1}, \ldots, a_r] + 2\eta^{m+1}[a_1, \ldots, a_{i-1}, a_{i+1}, \ldots, a_r] = 0$.

We set $j_n(\eta^m[a_1, \ldots, a_r]) = \langle\langle a_1, \ldots, a_r\rangle\rangle \in \mathrm{I}^r(F) \subset \mathrm{I}^n(F)$, and we check that the three above relations are satisfied. If F is of characteristic different from 2, this is essentially [**22**, Lemma 2.3]. In characteristic 2, the proof is the same but we spell it out for the convenience of the reader. We may use the presentation of $\mathrm{GW}(F)$ given in [**21**, Chapter IV, Lemma 1.1]. One of the relation reads as

$$\langle u\rangle + \langle v\rangle = \langle u + v\rangle + \langle (u+v)uv\rangle$$

for any $u, v \in F^\times$ with $u + v \neq 0$. Applying to $a, 1-a$, we obtain $\langle a\rangle + \langle 1-a\rangle = \langle 1\rangle + \langle a(1-a)\rangle$ showing that $\langle\langle a, 1-a\rangle\rangle = h^2$ in $\mathrm{GW}(F)$. Consequently, the first relation holds in $\mathrm{I}^n(F)$. The third relation holds by definition, and we are left with the second one. It suffices to prove that

$$\langle\langle ab\rangle\rangle = \langle\langle a\rangle\rangle + \langle\langle b\rangle\rangle + \langle\langle a, b\rangle\rangle \in \mathrm{I}(F).$$

Using $\langle u\rangle + \langle -u\rangle = h$ for any $u \in F^\times$, we get

$$\langle -1, a\rangle + \langle -1, b\rangle + \langle -1, a\rangle \cdot \langle -1, b\rangle = \langle -1, ab\rangle + 3h$$

whence $\langle\langle a\rangle\rangle + \langle\langle b\rangle\rangle + \langle\langle a, b\rangle\rangle = \langle\langle ab\rangle\rangle$. Finally, it is easy to check that j_n is $\mathrm{K}_0^{\mathrm{MW}}(F)$-linear. □

The proof of the lemma also yields the following easy corollary that we state for further reference.

COROLLARY 1.3. *For any $n \geq 1$, the diagram*

$$\begin{array}{ccc} \mathrm{K}_{n+1}^{\mathrm{MW}}(F) & \xrightarrow{j_{n+1}} & \mathrm{I}^{n+1}(F) \\ {\scriptstyle\eta}\downarrow & & \downarrow \\ \mathrm{K}_n^{\mathrm{MW}}(F) & \xrightarrow[j_n]{} & \mathrm{I}^n(F), \end{array}$$

in which the right-hand vertical map is the inclusion, is commutative.

If F is of characteristic different from 2, recall from [**20**, Theorem 4.1] that for any $n \in \mathbb{Z}$ there is a unique homomorphism

$$s_n : \mathrm{K}_n^{\mathrm{M}}(F) \to \bar{\mathrm{I}}^n(F)$$

such that $s_n(\{a_1, \ldots, a_n\}) = \langle\langle a_1, \ldots, a_n\rangle\rangle \pmod{\mathrm{I}^{n+1}(F)}$. In characteristic 2, essentially the same proof as in Lemma 1.2 shows that this homomorphism is well-defined and surjective. By definition, we obtain a commutative square (of $\mathrm{K}_0^{\mathrm{MW}}(F)$-modules)

$$\begin{array}{ccc} \mathrm{K}_n^{\mathrm{MW}}(F) & \xrightarrow{j_n} & \mathrm{I}^n(F) \\ \downarrow & & \downarrow \\ \mathrm{K}_n^{\mathrm{M}}(F) & \xrightarrow[s_n]{} & \bar{\mathrm{I}}^n(F). \end{array} \tag{1.1}$$

It turns out that this square is actually cartesian (at least in characteristic different from 2), but this is a highly non trivial fact. To explain this, let us first observe that the homomorphism j_n induces a surjective homomorphism

$$\bar{j}_n : \mathrm{K}_n^{\mathrm{MW}}(F)/(h) \to \mathrm{I}^n(F)$$

by $\mathrm{K}_0^{\mathrm{MW}}(F)$-linearity.

THEOREM 1.4. *Suppose that F is of characteristic different from 2. Then, the diagram (1.1) is Cartesian.*

PROOF. We have a commutative diagram of exact sequences

$$\begin{array}{ccccccc} \mathrm{K}_{n+1}^{\mathrm{MW}}(F) & \xrightarrow{\eta} & \mathrm{K}_n^{\mathrm{MW}}(F) & \longrightarrow & \mathrm{K}_n^{\mathrm{M}}(F) & \longrightarrow & 0 \\ {\scriptstyle j_{n+1}}\downarrow & & {\scriptstyle j_n}\downarrow & & {\scriptstyle s_n}\downarrow & & \\ \mathrm{I}^{n+1}(F) & \longrightarrow & \mathrm{I}^n(F) & \longrightarrow & \bar{\mathrm{I}}^n(F) & \longrightarrow & 0. \end{array}$$

On the other hand, we know from [**22**, Theorem 2.4] that $\bar{j}_n$ is an isomorphism (this is a consequence of Voevodsky's affirmation of Milnor conjecture) for each $n \in \mathbb{Z}$. It follows that the map

$$\mathrm{K}_{n+1}^{\mathrm{MW}}(F)/(h) \xrightarrow{\eta} \mathrm{K}_n^{\mathrm{MW}}(F)/(h)$$

is injective. Consequently $(h) \cap (\eta) = (\eta h) = 0$ and $\bar{\mathrm{I}}^n(F) = \mathrm{K}_n^{\mathrm{MW}}(F)/(\eta, h)$. The claim follows. □

REMARK 1.5. I don't know if the statement is true if F is of characteristic 2 (see however [**23**, Remark 2.12]).

To conclude our elementary explorations of the properties of Milnor-Witt K-theory, let us observe that the multiplication by h induces a homomorphism

$$h_n : \mathrm{K}_n^{\mathrm{M}}(F) \to \mathrm{K}_n^{\mathrm{MW}}(F)$$

since $\eta h = 0$. This homomorphism can also be described by

$$h_n(\{a_1, \ldots, a_n\}) = [a_1^2, a_2, \ldots, a_n].$$

We conclude this section with some basic properties of Milnor-Witt K-theory that we'll use in the sequel ([**23**, Lemma 3.7, Corollary 3.8, Lemma 3.14]).

LEMMA 1.6. *Let* $\epsilon := -\langle -1 \rangle \in \mathrm{K}_0^{\mathrm{MW}}(F)$. *For any* $n \in \mathbb{Z}$, *let*

$$n_\epsilon = \begin{cases} \sum_{i=1}^n \langle (-1)^{i-1} \rangle & \text{if } n > 0. \\ 0 & \text{if } n = 0. \\ \epsilon \sum_{i=1}^{-n} \langle (-1)^{i-1} \rangle & \text{if } n < 0. \end{cases}$$

Then, the following properties are satisfied.

(1) *For any* $a \in F^\times$, *we have* $[a, a] = [-1, a] = [a, -1]$ *and* $[a, -a] = 0$.
(2) *For any* $\alpha \in \mathrm{K}_n^{\mathrm{MW}}(F)$ *and* $\beta \in \mathrm{K}_m^{\mathrm{MW}}(F)$, *we have* $\alpha\beta = \epsilon^{mn}\beta\alpha$.
(3) *For any* $n \in \mathbb{Z}$ *and* $a \in F^\times$, *we have* $[a^n] = n_\epsilon[a]$.

1.2. Residue homomorphisms. Let us now describe the fundamental tool used to define a Gersten-type complex with coefficients in Milnor-Witt K-theory: the residue homomorphism. We follow again closely [**23**, §3]. Let then F be a field and $v : F \to \mathbb{Z} \cup \{-\infty\}$ be a discrete valuation with residue field $\kappa(v)$ and valuation ring $\mathcal{O}_v$. We choose a uniformizing parameter $\pi = \pi_v$ of v.

THEOREM 1.7. *There is a unique homomorphism of graded abelian groups*

$$\partial_v^\pi : \mathrm{K}_*^{\mathrm{MW}}(F) \to \mathrm{K}_{*-1}^{\mathrm{MW}}(\kappa(v))$$

of degree -1 *such that* ∂_v^π *commutes with the multiplication by* η *and*

(1) $\partial_v^\pi([\pi, u_1, \ldots, u_n]) = [\overline{u}_1, \ldots, \overline{u}_n]$,
(2) $\partial_v^\pi([u_1, \ldots, u_n]) = 0$,

for any $u_1, \ldots, u_n \in \mathcal{O}_v^\times$. *We call it the* residue homomorphism.

PROOF. See [**23**, Theorem 3.15]. □

REMARK 1.8. Granted the existence of ∂_v^π, the unicity is easy. Indeed, let $[a_1, \ldots, a_n]$ be a symbol. Writing $a_i = u_i\pi^{n_i}$ with $u_i \in \mathcal{O}_v^\times$ and $n_i \in \mathbb{Z}$ for each i and using $[\pi^{n_i}] = (n_i)_\epsilon[\pi]$, $[\pi, \pi] = [\pi, -1]$ and the graded commutativity of Milnor-Witt K-theory, we can express $[a_1, \ldots, a_n]$ as a sum of symbols of the form $\eta^m[\pi, u_1, \ldots, u_{n+m-1}]$ and $\eta^m[u_1, \ldots, u_{n+m}]$ for some (non constant) $m \in \mathbb{N}$. The images of these symbols are determined by properties 1. and 2. above, yielding unicity.

REMARK 1.9. The homomorphism ∂_v^π depends not only on the valuation v, but also on the choice of the uniformizing parameter π. Indeed, suppose that $\pi' = u\pi$ for some $u \in \mathcal{O}_v^\times$. We then have $\partial_v^\pi([\pi,-1]) = [-1]$ by construction, while

$$\begin{aligned}
\partial_v^{\pi'}([\pi,-1]) &= \partial_v^{\pi'}([u^{-1}\pi',-1]) \\
&= \partial_v^{\pi'}([u^{-1},-1]+[\pi',-1]+\eta[u^{-1},\pi',-1]) \\
&= \partial_v^{\pi'}([u^{-1},-1]+[\pi',-1]+\eta[\pi',u^{-1},-1]) \\
&= [-1]+\eta[u^{-1},-1] \\
&= \langle u^{-1}\rangle[-1].
\end{aligned}$$

In general, $[-1] \neq \langle u^{-1}\rangle[-1]$. For instance, taking $\kappa(v) = \mathbb{R}$ and using the homomorphism $\mathrm{K}_1^{\mathrm{MW}}(\mathbb{R}) \to \mathrm{I}(\mathbb{R})$ we see that $[-1] \neq \langle -1\rangle[-1]$.

The fact that ∂_v^π depends on the choice of the uniformizing parameter is the reason for introducing twisted Milnor-Witt K-theory in Section 1.5 and the motivation for considering graded line bundles in the next section. To conclude the present section, we state a few fundamental results on the residue homomorphism ([**23**, Proposition 3.17, Theorem 3.24]) and introduce the geometric transfer.

LEMMA 1.10. *Let $\alpha \in \mathrm{K}_*^{\mathrm{MW}}(F)$ and let $u \in \mathcal{O}_v^\times$. Then*

$$\partial_v^\pi(\langle u\rangle\alpha) = \langle \overline{u}\rangle\partial_v^\pi(\alpha).$$

THEOREM 1.11. *For any $n \in \mathbb{Z}$, there is a split short exact sequence*

$$0 \to \mathrm{K}_n^{\mathrm{MW}}(F) \to \mathrm{K}_n^{\mathrm{MW}}(F(t)) \xrightarrow{\sum \partial_p^p} \bigoplus_p \mathrm{K}_{n-1}^{\mathrm{MW}}(F(p)) \to 0$$

where p runs through the set of monic irreducible polynomials of $F[t]$ and $F(p) = F[t]/(p)$. The left-hand homomorphism is induced by the field extension $F \subset F(t)$ and the right-hand homomorphism is the sum of the residues homomorphisms associated to the p-adic valuation and the uniformizing parameter p.

REMARK 1.12. There is an obvious retraction of $\mathrm{K}_n^{\mathrm{MW}}(F) \to \mathrm{K}_n^{\mathrm{MW}}(F(t))$. Indeed, for any $\alpha \in \mathrm{K}_n^{\mathrm{MW}}(F(t))$, we can consider $[t]\alpha \in \mathrm{K}_{n+1}^{\mathrm{MW}}(F(t))$ and then use the residue homomorphism ∂_t^t.

There is yet another valuation on $F(t)$ that we haven't used so far: the valuation at ∞. For any polynomials $f, g \in (F[t]\setminus 0)^2$, we set $v_\infty(f/g) = \deg(g) - \deg(f)$. We can choose $-\frac{1}{t}$ as a uniformizing parameter (we'll come back later on the reasons of this choice) and we obtain a residue homomorphism

$$\partial_\infty^{-\frac{1}{t}} : \mathrm{K}_n^{\mathrm{MW}}(F(t)) \to \mathrm{K}_{n-1}^{\mathrm{MW}}(F)$$

for any $n \in \mathbb{Z}$. Following [**23**, §4.2], we set the following definition.

DEFINITION 1.13. Let p be a monic irreducible polynomial in $F[t]$, the composite

$$\tau_F^{F(p)} : \mathrm{K}_{n-1}^{\mathrm{MW}}(F(p)) \subset \bigoplus_p \mathrm{K}_{n-1}^{\mathrm{MW}}(F(p)) \xrightarrow{s} \mathrm{K}_n^{\mathrm{MW}}(F(t)) \xrightarrow{-\partial_\infty^{-\frac{1}{t}}} \mathrm{K}_{n-1}^{\mathrm{MW}}(F),$$

where s is any section of $\sum \partial_p^p$, is called the *geometric transfer* (in weight $n-1$) associated to the extension $F \subset F(p)$.

REMARK 1.14. As the composite $\mathrm{K}_n^{\mathrm{MW}}(F) \to \mathrm{K}_n^{\mathrm{MW}}(F(t)) \xrightarrow{-\partial_\infty^{-\frac{1}{t}}} \mathrm{K}_{n-1}^{\mathrm{MW}}(F)$ is trivial, the geometric transfer is independent of the choice of the section s.

As a consequence, we obtain a useful unicity property, that is a consequence of the definition of a cokernel.

LEMMA 1.15. *The geometric transfers $\tau_F^{F(p)}$ are the unique homomorphisms which make the diagram*

$$\begin{array}{ccccccccc} 0 & \longrightarrow & \mathrm{K}_n^{\mathrm{MW}}(F) & \longrightarrow & \mathrm{K}_n^{\mathrm{MW}}(F(t)) & \xrightarrow{\sum \partial_p^p} & \bigoplus_p \mathrm{K}_{n-1}^{\mathrm{MW}}(F(p)) & \longrightarrow & 0 \\ & & & & \Big\downarrow{\scriptstyle -\partial_\infty^{-\frac{1}{t}}} & \swarrow{\scriptstyle \sum \tau_F^{F(p)}} & & & \\ & & & & \mathrm{K}_{n-1}^{\mathrm{MW}}(F) & & & & \end{array}$$

commutative. In other words, there are the unique homomorphisms $f_p : \mathrm{K}_{n-1}^{\mathrm{MW}}(F(p)) \to \mathrm{K}_{n-1}^{\mathrm{MW}}(F)$ with $\sum(f_p \circ \partial_p^p) + \partial_\infty^{-\frac{1}{t}} = 0$.

REMARK 1.16. This unicity property is useful when trying to identify the geometric transfers in the context of symmetric bilinear forms. Indeed, if $n \leq -1$, the geometric transfers give transfer maps

$$\mathrm{W}(F(p)) \to \mathrm{W}(F)$$

at the level of Witt groups. Now, transfers for Witt groups are usually obtained via the so-called Scharlau transfer map. For any nontrivial homomorphism of F-vector spaces $f_p : F(p) \to F$, one obtains a transfer homomorphism $(f_p)_* : \mathrm{W}(F(p)) \to \mathrm{W}(F)$ by considering the composite

$$V \times V \to F(p) \xrightarrow{f_p} F$$

for any symmetric bilinear map $V \times V \to F(p)$. In our case, $F(p)$ is a monogeneous field extension, with basis $\{1, t, \ldots, t^{d-1}\}$ where $d = \deg(p)$ and we can consider the homomorphism $f_p : F(p) \to F$ defined by $f(t^i) = 0$ for $i = 0, \ldots, d-2$ and $f(t^{d-1}) = 1$. In characteristic different from 2, we may use [**28**, Theorem 4.1] to see that the geometric transfers are the ones obtained using the homomorphisms f_p we just defined. The reader can adapt the proof of *loc. cit.* to the case of characteristic 2 as well.

REMARK 1.17. The main problem with the geometric transfers is that they actually depend on choices. This is one of the main reasons for introducing a slightly different (and more subtle) transfer in Section 1.5.

1.3. Graded line bundles. Let X be a connected scheme, and let $\mathcal{P}(X)$ be the category of invertible $\mathcal{O}_X$-modules (or, equivalently, the category of line bundles over X). The category $\mathcal{P}(X)$ is a symmetric monoidal category (in the sense of [**19**, Chapter VII, §7]), where the operation is the tensor product of invertible $\mathcal{O}_X$-modules

$$(L, N) \mapsto L \otimes_{\mathcal{O}_X} N,$$

the unit is the module $\mathcal{O}_X$ and the associativity and commutativity constraints are the usual ones. We'll drop the subscript on the tensor product in the sequel. We have three useful isomorphisms, namely the switch isomorphisms $s : L \otimes N \to N \otimes L$ (which satisfies $s \otimes t \mapsto t \otimes s$ on sections), the isomorphism $u : L \otimes L^\vee \simeq \mathcal{O}_X$ given on sections by $l\otimes\varphi\mapsto\varphi(l)$ and the isomorphism $us: L^\vee\otimes L\simeq\mathcal{O}_X$. The category $\mathcal{G}(X)$ of graded line bundles on X is the category whose objects are pairs (L,a) where L is an object of $\mathcal{P}(X)$ and $a\in\mathbb{Z}$ and whose morphisms are of the form

$$\hom_{\mathcal{G}(X)}((L,a),(L',a'))=\begin{cases}\emptyset & \text{if } a\neq a'.\\ \mathrm{Isom}_{\mathcal{P}(X)}(L,L') & \text{if } a=a'.\end{cases}$$

In particular, all morphisms in this category are isomorphisms. We may enrich $\mathcal{G}(X)$ with a symmetric monoidal structure whose tensor product is defined by

$$(L,a)\otimes(L',a'):=(L\otimes L',a+a').$$

The associativity relation is induced by the associativity relation in $\mathcal{P}(X)$. The unit is the pair $(\mathcal{O}_X,0)$ with isomorphisms $(L,a)\otimes(\mathcal{O}_X,0)\simeq(L,a)$ and $(\mathcal{O}_X,0)\otimes(L,a)\simeq(L,a)$ induced by the isomorphisms $L\otimes\mathcal{O}_X\simeq L$ given on sections by $l\otimes a\mapsto al$ (and $a\otimes l\mapsto al$ for the other isomorphism). Finally, the symmetry isomorphism $(L,a)\otimes(L',a')\simeq(L',a')\otimes(L,a)$ is $(-1)^{aa'}s$ (where s is the switch isomorphism defined above). In $\mathcal{G}(X)$, each object is invertible. Indeed, we note that $(L,a)\otimes(L^\vee,-a)=(L\otimes L^\vee,0)\xrightarrow{u}(\mathcal{O}_X,0)$. By symmetry, we see that

$$(L^\vee,-a)\otimes(L,a)\simeq(L,a)\otimes(L^\vee,-a)\overset{u}{\simeq}(\mathcal{O}_X,0)$$

showing that (L,a) has also a right inverse. Note that the above isomorphism is not induced by us, but by $(-1)^a us$. All in all, $\mathcal{G}(X)$ is a Picard category in the sense of [**10**, §4].

The main purpose of $\mathcal{G}(X)$ is to understand the properties of the determinant of a vector bundle over X. Indeed, let $\mathcal{V}(X)$ be the category of vector bundles on X (with only isomorphisms as morphisms). There is a functor

$$D:\mathcal{V}(X)\to\mathcal{G}(X)$$

given on objects by $V\mapsto(\det V,\mathrm{rk}(V))$ and on morphisms by $f\mapsto\det f$. For each exact sequence (V) of the form

$$0\to V_1\xrightarrow{i}V_2\xrightarrow{p}V_3\to 0$$

with $\mathrm{rk}(V_i)=a_i$, we have an isomorphism

$$\varphi_{(V)}:\det V_1\otimes\det V_3\to\det V_2$$

defined as follows. Locally, the sequence splits, i.e. there exists a monomorphism $s:V_3\to V_2$ such that $ps=\mathrm{id}$. We may then define locally $\varphi_{(V)}$ by

$$(v_1\wedge\ldots\wedge v_{a_1})\otimes(w_1\wedge\ldots\wedge w_{a_3})\mapsto v_1\wedge\ldots\wedge v_{a_1}\wedge s(w_1)\wedge\ldots\wedge s(w_{a_3}).$$

It is straightforward to check that this definition doesn't depend on the choice of the section s and thus that one can glue the local definitions to get a global version. The isomorphism $\varphi_{(V)}$ readily gives an isomorphism $D(V_1)\otimes D(V_3)\simeq D(V_2)$ that we still denote by $\varphi_{(V)}$. This isomorphism is functorial in the sense that if we have a commutative diagram

$$\begin{array}{ccccccccc}0&\longrightarrow&V_1&\longrightarrow&V_2&\longrightarrow&V_3&\longrightarrow&0\\ &&\downarrow&&\downarrow&&\downarrow&&\\ 0&\longrightarrow&W_1&\longrightarrow&W_2&\longrightarrow&W_3&\longrightarrow&0\end{array}$$

of exact sequences in which the vertical arrows are isomorphisms, then the diagram

$$\begin{array}{ccc} D(V_1)\otimes D(V_3) & \xrightarrow{\varphi(V)} & D(V_2) \\ \downarrow & & \downarrow \\ D(W_1)\otimes D(W_3) & \xrightarrow[\varphi(W)]{} & D(W_2) \end{array}$$

in which the vertical arrows are the isomorphisms induced by the vertical arrows above, is also commutative. Moreover, we canonically have that $D(0) = (\mathcal{O}_X, 0)$. This datum satisfies the conditions of [**10**, §4.3]. In particular, suppose that we have a sequence of (admissible) monomorphisms of vector bundles

$$0 \subset W \subset V \subset E.$$

Then, the following diagram

$$\begin{array}{ccc} D(W)\otimes D(V/W)\otimes D(E/V) & \longrightarrow & D(V)\otimes D(E/V) \\ \downarrow & & \downarrow \\ D(W)\otimes D(E/W) & \longrightarrow & D(E) \end{array}$$

is commutative. Finally, let us mention that the isomorphism $\varphi_{(V)}$ is compatible with the commutativity of $\mathcal{G}(X)$ in the sense that if we can write $V_2 = V_1 \oplus V_3$ then the two exact sequences

$$(V) \qquad 0 \longrightarrow V_1 \longrightarrow V_2 \longrightarrow V_3 \longrightarrow 0$$

and

$$(V') \qquad 0 \longrightarrow V_3 \longrightarrow V_2 \longrightarrow V_1 \longrightarrow 0$$

give a commutative diagram

$$\begin{array}{ccc} D(V_1)\otimes D(V_3) & \longrightarrow & D(V_3)\otimes D(V_1) \\ & {\scriptstyle \varphi(V)}\searrow \quad \swarrow{\scriptstyle \varphi(V')} & \\ & D(V_2) & \end{array}$$

in which the horizontal arrow is the commutativity isomorphism in $\mathcal{G}(X)$.

As a result of this general formalism, it is in principle possible to understand (sometimes at the cost of cumbersome computations) the isomorphisms of determinant bundles associated to various exact sequences of vector bundles. This will be particularly useful when dealing with twisted Milnor-Witt K-theory. Before spending a section introducing the exact sequences we will use in this work, we conclude this section by mentioning various additional properties of $\mathcal{G}(X)$. First, we supposed at the beginning of this section that X was connected. When dealing with non-connected schemes, it is convenient to replace the integer appearing in the definition of a graded line bundle with a locally constant integer. The theory and constructions are evidently the same in this slightly more general context. Second, suppose that we have two (connected) schemes X and Y together with a morphism of schemes $f : X \to Y$. Given a graded line bundle (L, a) on Y, we may consider the graded line bundle (f^*L, a) on X. This defines a functor $f^* : \mathcal{G}(Y) \to \mathcal{G}(X)$ that we will use frequently in the rest of these notes.

1.4. Useful exact sequences. For any scheme X, we denote by $\Omega_{X/k}$ the sheaf of differentials of X over k. In case X is smooth and connected, then $\Omega_{X/k}$ is a vector bundle of rank $d_X := \dim(X)$. We may then consider the graded line bundle $D(\Omega_{X/k}) = (\omega_{X/k}, d_X)$ in $\mathcal{G}(X)$ (where $\omega_{X/k} = \det \Omega_{X/k}$). In case X is not connected, then the rank of $\Omega_{X/k}$ is a locally constant integer and we may still consider $D(\Omega_{X/k})$ in $\mathcal{G}(X)$ in the slightly more general sense we discussed above. We always assume that the schemes we use are connected and let the reader make the necessary changes in the general situation.

Let now $f : X \to Y$ be a smooth morphism of smooth schemes. Then, we have an exact sequence ([**16**, II.8])

$$f^*\Omega_{Y/k} \to \Omega_{X/k} \to \Omega_{X/Y} \to 0$$

which is also exact on the left as f is smooth. Consequently, we obtain an isomorphism of graded line bundles

$$D(f^*\Omega_{Y/k}) \otimes D(\Omega_{X/Y}) \simeq D(\Omega_{X/k})$$

Using the functor $\mathcal{G}(Y) \to \mathcal{G}(X)$ defined in the previous section, we may write the isomorphism as an isomorphism

$$f^*D(\Omega_{Y/k}) \otimes D(\Omega_{X/Y}) \simeq D(\Omega_{X/k}). \tag{1.2}$$

We will frequently use this isomorphism, or the isomorphisms associated to it after performing elementary operations involving taking inverses and permutations. For instance, we can transform (1.2) into an isomorphism

$$(\omega_{X/Y}, d_X - d_Y) \simeq (\omega_{X/k}, d_X) \otimes (\omega_{Y/k}^\vee, -d_Y)$$

multiplying first on the left by $(\omega_{Y/k}^\vee, -d_Y)$ and then permuting the latter with $(\omega_{X/k}, d_X)$.

Suppose now that we have a Cartesian square of smooth schemes

$$\begin{array}{ccc} X' & \xrightarrow{v} & X \\ {\scriptstyle g}\downarrow & & \downarrow{\scriptstyle f} \\ Y' & \xrightarrow[u]{} & Y \end{array}$$

with u (and thus also v) smooth. We then have a canonical isomorphism ([**18**, Corollary 4.3])

$$g^*D(\Omega_{Y'/Y}) = D(g^*\Omega_{Y'/Y}) \simeq D(\Omega_{X'/X}) \tag{1.3}$$

Finally, let us consider the situation where $i : X \to Y$ is a regular embedding of smooth schemes. Let C_XY be the conormal bundle to X in Y and let N_XY be the normal cone to X in Y. We then have an exact sequence

$$C_XY \to i^*\Omega_{Y/k} \to \Omega_{X/k} \to 0$$

which is in fact exact on the left. Therefore, we obtain a canonical isomorphism

$$D(C_XY) \otimes D(\Omega_{X/k}) \simeq i^*D(\Omega_{Y/k}) \tag{1.4}$$

As above, we may modify this isomorphism into an isomorphism

$$D(\Omega_{X/k}) \simeq (\det N_XY, d_X - d_Y) \otimes i^*D(\Omega_{Y/k})$$

by tensoring on the left by $(\det N_XY, d_X - d_Y)$.

1.5. Twisted Milnor-Witt K-theory. Let F be a field (which is a finitely generated field extension of k) and let $(L,a)\in\mathcal{G}(F)$ be a graded line bundle. We set L^0 for the set of nonzero elements of L and we observe that $F^\times$ acts on L^0 in an obvious way. It follows that the free abelian group $\mathbb{Z}[L^0]$ on the set L^0 is endowed with an action of the group algebra $\mathbb{Z}[F^\times]$. On the other hand, we know that $\mathrm{K}_*^{\mathrm{MW}}(F)$ is a $\mathrm{K}_0^{\mathrm{MW}}(F)$-algebra. Besides, we have a map $F^\times\to\mathrm{K}_0^{\mathrm{MW}}(F)$ given by $u\mapsto\langle u\rangle$. This induces a ring homomorphism $\mathbb{Z}[F^\times]\to\mathrm{K}_0^{\mathrm{MW}}(F)$ and it follows that $\mathrm{K}_*^{\mathrm{MW}}(F)$ is a $\mathbb{Z}[F^\times]$-module. The following definition is due to Morel.

DEFINITION 1.18. Let $(L,a)\in\mathcal{G}(F)$. We define

$$\mathrm{K}_*^{\mathrm{MW}}(F,L,a):=\mathrm{K}_*^{\mathrm{MW}}(F)\otimes_{\mathbb{Z}[F^\times]}\mathbb{Z}[L^0]$$

which we call the *Milnor-Witt K-theory of F twisted by (L,a).*

Even though a doesn't appear in the above definition, it is important to keep track of it. Indeed, we'll sometimes have to compare Milnor-Witt K-theory twisted by $(L,a)\otimes(L,a')$ with Milnor-Witt K-theory twisted by $(L,a')\otimes(L,a)$. We'll then use the commutativity isomorphism in $\mathcal{G}(X)$, which makes use of both a and a'. Another important thing to note is that $\mathrm{K}_*^{\mathrm{MW}}(F,L,a)$ is not a ring anymore, but a priori merely a graded abelian group. In fact, it has the structure of a $\mathrm{K}_*^{\mathrm{MW}}(F)$-module as we will shortly see. Before, let us observe that the action of $F^\times$ on L^0 is simply transitive. It follows that the choice of any non trivial element $l\in L^0$ yields an isomorphism

$$\mathrm{K}_*^{\mathrm{MW}}(F)\simeq\mathrm{K}_*^{\mathrm{MW}}(F,L,a)$$

given by $\alpha\mapsto\alpha\otimes l$. In view of this fact, we will often denote an element of $\mathrm{K}_*^{\mathrm{MW}}(F,L,a)$ by $\alpha\otimes l$ with $\alpha\in\mathrm{K}_*^{\mathrm{MW}}(F)$ and $l\in L^0$. Using this notation, we may write the $\mathrm{K}_*^{\mathrm{MW}}(F)$-module structure of $\mathrm{K}_*^{\mathrm{MW}}(F,L,a)$ as

$$\mathrm{K}_*^{\mathrm{MW}}(F)\times\mathrm{K}_*^{\mathrm{MW}}(F,L,a)\to\mathrm{K}_*^{\mathrm{MW}}(F,L,a)$$

mapping $(\beta,\alpha\otimes l)$ to $(\beta\alpha)\otimes l$. More generally, let (L,a) and (L',a') be two graded line bundles over F. Then, we define a pairing

$$\mathrm{K}_*^{\mathrm{MW}}(F,L,a)\times\mathrm{K}_*^{\mathrm{MW}}(F,L',a')\to\mathrm{K}_*^{\mathrm{MW}}(F,(L,a)\otimes(L',a'))$$

given by $(\alpha\otimes l,\beta\otimes l')\mapsto(\alpha\beta)\otimes(l\otimes l')$. We let the reader check that the diagram

$$\begin{array}{ccc}
\mathrm{K}_n^{\mathrm{MW}}(F,L,a)\otimes\mathrm{K}_m^{\mathrm{MW}}(F,L',a') & \longrightarrow & \mathrm{K}_{m+n}^{\mathrm{MW}}(F,(L,a)\otimes(L',a')) \\
\downarrow & & \downarrow \\
\mathrm{K}_m^{\mathrm{MW}}(F,L',a')\otimes\mathrm{K}_n^{\mathrm{MW}}(F,L,a) & \longrightarrow & \mathrm{K}_{m+n}^{\mathrm{MW}}(F,(L',a')\otimes(L,a)),
\end{array}$$

in which the left vertical map is the switch homomorphism and the right vertical map is the isomorphism induced by the commutativity rule in $\mathcal{G}(F)$, is $\langle(-1)^{aa'}\rangle\epsilon^{mn}$-commutative.

The main idea behind considering twisted Milnor-Witt K-theory is the definition of a residue homomorphism as in Theorem 1.7 which is independent of the choice of the uniformizing parameter. Let then F be a field with a discrete valuation $v:F\to\mathbb{Z}\cup\{-\infty\}$. Let $\mathcal{O}_v$ be the associated valuation ring with maximal ideal $\mathfrak{m}_v$ and residue field $\kappa(v)$, and let (L,a) be a graded line bundle on $\mathcal{O}_v$. We define the *twisted residue map*

$$(1.5)\qquad \partial_v:\mathrm{K}_*^{\mathrm{MW}}(F,L_F,a)\to\mathrm{K}_{*-1}^{\mathrm{MW}}(\kappa(v),((\mathfrak{m}_v/\mathfrak{m}_v^2)^*,-1)\otimes(L_{\kappa(v)},a))$$

by $\partial_v(\alpha\otimes l)=\partial_v^\pi(\alpha)\otimes\overline{\pi}^*\otimes l$. Here, $(\mathfrak{m}_v/\mathfrak{m}_v^2)^*$ is the dual (as $\kappa(v)$-vector spaces) of $\mathfrak{m}_v/\mathfrak{m}_v^2$, L_F and L_{κ_v} are the respective restrictions of L to F and $\kappa(v)$, π is a uniformizing parameter and $\overline{\pi}^*$ is the dual of the class $\overline{\pi}$ of π in $\mathfrak{m}_v/\mathfrak{m}_v^2$.

LEMMA 1.19. *The homomorphism ∂_v is well-defined and doesn't depend on the choice of a uniformizing parameter.*

PROOF. We prove the result for $(L,a)=(\mathcal{O}_v,0)$ for simplicity. The general case is similar. We are then left to show that the association $\alpha\mapsto\partial_v^\pi(\alpha)\otimes\overline{\pi}^*$ is well-defined and independent of π. The first assertion is clear as ∂_v^π is well-defined. Let now π' be another uniformizing parameter. There exists then $u\in\mathcal{O}_v^\times$ such that $u\pi=\pi'$. Arguing as in Remark 1.9, we see that $\partial_v^\pi=\langle u\rangle\partial_v^{\pi'}$. It follows thus from $\overline{\pi}^*=u(\overline{\pi}')^*$ that

$$\partial_v^\pi(\alpha)\otimes\overline{\pi}^*=\langle u\rangle\partial_v^{\pi'}(\alpha)\otimes\overline{\pi}^*=\langle u\rangle\partial_v^{\pi'}(\alpha)\otimes u(\overline{\pi}')^*=\partial_v^{\pi'}(\alpha)\otimes(\overline{\pi}')^*.$$

□

Having this twisted residue homomorphism in the pocket, we now turn to the task to define the transfer morphisms for twisted groups. Let then F/k be a finitely generated field extension. We consider the $F[t]$-module $\Omega_{F[t]/k}$ (which is free of rank equal to $\operatorname{tr.deg}(F/k)+1$ as k is perfect). For any monic irreducible polynomial $p\in F[t]$, we can consider the p-adic valuation and we obtain a twisted residue homomorphism

$$\partial_p:\mathrm{K}_*^{\mathrm{MW}}(F(t),D(\Omega_{F(t)/k}))\to\mathrm{K}_{*-1}^{\mathrm{MW}}(F(p),((\mathfrak{m}_p/\mathfrak{m}_p^2)^*,-1)\otimes_{F[t]}D(\Omega_{F[t]/k})).$$

Now, the canonical isomorphism (1.4) reads in this case as

$$(\mathfrak{m}_p/\mathfrak{m}_p^2,1)\otimes D(\Omega_{F(p)/k})\simeq F(p)\otimes D(\Omega_{F[t]/k})$$

and we get a canonical isomorphism $D(\Omega_{F(p)/k})\simeq((\mathfrak{m}_p/\mathfrak{m}_p^2)^*,-1)\otimes_{F[t]}D(\Omega_{F[t]/k})$. The above residue map can then be written

$$\partial_p:\mathrm{K}_*^{\mathrm{MW}}(F(t),D(\Omega_{F(t)/k}))\to\mathrm{K}_{*-1}^{\mathrm{MW}}(F(p),D(\Omega_{F(p)/k})).$$

It follows that we get a total residue homomorphism

$$d:\mathrm{K}_*^{\mathrm{MW}}(F(t),D(\Omega_{F(t)/k}))\to\bigoplus_p\mathrm{K}_{*-1}^{\mathrm{MW}}(F(p),D(\Omega_{F(p)/k}))$$

where the index set on the right-hand side runs through the set of monic irreducible polynomials in $F[t]$. On the other hand, the sequence $k\subset F\subset F[t]$ gives a version of the canonical isomorphism (1.2) of the form

$$D(\Omega_{F/k})\otimes D(\Omega_{F[t]/F})\simeq D(\Omega_{F[t]/k}).$$

Using the pairing with $\langle 1\rangle\otimes dt\in\mathrm{K}_0^{\mathrm{MW}}(F[t],D(\Omega_{F[t]/F}))$ and the field extension $F\subset F(t)$, we then obtain a homomorphism

$$\mathrm{K}_*^{\mathrm{MW}}(F,D(\Omega_{F/k}))\to\mathrm{K}_*^{\mathrm{MW}}(F(t),D(\Omega_{F(t)/k})).$$

PROPOSITION 1.20. *The sequence*

$$\begin{aligned}0\to\mathrm{K}_*^{\mathrm{MW}}(F,D(\Omega_{F/k}))\to&\mathrm{K}_*^{\mathrm{MW}}(F(t),D(\Omega_{F(t)/k}))\\&\xrightarrow{d}\bigoplus_p\mathrm{K}_{*-1}^{\mathrm{MW}}(F(p),D(\Omega_{F(p)/k}))\to 0\end{aligned}$$

is split exact.

PROOF. Any choice of a generator l of $\omega_{F/k}$ (which is the determinant of $\Omega_{F/k}$ and thus a F-vector space of dimension 1) will give a generator $l \wedge dt$ of $\omega_{F[t]/k}$ (by the isomorphism (1.2)). Using these generators together with the generators $\overline{p}^*$ of $(\mathfrak{m}_p/\mathfrak{m}_p^2)^*$, we obtain an isomorphism of sequences between the sequence of the statement and the sequence of Theorem 1.11. The claim follows then immediately. □

REMARK 1.21. Of course, one may further twist this sequence with a graded line bundle (L, a) over F and obtain the same result.

This exact sequence allows us to define transfers as in Definition 1.13.

DEFINITION 1.22. Let p be a monic irreducible polynomial in $F[t]$, we denote by

$$\mathrm{Tr}_F^{F(p)} : \mathrm{K}_{n-1}^{\mathrm{MW}}(F(p), D(\Omega_{F(p)/k})) \to \mathrm{K}_{n-1}^{\mathrm{MW}}(F, D(\Omega_{F/k}))$$

obtained mimicking the procedure of Definition 1.13. We call *canonical transfer* this homomorphism.

The canonical transfers satisfy the same universal property as the one for geometric transfers stated in Lemma 1.15. Moreover, they coincide with Morel's absolute transfers as defined in [**23**, §5.1] (in any characteristic) as evidenced by the following example which treats the separable case.

EXAMPLE 1.23. Suppose that the extension $F(p)/F$ is separable. Then, an instructing calculation shows that the transfer map $\mathrm{Tr}_F^{F(p)}$ can be described as follows. Let l be a generator of $\omega_{F/k}$. In view of the isomorphism $\Omega_{F(p)/k} \simeq \Omega_{F/k} \otimes F(p)$, we may consider l as a generator of $\omega_{F(p)/k}$ as well. For any $\alpha \in \mathrm{K}_*^{\mathrm{MW}}(F(p))$, we then find the formula

$$\mathrm{Tr}_F^{F(p)}(\alpha \otimes l) = \tau_F^{F(p)}(\langle p' \rangle \alpha) \otimes l.$$

As in the case of geometric transfers, we may have a look at the transfers on Witt groups $\mathrm{W}(F(p), \omega_{F(p)/k}) \to \mathrm{W}(F, \omega_{F/k})$ induced by the canonical transfers. In the separable case, the above example, Remark 1.16 and [**31**, III,§6, Lemme 2] show that this transfer coincides with the one obtained using the trace form. Moreover, the universal property shows that the canonical transfers actually coincide with the ones defined by M. Schmid in his thesis in characteristic different from 2 ([**30**, §2]).

Suppose now that $k \subset F \subset L$ are field extensions such that L/F is finite. We can find a filtration

$$F = F_0 \subset F_1 \subset F_2 \subset \ldots \subset F_n = L$$

such that F_i/F_{i-1} is monogeneous for $i = 1, \ldots, n$. In that case, we define a transfer homomorphism

$$\mathrm{Tr}_F^L : \mathrm{K}_*^{\mathrm{MW}}(L, D(\Omega_{L/k})) \to \mathrm{K}_*^{\mathrm{MW}}(F, D(\Omega_{F/k}))$$

as the composite $\mathrm{Tr}_F^L = \mathrm{Tr}_F^{F_1} \circ \ldots \circ \mathrm{Tr}_{F_{n-2}}^{F_{n-1}} \circ \mathrm{Tr}_{F_{n-1}}^L$. It is a priori far from clear that this transfer homomorphism is well-defined (i.e. that it doesn't depend on the choice of the filtration above). This is proved by F. Morel in [**23**, Theorem 4.27, §5] in full generality. Note however that in case the extension is separable one may use the above discussion (i.e. the fact that canonical transfers are actually induced by the trace form) to deduce that the transfer is well defined. In characteristic different from 2, we may also deduce the result using [**30**, Proposition 2.2.5]. In any case, we can write the following definition.

DEFINITION 1.24. Let $k \subset F \subset L$ be field extensions such that L/F is finite. We call the homomorphism

$$\mathrm{Tr}_F^L : \mathrm{K}_*^{\mathrm{MW}}(L, D(\Omega_{L/k})) \to \mathrm{K}_*^{\mathrm{MW}}(F, D(\Omega_{F/k}))$$

the *canonical transfer*.

As observed above, it's also possible to further twist Tr_F^L by a graded line bundle (L, a) in $\mathcal{G}(F)$. We leave the details to the reader. As a consequence of [**23**, Theorem 4.27, §5], we also note that the canonical transfers are functorial in the field extensions of k.

2. The Rost-Schmid complex and Chow-Witt groups

In the first lecture, we introduced canonical transfers for finite field extensions $F \subset L$ (both field extensions of k). These transfers, together with the twisted residue homomorphism (1.5) are the two main ingredients in the construction of the Rost-Schmid complex(es) that we now introduce (always following Morel).

2.1. The complexes. Let X be a (finite type, separated) scheme over k and let (L, a) be a graded line bundle over X. For any $j \in \mathbb{Z}$, the Rost-Schmid complex $\tilde{C}(X, j, L, a)$ in weight j is the complex (in homological dimension) whose term in degree i is

$$\tilde{C}_i(X, j, L, a) := \bigoplus_{x \in X_{(i)}} \mathrm{K}_{j+i}^{\mathrm{MW}}(k(x), D(\Omega_{k(x)/k}) \otimes (L, a)).$$

Let $x \in X_{(i)}$ and $y \in X_{(i-1)}$. The differential

$$d_y^x : \mathrm{K}_{j+i}^{\mathrm{MW}}(k(x), D(\Omega_{k(x)/k}) \otimes (L, a)) \to \mathrm{K}_{j+i-1}^{\mathrm{MW}}(k(y), D(\Omega_{k(y)/k}) \otimes (L, a))$$

is defined as follows. If $y \notin \overline{\{x\}}$, then we set $d_y^x = 0$. In the other case, let Z be the normalization of $\overline{\{x\}}$. We have a finite morphism $f : Z \to X$. Removing if necessary points of codimension ≥ 2 in Z and their images in X, we may assume that we have a finite morphism $Z \to X$ and that Z is smooth. In particular, we can consider its module of differentials $\Omega_{Z/k}$ (which is locally free of rank i). For any $z \in Z_{(i-1)}$ we then have a residue homomorphism (1.5)

$$d_z^x : \mathrm{K}_{j+i}^{\mathrm{MW}}(k(x), D(\Omega_{k(x)/k}) \otimes (L, a)) \\ \to \mathrm{K}_{j+i-1}^{\mathrm{MW}}(k(z), ((\mathfrak{m}_z/\mathfrak{m}_z^2)^*, -1) \otimes D(\Omega_{Z/k}) \otimes (L, a))$$

Using the canonical isomorphism (1.4)

$$((\mathfrak{m}_z/\mathfrak{m}_z^2)^*, -1) \otimes D(\Omega_{Z/k}) \simeq D(\Omega_{k(z)/k})$$

we obtain a homomorphism

$$d_z^x : \mathrm{K}_{j+i}^{\mathrm{MW}}(k(x), D(\Omega_{k(x)/k}) \otimes (L, a)) \to \mathrm{K}_{j+i-1}^{\mathrm{MW}}(k(z), D(\Omega_{k(z)/k}) \otimes (L, a))$$

If $z \mapsto y$ under the map $f : Z \to X$, the field extension $k(y) \subset k(z)$ is finite and there is a canonical transfer homomorphism

$$\mathrm{K}_{j+i-1}^{\mathrm{MW}}(k(z), D(\Omega_{k(z)/k}) \otimes (L, a)) \to \mathrm{K}_{j+i-1}^{\mathrm{MW}}(k(y), D(\Omega_{k(y)/k}) \otimes (L, a)).$$

Summing the composite maps over all $z \in Z$ mapping to y, we finally obtain a homomorphism

$$d_y^x : \mathrm{K}_{j+i}^{\mathrm{MW}}(k(x), D(\Omega_{k(x)/k}) \otimes (L, a)) \to \mathrm{K}_{j+i-1}^{\mathrm{MW}}(k(y), D(\Omega_{k(y)/k}) \otimes (L, a))$$

as required. For a given x and a given $\alpha \in \mathrm{K}^{\mathrm{MW}}_{j+i}(k(x), D(\Omega_{k(x)/k}) \otimes (L,a))$, there are only finitely many $y \in X_{(i-1)}$ such that $d^x_y(\alpha) \neq 0$ and we then obtain a well defined homomorphism

$$d_i : \tilde{C}_i(X,j,L,a) \to \tilde{C}_{i-1}(X,j,L,a).$$

In short, we have a graded abelian group $\tilde{C}(X,j,L,a)$ together with a homomorphism d of degree -1, i.e. a pair $(\tilde{C}(X,j,L,a),d)$. It is not clear at all that $d_{i-1}d_i = 0$. We will shortly state the theorem summarizing the situation, but before we focus on the Rost-Schmid complex in the special situation where X is smooth (and connected). In this case, we can consider the graded line bundle $D(\Omega_{X/k})$ on X and the pair $(\tilde{C}(X,j,D(\Omega_{X/k})^{-1}),d)$ where $D(\Omega_{X/k})^{-1}$ is the inverse of $D(\Omega_{X/k})$ in $\mathcal{G}(X)$ (more explicitly, $D(\Omega_{X/k})^{-1} = (\omega^{\vee}_{X/k}, -\dim(X))$). Using the canonical isomorphism (1.4), we see that the graded components are of the form

$$\tilde{C}_i(X,j,D(\Omega_{X/k})^{-1}) = \bigoplus_{x\in X_{(i)}} \mathrm{K}^{\mathrm{MW}}_{j+i}(k(x), D(\mathfrak{m}_x/\mathfrak{m}_x^2)^{-1}).$$

Setting $\dim(X) := d_X$, we may rewrite the above term as

$$\tilde{C}_i(X,j,D(\Omega_{X/k})^{-1}) = \bigoplus_{x\in X^{(d_X-i)}} \mathrm{K}^{\mathrm{MW}}_{j+i}(k(x), D(\mathfrak{m}_x/\mathfrak{m}_x^2)^{-1}).$$

Now, we can consider the graded abelian group $C(X,j)$ whose component of degree i is of the form

$$C(X,j)^i := \bigoplus_{x\in X^{(i)}} \mathrm{K}^{\mathrm{MW}}_{j-i}(k(x), D(\mathfrak{m}_x/\mathfrak{m}_x^2)^{-1})$$

and we have $\tilde{C}_i(X,j,D(\Omega_{X/k})^{-1}) = C(X,j+d_X)^{d_X-i}$. It follows that we obtain a homomorphism $d : C(X,j) \to C(X,j)$ of degree $+1$.

REMARK 2.1. Of course, we may also consider the complex $C(X,j,L,a)$ for any graded line bundle (L,a) over X.

THEOREM 2.2. *Let X be a finite type k-scheme, $j \in \mathbb{Z}$ be an integer and $(L,a) \in \mathcal{G}(X)$ be a graded line bundle over X. Then, $(\tilde{C}(X,j,L,a),d)$ is a (chain) complex provided k is of characteristic different from 2. If X is smooth, $(C(X,j,L,a),d)$ is a (cochain) complex for k of any characteristic.*

PROOF. If X is smooth, this is [**23**, Theorem 5.31]. In the general case, the proof can be found in [**9**, §4.2.8]. □

REMARK 2.3. The assumption that k is of characteristic different from 2 in the proof that $(\tilde{C}(X,j,L,a),d)$ is a complex is artificial, and will be removed in [**8**]. Further, one can relax the assumption on X by only supposing that X is essentially of finite type.

DEFINITION 2.4. Let X be a finite type scheme over k. We call $\tilde{C}(X,j,L,a)$ the *homological Rost-Schmid complex twisted by* (L,a). We write $\mathrm{H}_i(X,j,L,a)$ for the homology groups of $\tilde{C}(X,j,L,a)$. In case X is smooth, we call $C(X,j,L,a)$ the *cohomological Rost-Schmid complex twisted by* (L,a) and write $\mathrm{H}^i(X,j,L,a)$ for its cohomology groups. If $Y \subset X$ is a closed subset, we can also consider the subcomplexes of both complexes formed by points on Y. We respectively denote by $\tilde{C}_Y(X,j,L,a)$ and $C_Y(X,j,L,a)$ these complexes.

If X_{red} denotes the reduction of X, then we can observe that $\tilde{C}(X,j,L,a) = \tilde{C}(X_{\mathrm{red}},j,L,a)$ by very definition. Further, we can also observe that if $i : Y \subset X$ is a closed subscheme, then we also have an identification

$$\tilde{C}(Y,j,i^*(L,a)) = \tilde{C}_Y(X,j,L,a).$$

We finally come to the definition of Chow-Witt groups.

DEFINITION 2.5. Let X be a scheme (essentially of finite type over k, separated). For any $i \in \mathbb{N}$ and any graded line bundle (L,a) over X, we set

$$\widetilde{\mathrm{CH}}_i(X,L,a) = \mathrm{H}_i(X,-i,L,a).$$

We call this homology group the *homological Chow-Witt group of i-dimensional cycles twisted by* (L,a). For simplicity, we also sometimes simply say *homological i-th Chow-Witt group* or *Chow-Witt group.* If (L,a) is trivial, we omit it from the notation.

DEFINITION 2.6. Let X be an essentially smooth k-scheme. For any $i \in \mathbb{N}$ and any graded line bundle (L,a) over X, we set

$$\widetilde{\mathrm{CH}}^i(X,L,a) = \mathrm{H}^i(X,i,L,a).$$

We call this cohomology group the *cohomological Chow-Witt group of i-codimension cycles twisted by* (L,a). As above, we also sometimes simply say *cohomological i-th Chow-Witt group* or *Chow-Witt group* and we omit (L,a) from the notation in case it is trivial. If $Y \subset X$ is a closed subset, we may also consider the group

$$\widetilde{\mathrm{CH}}^i_Y(X,L,a) = \mathrm{H}^i(C_Y(X,i,L,a)).$$

and call it the *cohomological Chow-Witt group of i-codimension cycles supported on Y and twisted by* (L,a), or simply *Chow-Witt group supported on Y.*

2.2. Basic properties. We now state some basic properties of the Chow-Witt groups. First note that $\widetilde{\mathrm{CH}}_i(X,L,a) = 0$ and $\widetilde{\mathrm{CH}}^i(X,L,a) = 0$ for $i > \dim(X)$ or $i < 0$. If X is smooth of (constant) dimension d_X, the above identification of complexes yields an isomorphism

$$\widetilde{\mathrm{CH}}_i(X, D(\Omega_{X/k})^{-1} \otimes (L,a)) \simeq \widetilde{\mathrm{CH}}^{d_X - i}(X,L,a).$$

Next, observe that we may replace Milnor-Witt K-theory by Milnor K-theory in the definition of the Rost-Schmid complex. In a precise manner, this amounts to consider the quotient of each term by η. Since the differential maps commute with η, it follows that we get a differential at the level of the quotient complex. By definition, it coincides with the differential defined in [**27**, §2, (2.1.0)] and consequently we obtain homomorphisms

$$\widetilde{\mathrm{CH}}_i(X,L,a) \to \mathrm{CH}_i(X)$$

for any $i \in \mathbb{N}$ (the same applies for the cohomological versions). In general, this map is neither injective, nor surjective as shown by [**13**, Corollary 11.8]. Next, suppose that $\iota : Y \subset X$ is a closed subset. We may see Y as a scheme by endowing it with its reduced structure and we know that $\tilde{C}(Y,j,\iota^*(L,a)) = \tilde{C}_Y(X,j,L,a)$. Let $U \subset X$ be the complement of Y in X with embedding $u : U \to X$. It is an easy exercise to show that we have an exact sequence of complexes

$$0 \to \tilde{C}(Y,j,\iota^*(L,a)) \to \tilde{C}(X,j,L,a) \to \tilde{C}(U,j,u^*(L,a)) \to 0$$

and consequently a long exact sequence

$$\ldots \to \widetilde{\mathrm{CH}}_i(Y,\iota^*(L,a)) \to \widetilde{\mathrm{CH}}_i(X,L,a) \to \widetilde{\mathrm{CH}}_i(U,u^*(L,a)) \to \ldots$$

in homology. In contrast with the situation in the case of Chow groups, the right-hand arrow $\widetilde{\mathrm{CH}}_i(X,L,a) \to \widetilde{\mathrm{CH}}_i(U,u^*(L,a))$ is in general not surjective ([**13**]). In case the scheme X is smooth, and $\iota : Y \subset X$ is closed, we obtain the same sequence, with the distinction that $\widetilde{\mathrm{CH}}_i(Y,\iota^*(L,a))$ can be compared to the relevant cohomological group only if Y is also smooth. Letting $r = d_X - d_Y$ be the (constant) codimension of Y in X and considering the subcomplex $C_Y(X,j,L,a) \subset C(X,j,L,a)$ of points supported on Y, we have identifications

$$\begin{aligned} C_Y(X,j,L,a) &= \tilde{C}_Y(X,j-d_X,D(\Omega_{X/k})^{-1}\otimes(L,a)) \\ &= \tilde{C}(Y,j-d_X,\iota^*D(\Omega_{X/k})^{-1}\otimes\iota^*(L,a)) \\ &= C(Y,j+d_Y-d_X,D(\Omega_{Y/k})\otimes\iota^*D(\Omega_{X/k})^{-1}\otimes\iota^*(L,a)) \end{aligned}$$

Thus, the above exact sequence reads

$$\ldots \to \widetilde{\mathrm{CH}}^{i-r}(Y,D(\Omega_{Y/k})\otimes\iota^*D(\Omega_{X/k})^{-1}\otimes\iota^*(L,a)) \to \widetilde{\mathrm{CH}}^{i}(X,L,a) \\ \to \widetilde{\mathrm{CH}}^{i}(U,u^*(L,a)) \to \ldots$$

in this case. One may moreover use the canonical isomorphism (1.2) to identify $D(\Omega_{Y/k})\otimes\iota^*D(\Omega_{X/k})^{-1}$ with $D(\Omega_{X/Y})$.

2.3. Push-forwards. The purpose of this section is to show that given a proper morphism $f : X \to Y$ of finite type schemes over k there exists a morphism of complexes

$$f_* : \tilde{C}(X,j,f^*(L,a)) \to \tilde{C}(Y,j,L,a)$$

for any $j \in \mathbb{Z}$ and any graded line bundle (L,a) on Y. The association $f \mapsto f_*$ is functorial (i.e. respects compositions and identities) and therefore homological Chow-Witt groups define a covariant functor from the category of finite type k-schemes (with projective morphisms) to the category of abelian groups. The definition of f_* is quite straightforward, given the existence of transfer maps. Indeed, let $x \in X$ and let $y = f(x) \in Y$ be such that $x \in X_{(i)}$ and $y \in Y_{(i')}$. If the field extension $k(y) \subset k(x)$ is infinite, then we define

$$(f_*)^x_y : \mathrm{K}^{\mathrm{MW}}_{j+i}(k(x),D(\Omega_{k(x)/k})\otimes(L,a)) \to \mathrm{K}^{\mathrm{MW}}_{j+i'}(k(y),D(\Omega_{k(y)/k})\otimes(L,a))$$

to be trivial. If the extension $k(y) \subset k(x)$ is finite, then $i = i'$ and we define

$$(f_*)^x_y : \mathrm{K}^{\mathrm{MW}}_{j+i}(k(x),D(\Omega_{k(x)/k})\otimes(L,a)) \to \mathrm{K}^{\mathrm{MW}}_{j+i}(k(y),D(\Omega_{k(y)/k})\otimes(L,a))$$

to be the canonical transfer $\mathrm{Tr}^{k(x)}_{k(y)}$ of Definition 1.24. Summing up these maps, we obtain a morphism of graded abelian groups

$$f_* : \tilde{C}(X,j,f^*(L,a)) \to \tilde{C}(Y,j,L,a)$$

this is evidently functorial since transfer maps are functorial.

THEOREM 2.7. *Let $f : X \to Y$ be a morphism of k-schemes essentially of finite type. Then,*

$$f_* : \tilde{C}(X,j,f^*(L,a)) \to \tilde{C}(Y,j,L,a)$$

is a morphism of complexes provided one of the following conditions hold:

(1) *f is finite and both X and Y are essentially smooth.*
(2) *f is proper and k is of characteristic different from 2.*

PROOF. The result under the first assumption is proved in [**23**, Corollary 5.30]. For the proof of the statement under the second assumption, we refer to [**9**, §4.2.5, Theorem 4.2.7]. □

REMARK 2.8. As in Remark 2.3, the assumption that k is of characteristic different from 2 should be superfluous. We hope to lift it in [**8**].

As an obvious consequence, we see that for any $i \in \mathbb{N}$, the homological Chow group $\widetilde{\mathrm{CH}}_i$ is a functor from the category of finite-type schemes (with only projective morphisms as morphisms) to the category of abelian groups. Moreover, this allows to define a refinement of the usual degree map on Chow groups. Suppose that X is a projective k-scheme with structural morphism $p : X \to \operatorname{Spec} k$. We then have a push-forward map

$$p_* : \widetilde{\mathrm{CH}}_0(X) \to \widetilde{\mathrm{CH}}_0(k) = \mathrm{K}_0^{\mathrm{MW}}(k).$$

and a commutative diagram

$$\begin{array}{ccc} \widetilde{\mathrm{CH}}_0(X) & \xrightarrow{p_*} & \mathrm{K}_0^{\mathrm{MW}}(k) \\ \downarrow & & \downarrow \\ \mathrm{CH}_0(X) & \xrightarrow[\deg]{} & \mathbb{Z} \end{array}$$

in which the vertical homomorphisms are the ones from Chow-Witt groups to Chow groups and the bottom horizontal one is the degree map. Both vertical homomorphisms are surjective and we see that p_* is a refinement of the degree map. We denote it $\widetilde{\deg}$ below and call it the *Milnor-Witt degree* of X. Virtually, all the questions that are interesting for deg are also interesting for $\widetilde{\deg}$. For instance, it is clear (by functoriality) that if X has a rational point, then $\widetilde{\deg}$ is surjective. The converse statement is also interesting: Suppose that $\widetilde{\deg}$ is surjective, then does X have a rational point? Funnily, the answer is that $\widetilde{\deg}$ is no better than deg in this respect, i.e. it doesn't detect rational points. For more information on this story, we refer to [**2**]. This question set apart, we may observe that the image of $\widetilde{\deg}$ is harder to compute than the image of the degree. Indeed, $\mathrm{K}_0^{\mathrm{MW}}(k)$ is not a PID and the only a priori structure of the image is that it is a $\mathrm{K}_0^{\mathrm{MW}}(k)$-submodule of $\mathrm{K}_0^{\mathrm{MW}}(k)$. We refer to [**5**] for some computations in this direction, and we note that it would be interesting to compute some Milnor-Witt degrees of hypersurfaces in $\mathbb{P}^n$.

Finally, let's state the cohomological version of Theorem 2.7 which is obtained via identification of the cohomological and homological complexes.

THEOREM 2.9. *Let $f : X \to Y$ be a morphism of smooth k-schemes, and let $d = \dim(Y) - \dim(X)$. Then the morphism of degree d*

$$f_* : C(X, j, D(\omega_{X/k}) \otimes f^*(L, a)) \to C(Y, j + d, D(\Omega_{Y/k}) \otimes (L, a))$$

is a morphism of complexes provided one of the following conditions hold:

(1) *f is finite.*
(2) *f is proper and k is of characteristic different from 2.*

2.4. Flat pull-backs. We now pass to the notion of pull-back for Chow-Witt groups. As in the case of Chow groups, the construction of the pull-backs is more delicate than the one for push-forwards. The subtle point here is that we need a notion of a "quadratic" (or more precisely "symmetric bilinear") length in order to define explicitly the pull-back for flat morphisms. Our aim is then to construct a morphism of complexes

$$f^* : C(X, j, L, a) \to C(Y, j, f^*(L, a))$$

for any flat morphism $f : Y \to X$ between (essentially) smooth schemes and any graded line bundle (L, a) over X. We start by observing that the basic bricks of the complex $C(X, j, L, a)$ are of the form $\mathrm{K}_j^{\mathrm{MW}}(k(x), D(\mathfrak{m}_x/\mathfrak{m}_x^2)^{-1} \otimes (L, a))$. We omit (L, a) in the sequel, letting the reader making the necessary changes to keep track of this graded line bundle. The idea behind what follows is that the Milnor-Witt K-theory of $k(x)$ is deeply linked with Hermitian K-theory (aka higher Grothendieck-Witt groups) of the regular local ring $\mathcal{O}_{X,x}$.

Let then $x \in X^{(i)}$ and consider the category $\mathcal{O}_{X,x}^{\mathrm{fl}}$ of finite length $\mathcal{O}_{X,x}$-modules, which is an abelian category. To any finite length module M, we can associate a finite length module $M^\sharp := \mathrm{Ext}^i_{\mathcal{O}_{X,x}}(M, \mathcal{O}_{X,x})$, obtaining an exact functor

$$^\sharp : (\mathcal{O}_{X,x}^{\mathrm{fl}})^{\mathrm{op}} \to \mathcal{O}_{X,x}^{\mathrm{fl}}$$

Moreover, there is a canonical isomorphism $\varpi : \mathrm{id} \to ^{\sharp\sharp}$ ([**3**, §6]) turning $\mathcal{O}_{X,x}^{\mathrm{fl}}$ into an exact category with duality in the sense of [**29**, Definition 2.1]. To this category, we may then associate its Grothendieck-Witt group $\mathrm{GW}^{\mathrm{fl}}(\mathcal{O}_{X,x})$ in the sense of [**29**, §2.2]. Roughly speaking, this group is the Grothendieck group of the set of isometry classes of pairs (M, φ) (where M is a finite length module and $\varphi : M \to M^\sharp$ is a symmetric isomorphism) endowed with the orthogonal sum as operation, modulo an extra relation identifying a pair (M, φ) with a totally isotropic submodule with an hyperbolic module. We now explain how to compute $\mathrm{GW}^{\mathrm{fl}}(\mathcal{O}_{X,x})$. Let $\mathcal{O}_{X,x}^0 \subset \mathcal{O}_{X,x}^{\mathrm{fl}}$ be the subcategory of semi-simple objects. It inherits the duality $^\sharp$ from the latter and is in turn an exact category with duality. The induced homomorphism between Grothendieck-Witt groups

$$\mathrm{GW}(\mathcal{O}_{X,x}^0) \to \mathrm{GW}^{\mathrm{fl}}(\mathcal{O}_{X,x})$$

is actually an isomorphism by [**26**, Theorem 6.10]. We now identify the left-hand side with a more familiar object. Let V be a (finite dimensional) $k(x)$-vector space. We claim that there is an isomorphism of $k(x)$-vector spaces

$$\hom_{k(x)}(V, \mathrm{Ext}^i_{\mathcal{O}_{X,x}}(k(x), \mathcal{O}_{X,x})) \to \mathrm{Ext}^i_{\mathcal{O}_{X,x}}(V, \mathcal{O}_{X,x}).$$

Indeed, one can choose a projective $\mathcal{O}_{X,x}$-resolution of $k(x)$

$$0 \to P_i \to \ldots \to P_1 \to P_0 \to k(x) \to 0$$

Dualizing (and setting $P_j^\vee = \hom_{\mathcal{O}_{X,x}}(P, \mathcal{O}_{X,x})$), we obtain a projective resolution

$$0 \to P_0^\vee \to P_1^\vee \to \ldots \to P_i^\vee \to \mathrm{Ext}^i_{\mathcal{O}_{X,x}}(k(x), \mathcal{O}_{X,x}) \to 0$$

We can pull-back this exact sequence along any homomorphism of $\mathcal{O}_{X,x}$-modules $V \to \mathrm{Ext}^i_{\mathcal{O}_{X,x}}(k(x), \mathcal{O}_{X,x})$ and get a projective resolution of V as an $\mathcal{O}_{X,x}$-module, which is nothing else than an extension of V by $P_0^\vee$. Now, it is clear that we can choose $P_0 = \mathcal{O}_{X,x}$ and we obtain a map

$$\hom_{k(x)}(V, \mathrm{Ext}^i_{\mathcal{O}_{X,x}}(k(x), \mathcal{O}_{X,x})) \to \mathrm{Ext}^i_{\mathcal{O}_{X,x}}(V, \mathcal{O}_{X,x}).$$

as required. We let the reader check that it is $k(x)$-linear, functorial in V and that it respects direct sums of $k(x)$-vector spaces. Since it is obviously an isomorphism for $V = k(x)$, it is an isomorphism. In more sophisticated terms, the above homomorphism defines a duality preserving functor between the category of $k(x)$-vector spaces, endowed with the duality $\hom_{k(x)}(\_, \mathrm{Ext}^i_{\mathcal{O}_{X,x}}(k(x), \mathcal{O}_{X,x}))$ and the category $\mathcal{O}^0_{X,x}$. This functor is an equivalence, and therefore we obtain an isomorphism of Grothendieck-Witt groups

$$\mathrm{GW}(k(x), \mathrm{Ext}^i_{\mathcal{O}_{X,x}}(k(x), \mathcal{O}_{X,x})) \simeq \mathrm{GW}(\mathcal{O}^0_{X,x}).$$

Finally, we can state the following proposition, which is the basis of the definition of flat pull-backs.

PROPOSITION 2.10. *Let X be an essentially smooth scheme and let $x \in X^{(i)}$. Then, we have a canonical isomorphism*

$$\mathrm{K}_0^{\mathrm{MW}}(k(x), D(\mathfrak{m}_x/\mathfrak{m}_x^2)^{-1}) \to \mathrm{GW}^{\mathrm{fl}}(\mathcal{O}_{X,x}).$$

PROOF. First, we know from Lemma 1.1 that we have an isomorphism

$$\mathrm{K}_0^{\mathrm{MW}}(k(x)) \to \mathrm{GW}(k(x))$$

such that $\langle u \rangle \mapsto \langle u \rangle$ for any $u \in k(x)^\times$. For any $k(x)$-vector space L of dimension 1, we define a map

$$\mathrm{K}_0^{\mathrm{MW}}(k(x)) \otimes_{\mathbb{Z}[k(x)^\times]} \mathbb{Z}[L^0] \to \mathrm{GW}(k(x), L)$$

by $\langle u \rangle \otimes l \mapsto \varphi_{ul}$, where $\varphi_l : k(x) \times k(x) \to L$ is defined by $\varphi_l(x, y) = xyl$. We let the reader check (for instance, using the isomorphisms between untwisted and twisted groups induced by the choice of l) that this map is in fact an isomorphism of abelian groups. Consequently, we have a sequence of isomorphisms

$$\begin{aligned}&\mathrm{K}_0^{\mathrm{MW}}(k(x), a, \mathrm{Ext}^i_{\mathcal{O}_{X,x}}(k(x), \mathcal{O}_{X,x}))\\ &\qquad \simeq \mathrm{GW}(k(x), \mathrm{Ext}^i_{\mathcal{O}_{X,x}}(k(x), \mathcal{O}_{X,x})) \simeq \mathrm{GW}^{\mathrm{fl}}(\mathcal{O}_{X,x})\end{aligned}$$

for any integer $a \in \mathbb{Z}$. The claim then follows using the canonical isomorphism $(\mathfrak{m}_x/\mathfrak{m}_x^2)^* \simeq \mathrm{Ext}^i_{\mathcal{O}_{X,x}}(k(x), \mathcal{O}_{X,x})$ deduced from [**6**, A X.165, Proposition 9] and setting $a = -1$. □

One of the interesting features about Grothendieck-Witt groups of finite length modules is that they are functorial for flat morphisms between rings of the same Krull dimension. Indeed, let A, B be regular local rings of the same Krull dimension d, and let $f : A \to B$ be a flat homomorphism of local rings. We have a functor $f^* : A^{\mathrm{fl}} \to B^{\mathrm{fl}}$ between the respective categories of finite length modules given by $M \mapsto B \otimes_A M$. For any projective A-module P, we have isomorphisms

$$B \otimes \mathrm{Hom}_A(P, A) \simeq \mathrm{Hom}_B(B \otimes P, B)$$

and it follows that we have an isomorphism $B\otimes_A \mathrm{Ext}^d_A(M,A)\simeq \mathrm{Ext}^d_B(B\otimes_A M,B)$ which is functorial in M. Consequently, we obtain a homomorphism

$$f^*:\mathrm{GW}^{\mathrm{fl}}(A)\to \mathrm{GW}^{\mathrm{fl}}(B)$$

which is actually functorial.

Let now X and Y be essentially smooth schemes and let $f:Y\to X$ be a flat morphism. Let $x\in X^{(i)}$ and let $y\in Y^{(i)}$ be such that $f(y)=x$ (in fact, y is a minimal point of the fiber of x). The induced homomorphism

$$f:\mathcal{O}_{X,x}\to \mathcal{O}_{Y,y}$$

is then a flat homomorphism of regular local rings and we get an induced homomorphism f^* on the respective Grothendieck-Witt groups of finite length modules. In view of Proposition 2.10, we obtain a homomorphism $m(x,y,f)$ of the form

$$\begin{aligned}\mathrm{K}_0^{\mathrm{MW}}(k(x),D(\mathfrak{m}_x/\mathfrak{m}_x^2)^{-1})&\simeq \mathrm{GW}^{\mathrm{fl}}(\mathcal{O}_{X,x})\stackrel{f^*}{\to}\mathrm{GW}^{\mathrm{fl}}(\mathcal{O}_{Y,y})\\ &\simeq \mathrm{K}_0^{\mathrm{MW}}(k(y),D(\mathfrak{m}_y/\mathfrak{m}_y^2)^{-1})\end{aligned}$$

which we call *local multiplicity along* f. This terminology can be justified as follows. If R is a (noetherian) regular local ring with maximal ideal $\mathfrak{m}$, then one can consider its completion $\hat{R}$ with respect to the $\mathfrak{m}$-adic valuation. This is still a regular local ring with the same residue field $\kappa(m)$ as R and the extension $R\to\hat{R}$ is flat. As a consequence of Proposition 2.10, we see that the induced map

$$\mathrm{GW}^{\mathrm{fl}}(R)\to \mathrm{GW}^{\mathrm{fl}}(\hat{R})$$

is an isomorphism. We may use the Cohen structure theorem to find a presentation $\hat{R}=\kappa(x)[[x_1,\ldots,x_d]]$ and finally we see that what we are looking at is in fact the group $\mathrm{GW}^{\mathrm{fl}}(\kappa(x)[[t_1,\ldots,t_d]])$. A flat homomorphism as above then induces a homomorphism of the form

$$\mathrm{GW}^{\mathrm{fl}}(k(x)[[t_1,\ldots,t_i]])\to \mathrm{GW}^{\mathrm{fl}}(k(y)[[t_1,\ldots,t_i]])$$

which is in some sense the multiplication by the (motivic) Brouwer degree of the homomorphism $k(x)[[t_1,\ldots,t_i]]\to k(y)[[t_1,\ldots,t_i]]$.

The local multiplicity along f allows us to define homomorphisms

$$(f^*)^y_x:\mathrm{K}_{j-i}^{\mathrm{MW}}(k(x),D(\mathfrak{m}_x/\mathfrak{m}_x^2)^{-1})\to \mathrm{K}_{j-i}^{\mathrm{MW}}(k(y),D(\mathfrak{m}_y/\mathfrak{m}_y^2)^{-1})$$

for any $j\in\mathbb{Z}$ as follows. Let $\iota:\mathrm{K}_{j-i}^{\mathrm{MW}}(k(x))\to \mathrm{K}_{j-i}^{\mathrm{MW}}(k(y))$ be the homomorphism induced by the field extension. We set

$$(f^*)^y_x(\alpha\otimes\overline{x}_1^*\wedge\ldots\wedge\overline{x}_i^*)=\iota(\alpha)\cdot m(x,y,f)(\langle 1\rangle\otimes x_1^*\wedge\ldots\wedge\overline{x}_i^*)$$

for any generators $\overline{x}_1,\ldots,\overline{x}_i$ of $\mathfrak{m}_x/\mathfrak{m}_x^2$. Of course, we may further twist by any graded line bundle (L,a) and we leave as usual the details to the reader.

EXAMPLE 2.11. Suppose that $f:Y\to X$ is smooth and that $x\in X^{(i)}$ and $y\in Y^{(i)}$ are such that $f(y)=x$. In that case, f induces an isomorphism $\overline{f}:(\mathfrak{m}_x/\mathfrak{m}_x^2)\otimes k(y)\to \mathfrak{m}_y/\mathfrak{m}_y^2$ and the local multiplicity $m(x,y,f)$ is easily seen to satisfy

$$m(x,y,f)(\langle 1\rangle\otimes\overline{x}_1^*\wedge\ldots\wedge\overline{x}_i^*)=\langle 1\rangle\otimes\overline{f}(\overline{x}_1)^*\wedge\ldots\wedge\overline{f}(\overline{x}_i)^*$$

showing that in that case $(f^*)^y_x$ is essentially determined by $\iota:\mathrm{K}_{j-i}^{\mathrm{MW}}(k(x))\to \mathrm{K}_{j-i}^{\mathrm{MW}}(k(y))$.

All in all, we see that if $f : Y \to X$ is a flat morphism, we can define a pull-back homomorphism

$$f^* : C(X, j, L, a) \to C(Y, j, f^*(L, a))$$

by summing up the homomorphisms $(f^*)_x^y$. The association $f \mapsto f^*$ is functorial.

THEOREM 2.12. *Let $f : Y \to X$ be a flat morphism of essentially smooth k-schemes and let (L, a) be a graded line bundle over X. Then,*

$$f^* : C(X, j, L, a) \to C(Y, j, f^*(L, a))$$

is a morphism of complexes provided one of the following conditions hold:

(1) *f is smooth.*
(2) *k is of characteristic different from 2.*

PROOF. If f is smooth, then this is [**23**, Lemma 5.16]. If f is flat, then the proof goes back to [**12**, Corollaire 10.4.2]. □

REMARK 2.13. As for push-forwards, the assumption on the base field is irrelevant and should be dropped in the near future.

As a consequence, we see that the association $X \mapsto \widetilde{\mathrm{CH}}^i(X)$ induces a functor from the category of (essentially) smooth k-schemes with flat morphisms to the category of abelian groups. Interestingly, we can also define pull-backs for homological Chow-Witt groups, provided the morphisms are smooth. Let then $f : Y \to X$ be a smooth morphism between schemes which are essentially of finite type. For simplicity, we suppose that the relative dimension $d := d_Y - d_X$ is constant (for instance X, Y are connected). Let $x \in X_{(i)}$ and let $y \in Y_{(d+i)}$ be such that $f(x) = y$. In that case, we have a commutative diagram of exact sequences

$$\begin{array}{ccccccccc} 0 & \longrightarrow & (\mathfrak{m}_x/\mathfrak{m}_x^2) \otimes k(y) & \longrightarrow & \Omega_{X/k} \otimes k(y) & \longrightarrow & \Omega_{k(x)/k} \otimes k(y) & \longrightarrow & 0 \\ & & \downarrow & & \downarrow & & \downarrow & & \\ 0 & \longrightarrow & \mathfrak{m}_y/\mathfrak{m}_y^2 & \longrightarrow & \Omega_{Y/k} \otimes k(y) & \longrightarrow & \Omega_{k(y)/k} & \longrightarrow & 0 \end{array}$$

in which the vertical arrows are induced by f. Since f is smooth, the left-hand vertical arrow is an isomorphism and the other two arrows are injective. The cokernel of the middle arrow is $\Omega_{Y/X} \otimes k(y)$ and we obtain an exact sequence

$$0 \to \Omega_{k(x)/k} \otimes k(y) \to \Omega_{k(y)/k} \to \Omega_{Y/X} \otimes k(y) \to 0$$

and a canonical isomorphism $D(\Omega_{k(y)/k}) \simeq D(\Omega_{k(x)/k} \otimes k(y)) \otimes D(\Omega_{Y/X} \otimes k(y))$. Using the homomorphism $\iota : \mathrm{K}_{j+i}^{\mathrm{MW}}(k(x)) \to \mathrm{K}_{j+i}^{\mathrm{MW}}(k(y))$, we can then define

$$\mathrm{K}_{j+i}^{\mathrm{MW}}(k(x), D(\Omega_{k(x)/k})) \to \mathrm{K}_{j+i}^{\mathrm{MW}}(k(y), D(\Omega_{k(y)/k}) \otimes D(\Omega_{Y/X} \otimes k(y))^{-1}).$$

Summing up these homomorphisms, we obtain a homomorphism (of degree d)

$$f^* : \tilde{C}(X, j, L, a) \to \tilde{C}(Y, j - d, D(\Omega_{Y/X})^{-1} \otimes f^*(L, a))$$

which coincides with the pull-back for cohomological Chow-Witt groups in case X and Y are smooth. We then have the following result (with the usual remark about the characteristic of the base field).

THEOREM 2.14. *Let $f : Y \to X$ be a smooth morphism of k-schemes essentially of finite type over k, and let (L, a) be a graded line bundle over X. Then, the pull-back homomorphism (of degree $d = d_Y - d_X$)*

$$f^* : \tilde{C}(X, j, L, a) \to \tilde{C}(Y, j - d, D(\Omega_{Y/X})^{-1} \otimes f^*(L, a))$$

is a morphism of complexes if either of the following two conditions holds:

(1) *both schemes are essentially smooth.*
(2) *k is of characteristic different from 2.*

PROOF. The first case is an immediate consequence of the previous theorem. For the second case, see [**9**, §4.2.4]. □

THEOREM 2.15 (Homotopy invariance). *Let X be a scheme which is essentially of finite type over k and let $p : X \times \mathbb{A}^1 \to X$ be the projection. Suppose that either k is of characteristic different from 2 or that X is smooth. Then, the homomorphism*

$$p^* : \tilde{C}(X, j, L, a) \to \tilde{C}(X \times \mathbb{A}^1, j - 1, D(\Omega_{X \times \mathbb{A}^1/X})^{-1} \otimes p^*(L, a))$$

is a quasi-isomorphism for any graded line bundle (L, a) over X.

PROOF. We observe that in the first case the extra assumption is only to make sure that the Rost-Schmid complexes are indeed complexes (which we know without extra conditions if X is smooth). In any case, the proof is a verbatim of [**30**, Satz 6.1.1] with input the exact sequences of Proposition 1.20. □

As a corollary of this theorem and of the existence of exact sequences of localization discussed in Section 2.2, we obtain the following result.

COROLLARY 2.16. *Let $p : Y \to X$ be an affine bundle over a scheme X essentially of finite type over k and let $r = d_Y - d_X$. Then the pull-back homomorphism*

$$p^* : \tilde{C}(X, j, L, a) \to \tilde{C}(Y, j - r, D(\Omega_{Y/X})^{-1} \otimes p^*(L, a))$$

is a quasi-isomorphism for any graded line bundle (L, a) over X.

2.5. Euler classes. Let X be a scheme (essentially) of finite type over k, and let $p : E \to X$ be a vector bundle of rank r. Let $s : X \to E$ be the zero section of E. For any graded line bundle (L, a) over X, we have two morphisms of complexes (the latter being of degree r)

$$s_* : \tilde{C}(X, j, s^*(D(\Omega_{E/X})^{-1}) \otimes (L, a)) \to \tilde{C}(E, j, D(\Omega_{E/X})^{-1} \otimes p^*(L, a))$$

and

$$p^* : \tilde{C}(X, j + r, L, a) \to \tilde{C}(E, j, D(\Omega_{E/X})^{-1} \otimes p^*(L, a))$$

with $\Omega_{E/X} \simeq p^* E^\vee$. This induces a homomorphism for any $i \in \mathbb{N}$

$$e(E) : \mathrm{H}_i(X, j, D(E) \otimes (L, a)) \to \mathrm{H}_{i-r}(X, j + r, L, a)$$

that we call *the (homological) Euler homomorphism.* Supposing that X is smooth, we may replace (L, a) by the graded line bundle $D(E^\vee) \otimes D(\Omega_{X/k})^{-1} \otimes (L, a)$ to get a homomorphism

$$e(E) : \mathrm{H}_i(X, j, D(\Omega_{X/k})^{-1} \otimes (L, a)) \to \mathrm{H}_{i-r}(X, j + r, D(E^\vee) \otimes D(\Omega_{X/k})^{-1} \otimes (L, a))$$

In view of the identification $\tilde{C}_i(X, j, D(\Omega_{X/k})^{-1}) = C(X, j + d_X)^{d_X - i}$, this translates into a homomorphism

$$e(E) : \mathrm{H}^i(X, j, L, a) \to \mathrm{H}^{i+r}(X, j + r, D(E^\vee) \otimes (L, a)).$$

called *the (cohomological) Euler homomorphism.* Taking $i = j = 0$, we can consider $e(E)(\langle 1\rangle) \in \mathrm{H}^r(X, r, D(E^\vee)) = \widetilde{\mathrm{CH}}^r(X, D(E^\vee))$. This is the *Euler class* of E. As a consequence of the projection formula we'll see later, the cohomological Euler homomorphism is the multiplication by the Euler class.

3. Products and general pull-backs

3.1. Exterior product. In this section, we show that the cohomological Chow-Witt groups admit exterior products and pull-backs with respect to arbitrary morphisms of (essentially) smooth schemes. Let then X, Y be essentially smooth schemes and let $x \in X^{(i)}$ and $y \in Y^{(i')}$. The base field being perfect, the product $k(x) \otimes_k k(y)$ is actually a product of fields, say $k(x) \otimes_k k(y) = k(u_1) \times \ldots \times k(u_n)$ for $u_1, \ldots, u_n \in (X \times Y)^{(i+i')}$. Writing $X_x := \mathrm{Spec}(\mathcal{O}_{X,x})$ and $Y_y := \mathrm{Spec}(\mathcal{O}_{Y,y})$, we get a Cartesian square

$$\begin{array}{ccc} \mathrm{Spec}(k(x) \otimes k(y)) & \longrightarrow & X_x \times \mathrm{Spec}(k(y)) \\ \downarrow & & \downarrow \\ \mathrm{Spec}(k(x)) \times Y_y & \longrightarrow & X_x \times Y_y \end{array}$$

The morphisms in the diagram are all complete intersection, and we then obtain a decomposition ([**15**, Appendix B.7.4])

$$\big((\mathfrak{m}_x/\mathfrak{m}_x^2)^* \otimes k(u)\big) \oplus \big((\mathfrak{m}_y/\mathfrak{m}_y^2)^* \otimes k(u)\big) = (\mathfrak{m}_u/\mathfrak{m}_u^2)^*$$

for each $u = u_1, \ldots, u_n$. Writing $i_{x,u} : k(x) \to k(u)$ and $i_{y,u} : k(y) \to k(u)$ for the field extensions, we then obtain a canonical isomorphism

$$i_{x,u}^* D(\mathfrak{m}_x/\mathfrak{m}_x^2)^{-1} \otimes i_{y,u}^* D(\mathfrak{m}_y/\mathfrak{m}_y^2)^{-1} \simeq D(\mathfrak{m}_u/\mathfrak{m}_u^2)^{-1}. \tag{3.1}$$

DEFINITION 3.1. Let $x \in X^{(i)}$, $y \in Y^{(i')}$ and $u \in (X \times Y)^{(i+i')}$ be as above. We write $\mu(x, y; u)$:

$$\mathrm{K}_*^{\mathrm{MW}}(k(x), D(\mathfrak{m}_x/\mathfrak{m}_x^2)^{-1}) \times \mathrm{K}_*^{\mathrm{MW}}(k(y), D(\mathfrak{m}_y/\mathfrak{m}_y^2)^{-1}) \to \mathrm{K}_*^{\mathrm{MW}}(k(u), D(\mathfrak{m}_u/\mathfrak{m}_u^2)^{-1})$$

for the composite of the product of twisted Milnor-Witt K-theory groups and the isomorphism induced by (3.1).

Explicitly, let $\alpha \in \mathrm{K}_*^{\mathrm{MW}}(k(x))$, $\overline{x}_1, \ldots, \overline{x}_i$ be generators of $\mathfrak{m}_x/\mathfrak{m}_x^2$, $\beta \in \mathrm{K}_*^{\mathrm{MW}}(k(y))$ and $\overline{y}_1, \ldots, \overline{y}_{i'}$ be generators of $\mathfrak{m}_y/\mathfrak{m}_y^2$, then

$$\mu(x, y; u)(\alpha \otimes \overline{x}_1^* \wedge \ldots \wedge \overline{x}_i^*, \beta \otimes \overline{y}_1^* \wedge \ldots \wedge \overline{y}_{i'}^*) = \alpha \cdot \beta \otimes \overline{x}_1^* \wedge \ldots \wedge \overline{x}_i^* \wedge \overline{y}_1^* \wedge \ldots \wedge \overline{y}_{i'}^*.$$

We let the reader prove the following result, which will imply the graded commutativity of the exterior product on Chow-Witt groups.

LEMMA 3.2. *Let $r, s \in \mathbb{Z}$. The following diagram, in which the left-hand map is the switch,*

$$\begin{array}{ccc} \mathrm{K}_r^{\mathrm{MW}}(k(x), D(\mathfrak{m}_x/\mathfrak{m}_x^2)^{-1}) \otimes \mathrm{K}_s^{\mathrm{MW}}(k(y), D(\mathfrak{m}_y/\mathfrak{m}_y^2)^{-1}) & \xrightarrow{\mu(x,y;u)} & \mathrm{K}_{r+s}^{\mathrm{MW}}(k(u), D(\mathfrak{m}_u/\mathfrak{m}_u^2)^{-1}) \\ \downarrow & & \downarrow{\scriptstyle \epsilon^{rs}\langle(-1)^{ii'}\rangle} \\ \mathrm{K}_s^{\mathrm{MW}}(k(y), D(\mathfrak{m}_y/\mathfrak{m}_y^2)^{-1}) \otimes \mathrm{K}_r^{\mathrm{MW}}(k(x), D(\mathfrak{m}_x/\mathfrak{m}_x^2)^{-1}) & \xrightarrow[\mu(y,x;u)]{} & \mathrm{K}_{r+s}^{\mathrm{MW}}(k(u), D(\mathfrak{m}_u/\mathfrak{m}_u^2)^{-1}) \end{array}$$

is commutative.

We may assemble the homomorphisms $\mu(x,y;u)$ to obtain a homomorphism of graded abelian groups

$$\mu : C(X,j,L,a) \times C(Y,j',L',a') \to C(X \times Y, j+j', (L,a) \otimes (L',a')).$$

We also often denote μ by $\times$. Note also the extra subtlety induced by the presence of graded line bundles. Indeed, in that case, one has to use the canonical isomorphism of graded line bundles over $k(u)$

$$D(\mathfrak{m}_x/\mathfrak{m}_x^2)^{-1} \otimes (L,a) \otimes D(\mathfrak{m}_y/\mathfrak{m}_y^2)^{-1} \otimes (L',a') \\ \to D(\mathfrak{m}_x/\mathfrak{m}_x^2)^{-1} \otimes D(\mathfrak{m}_y/\mathfrak{m}_y^2)^{-1} \otimes (L,a) \otimes (L',a')$$

and proceed as above. This being said, we have the following result (written as for now only in characteristic different from 2) whose proof can be found in [**11**, Corollary 4.12].

PROPOSITION 3.3. *Let X and Y be essentially smooth schemes over k. Then, the homomorphism*

$$\mu : C(X,j,L,a) \times C(Y,j',L',a') \to C(X \times Y, j+j', (L,a) \otimes (L',a')).$$

induces a well-defined product on cohomology provided the base field is of characteristic different from 2.

The above proposition is the result of putting together two Leibniz-type formulas (one for Witt groups, the other for K^{M}-cohomology). We refer the reader to [**8**] for the following more precise result (without any extra assumption on the base field).

PROPOSITION 3.4 (Leibniz formula). *Let $\alpha \in C^i(X,j,L,a)$ and*

$$\beta \in C^{i'}(Y,j',L',a').$$

Then

$$d^{i+i'}_{X\times Y}(\alpha \times \beta) = (d^i_X(\alpha) \times \beta) + (-1)^i \epsilon^j \langle (-1)^a \rangle (\alpha \times d^{i'}_Y(\beta)).$$

3.2. Cartier divisors. In this section, we explain how to pull-back along a (principal) Cartier divisor, starting with a few definitions and basic facts on the latter following [**15**, Appendix B.4]. Let X be an integral scheme and let $C = \{U_i, f_i\}$ be a Cartier divisor on X. Here $U_i \subset X$ is an open subscheme, and the functions $f_i \in k(U_i) = k(X)$ are such that for each i,j we have $f_i/f_j \in \mathcal{O}^{\times}_{U_i \cap U_j}$. The support $|C|$ of C is the closed subset of X for which a local equation of C is not invertible. To C, we can associate a line bundle $\mathcal{O}(C)$ which is defined as the subsheaf of $k(X)$ generated on U_i by f_i^{-1}. Recall finally that C is said to be effective if local equations are regular. An effective Cartier divisor always has a canonical section s_C, which corresponds to $1 \in k(X)$. The vanishing locus of the canonical section is precisely $|C|$.

Our first goal is to associate to a Cartier divisor an explicit canonical class in the group $\widetilde{\mathrm{CH}}^1_{|C|}(X, D(\mathcal{O}(C))^{-1})$. We require here X to be smooth (and integral). Let $x \in X^{(1)}$. There exists i such that $x \in U_i$ and we can consider the residue homomorphism

$$\partial_x : \mathrm{K}^{\mathrm{MW}}_1(k(X), D(\mathcal{O}(C))^{-1}) \to \mathrm{K}^{\mathrm{MW}}_0(k(x), D(\mathfrak{m}_x/\mathfrak{m}_x^2)^{-1} \otimes D(\mathcal{O}(C))^{-1}).$$

We set $\widetilde{\mathrm{ord}}_x(C) := \partial_x([f_i] \otimes f_i)$ and observe that this definition depends a priori on the choice of i such that $x \in U_i$ (we have identified $k(X)$ with $k(U_i)$ on the left-hand side). However, the following computation shows that $\widetilde{\mathrm{ord}}_x(C)$ is actually independent of such a choice. Indeed, suppose that $x \in U_i \cap U_j$. Then

$$\begin{aligned}[f_i] \otimes f_i &= [(f_i/f_j) \cdot f_j] \otimes (f_i/f_j) \cdot f_j \\ &= ([f_i/f_j] + \langle f_i/f_j \rangle [f_j]) \otimes (f_i/f_j) \cdot f_j = [f_i/f_j] \otimes f_i + [f_j] \otimes f_j\end{aligned}$$

and we can use the fact that f_i/f_j is invertible to see that the residue of both $[f_i] \otimes f_i$ and $f_j \otimes f_j$ are the same. This leads to the following definition.

DEFINITION 3.5. We set

$$\widetilde{\mathrm{ord}}(C) = \sum_{x \in X^{(1)} \cap |C|} \widetilde{\mathrm{ord}}_x(C) \in C(X, 1, D(\mathcal{O}(C))^{-1}).$$

A simple computation shows that in fact $\widetilde{\mathrm{ord}}(C)$ is a cocycle as could be seen by repeating the proof of [**14**, Lemma 2.1.3] (and noting that the assumption on the characteristic of k is not used there). If now $\alpha \in C(X, j, L, a)^i$, the question is to know if one can define an element $C \cdot \alpha \in C_{|C|}(X, j, D(\mathcal{O}(C))^{-1} \otimes (L, a))^{i+1}$ which coincides with $\widetilde{\mathrm{ord}}(C)$ in case $\alpha = \langle 1 \rangle \in C(X, 0)^0$. It is possible to do it in fairly general situations, but only for cocycles as explained in [**8**].

We now pass to the definition of a partial pull-back which will be needed in the sequel. For $f \in \mathcal{O}(X)$, we denote by $V(f)$ the zero locus of f, by $U(f)$ its complement and by $\iota : U(f) \to X$ the inclusion. We sometimes simply write U and V when the context is clear. We can also observe that the codomain of the differential

$$d^i : C(U(f), j, \iota^*(L, a))^i \to C(X, j, L, a)^{i+1}$$

can be split into $C(X, j, L, a)^{i+1} = C(U(f), j, \iota^*(L, a))^{i+1} \oplus C_{V(f)}(X, j, L, a)^{i+1}$. Consequently, we can write $d^i = d^i_U + d(f)^i$, where d^i_U is the differential in the Rost-Schmid complex of U and $d(f)^i$ is the component supported on $V(f)$. We define a homomorphism

$$\mu^i_f : C(U(f), j, \iota^*(L, a))^i \to C_{V(f)}(X, j+1, D(f)^{-1} \otimes (L, a))^{i+1}$$

as follows, where $D(f)$ is the Cartier divisor associated to f.

Let $x \in U(f)^{(i)}$ and let $\alpha' = \alpha \otimes (\overline{x}^*_1 \wedge \ldots \wedge \overline{x}^*_i) \otimes l \in \mathrm{K}^{\mathrm{MW}}_{j-i}(k(x), D(\mathfrak{m}_x/\mathfrak{m}_x^2)^{-1} \otimes (L, a))$. We set

$$\mu^i_f(\alpha') = d(f)^i(\langle (-1)^i \rangle [f] \alpha \otimes (\overline{x}^*_1 \wedge \ldots \wedge \overline{x}^*_i) \otimes f \otimes l).$$

LEMMA 3.6. *The homomorphisms*

$$\mu^i_f : C(U(f), j, \iota^*(L, a))^i \to C_{V(f)}(X, j+1, D(f)^{-1} \otimes (L, a))^{i+1}$$

induce a morphism of complexes (of degree 1*)*

$$\mu_f : C(U(f), j, \iota^*(L, a)) \to C_{V(f)}(X, j+1, D(f)^{-1} \otimes (L, a)).$$

PROOF. See [**1**, §2.3.1]. □

In case $V(f)$ is smooth, we may use the identification of Section 2.2 to see μ_f as a homomorphism (of degree 0)

$$C(U(f), j, \iota^*(L, a)) \to C(V(f), j, (L, a)_{|V(f)})$$

that we still denote by μ_f.

REMARK 3.7. Let $f \in \mathcal{O}(X)$ be a global section as above, and let $\lambda \in \mathcal{O}(X)^\times$ be an invertible global section. In general, the homomorphisms μ_f and $\mu_{\lambda f}$ are different, even for cocycles on $U(f) = U(\lambda f)$. However, they coincide for cocycles on X and form a special case of the pull-back along a Cartier divisor hinted at above. To see that μ_f and $\mu_{\lambda f}$ are in general different, one may consider $X = \mathbb{A}^1$, $f = t$ and $\lambda = -1$. For $\alpha = [t] \in C(\mathbb{A}^1, 1)^0$, we find

$$\mu_t([t]) = [-1] \otimes \overline{t}^* \otimes t \in C_{V(t)}(\mathbb{A}^1, 2, D(t)^{-1})^1$$

On the other hand, $\mu_{-t}([t]) = 0$ as $[-t, t] = 0$. More generally, we have $\mu_{\lambda t}([t]) = \epsilon[-\lambda] \otimes \overline{t}^* \otimes t$ for $\lambda \in k^\times$.

3.3. General pull-backs. In this section, we construct pull-backs for arbitrary morphisms following [**27**] and [**11**]. We start by briefly recalling the deformation to the normal cone construction (as explained for instance in [**27**, §10]). This is a fundamental object in the definition of the pull-back for Chow-Witt groups (and more generally a fundamental object for any reasonable cohomology theory). Let $i : Y \subset X$ be an embedding of smooth schemes (say of constant relative dimension $r = d_X - d_Y$) with normal (vector) bundle $\pi : N_Y X \to Y$. The deformation to the normal cone is a commutative diagram

$$\begin{array}{ccc} Y \times \mathbb{A}^1 & \xrightarrow{\iota} & D(Y, X) \\ & \searrow^{i \times 1} & \downarrow p \\ & & X \times \mathbb{A}^1 \end{array}$$

such that:

(1) $D(Y, X)$ is smooth.
(2) The restriction of p to $X \times (\mathbb{A}^1 \setminus \{0\})$ is an isomorphism.
(3) The restriction of p to $X \times \{0\}$ is the composite $N_Y X \xrightarrow{\pi} Y \xrightarrow{i} X$ and $\iota(0) : Y \to N_Y X$ is the zero section.

If t denotes a coordinate of $\mathbb{A}^1$, we then have a global section $t \in \mathcal{O}(D(Y, X))$ whose zero locus is $N_Y X$ (which is a vector bundle over Y). Any graded line bundle (L, a) on X yields a graded line bundle on $D(Y, X)$ whose restriction to $U := D(Y, X) \setminus N_Y X$ is isomorphic via p to the pull-back (via the projection) of (L, a) to $X \times (\mathbb{A}^1 \setminus \{0\})$. Besides, the restriction to $N_Y X$ of this graded line bundle is $\pi^*(L, a)_{|Y}$ ([**24**, §3.1]). We then have morphisms of complexes (of degree 0)

$$C(X, j, L, a) \to C(X \times (\mathbb{A}^1 \setminus \{0\}), j, L, a) \xrightarrow{\mu_t} C(N_Y X, j, \pi^*(L, a)_{|Y})$$

where the first map is the pull-back along the projection $X \times (\mathbb{A}^1 \setminus \{0\}) \to X$. Let us note that this morphism is very explicit. However, we would like to replace the right-hand term with the Rost-Schmid complex of Y. This requires to pass to cohomology and use the homotopy invariance property described in Corollary 2.16. Unfortunately, it is in general very hard to compute the inverse (in cohomology) of π^*, making pull-backs usually hard to compute. Note however that in case $N_Y X \simeq \mathbb{A}^n_Y$, then there is a canonical homotopy using iterations of the method described in [**23**, proof of Theorem 5.38]. In any case, we may now define the pull-back along $i : Y \to X$ as follows.

DEFINITION 3.8. Let $i : Y \to X$ be a closed embedding of smooth schemes. Let further $r, j \in \mathbb{Z}$ and (L, a) be a graded line bundle on X. We define

$$i^* : \mathrm{H}^r(X, j, L, a) \to \mathrm{H}^r(Y, j, (L, a)_{|Y})$$

as $i^*(\alpha) = (\pi^*)^{-1}(\mu_t(\alpha))$.

This pull-back allows us to define the pull-back along any morphism $f : Y \to X$. Indeed, we can factorize f using its graph

$$\begin{array}{ccc} Y & \xrightarrow{\Gamma_f} & Y \times X \\ & \searrow^{f} & \downarrow{\scriptstyle p} \\ & & X \end{array}$$

and observe that p is smooth and Γ_f is an embedding of smooth schemes.

DEFINITION 3.9. For any $r, j \in \mathbb{Z}$ and $(L, a) \in \mathcal{G}(X)$, we define

$$f^* : \mathrm{H}^r(X, j, L, a) \to \mathrm{H}^r(Y, j, f^*(L, a))$$

as $f^* := \Gamma^* p^*$ and call f^* the *pull-back along* f.

THEOREM 3.10. *The association $f \mapsto f^*$ is well-defined and functorial in case the base field k is of characteristic different from* 2.

PROOF. This is [**1**, Theorem 2.3]. □

REMARK 3.11. As usual, the assumption on k will be dropped in [**8**].

REMARK 3.12. The reader may have observed that the method above also allows us to define pull-backs along regular embeddings $f : Y \to X$ for which X and Y are not smooth. This induces a (partial) pull-back at the level of the homology of the Rost-Schmid complexes. More precisely, we obtain pull-backs

$$f^* : \mathrm{H}_i(X, j, L, a) \to \mathrm{H}_{i-r}(Y, j + r, D(N_Y X)^{-1} \otimes (L, a))$$

where $r = d_X - d_Y$. We'll come back to this in [**8**].

3.4. The ring structure. The general pull-back map introduced in the previous section allows to define a product on the cohomology of a smooth scheme X. Indeed, we have an exterior product (well-defined on cohomology groups)

$$\mu : C(X, j, L, a) \times C(X, j', L', a') \to C(X \times X, j + j', (L, a) \otimes (L', a')).$$

that we can compose with the pull-back map associated to the diagonal $\Delta : X \to X \times X$. This yields a well-defined product

$$\mathrm{H}^i(X, j, L, a) \otimes \mathrm{H}^{i'}(X, j', L', a') \to \mathrm{H}^{i+i'}(X, j + j', (L, a) \otimes (L', a'))$$

which is associative ([**11**, Proposition 6.6]). The unit is given by the class of $\langle 1 \rangle \in \mathrm{K}_0^{\mathrm{MW}}(k)$ (or rather its pull-back to X). More precisely, the product defined above yields a structure of a graded $\mathrm{K}_0^{\mathrm{MW}}(k)$-algebra. The graded commutativity is essentially given by Lemma 3.2. We find that the diagram

$$\begin{array}{ccc} \mathrm{H}^i(X, j, L, a) \otimes \mathrm{H}^{i'}(X, j', L', a') & \longrightarrow & \mathrm{H}^{i+i'}(X, j + j', (L, a) \otimes (L', a')) \\ \downarrow & & \downarrow{\scriptstyle \langle (-1)^{(a+i)(a'+i')} \rangle \epsilon^{(j-i)(j'-i')}} \\ \mathrm{H}^{i'}(X, j', L', a') \otimes \mathrm{H}^i(X, j, L, a) & \longrightarrow & \mathrm{H}^{i+i'}(X, j + j', (L, a) \otimes (L', a')) \end{array}$$

is commutative. In particular, the product

$$\widetilde{\mathrm{CH}}^i(X, L, a) \otimes \widetilde{\mathrm{CH}}^{i'}(X, L', a') \to \widetilde{\mathrm{CH}}^{i+i'}(X, (L, a) \otimes (L', a'))$$

is $\langle(-1)^{(a+i)(a'+i')}\rangle$-commutative.

PROPOSITION 3.13. *Let $f : X \to Y$ be a morphism of smooth schemes. Then the pull-back map f^* constructed in the previous section is a ring homomorphism.*

PROOF. See [**11**, Proposition 7.2]. □

3.5. Important properties. In this section, we finally list the important properties of the (co-)homology of the Rost-Schmid complexes, starting with the homological story. We assume that k is of characteristic different from 2 for simplicity, with the usual remark about this assumption. We refer to [**8**] for the first result (alternatively, this follows from the smooth base change of [**9**, §1.1.7]).

THEOREM 3.14 (Homological base change). *Let*

$$\begin{array}{ccc} X' & \xrightarrow{f'} & Y' \\ {\scriptstyle p'}\downarrow & & \downarrow{\scriptstyle p} \\ X & \xrightarrow[f]{} & Y \end{array}$$

be a Cartesian square of schemes (essentially) of finite type over k and let $d = \dim(Y') - \dim(Y)$. Suppose that f is proper and that p is smooth. Then, the diagram

$$\begin{array}{ccc} \tilde{C}(X', j-d, D(\Omega_{X'/X})^{-1} \otimes (p')^* f^*(L,a)) & \xrightarrow{(f')_* \mathrm{can}} & \tilde{C}(Y', j-d, D(\Omega_{Y'/Y})^{-1} \otimes p^*(L,a)) \\ {\scriptstyle (p')^*}\uparrow & & \uparrow{\scriptstyle p^*} \\ \tilde{C}(X, j, f^*(L,a)) & \xrightarrow[f_*]{} & \tilde{C}(Y, j, L, a) \end{array}$$

is commutative. Here, can *is the isomorphism induced by the canonical isomorphism $\Omega_{X'/X} \simeq (f')^*\Omega_{Y'/Y}$ of* [**18**, Corollary 4.3].

THEOREM 3.15 (cdh descent). *Let*

$$\begin{array}{ccc} W & \xrightarrow{s} & V \\ {\scriptstyle g}\downarrow & & \downarrow{\scriptstyle f} \\ Y & \xrightarrow[r]{} & X \end{array}$$

be a Cartesian square of schemes. Assume that r is a closed immersion, f is proper and the induced morphism $(V \setminus W) \to (X \setminus Y)$ is an isomorphism. Then, we have a long exact sequence

$$\ldots \to \mathrm{H}_i(W, j, (ig)^*(L, a)) \xrightarrow{s_* - g_*} \mathrm{H}_i(V, j, f^*(L, a)) \oplus \mathrm{H}_i(Y, j, r^*(L, a)) \xrightarrow{f_* + r_*}$$

$$\xrightarrow{f_* + r_*} \mathrm{H}_i(X, j, L, a) \xrightarrow{\delta} \mathrm{H}_{i-1}(W, j, (ig)^*(L, a)) \to \ldots$$

PROOF. The proof relies on the exact triangle [**9**, 1.1.12.c] specialized to the case of the spectrum representing Borel-Moore motivic MW-cohomology ([**9**, §4.2.2]). The coniveau spectral sequence associated to this homology theory is preserved by proper push-forwards and one of the lines at page 2 in the spectral sequence coincides with the Rost-Schmid complex ([**9**, §4.2.8]).

Alternatively, one may use the localization sequences associated to the open embeddings $(V \setminus W) \subset V$ and $(X \setminus Y) \subset X$ and prove by hand that the Cartesian square of schemes induces a Cartesian square of Rost-Schmid complexes. □

THEOREM 3.16 (Nisnevich descent). *Let*

$$\begin{array}{ccc} W & \xrightarrow{s} & V \\ {\scriptstyle g}\downarrow & & \downarrow{\scriptstyle f} \\ Y & \xrightarrow[r]{} & X \end{array}$$

be a Cartesian square of schemes. Assume that r is an open immersion, that f is étale and that the induced morphism (of reduced schemes) $(V \setminus W) \to (X \setminus Y)$ is an isomorphism. Then, we have a long exact sequence

$$\ldots \to \mathrm{H}_i(X, j, L, a) \xrightarrow{r^*+f^*} \mathrm{H}_i(Y, j, r^*(L, a)) \oplus \mathrm{H}_i(V, j, f^*(L, a)) \xrightarrow{s^*-g^*}$$

$$\xrightarrow{s^*-g^*} \mathrm{H}_i(W, j, (fs)^*(L, a)) \xrightarrow{\delta} \mathrm{H}_{i-1}(X, j, L, a) \to \ldots$$

PROOF. This is an easy consequence of the localization sequences of Section 2.2 associated to the open embeddings $Y \to X$ and $W \to V$, together with the fact that the Rost-Schmid complex of a scheme coincides with the Rost-Schmid complex of its associated reduced scheme. □

Let's now pass to the properties of the cohomology of the Rost-Schmid complex. We first note that both cohomological Nisnevich descent and cohomological cdh descent are special cases of the theorems above, taking only smooth schemes. The only difference with the homological situation is that we have more general pullbacks and that the base change formula is therefore more general. Before stating the result, we start with a few considerations on virtual relative bundles. Let $f : X \to Y$ be a morphism of schemes of finite type. We say that f is an l.c.i. morphism if it admits a factorization

$$X \xrightarrow{i} Z \xrightarrow{p} Y$$

such that p is smooth and i is a regular embedding. For instance, any morphism of smooth schemes is l.c.i. (taking the factorization given by the graph). Then, we have two vector bundles on X, namely $i^*\Omega_{Z/Y}$ and $C_X Z$. We can define the virtual vector bundle $\tau_{pi} := C_X Z - i^*\Omega_{Z/Y}$ (in the sense of [**10**, §4.2]), which is called *the relative bundle (to f) with respect to the factorization $f = pi$*. Suppose next that

we have a commutative diagram

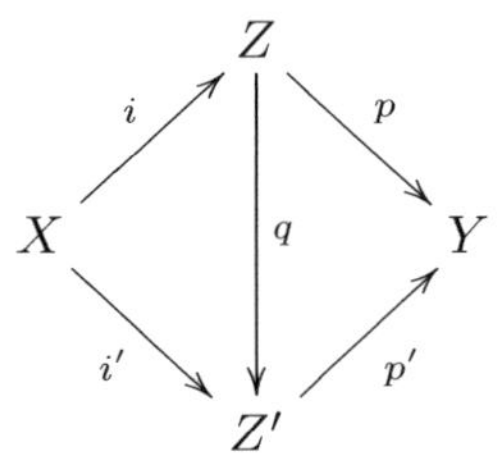

where $pi = p'i' = f$ and q is *smooth*. In that case, we have exact sequences (e.g. [**15**, Appendix B.7.5])

$$0 \to q^*\Omega_{Z'/Y} \to \Omega_{Z/Y} \to \Omega_{Z/Z'} \to 0$$

and

$$0 \to C_X Z' \to C_X Z \to i^*\Omega_{Z/Z'} \to 0.$$

Therefore, we obtain isomorphisms

$$\begin{array}{rcl} C_X Z - i^*\Omega_{Z/Y} & \to & C_X Z' + i^*\Omega_{Z/Z'} - i^*\Omega_{Z/Z'} - (i')^*\Omega_{Z'/Y} \\ & \to & C_X Z' - (i')^*\Omega_{Z'/Y}, \end{array}$$

i.e. a canonical isomorphism $\tau_{pi} \to \tau_{p'i'}$. If now we only have a diagram

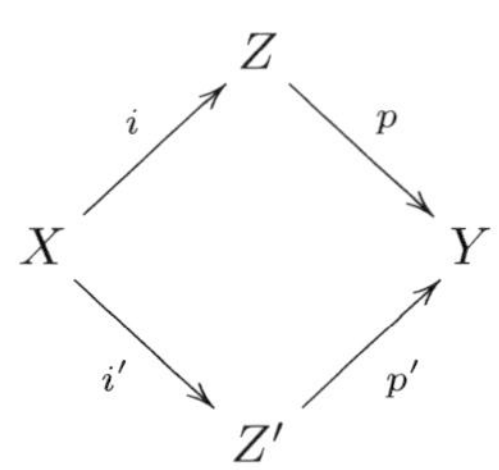

with $pi = p'i' = f$, then we can consider the factorization

$$X \xrightarrow{i \times i'} Z \times_Y Z' \xrightarrow{\pi} Y$$

where π is the obvious projection from the fiber product to Y. It is again a factorization of f, and the respective projections to Z and Z' induce morphisms of factorizations as above. We then get a diagram

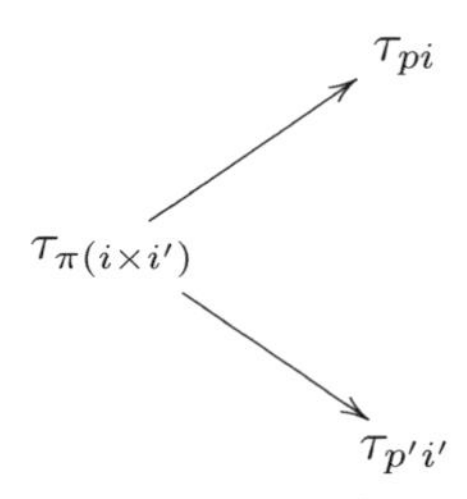

in the category of virtual vector bundles (or the same in the category of graded line bundles). This construction allows to define τ_f as the limit (over all factorizations) of τ_{pi} and obtain a well-defined $D(\tau_f) \in G(X)$.

In case both schemes X and Y are essentially smooth, we can use the canonical isomorphisms (1.2)

$$p^*D(\Omega_{Y/k}) \simeq D(\Omega_{Z/k}) \otimes D(\Omega_{Z/Y})^{-1}$$

and (1.4)

$$D(C_X Z) \otimes D(\Omega_{X/k}) \simeq i^* D(\Omega_{Z/k})$$

for any factorization $f = pi$ to find a canonical isomorphism

$$\begin{aligned} D(\tau_{pi}) &= D(C_X Z) \otimes i^* D(\Omega_{Z/Y})^{-1} \\ &\simeq i^* D(\Omega_{Z/k}) \otimes D(\Omega_{X/k})^{-1} \otimes i^* D(\Omega_{Z/k})^{-1} \otimes f^* D(\Omega_{Y/k}) \end{aligned}$$

Switching the two middle factors and cancelling, we find an isomorphism

$$\alpha(X,Y) : D(\tau_{pi}) \simeq D(\Omega_{X/k})^{-1} \otimes f^* D(\Omega_{Y/k}).$$

We let the reader check that this induces an isomorphism $D(\tau_f) \simeq D(\Omega_{X/k})^{-1} \otimes f^* D(\Omega_{Y/k})$ that we still denote by $\alpha(X,Y)$.

Recall next that a Cartesian square of schemes (of finite type)

$$\begin{array}{ccc} X' & \xrightarrow{f'} & Y' \\ {\scriptstyle g'}\downarrow & & \downarrow{\scriptstyle g} \\ X & \xrightarrow[f]{} & Y \end{array}$$

is *tor-independent* if for any $x \in X$ and $y' \in Y$ with $y = f(x) = g(y')$ we have

$$\mathrm{Tor}_i^{\mathcal{O}_{Y,y}}(\mathcal{O}_{X,x}, \mathcal{O}_{Y',y'}) = 0$$

for any $i > 0$. If g is l.c.i., we can form the following diagram in which all squares are Cartesian

$$\begin{array}{ccc} X' & \xrightarrow{f'} & Y' \\ {\scriptstyle i'}\downarrow & & \downarrow{\scriptstyle i} \\ X'' & \xrightarrow{f''} & Y'' \\ {\scriptstyle p'}\downarrow & & \downarrow{\scriptstyle p} \\ X & \xrightarrow[f]{} & Y \end{array}$$

and p, p' are smooth, while i is a regular embedding.

LEMMA 3.17. *The Cartesian square*

$$\begin{array}{ccc} X' & \xrightarrow{f'} & Y' \\ {\scriptstyle i'}\downarrow & & \downarrow{\scriptstyle i} \\ X'' & \xrightarrow[f'']{} & Y'' \end{array}$$

is tor-independent. Consequently, i' is a regular embedding and the homomorphism

$$(f')^* C_{Y'} Y'' \to C_{X'} X''$$

is an isomorphism.

PROOF. The property of being tor-independent being local, we may restrict to the affine case and moreover suppose that i is complete intersection. We have a commutative diagram of rings

$$\begin{array}{ccc} B' & \longleftarrow & A' \\ \uparrow & & \uparrow \\ B'' & \longleftarrow & A'' \\ \uparrow & & \uparrow \\ B & \longleftarrow & A \end{array}$$

in which the squares are cocartesian, A'' (resp. B'') is a smooth A-algebra (resp. smooth B-algebra) and $A' = A''/I$ where I is a complete intersection ideal. Let $P_\bullet \to B$ be a projective resolution of B as an A-module. We have a commutative square of complexes of A'-modules

$$\begin{array}{ccc} ((P_\bullet) \otimes_A A'') \otimes_{A''} A' & \longrightarrow & (B \otimes_A A'') \otimes_{A''} A' \\ \downarrow & & \downarrow \\ (P_\bullet) \otimes_A A' & \longrightarrow & B \otimes_A A' \end{array}$$

in which the vertical arrows are isomorphisms. Since $\mathrm{Tor}_i^A(B, A') = 0$ for $i > 0$, the bottom arrow is a quasi-isomorphism and it follows that the top arrow is also a quasi-isomorphism. Since A'' is in particular flat over A, $(P_\bullet) \otimes_A A'' \to B \otimes_A A'' = B''$ is a projective resolution and we conclude that $\mathrm{Tor}_i^{A''}(B'', A') = 0$ for $i > 0$, i.e. that the top square is tor-independent. Since $A' = A''/I$ is defined by a regular sequence, we see that the Koszul complex K associated to I is a free resolution of A' as A''-module. Since $\mathrm{Tor}_i^{A''}(B'', A') = 0$ for $i > 0$, we see that $K \otimes_{A''} B'' \to A' \otimes_{A''} B'' = B'$ is a quasi-isomorphism. It follows that B' is also defined by a regular sequence and that

$$B'' \otimes_{A''} (I/I^2) \simeq IB''/(IB'')^2.$$

$\square$

As a result, a tor-independent Cartesian square

$$\begin{array}{ccc} X' & \xrightarrow{f'} & Y' \\ {\scriptstyle g'}\downarrow & & \downarrow{\scriptstyle g} \\ X & \xrightarrow[f]{} & Y \end{array}$$

in which g is l.c.i. comes equipped with a morphism of virtual vector bundles $(f')^*(\tau_g) \to \tau_{g'}$, which is also independent of the choice of the factorization of g. This yields a well-defined isomorphism

$$\mathrm{can} : (f')^* D(\tau_g) \simeq D(\tau_{g'})$$

which is the one needed in the base change formula that we can finally state.

THEOREM 3.18 (Cohomological base change). *Let*

$$\begin{array}{ccc} X' & \xrightarrow{f'} & Y' \\ {\scriptstyle g'}\downarrow & & \downarrow{\scriptstyle g} \\ X & \xrightarrow[f]{} & Y \end{array}$$

be a tor-independent Cartesian square of smooth schemes over a field of characteristic different from 2. Suppose that f is proper of relative dimension $d = \dim(Y) - \dim(X)$. Then, the following diagram commutes

$$\begin{array}{ccc} \mathrm{H}^i(X', j, D(\tau_{g'})^{-1} \otimes D(\Omega_{X'/k}) \otimes (fg')^*(L,a)) & \xrightarrow{f'_*\mathrm{can}} & \mathrm{H}^{i+d}(Y', j+d, D(\tau_g)^{-1} \otimes D(\Omega_{Y'/k}) \otimes g^*(L,a)) \\ {\scriptstyle \alpha(X',X)}\uparrow & & \uparrow{\scriptstyle \alpha(Y',Y)} \\ \mathrm{H}^i(X', j, (g')^* D(\Omega_{X/k}) \otimes (fg')^*(L,a)) & & \mathrm{H}^{i+d}(Y', j, g^* D(\Omega_{Y/k}) \otimes g^*(L,a)) \\ {\scriptstyle (g')^*}\uparrow & & \uparrow{\scriptstyle g^*} \\ \mathrm{H}^i(X, j, D(\Omega_{X/k}) \otimes f^*(L,a)) & \xrightarrow[f_*]{} & \mathrm{H}^{i+d}(Y, j+d, D(\Omega_{Y/k}) \otimes (L,a)) \end{array}$$

for any $i, j \in \mathbb{Z}$ and any graded line bundle (L,a) over Y.

We refer the reader to [**8**] for a (characteristic free) proof. In fact, the result holds more generally if we assume that the schemes are of finite type over a field, and that the vertical map are l.c.i.

One of the main consequences of the base change theorem is a projection formula for the cohomology of the Rost-Schmid complex.

THEOREM 3.19 (Projection formula). *Let $f : X \to Y$ be a proper morphism of essentially smooth schemes over k. Let $d = \dim(Y) - \dim(Y)$ and let (L,a) and (L',a') be graded line bundles over Y. For any $\alpha \in \mathrm{H}^i(Y,j,L,a)$ and $\beta \in \mathrm{H}^{i'}(X, j', D(\Omega_{X/k}) \otimes f^*(L',a'))$ we have*

$$f_*(f^*(\alpha) \cdot \beta) = \alpha \cdot f_*(\beta) \in \mathrm{H}^{i+i'+d}(Y, j+j'+d, D(\Omega_{Y/k}) \otimes (L,a) \otimes (L',a'))$$

and

$$f_*(\beta \cdot f^*(\alpha)) = f_*(\beta) \cdot \alpha \in \mathrm{H}^{i+i'+d}(Y, j+j'+d, D(\Omega_{Y/k}) \otimes (L',a') \otimes (L,a)).$$

PROOF. We can consider the Cartesian square

$$\begin{array}{ccc} X & \xrightarrow{\Gamma_f} & X \times Y \\ {\scriptstyle f}\downarrow & & \downarrow{\scriptstyle f \times \mathrm{id}} \\ Y & \xrightarrow[\Delta_Y]{} & Y \times Y \end{array}$$

where Δ_Y is the diagonal morphism and Γ_f is the graph of f. If Δ_X is the diagonal morphism of X, we have by definition $\Gamma_f = (\mathrm{id} \times f) \circ \Delta_X$. By base change, we have (using the relevant canonical isomorphisms)

$$\Delta_Y^*(f \times \mathrm{id})_* = f_* \Gamma_f^* = f_* \Delta_X^* (\mathrm{id} \times f)^*.$$

We let the reader check that $(f \times \mathrm{id})_* = f_* \times \mathrm{id}$, which gives immediately the second formula. The same argument switching X and Y on the right yields the other statement (see also [**7**, Remark 3.6] for an alternative argument). □

4. Some computations

The theory of Chow-Witt groups and more generally of the (co-)homology of the Rost-Schmid complex being relatively new, there are few computations available in the literature, especially compared to the amount of computations involving Chow groups. One primary difficulty is already to compute the Chow-Witt group of the base field, i.e. its Grothendieck-Witt group $\mathrm{GW}(k)$. For a general field, this is a very interesting but quite difficult question, that we completely avoid here. So, all the computations below will be in term of $\mathrm{GW}(k)$, not bothering about its actual shape.

Perhaps the most basic computation of the Chow ring is the one of the projective space $\mathbb{P}^n_k$. It is a truncated polynomial algebra generated by the first Chern class of the tautological line bundle. If one can define the Euler class of the tautological bundle, it doesn't live in the regular Chow-Witt group, but rather on its version twisted by the graded line bundle $D(\mathcal{O}(-1))$. This apparently innocuous difference leads to a completely different result as the ordinary one ([**13**, Corollary 11.8]).

THEOREM 4.1. *We have*

$$\widetilde{\mathrm{CH}}^i(\mathbb{P}^n_k) = \begin{cases} \mathrm{GW}(k) & \textit{if } i=0 \textit{ or } i=n \textit{ with } n \textit{ odd.} \\ \mathbb{Z} & \textit{if } i \textit{ is even and } i\neq 0. \\ 2\mathbb{Z} & \textit{if } i \textit{ is odd and } i\neq n. \end{cases}$$

$$\widetilde{\mathrm{CH}}^i(\mathbb{P}^n_k, D(\mathcal{O}(-1))) = \begin{cases} 2\mathbb{Z} & \textit{if } i \textit{ is even and } i\neq n. \\ \mathbb{Z} & \textit{if } i \textit{ is odd.} \\ \mathrm{GW}(k) & \textit{if } i=n \textit{ and } n \textit{ is even.} \end{cases}$$

In the statement, both the factors $\mathbb{Z}$ and $2\mathbb{Z}$ indicate that the relevant group is $\mathbb{Z}$, but that its image under the projection $\widetilde{\mathrm{CH}}^i(\mathbb{P}^n_k) \to \mathrm{CH}^i(\mathbb{P}^n_k) = \mathbb{Z}$ is either everything or of index 2. More generally, one can compute the whole cohomology of the Rost-Schmid complex of $\mathbb{P}^n_k$ ([**13**, Theorem 11.7]). As an interesting byproduct, we can compute the Chow-Witt group (or actually also the ring structure) of the classifying space $\mathrm{B}\mathbb{G}_m$. To define the latter, consider for $n\in\mathbb{N}$ the embedding $k^{n+1}\to k^{n+2}$ as the first $n+1$ factors. It induces a closed immersion $b_n:\mathbb{P}^n\to\mathbb{P}^{n+1}$ and we obtain a sequence

$$\operatorname{Spec} k \xrightarrow{b_0} \mathbb{P}^1 \to \ldots \to \mathbb{P}^n \xrightarrow{b_n} \mathbb{P}^{n+1} \to \ldots$$

For any $i\in\mathbb{N}$, this yields a sequence

$$\widetilde{\mathrm{CH}}^i(\operatorname{Spec} k) \xleftarrow{b_1^*} \widetilde{\mathrm{CH}}^i(\mathbb{P}^1) \leftarrow \ldots \leftarrow \widetilde{\mathrm{CH}}^i(\mathbb{P}^n) \xleftarrow{b_n^*} \widetilde{\mathrm{CH}}^i(\mathbb{P}^{n+1}) \leftarrow \ldots$$

This sequence eventually stabilizes ([**1**, Theorem 3.3]), giving a "concrete" model for the limit, which is by definition the group $\widetilde{\mathrm{CH}}^i(\mathrm{B}\mathbb{G}_m)$. In short, we have $\widetilde{\mathrm{CH}}^i(\mathrm{B}\mathbb{G}_m) = \widetilde{\mathrm{CH}}^i(\mathbb{P}^n)$ for n large enough. We can define the group $\widetilde{\mathrm{CH}}^i(\mathrm{B}\mathbb{G}_m, D(\mathcal{O}(-1)))$ in a similar fashion ([**1**, Theorem 3.3]). As a straightforward consequence of the previous theorem, we obtain the following result.

THEOREM 4.2. *We have*

$$\widetilde{\mathrm{CH}}^i((\mathrm{B}\mathbb{G}_m) = \begin{cases} \mathrm{GW}(k) & \text{if } i = 0. \\ \mathbb{Z} & \text{if } i \neq 0 \text{ is even.} \\ 2\mathbb{Z} & \text{if } i \text{ is odd.} \end{cases}$$

while

$$\widetilde{\mathrm{CH}}^i((\mathrm{B}\mathbb{G}_m, D(\mathcal{O}(-1))) = \begin{cases} \mathbb{Z} & \text{if } i \text{ is odd.} \\ 2\mathbb{Z} & \text{if } i \text{ is even.} \end{cases}$$

In terms of characteristic classes, the total ring

$$\widetilde{\mathrm{CH}}^{\mathrm{tot}}(\mathrm{B}\mathbb{G}_m) := \widetilde{\mathrm{CH}}^*(\mathrm{B}\mathbb{G}_m) \oplus \widetilde{\mathrm{CH}}^*((\mathrm{B}\mathbb{G}_m, D(\mathcal{O}(-1)))$$

is generated in degree 1 by the class $e := e(\mathcal{O}(1)) \in \widetilde{\mathrm{CH}}^1((\mathrm{B}\mathbb{G}_m, D(\mathcal{O}(-1)))$ and by another class that we now describe. Recall from Section 1 that we have a homomorphism

$$h_n : \mathrm{K}_n^{\mathrm{M}}(F) \to \mathrm{K}_n^{\mathrm{MW}}(F)$$

defined by $\{a_1, \ldots, a_n\} \mapsto [a_1^2, a_2, \ldots, a_n]$. This induces a homomorphism

$$H_i : \mathrm{CH}^i(X) \to \widetilde{\mathrm{CH}}^i(X)$$

for any smooth scheme X. In particular, we can consider for any $i \in \mathbb{N}$ the class

$$c := H_1(c_1(\mathcal{O}(1))) \in \widetilde{\mathrm{CH}}^1(\mathrm{B}\mathbb{G}_m).$$

It turns out that

$$\widetilde{\mathrm{CH}}^{\mathrm{tot}}(\mathrm{B}\mathbb{G}_m) = \mathrm{GW}(k)[e, c]/J$$

where J is the ideal generated by the relations:

(1) $\mathrm{I}(k)e = 0 = \mathrm{I}(k)c$.
(2) $ec = ce$.
(3) $c^2 = 2he^2$.

The interest of computing the total Chow-Witt ring of $\mathrm{B}\mathbb{G}_m$ lies in the fact that the latter gives a complete set of characteristic classes attached to line bundles ([**32**, Theorem 1.3] and [**17**, Proposition 3.3]). Thus, the outcome of our computation is that there are essentially two characteristic classes attached to a line bundle: the Euler class (which lives in a twisted group) and a class which is the image of the first Chern class.

Actually, the situation we just described can be generalized to more general (reasonable) group schemes G over k. Hornbostel-Wendt computed recently the total Chow-Witt ring of both BSL_n and BSp_{2n} (no twist by a line bundle is needed here). We refer the reader to their paper ([**17**]) for a precise formulation, the essential fact being that there is essentially one new kind of characteristic class: the Borel classes of Panin-Walter ([**25**]); the other classes that appear are roughly defined through Chow groups and Chow groups modulo 2. Their computation was later extended to BGL_n by Wendt in [**33**]. The reader can compare with our computation above in case $n = 1$. The obvious question is now to understand the classifying spaces of orthogonal groups. It is not clear to me what can be expected.

References

[1] A. Asok and J. Fasel, *Comparing Euler classes*, Q. J. Math. **67** (2016), no. 4, 603–635. MR3609848

[2] A. Asok and C. Haesemeyer, *The 0-th stable $\mathbb{A}^1$-homotopy sheaf and quadratic zero cycles*, arXiv:1108.3854, 2011.

[3] Paul Balmer and Charles Walter, *A Gersten-Witt spectral sequence for regular schemes* (English, with English and French summaries), Ann. Sci. École Norm. Sup. (4) **35** (2002), no. 1, 127–152, DOI 10.1016/S0012-9593(01)01084-9. MR1886007

[4] Jean Barge and Fabien Morel, *Groupe de Chow des cycles orientés et classe d'Euler des fibrés vectoriels* (French, with English and French summaries), C. R. Acad. Sci. Paris Sér. I Math. **330** (2000), no. 4, 287–290, DOI 10.1016/S0764-4442(00)00158-0. MR1753295

[5] S. M. Bhatwadekar, J. Fasel, and S. Sane, *Euler class groups and 2-torsion elements*, J. Pure Appl. Algebra **218** (2014), no. 1, 112–120, DOI 10.1016/j.jpaa.2013.04.016. MR3120613

[6] Nicolas Bourbaki, *Éléments de mathématique* (French), Masson, Paris, 1980. Algèbre. Chapitre 10. Algèbre homologique. [Algebra. Chapter 10. Homological algebra]. MR610795

[7] B. Calmès and J. Fasel, *The category of finite Chow-Witt correspondences*, arXiv:1412.2989, 2014.

[8] B. Calmès and J. Fasel, *A primer for Milnor-Witt K-theory sheaves and their cohomology*, in preparation, 2017.

[9] F. Déglise and J. Fasel, *The Milnor-Witt motivic ring spectrum and its associated theories*, arXiv:1708.06102, 2017.

[10] P. Deligne, *Le déterminant de la cohomologie* (French), Current trends in arithmetical algebraic geometry (Arcata, Calif., 1985), Contemp. Math., vol. 67, Amer. Math. Soc., Providence, RI, 1987, pp. 93–177, DOI 10.1090/conm/067/902592. MR902592

[11] Jean Fasel, *The Chow-Witt ring*, Doc. Math. **12** (2007), 275–312. MR2350291

[12] Jean Fasel, *Groupes de Chow-Witt*, Mém. Soc. Math. Fr. (N.S.) **113** (2008), viii+197. MR2542148

[13] Jean Fasel, *The projective bundle theorem for $\mathbf{I}^j$-cohomology*, J. K-Theory **11** (2013), no. 2, 413–464, DOI 10.1017/is013002015jkt217. MR3061003

[14] J. Fasel and P. A. Østvær, *A cancellation theorem for MW-motives*, arXiv:1708.06098, 2016.

[15] William Fulton, *Intersection theory*, 2nd ed., Ergebnisse der Mathematik und ihrer Grenzgebiete. 3. Folge. A Series of Modern Surveys in Mathematics [Results in Mathematics and Related Areas. 3rd Series. A Series of Modern Surveys in Mathematics], vol. 2, Springer-Verlag, Berlin, 1998. MR1644323

[16] Robin Hartshorne, *Algebraic geometry*, Springer-Verlag, New York-Heidelberg, 1977. Graduate Texts in Mathematics, No. 52. MR0463157

[17] Jens Hornbostel, *Oriented Chow groups, Hermitian K-theory and the Gersten conjecture*, Manuscripta Math. **125** (2008), no. 3, 273–284, DOI 10.1007/s00229-007-0148-5. MR2373061

[18] Ernst Kunz, *Kähler differentials*, Advanced Lectures in Mathematics, Friedr. Vieweg & Sohn, Braunschweig, 1986. MR864975

[19] Saunders Mac Lane, *Categories for the working mathematician*, 2nd ed., Graduate Texts in Mathematics, vol. 5, Springer-Verlag, New York, 1998. MR1712872

[20] John Milnor, *Algebraic K-theory and quadratic forms*, Invent. Math. **9** (1969/1970), 318–344, DOI 10.1007/BF01425486. MR0260844

[21] J. Milnor and D. Husemoller, *Symmetric bilinear forms*, Ergebnisse der Mathematik und ihrer Grenzgebiete. 3. Folge. A Series of Modern Surveys in Mathematics [Results in Mathematics and Related Areas. 3rd Series. A Series of Modern Surveys in Mathematics], vol. 73, Springer, Berlin, 1973.

[22] Fabien Morel, *Sur les puissances de l'idéal fondamental de l'anneau de Witt* (French, with English and French summaries), Comment. Math. Helv. **79** (2004), no. 4, 689–703, DOI 10.1007/s00014-004-0815-z. MR2099118

[23] Fabien Morel, *$\mathbb{A}^1$-algebraic topology over a field*, Lecture Notes in Mathematics, vol. 2052, Springer, Heidelberg, 2012. MR2934577

[24] Alexander Nenashev, *Gysin maps in Balmer-Witt theory*, J. Pure Appl. Algebra **211** (2007), no. 1, 203–221, DOI 10.1016/j.jpaa.2007.01.006. MR2333767

[25] I. Panin and C. Walter, *Quaternionic grassmannians and pontryagin classes in algebraic geometry*, arXiv:1011.0649, 11 2010.

[26] H.-G. Quebbemann, W. Scharlau, and M. Schulte, *Quadratic and Hermitian forms in additive and abelian categories*, J. Algebra **59** (1979), no. 2, 264–289, DOI 10.1016/0021-8693(79)90126-1. MR543249

[27] Markus Rost, *Chow groups with coefficients*, Doc. Math. **1** (1996), No. 16, 319–393. MR1418952

[28] Winfried Scharlau, *Quadratic reciprocity laws*, J. Number Theory **4** (1972), 78–97, DOI 10.1016/0022-314X(72)90012-1. MR0292753

[29] Marco Schlichting, *Hermitian K-theory of exact categories*, J. K-Theory **5** (2010), no. 1, 105–165, DOI 10.1017/is009010017jkt075. MR2600285

[30] M. Schmid, *Wittringhomologie*, Ph.D. thesis, Universität Regensburg, 1998.

[31] Jean-Pierre Serre, *Corps locaux* (French), Hermann, Paris, 1968. Deuxième édition; Publications de l'Université de Nancago, No. VIII. MR0354618

[32] Burt Totaro, *The Chow ring of a classifying space*, Algebraic K-theory (Seattle, WA, 1997), Proc. Sympos. Pure Math., vol. 67, Amer. Math. Soc., Providence, RI, 1999, pp. 249–281, DOI 10.1090/pspum/067/1743244. MR1743244

[33] M. Wendt, *Chow-Witt rings of Grassmannians*, arXiv:1805.06142, 2018.

Institut Fourier - UMR 5582, Université Grenoble Alpes, CS 40700, 38058 Grenoble Cedex 9; France

Email address: `jean.fasel@gmail.com`

Contemporary Mathematics
Volume **745**, 2020
https://doi.org/10.1090/conm/745/15024

Chow-Witt rings of split quadrics

Jens Hornbostel, Heng Xie, and Marcus Zibrowius

Abstract. We compute the Chow-Witt rings of split quadrics over a field of characteristic not two. We even determine the full bigraded **I**-cohomology and Milnor-Witt cohomology rings, including twists by line bundles. The results on **I**-cohomology corroborate the general philosophy that **I**-cohomology is an algebro-geometric version of singular cohomology of real varieties: our explicit calculations confirm that the **I**-cohomology ring of a split quadric over the reals is isomorphic to the singular cohomology ring of the space of its real points.

Contents

1. Introduction

Chow-Witt groups were introduced by Barge and Morel [**BM00**] as a cohomology theory containing an *Euler class*, detecting if a vector bundle over a smooth affine scheme over a field of characteristic $\neq 2$ splits off a trivial line bundle in the critical range. This generalized a previous result of Murthy [**Mur94**] for Chow groups of smooth affine schemes over algebraically closed fields. In recent years,

2010 *Mathematics Subject Classification.* 14F42, 14C15, 19G12, 55N25.

Research for this publication was conducted in the framework of the DFG Research Training Group 2240: Algebro-Geometric Methods in Algebra, Arithmetic and Topology.

The second author was supported by EPSRC Grant EP/M001113/1. He would like to thank the Max-Planck-Institut in Bonn for its hospitality. The authors where supported by DFG grants HO 4279/3-1 and ZI 1460/1-1 within the DFG Priority Programme 1786: Homotopy Theory and Algebraic Geometry.

subsequent work of Morel [**Mor12**], Fasel [**Fas08**], Fasel-Srinivas [**FS09**] and Asok-Fasel [**AF16**] has completed the picture if two is invertible. As Chow-Witt groups are a quadratic refinement of Chow groups, one might hope for generalizations of other results using Chow-Witt groups in the future.

In this paper, we determine the Chow-Witt ring of a split quadric Q_n of dimension $n \geq 3$ over a field F of characteristic $\neq 2$. Our computation follows the general strategies laid out by Fasel, Wendt and the first author in the computations for projective spaces, Grassmannians and classifying spaces [**Fas13, HW19, Wen18**]. The main step (in both descriptions) is thus the computation of **I**-cohomology. We compute these cohomology groups in all bidegrees, and over an arbitrary smooth base (not just a field). That is, we compute the group

$$\mathrm{H}^i(Q_n, \mathbf{I}^j, \mathcal{O}(l))$$

for all values of i, j and l, where Q_n is a split quadric over a smooth scheme, and where $\mathcal{O}(l)$ denotes the lth tensor power of $\mathcal{O}_Q(1)$ (see Section 2 for precise definitions). This additive result is summarized in Theorem 5.5 below.

For the computation of the Chow-Witt ring, we need to understand the **I**-cohomology groups in "geometric" bidegrees. With untwisted coefficients, their direct sum yields the graded ring $\mathrm{H}^\bullet(Q_n, \mathbf{I}^\bullet) := \bigoplus_{i\geq 0} \mathrm{H}^i(Q_n, \mathbf{I}^i)$, which is a graded commutative algebra over the Witt ring $\mathrm{W}(F)$ of the base field. Adding twisted coefficients, this extends to a $\mathbb{Z} \oplus \mathbb{Z}/2$-graded $\mathrm{W}(F)$-algebra

$$\mathrm{H}^\bullet(Q_n, \mathbf{I}^\bullet, \mathcal{O} \oplus \mathcal{O}(1)) := \bigoplus_{i\geq 0} \left(\mathrm{H}^i(Q_n, \mathbf{I}^i, \mathcal{O}) \oplus \mathrm{H}^i(Q_n, \mathbf{I}^i, \mathcal{O}(1))\right).$$

The explicit ring structure can often be written down more concisely when twisted coefficients are taken into account. Indeed, this is already true for projective spaces. The following result is Proposition 4.5 in a recent preprint of Matthias Wendt [**Wen18**], based on the additive computations of Fasel [**Fas13**].

THEOREM (Fasel/Wendt). *Let F be field of characteristic $\neq 2$, with Witt ring $\mathrm{W}(F)$ and fundamental ideal $\mathrm{I}(F) \subset \mathrm{W}(F)$. Let $\mathbb{P}^p$ denote the p-dimensional projective space over F. For any $p \geq 1$, we have a $\mathbb{Z}\oplus\mathbb{Z}/2$-graded ring isomorphism*

$$\mathrm{H}^\bullet(\mathbb{P}^p, \mathbf{I}^\bullet, \mathcal{O} \oplus \mathcal{O}(1)) = \mathrm{W}(F)[\xi, \alpha]/(\mathrm{I}(F)\xi, \xi^{p+1}, \xi\alpha, \alpha^2) \quad \textit{with} \begin{cases} |\xi| = (1, \overline{1}) \\ |\alpha| = (p, \overline{p+1}) \end{cases}$$

We generalize this result for projective spaces to the full $\mathbb{Z} \oplus \mathbb{Z} \oplus \mathbb{Z}/2$-graded setting in Theorem 6.1 below. For split quadrics, we obtain the following corresponding result (Theorems 6.8 and 6.9):

THEOREM A. *For any $n \geq 3$, we have a $\mathbb{Z} \oplus \mathbb{Z}/2$-graded ring isomorphism*

$$\mathrm{H}^\bullet(Q_n, \mathbf{I}^\bullet, \mathcal{O} \oplus \mathcal{O}(1)) \cong \begin{cases} \mathrm{W}(F)[\xi, \alpha, \beta]/(\mathrm{I}(F)\xi, \xi^{p+1}, \xi\alpha + \xi\beta, \alpha^2 + \beta^2, \alpha\beta) & \textit{if } n = 2p \textit{ and } p \textit{ is even} \\ \mathrm{W}(F)[\xi, \alpha, \beta]/(\mathrm{I}(F)\xi, \xi^{p+1}, \xi\alpha + \xi\beta, \alpha^2, \beta^2) & \textit{if } n = 2p \textit{ and } p \textit{ is odd} \\ \mathrm{W}(F)[\xi, \alpha, \beta]/(\mathrm{I}(F)\xi, \xi^{p+1}, \xi\alpha, \alpha^2, \beta^2) & \textit{if } n = 2p + 1 \end{cases}$$

Here, the generators are of degrees $|\xi| = (1, \overline{1})$, $|\alpha| = (p, \overline{p+1})$, $|\beta| = (q, \overline{q+1})$ where $q = p$ if $n = 2p$ is even and $q = p + 1$ if $n = 2p + 1$ is odd.

In fact, Theorems 6.8 and 6.9 again describe the full $\mathbb{Z} \oplus \mathbb{Z} \oplus \mathbb{Z}/2$-graded $\mathbf{I}$-cohomology of Q_n. Finally, for the Chow-Witt ring, we arrive at two related presentations. The first presentation, discussed in Section 8, looks as follows.

THEOREM B. *The* $\mathbb{Z}\oplus\mathbb{Z}/2$*-graded* $\mathrm{GW}(F)$*-algebra* $\widetilde{\mathrm{CH}}^\bullet(Q_n, \mathcal{O}\oplus\mathcal{O}(1))$ *can be described as a fibre product of* $\mathbb{Z}\oplus\mathbb{Z}/2$*-graded rings of the following form:*

$$\widetilde{\mathrm{CH}}^\bullet(Q_n, \mathcal{O}\oplus\mathcal{O}(1)) \xrightarrow{\cong} \mathrm{H}^\bullet(Q_n, \mathbf{I}^\bullet, \mathcal{O}\oplus\mathcal{O}(1)) \times_{\mathrm{Ch}^\bullet(Q_n,\mathcal{O}\oplus\mathcal{O}(1))} (\ker\partial \oplus \ker\partial_{\mathcal{O}(1)}).$$

Here, the second factor $\ker\partial \oplus \ker\partial_{\mathcal{O}(1)}$ *is described explicitly in Theorem* 8.3. *The fibre product is taken over the* $\mathbb{Z}\oplus\mathbb{Z}/2$*-graded ring* $\mathrm{Ch}^\bullet(Q_n, \mathcal{O}\oplus\mathcal{O}(1)) := \mathrm{Ch}^\bullet(Q_n)[\tau]/(\tau^2-1)$*, where* $\mathrm{Ch}^\bullet(Q_n)$ *is the (well-known) Chow-ring modulo two of* Q_n*, concentrated in degrees* $(*,0)$*, and* τ *is an artificial generator of bidegree* $(0,1)$.

The second presentation of the Chow-Witt ring is discussed in the final section. It gives a full additive description of the $\mathbb{Z}\oplus\mathbb{Z}\oplus\mathbb{Z}/2$-graded Milnor-Witt cohomology $\mathrm{H}^{\mathrm{tot}}(Q_n, \mathbf{K}^{MW})$ including all twists, refining [**Fas13**, Theorem 11.7] from projective spaces to split quadrics. The arguments presented in Section 6 can also be refined to obtain a description of $\mathrm{H}^{\mathrm{tot}}(Q_n, \mathbf{K}^{MW})$ as an algebra in terms of generators and relations over the coefficient ring $\mathrm{H}^{\mathrm{tot}}(F, \mathbf{K}^{MW})$, but we do not make this explicit here.

For a real variety X, $\mathbf{I}$-cohomology is closely related to the singular cohomology of the real points $X(\mathbb{R})$ equipped with the analytic topology. The strongest general result in this direction is the following theorem of Jacobson [**Jac17**]:

THEOREM (Jacobson). *For a real variety* X*, the group* $\mathrm{H}^i(X, \mathbf{I}^j)$ *is isomorphic to the singular cohomology group* $\mathrm{H}^i(X(\mathbb{R}), \mathbb{Z})$ *in all bidegrees with* $j > \dim(X)$.

Note that this general result does not apply to the interesting geometric bidegrees $i = j$ with $i \leq \dim(X)$. However, the known computations for projective spaces and Grassmannians indicate that the situation for cellular varieties is even better. In forthcoming work with Wendt [**HWXZ**], we will show that realization induces a ring isomorphism $\mathrm{H}^\bullet(X, \mathbf{I}^\bullet) \cong \mathrm{H}^\bullet(X(\mathbb{R}), \mathbb{Z})$ for all smooth cellular real varieties X. This isomorphism also extends to twisted coefficients. Here, we verify this result for split real quadrics, through explicit computations.

Unfortunately, we could not find the integral cohomology ring of a real split quadric in the literature. All we found was the additive structure of cohomology with $\mathbb{Z}/2$-coefficients, which can be computed using the Gysin sequence, see [**Die82**, § 24.39, problem 10], [**WL62**] or [**CGT**]. We therefore include computations of the integral cohomology rings of real quadrics as a preliminary section in this paper. These computations cover all real quadrics, not just the split ones, and also include twisted coefficients. They are completely independent of the computations of $\mathbf{I}$-cohomology that follow, thus providing a baseline. The results are summarized in Theorem 3.20. The computations for split quadrics extend from Chow groups not only to Milnor cohomology (Corollary 6.4), but also to full bigraded motivic cohomology [**DI07**, Prop. 4.3]. See [**Ker09**] or [**RS18**, Section 3] for a comparison of these two bigraded theories. We expect that our techniques and results on Milnor-Witt cohomology groups of split quadrics also extend to the full bigraded "Milnor-Witt motivic cohomology" (also known as "generalized motivic cohomology") as developed by Fasel et al., see e.g. [**DF17**, Theorem 4.2.4] and [**CF17**], but we have not pursued this.

There are many partial results on the Chow groups of quadrics by Swan, Karpenko, Merkurjev, Rost, Vishik and others. Partial results on Witt groups of split quadrics of Nenashev **[Nen09]** can be completed using the blowup setup of Balmer-Calmès that we also use in our computations here **[Cal08]**. Witt groups of split quadrics over the complex numbers have been computed by the third author via another method, see **[Zib11]**. We hope to return to computations of the **I**-cohomology of other kinds of quadrics in future work. The trace map on Clifford algebras has been found very helpful for studying Witt groups of general quadrics, cf. the recent work of the second author **[Xie19]**.

Acknowledgements. We would like to thank the anonymous referee for carefully reading our manuscript and suggesting many improvements.

2. Conventions and notation

Topology. We fix the following notation:

- $Q_{p,q}$ is the $(p+q)$-dimensional **real quadric** defined as a subspace of $\mathbb{RP}^{p+q+1}$ by the equation

$$x_0^2 + x_1^2 + \cdots + x_p^2 = y_0^2 + y_1^2 + \cdots + y_q^2, \tag{2.1}$$

  where $[x_0 : \ldots : x_p : y_0 : \ldots : y_q]$ are homogeneous coordinates on $\mathbb{RP}^{p+q+1}$. Note that $Q_{p,q} \cong Q_{q,p}$. We always assume $1 \leq p \leq q$.

Additional notation is introduced at the beginning of Section 3.

Algebraic geometry. We fix a base scheme S which is smooth over a field F of characteristic $\neq 2$. More precisely, S is assumed to be noetherian, separated, smooth and of finite type over F. We may and will moreover assume without loss of generality that S is connected. These are precisely the assumptions on the base scheme S used in **[Fas13]**. The following schemes and morphisms are all defined over S:

- Q_n denotes the n-dimensional **split quadric**, i.e. the closed subvariety of $\mathbb{P}^{n+1}$ defined by the equation

$$\begin{cases} x_0y_0 + \ldots + x_py_p = 0 & \text{when } n = 2p \\ x_1y_1 + \ldots + x_py_p + z^2 = 0 & \text{when } n = 2p+1 \end{cases} \tag{2.2}$$

  When $n = 2p$, Q_n is isomorphic to the subvariety $Q_{p,p}$ defined by equation (2.1). When $n = 2p+1$, it is isomorphic to $Q_{p,p+1}$.
- $\mathbb{P}^p_x$, $\mathbb{P}^p_y$, $\mathbb{P}^p_{x'}$ and $\mathbb{P}^p_{y'}$ denote the following subvarieties of Q_n:

  When $n = 2p$, we define $\mathbb{P}^p_x, \mathbb{P}^p_y \subset \mathbb{P}^{2p+1}$ by the equations $y_0 = y_1 = \ldots = y_p = 0$ and $x_0 = x_1 = \ldots = x_p = 0$ respectively. We define $\mathbb{P}^p_{x'}, \mathbb{P}^p_{y'} \subset \mathbb{P}^{2p+1}$ by the equations $x_0 = y_1 = \ldots = y_p = 0$ and $y_0 = x_1 = \ldots = x_p = 0$.

  When $n = 2p+1$, we define $\mathbb{P}^p_x, \mathbb{P}^p_y \subset \mathbb{P}^{2p+2}$ by the equations $y_0 = y_1 = \ldots = y_p = z = 0$ and $x_0 = x_1 = \ldots = x_p = z = 0$, respectively.
- $q\colon Q_n \to S$ and $p\colon \mathbb{P}^p \to S$ denote the projection maps
- $i_y\colon \mathbb{P}^p_y \hookrightarrow Q_n - \mathbb{P}^p_x$, $\iota_y\colon \mathbb{P}^p_y \hookrightarrow Q_n$, $j\colon Q_n - \mathbb{P}^p_x \hookrightarrow Q_n$ and $i\colon Q_n \hookrightarrow \mathbb{P}^{n+1}$ are the obvious embeddings; likewise for i_x, ι_x, $i_{x'}$, $\iota_{x'}$, $i_{y'}$ and $\iota_{y'}$

- $\rho\colon Q_n - \mathbb{P}^p_x \to \mathbb{P}^p_y$ is the morphism given by

$$\begin{cases} \rho\colon [x_0 : \ldots : x_p : y_0 : \ldots : y_p] \mapsto [0 : \ldots : 0 : y_0 : \ldots : y_p] & \text{if } n = 2p \\ \rho\colon [x_0 : \ldots : x_p : y_0 : \ldots : y_p : z] \mapsto [0 : \ldots : 0 : y_0 : \ldots : y_p : 0] & \text{if } n = 2p+1 \end{cases}$$

  (Some of this notation is summarized in Figure 1.)
- $\mathcal{O}_Q(1)$ is defined as the restriction along $i\colon Q_n \hookrightarrow \mathbb{P}^{n+1}$ of the canonical line bundle $\mathcal{O}(1)$ on $\mathbb{P}^{n+1}$. When there is no danger of confusion, we simply denote this restriction again by $\mathcal{O}(1)$.

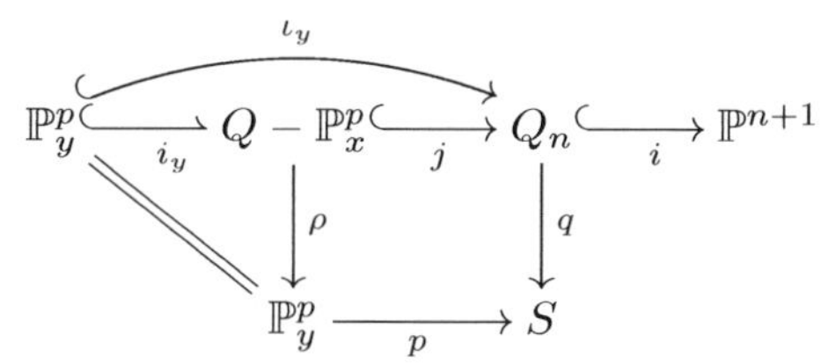

FIGURE 1. Various morphisms used in the algebro-geometric calculations

3. Singular cohomology of real quadrics

In this section, we study the real quadrics $Q_{p,q} \subset \mathbb{RP}^{p+q+1}$ (with the analytic topology). Note first that $Q_{p,q}$ is a two-fold covering space of $\mathbb{RP}^p \times \mathbb{RP}^q$, via the obvious map that takes $[x : y]$ to $([x], [y])$. Write π_1 and π_2 for the two components of this map. A two-fold cover of the quadric itself is given by $S^p \times S^q$: the real quadric $Q_{p,q}$ is the quotient of $S^p \times S^q$ modulo the involution $\tau := \tau_p \times \tau_q$, where τ_n denotes the involution of S^n sending x to $-x$. The composition of the two covering maps is the canonical four-fold cover of $\mathbb{RP}^p \times \mathbb{RP}^q$. The situation is summarized by the central vertical column of Figure 2. Also displayed there are the "diagonal" embedding $\Delta^S\colon S^p \hookrightarrow S^p \times S^q$ that sends x to $(x, (x, 0))$ and the two induced maps $\Delta\colon \mathbb{RP}^p \to Q_{p,q}$ and $\mathbb{RP}^p \to \mathbb{RP}^p \times \mathbb{RP}^q$, respectively. Note that Δ splits the projection $\pi_1\colon Q_{p,q} \to \mathbb{RP}^p$.

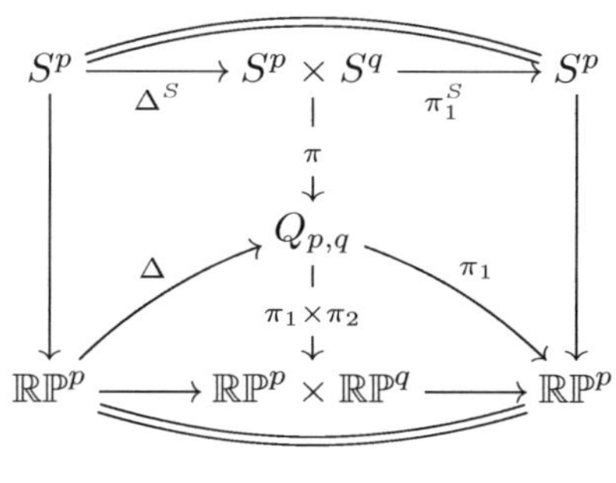

FIGURE 2. The various continuous maps relating spheres, projective spaces and real quadrics

Twisted coefficients. A free rank one local coefficient system over a path-connected space X can be defined as an action of the fundamental group on $\mathbb{Z}$, i.e. as a group homomorphism $\pi_1(X,x) \to \mathrm{Aut}(\mathbb{Z}) = \{\pm 1\}$, for any choice of base point x [**DK01**, § 5.1][**Spa81**, Ex. I.F]. As is customary, we denote the trivial rank one local coefficient system corresponding to the constant group homomorphism $\pi_1(X,x) \to \mathrm{Aut}(\mathbb{Z})$ simply by $\mathbb{Z}$. For semi-locally simply connected X, such coefficient systems correspond to fibre bundles with typical fibre $\mathbb{Z}$ and structure group $\{\pm 1\}$ [**DK01**, Lemma 4.7]. Equivalently, we may identify them with principal $\{\pm 1\}$-bundles, i.e. with two-fold coverings of X.

We write $\mathbb{Z}(1)$ for the rank one local coefficient system $\rho\colon \pi_1(\mathbb{RP}^p) \to \{\pm 1\}$ on $\mathbb{RP}^p$ corresponding to the two-fold cover $S^p \to \mathbb{RP}^p$. For $p \geq 2$, the two-fold cover is the universal cover, and ρ is an isomorphism, but the system $\mathbb{Z}(1)$ also exists for $p = 1$. The situation for the real quadrics is similar. We have a canonical two-fold cover $S^p \times S^q \to Q_{p,q}$, and we again write $\mathbb{Z}(1)$ for the corresponding local coefficient system $\pi_1(Q_{p,q}) \to \{\pm 1\}$. Again, $S^p \times S^q$ is the universal cover for $q \geq p \geq 2$. Note that $\mathbb{Z}(1)$ pulls back to $\mathbb{Z}(1)$ under the maps π_1 and Δ in Figure 2, so the notation is consistent.

More generally, given a coefficient ring R and an integer s, we define $R(s) := R \otimes_{\mathbb{Z}} \mathbb{Z}(1)^{\otimes s}$ with the action of the fundamental group induced by the trivial action on R and the above action on $\mathbb{Z}(1)$. Of course, this really only depends on R and the value of s mod 2. Also note that $\mathbb{Z}/2(s) = \mathbb{Z}/2$ for all s. The direct sum $R \oplus R(1)$ is a $\mathbb{Z}/2$-graded ring, and hence cohomology with local coefficients in $R \oplus R(1)$ is a $\mathbb{Z} \oplus \mathbb{Z}/2$-graded ring, which decomposes additively as

$$\mathrm{H}^\bullet(Q_{p,q}, R \oplus R(1)) = \mathrm{H}^\bullet(Q_{p,q}, R) \oplus \mathrm{H}^\bullet(Q_{p,q}, R(1)).$$

Elements $\alpha \in \mathrm{H}^i(Q_{p,q}, R)$ have degree $|\alpha| = (i, \overline{0})$; elements $\beta \in \mathrm{H}^i(Q_{p,q}, R(1))$ have degree $|\beta| = (i, \overline{1})$. We sometimes also view $R \oplus R$ as a $\mathbb{Z}/2$-graded coefficient ring for $S^p \times S^q$ and

$$\mathrm{H}^\bullet(S^p \times S^q, R \oplus R) = \mathrm{H}^\bullet(S^p \times S^q, R) \oplus \mathrm{H}^\bullet(S^p \times S^q, R)$$

as a $\mathbb{Z} \oplus \mathbb{Z}/2$-graded ring, in the evident way.

Rational cohomology ring. The description of $Q_{p,q}$ as a quotient of $S^p \times S^q$ implies that $\mathrm{H}^\bullet(Q_{p,q}, \mathbb{Q})$ is isomorphic to the subring of $\mathrm{H}^\bullet(S^p \times S^q, \mathbb{Q})$ fixed by τ^* [**Hat02**, Proposition 3G.1]. Similar arguments show that, more generally, $\mathrm{H}^\bullet(Q_{p,q}, \mathbb{Q} \oplus \mathbb{Q}(1))$ is isomorphic to the subring of the $\mathbb{Z} \oplus \{\pm 1\}$-graded ring $\mathrm{H}^\bullet(S^p \times S^q, \mathbb{Q} \oplus \mathbb{Q})$ fixed by the action of τ^* on homogeneous elements in the $+1$-graded part and the action of $-\tau^*$ on the (-1)-graded part. As $\tau_n^*\colon \mathrm{H}^n(S^n) \to \mathrm{H}^n(S^n)$ is given by multiplication with $(-1)^{n+1}$, we find:

$$\mathrm{H}^\bullet(Q_{p,q}, \mathbb{Q} \oplus \mathbb{Q}(1)) \cong \mathbb{Q}[\alpha, \beta]/(\alpha^2, \beta^2) \quad \text{with} \begin{cases} |\alpha| = (p, \overline{p+1}) \\ |\beta| = (q, \overline{q+1}) \end{cases} \tag{3.1}$$

In particular, $Q_{p,q}$ is orientable if and only if $p + q$ is even (use [**Hat02**, Theorem 3.26]).

A (minimal) cell structure. In order to compute the twisted (co)homology of the real quadric $Q_{p,q}$ with integral coefficients, we need to study its two-fold cover $S^p \times S^q$ in more detail.

PROPOSITION 3.2. *Assume $p \leq q$. There exists a cell structure on $S^p \times S^q$ consisting of $4p+4$ cells for which the action of $\tau := \tau_p \times \tau_q$ is cellular: we have characteristic maps $j^{\pm}_{0,0}, \ldots, j^{\pm}_{p,0}$ for cells in dimensions $0, \ldots, p$ and characteristic maps $j^{\pm}_{0,q}, \ldots, j^{\pm}_{p,q}$ for cells in dimensions $q, \ldots, p+q$ such that $\tau \circ j^+_{k,i} = j^-_{k,i}$. The skeleta of $X = S^p \times S^q$ with respect to this cell structure are given by:*

$$\begin{aligned} X^k &= \Delta^S(S^k) && \text{for } 0 \leq k < p \\ X^k &= \Delta^S(S^p) && \text{for } p \leq k < q \\ X^{k+q} &= \Delta^S(S^p) \cup (S^k \times S^q) && \text{for } q \leq k+q \leq p+q \end{aligned}$$

Here, $\Delta^S \colon S^p \to S^p \times S^q$ denotes the map $x \mapsto (x, (x, 0))$, and S^k is viewed as an "equator" of S^p for $k \leq p$, embedded via the first $k+1$ coordinates.

PROOF. We will give an explicit description of this cell structure on $S^p \times S^q$. First, let us recall and make explicit one common cell structure on a single sphere S^n. For $k \in \{0, \ldots, n\}$, let j^+_k denote the following embedding of D^k into S^n:

$$\begin{aligned} j^+_k \colon D^k &\longrightarrow S^n \\ (x_0, \ldots, x_{k-1}) &\mapsto (x_0, \ldots, x_{k-1}, \sqrt{1 - \|x\|^2}, 0, \ldots, 0) \end{aligned}$$

Another embedding of D^k into S^n is given by $j^-_k := \tau_n \circ j^+_k$. Together, these maps j^+_k and j^-_k for $k \in \{0, \ldots, n\}$ define a cell structure on S^n with two cells in each dimension.

Now take $n = p$. We give $\Delta^S(S^p) \subset S^p \times S^q$ the cell structure induced by the cell structure on S^p just described. That is, we consider the characteristic maps

$$j^{\pm}_{k,0} := \Delta^S \circ j^{\pm}_k \colon D^k \to S^p \times S^q. \tag{3.3}$$

Note that $\tau \circ j^+_{k,0} = j^-_{k,0}$ as desired.

The second family of $2p+2$ cells lives in dimensions $q, \ldots, p+q$. In order to define these cells, we need to fix some further notation. Consider the "northern hemisphere" $j^+_q(D^q) \subset S^q$, i.e. the set of all points $x = (x_0, \ldots, x_q) \in S^q$ with $x_q \geq 0$. Let $e_q := (0, \ldots, 0, 1)$ be the "north pole". Choose and fix a continuous map

$$\begin{aligned} R \colon j^+_q(D^q) &\to \mathrm{SO}(q+1) \\ x &\mapsto R_x \end{aligned}$$

with the property that $R_x \cdot e_q = x$. Such a map exists by Lemma 3.7 below.

Also, we fix for each dimension n a map that "wraps the n-disk around the n-sphere", i.e. a surjection $w_n \colon D^n \to S^n$ that sends $0 \in D^n$ to the north pole $e_n \in S^n$ and every point on the boundary ∂D^n to the south pole $-e_n \in S^n$, and that induces a homeomorphism $\bar{w}_n \colon D^n/\partial D^n \cong S^n$. We choose w_n as follows:

$$\begin{aligned} w_n \colon D^n &\to S^n \\ x &\mapsto (s(x) \cdot x, 1 - 2\|x\|), \end{aligned} \tag{3.4}$$

where $s(x)$ is a non-negative scalar determined by the requirement that $w_n(x) \in S^n$. Finally, fix homeomorphisms $\phi_{m,n}$ between products of disks $D^m \times D^n$ and the disks D^{m+n} of the appropriate dimensions:

$$\begin{aligned} \phi_{m,n} \colon D^m \times D^n &\xrightarrow{\cong} D^{m+n} \\ (x, y) &\mapsto \tfrac{\max(\|x\|, \|y\|)}{\|(x,y)\|}(x, y) \end{aligned} \tag{3.5}$$

In view of the homeomorphisms $\phi_{k,q}$, we may define the characteristic maps for the $(k+q)$-cells of $S^p \times S^q$ on the products $D^k \times D^q$. For every $k \in \{0, \dots, p\}$, we define one such characteristic map as follows:

$$\begin{aligned} j^+_{k,q} \colon D^k \times D^q &\to S^p \times S^q \\ (x, y) &\mapsto (j^+_k(x), -R_{(j^+_k(x),0)}(w_q y)) \end{aligned} \tag{3.6}$$

Here, the entry $j^+_k(x)$ in the first coordinate denotes the embedding $j^+_k : D^k \to S^p$, and the entry $(j^+_k(x), 0)$ in the second coordinate denotes the composition of j^+_k with the embedding $x \mapsto (x, 0)$ of S^p into S^q as an "equator". We define a second characteristic map for a $(k+q)$-cell as $j^-_{k,q} := \tau \circ j^+_{k,q}$.

When restricted to the inner points of $D^k \times D^q$, the characteristic maps $j^\pm_{k,q}$ induce homeomorphisms:

$$\overset{\circ}{D}{}^k \times \overset{\circ}{D}{}^q \xrightarrow{\cong} (j^\pm_k(\overset{\circ}{D}{}^k) \times S^q) \setminus \Delta^S(S^p)$$

The images of these homeomorphisms clearly constitute a cover of $(S^p \times S^q) \setminus \Delta^S(S^p)$ by disjoint sets. Thus, altogether the maps $j^\pm_{k,i}$ for $i \in \{0, q\}$ and $k \in \{0, \dots, p\}$ defined in (3.3) and (3.6) constitute a cell structure on $S^p \times S^q$. $\square$

LEMMA 3.7. *There exists a continuous map $R \colon j^+_q(D^q) \to \mathrm{SO}(q+1), x \mapsto R_x$ such that $R_x \cdot e_q = x$ for all x.*

PROOF. Consider the evaluation map $\mathrm{SO}(q+1) \to S^q$ that takes a matrix R to $R \cdot e_q$. This evaluation map constitutes a principal $\mathrm{SO}(q)$-bundle. Over the contractible space $j^+_q(D^q) \cong D^q$, this bundle is necessarily trivial, hence it admits a section. $\square$

Cellular (co)homology. Let $\mathbb{Z}[\tau]$ denote the group ring associated with the group with two elements, i.e. $\mathbb{Z}[\tau] := \mathbb{Z} \oplus \mathbb{Z}\tau$ with $\tau^2 = 1$. Let $\widetilde{C}^{(n)}(\overline{0})$ denote the chain complex of free $\mathbb{Z}[\tau]$-modules

$$\widetilde{C}^{(n)}(\overline{0}) \colon \mathbb{Z}[\tau] \xleftarrow{1-\tau} \mathbb{Z}[\tau] \xleftarrow{1+\tau} \mathbb{Z}[\tau] \xleftarrow{1-\tau} \mathbb{Z}[\tau] \xleftarrow{1+\tau} \cdots \leftarrow \mathbb{Z}[\tau]$$

concentrated in degrees 0 to n. This is the cellular chain complex of S^n with respect to the cell structure arising from the two-fold covering map $S^n \to \mathbb{RP}^n$ and the standard cell structure on $\mathbb{RP}^n$. Multiplication by τ on the complex is the chain map induced by the involution τ_n of S^n. The usual homology of $\mathbb{RP}^n$ with coefficients in $\mathbb{Z}$ is the homology of $\widetilde{C}^{(n)}(\overline{0}) \otimes_{\mathbb{Z}[\tau]} \mathbb{Z}$, where $\mathbb{Z}$ is viewed as trivial $\mathbb{Z}[\tau]$-module; if instead we tensor with the $\mathbb{Z}[\tau]$-module $\mathbb{Z}(1)$ on which τ acts as -1, we obtain the homology of $\mathbb{RP}^n$ with twisted coefficients.

Let $\widetilde{C}^{(n)}(\overline{1})$ denote the complex with the same groups in each degree, but with the roles of multiplication by $1-\tau$ and multiplication by $1+\tau$ reversed:

$$\widetilde{C}^{(n)}(\overline{1}) \colon \mathbb{Z}[\tau] \xleftarrow{1+\tau} \mathbb{Z}[\tau] \xleftarrow{1-\tau} \mathbb{Z}[\tau] \xleftarrow{1+\tau} \mathbb{Z}[\tau] \xleftarrow{1-\tau} \cdots \leftarrow \mathbb{Z}[\tau]$$

PROPOSITION 3.8. *For $p \leq q$, the cellular chain complex associated with the cell structure on $S^p \times S^q$ described in Proposition 3.2 has the form $C(S^p \times S^q) \cong \widetilde{C}^{(p)}(\overline{0}) \oplus \widetilde{C}^{(p)}(\overline{q+1})[q]$, where $\overline{q+1}$ indicates the value of $q+1$ modulo two, and $[q]$ indicates that the second summand is shifted q degrees to the right. Multiplication by τ on the complex is the chain map induced by the involution τ on $S^p \times S^q$.*

PROOF SKETCH. We clearly have a decomposition

$$C(S^p \times S^q) = C(\Delta^S(S^p)) \oplus C'$$

for some complex C' concentrated in degrees $q, \dots, p+q$, except that, a priori, there might be some differentials from C' to $C(\Delta(S^p))$. However, there are no such differentials. This can easily be seen by considering the map on chain complexes induced by the projection $\pi_1^S \colon S^p \times S^q \to S^p$: this map is zero on C' and maps $C(\Delta^S(S^p))$ isomorphically to $C(S^p)$. Thus, the above decomposition is an honest decomposition of chain complexes.

We may identify $C(\Delta^S(S^p))$ with $C(S^p)$ with respect to the usual cell structure on S^p consisting of two cells in each dimension. This is precisely the cell structure used in the usual computation of the homology of $\mathbb{RP}^p$ (cf. [**DK01**, 5.2.1]). The complex C' can be computed in an analogous fashion. The non-zero boundary maps d of C' are $\mathbb{Z}[\tau]$-linear maps between free $\mathbb{Z}[\tau]$-modules of rank one, each with a generator $\{j^+_{k,q}\}$ represented by one of the characteristic maps $j^+_{k,q}$. Write $d\colon C'_{k+q} \to C'_{k-1+q}$ as

$$d\{j^+_{k,q}\} = d_+\{j^+_{k-1,q}\} + d_-\tau\{j^+_{k-1,q}\}.$$

Both coefficients d_+ and d_- can easily be seen to be ± 1. The crucial value we need to know for the homology/cohomology calculations is the *relative* sign $d_+/d_- \in \{\pm 1\}$.

Recall how the coefficients of the boundary map d in the cellular chain complex of a CW complex X are defined. Let X^i denote the i-skeleton. Given a characteristic map $j\colon D^i \to X^i$, write $\pi(j)\colon X^i \twoheadrightarrow S^i$ for the map that sends the complement of the open cell defined by j to the "south pole" $-e_i$ and restricts to the "wrapping map" defined in (3.4) on the open cell itself: $\pi(j)\circ j = w_i$. Given an $(i-1)$-cell with characteristic map j', the coefficient of $\{j'\}$ in $d\{j\}$ is the degree of the composition $\pi(j') \circ j_{|\partial D^i}$.

So consider $X = S^p \times S^q$ with our chosen cell structure. The coefficients $d_\pm$ are the degrees of the following two compositions:

$$f^{++}\colon \partial(D^k \times D^q) \xrightarrow{j^+_{k,q|\partial(D^k\times D^q)}} X^{k-1+q} \xrightarrow{\pi(j^+_{k-1,q})} S^{k-1+q} \tag{3.9}$$

$$f^{+-}\colon \partial(D^k \times D^q) \xrightarrow{j^+_{k,q|\partial(D^k\times D^q)}} X^{k-1+q} \xrightarrow{\pi(j^-_{k-1,q})} S^{k-1+q} \tag{3.10}$$

In order to compare the degrees of these maps, we first describe them more explicitly. Recall from (3.5) our notation $\phi_{m,n}$ for the homeomorphism $D^m \times D^n \cong D^{m+n}$, and let $\bar{\phi}_{m,n}$ denote the induced homeomorphism

$$\bar{\phi}_{m,n}\colon \frac{D^m \times D^n}{\partial(D^m \times D^n)} \xrightarrow{\quad\cong\quad} \frac{D^{m+n}}{\partial D^{m+n}}.$$

Also recall that $\bar{w}_n$ denotes the homeomorphism $D^n/\partial D^n \to S^n$ induced by the wrapping map w_n. Let $w_{m,n}\colon D^m \times D^n \to S^{n+m}$ denote the composition $w_{m,n} := w_{m+n} \circ \phi_{m,n}$, and let $\bar{w}_{m,n}$ denote the following composition in which the first map is the canonical surjection from the product to the smash product:

$$\bar{w}_{m,n}\colon \frac{D^m}{\partial D^m} \times \frac{D^n}{\partial D^n} \to \frac{D^m}{\partial D^m} \wedge \frac{D^n}{\partial D^n} = \frac{D^m \times D^n}{\partial(D^m \times D^n)} \xrightarrow[\bar{\phi}_{m,n}]{\cong} \frac{D^{m+n}}{\partial D^{m+n}} \xrightarrow[\bar{w}_{m+n}]{\cong} S^{m+n}$$

The maps $\pi(j^{\pm}_{k,q})\colon X^{k+q} \to S^{k+q}$ can be described as follows:

$$\pi(j^{+}_{k,q})(x,y) = \begin{cases} \bar{w}_{k,q}\left(x_0,\dots,x_{k-1},\bar{w}_q^{-1}\left(-R^{-1}_{(x,0)}(y)\right)\right) & \text{if } x_k \geq 0 \text{ and } (x,0)\neq y \\ -e_{k+q} & \text{otherwise} \end{cases}$$

$$\pi(j^{-}_{k,q})(x,y) = \begin{cases} \bar{w}_{k,q}\left(-(x_0,\dots,x_{k-1}),\bar{w}_q^{-1}\left(-R^{-1}_{-(x,0)}(y)\right)\right) & \text{if } x_k \leq 0 \text{ and } (x,0)\neq y \\ -e_{k+q} & \text{otherwise} \end{cases}$$

In either line, the coordinates are $x=(x_0,\dots,x_k,0,\dots,0)\in S^p$ and $y=(y_0,\dots,y_q) \in S^q$. Note that only the first k coordinates of x, i.e. the coordinates $(x_0,\dots,x_{k-1})\in D^k$, appear as arguments of $\bar{w}_{k,q}$. For the maps $f^{+\pm}\colon \partial(D^k\times D^q)\to S^{k+q-1}$ in (3.9) and (3.10), we obtain the following description:

(3.11)
$$f^{++}(x,y) = \begin{cases} \bar{w}_{k-1,q}\left(x_0,\dots,x_{k-2},y\right) & \text{if } x_{k-1}\geq 0 \text{ and } \|x\|=1 \\ -e_{k-1+q} & \text{otherwise} \end{cases}$$

(3.12)
$$f^{+-}(x,y) = \begin{cases} \bar{w}_{k-1,q}\big(-(x_0,\dots,x_{k-2}), \\ \qquad \bar{w}_q^{-1}\left(-R^{-1}_{-(x,0)}R_{(x,0)}w_q(y)\right)\big) & \text{if } x_{k-1}\leq 0 \text{ and } \|x\|=1 \\ -e_{k-1+q} & \text{otherwise} \end{cases}$$

Here, the coordinates in either line are $x=(x_0,\dots,x_{k-1})\in D^k$ and $y=(y_0,\dots,y_{q-1}) \in D^q$.

The map f^{+-} we have just computed is homotopic to the following simpler map:

(3.13)
$$\widetilde{f}^{+-}(x,y) = \begin{cases} \bar{w}_{k-1,q}\left(-(x_0,\dots,x_{k-2}),(-1)^{q+1}y_0,y_1,\dots,y_{q-1}\right) & \text{if } x_{k-1}\leq 0 \text{ and } \|x\|=1 \\ -e_{k-1+q} & \text{otherwise} \end{cases}$$

To construct a homotopy $f^{+-}\rightsquigarrow \widetilde{f}^{+-}$, first consider the following map:

$$\begin{aligned} S\colon j^{-}_{q-1}(D^{q-1}) &\to \mathrm{O}(q+1) \\ x &\mapsto S_x := -R^{-1}_{-x}R_x \end{aligned}$$

Here, x has coordinates $(x_0,\dots,x_{q-1},0)\in D^q$ with $x_{q-1}\leq 0$. Note that both x and $-x$ lie in $j^{+}_q(D^q)$, so that both R_x and R_{-x} are defined. It follows from our description of R in Lemma 3.7 that $S_xe_q=e_q$ for each x. So the map S factors through the inclusion of $\mathrm{O}(q)$ into $O(q+1)$ as "upper left block". Let $x\mapsto S'_x\in\mathrm{O}(q)$ denote the restricted map:

$$j^{-}_{q-1}(D^{q-1}) \xrightarrow{S'} \mathrm{O}(q) \hookrightarrow \mathrm{O}(q+1), \qquad \text{composite } S$$

As $j_{q-1}^-(D^{q-1})$ is contractible, the map S' is nullhomotopic. As R_x always has determinant one, the determinant of S'_x can be computed as $\det(S'_x) = \det(S_x) = \det(-\mathrm{id}_{\mathbb{R}^{q+1}}) = (-1)^{q+1}$. We may therefore choose a homotopy from S' to the constant map $j_{q-1}^-(D^{q-1}) \to \mathrm{O}(q)$ with value the $(q \times q)$-diagonal matrix T_q^{q+1}, where

$$T_q^i := \begin{pmatrix} (-1)^i & & & \\ & 1 & & \\ & & \ddots & \\ & & & 1 \end{pmatrix}.$$

Consequently, S is homotopic to the map $j_{q-1}^-(D^{q-1}) \to \mathrm{O}(q+1)$ with constant value $(T_{q+1})^{q+1}$. Denote this homotopy by $S(t)$:

$$\begin{aligned} [0,1] \times j_{q-1}^-(D^{q-1}) &\to \mathrm{O}(q+1) \\ (x,t) &\mapsto S_x(t) \end{aligned}$$

Then $S_x(0) = S_x$, $S_x(1) = (T_{q+1})^{q+1}$, and $S_x(t)e_q = e_q$ for all x and all t. Now consider the homotopy $f\colon [0,1] \times \partial(D^k \times D^q) \to S^{k-1+q}$ defined as follows:

$$f(t)(x,y) := \begin{cases} \bar{w}_{k-1,q}(-(x_0,\dots,x_{k-2}), & \\ \qquad \bar{w}_q^{-1}(S_{(x,0)}(t)\cdot w_q(y))) & \text{if } x_{k-1} \le 0 \text{ and } \|x\| = 1 \\ -e_{k-1+q} & \text{otherwise} \end{cases}$$

This is the homotopy from f^{+-} to $\widetilde{f}^{+-}$ that we need. Indeed, the equality $f(0) = f^{+-}$ is obvious, and the equality $f(1) = \widetilde{f}^{+-}$ easily follows once we observe that $T_{q+1}w_q(y) = w_q(T_q y)$ for our choice of wrapping map w_q (see (3.4)).

It follows in particular that f^{+-} and $\widetilde{f}^{+-}$ have the same degree. On the other hand, we see from the formulas given in (3.11) and (3.13) that $\widetilde{f}^{+-} = f^{++}\sigma$ for the involution $\sigma\colon \partial(D^k \times D^q) \to \partial(D^k \times D^q)$ given by

$$\sigma(x_0,\dots,x_{k-1},y_0,\dots,y_{q-1}) := (-x_0,-x_1,\dots,-x_{k-1},(-1)^{q+1}y_0,y_1,\dots,y_{q-1}).$$

So altogether we find that $d_+/d_- = \deg(f^{++})/\deg(\widetilde{f}^{+-}) = \deg(\sigma) = (-1)^{k+q+1}$. This shows that the cellular chain complex of $S^p \times S^q$ has the form claimed. □

Now let $\mathbb{Z}(0) = \mathbb{Z}$ and $\mathbb{Z}(1)$ denote the trivial $\mathbb{Z}[\tau]$-module and the $\mathbb{Z}[\tau]$-module on which τ acts as multiplication by -1, respectively. Let $C^{(n)}(\bar{t})$ denote the following chain complexes of abelian groups:

$$\begin{aligned} C^{(n)}(\bar{0})&\colon \mathbb{Z} \xleftarrow{0} \mathbb{Z} \xleftarrow{2} \mathbb{Z} \xleftarrow{0} \mathbb{Z} \xleftarrow{2} \cdots \leftarrow \mathbb{Z} \\ C^{(n)}(\bar{1})&\colon \mathbb{Z} \xleftarrow{2} \mathbb{Z} \xleftarrow{0} \mathbb{Z} \xleftarrow{2} \mathbb{Z} \xleftarrow{0} \cdots \leftarrow \mathbb{Z} \end{aligned}$$

These are related to the chain complexes of $\mathbb{Z}[\tau]$-modules considered above by canonical isomorphisms $\widetilde{C}^{(n)}(\bar{t}) \otimes_{\mathbb{Z}[\tau]} \mathbb{Z}(s) \cong C^{(n)}(\overline{s+t})$.

COROLLARY 3.14. *Assume $p \le q$. There exists a cell structure on $Q_{p,q}$ with $2p+2$ cells: one i-cell for each $i \in \{0,\dots,p\}$, and one i-cell for each $i \in \{q,\dots,p+q\}$. The associated chain complexes with coefficients in $\mathbb{Z}(s)$ are given by*

$$C(Q_{p,q},\mathbb{Z}(s)) \cong C^{(p)}(\bar{s}) \oplus C^{(p)}(\overline{s+q+1})[q].$$

This description of $C(Q_{p,q}, \mathbb{Z}(s))$ immediately gives us additive descriptions of the homology and cohomology of $Q_{p,q}$ with arbitrary – untwisted and twisted – coefficients. For example, $\mathrm{H}_\bullet(Q_{p,q}, \mathbb{Z}/2)$ is just a direct sum of two (shifted) copies of $\mathrm{H}_\bullet(\mathbb{RP}^p, \mathbb{Z}/2)$. Table 1 displays the cohomology groups of some split real quadrics with untwisted and twisted integral coefficients.

COROLLARY 3.15. *For coefficients $R \in \{\mathbb{Z}, \mathbb{Z}(1), \mathbb{Z}/2\}$, the embedding $\Delta\colon \mathbb{RP}^p \hookrightarrow Q_{p,q}$ and the projection $\pi_1\colon Q_{p,q} \twoheadrightarrow \mathbb{RP}^p$ induce mutually inverse isomorphisms $\mathrm{H}^i(Q_{p,q}, R) \cong \mathrm{H}^i(\mathbb{RP}^p, R)$ in all degrees $i < p$. In degree p, Δ^* is an epimorphism split by the monomorphism π_1^*.*

PROOF. This is clear from Figure 2 and the fact that any split monomorphism or epimorphism $\mathbb{Z} \to \mathbb{Z}$ or $\mathbb{Z}/2 \to \mathbb{Z}/2$ is an isomorphism. □

COROLLARY 3.16. *For $p < q$ and $s \in \mathbb{Z}$, each cohomology group $\mathrm{H}^i(Q_{p,q}, \mathbb{Z}(s))$ is isomorphic to one of the groups 0, $\mathbb{Z}/2$ or $\mathbb{Z}$. In all degrees i in which $\mathrm{H}^i(Q_{p,q}, \mathbb{Z}(s)) \cong \mathbb{Z}$, the mod-2-reduction map $\mathrm{H}^i(Q_{p,q}, \mathbb{Z}(s)) \to \mathrm{H}^i(Q_{p,q}, \mathbb{Z}/2)$ is a surjection $\mathbb{Z} \twoheadrightarrow \mathbb{Z}/2$. In all degrees i in which $\mathrm{H}^i(Q_{p,q}, \mathbb{Z}(s)) \cong \mathbb{Z}/2$, the mod-2-reduction map is an isomorphism.*

The same assertions are also true in the case $p = q$ in all degrees $i \neq p$. The reduction maps $\mathrm{H}^p(Q_{p,p}, \mathbb{Z}(s)) \to \mathrm{H}^p(Q_{p,p}, \mathbb{Z}/2)$ are epimorphisms or monomorphisms as follows:

$$\begin{aligned} \mathbb{Z} \oplus \mathbb{Z} &\twoheadrightarrow \mathbb{Z}/2 \oplus \mathbb{Z}/2 \quad \text{when } p+s \text{ is odd;} \\ \mathbb{Z}/2 &\hookrightarrow \mathbb{Z}/2 \oplus \mathbb{Z}/2 \quad \text{when } p+s \text{ is even.} \end{aligned}$$

PROOF. This is immediate from the cellular chain complexes. □

REMARK 3.17 (Alternative additive computations). The additive structure of $\mathrm{H}^\bullet(Q_{p,q}, \mathbb{Z} \oplus \mathbb{Z}(1))$ can also be computed using localization sequences, (twisted) Thom isomorphisms and a topological version of the blow-up setup described in Section 4. That is, there is a topological variant of the algebro-geometric computations that will follow, which we leave as an exercise to the diligent reader. A third approach would be to derive the topological results from the geometric ones, using the results of the forthcoming article **[HWXZ]**, which will establish an isomorphism given by the *real realization functor* between the **I**-cohomology ring of any smooth real cellular variety and the singular cohomology ring of its real points (see Theorem 7.1 below). The topological computations presented here are intentionally independent of these considerations.

Mod-two cohomology ring.

PROPOSITION 3.18. *The cohomology ring of the real quadric $Q_{p,q}$ with coefficients in $\mathbb{Z}/2$ has the form*

$$\mathrm{H}^\bullet(Q_{p,q}, \mathbb{Z}/2) = \begin{cases} \mathbb{Z}/2[\xi,\zeta]/(\xi^{p+1}, \zeta^2 + \xi^p\zeta) & \text{if } p = q \text{ and } p \text{ is even} \\ \mathbb{Z}/2[\xi,\zeta]/(\xi^{p+1}, \zeta^2) & \text{in all other cases} \end{cases}$$

with $|\xi| = 1$ and $|\zeta| = q$. For $p < q$, the generators ξ and ζ are the unique non-zero elements of the respective degrees. In the case $p = q$, ξ is the unique non-zero element of degree 1, and ζ is the unique non-zero element in the kernel of $\Delta^*\colon \mathrm{H}^p(Q_{p,p}, \mathbb{Z}/2) \to \mathrm{H}^p(\mathbb{RP}^p, \mathbb{Z}/2)$ (cf. Figure 2).

Proof of 3.18, apart from the computation of ζ^2 when $p = q$. Write h^i for $\mathrm{H}^i(Q_{p,q}, \mathbb{Z}/2)$. Let ξ be the generator of h^1. By Corollary 3.15 and the known cohomology of $\mathbb{RP}^p$, each of the powers ξ^i generates h^i for $i \in \{0, \dots, p-1\}$. Moreover, $\xi^p \in h^p$ is non-zero, and $\xi^{p+1} = 0$. Let $\zeta = \zeta_0$ denote the generator of the kernel of Δ^* on h^q. When $p < q$, $\ker(\Delta^*) = h^q$, so ζ_0 is simply a generator of h^q. When $p = q$, the elements ξ^p and ζ_0 together form a basis of h^p. Pick generators $\zeta_1, \dots, \zeta_p$ in the remaining degrees $h^{1+q}, \dots, h^{p+q}$, so that $|\zeta_i| = i + q$. Poincaré duality implies $\xi^{p-i}\zeta_i = \zeta_p$ for all $i \in \{0, \dots, p\}$. It follows that $\xi^i\zeta_0 = \zeta_i$ for $i \in \{0, \dots, p\}$. In all cases with $p < q$, it is moreover clear for degree reasons that $\zeta^2 = 0$, so in these cases the proof is complete. In the case $p = q$, it remains to compute ζ^2. This will be done in the course of the proof of Theorem 3.20 below. □

Integral cohomology ring. The integral cohomology ring of $\mathbb{RP}^p$ with twisted integral coefficients is as follows:

$$\mathrm{H}^\bullet(\mathbb{RP}^p, \mathbb{Z} \oplus \mathbb{Z}(1)) = \mathbb{Z}[\xi, \alpha]/(2\xi, \xi^{p+1}, \xi\alpha, \alpha^2) \quad \text{with} \begin{cases} |\xi| = (1, \bar{1}) \\ |\alpha| = (p, \overline{p+1}) \end{cases} \tag{3.19}$$

Mod-2-reduction to $\mathrm{H}^\bullet(\mathbb{RP}^p, \mathbb{Z}/2) = \mathbb{Z}/2[\xi]/\xi^{p+1}$ is determined by $\xi \mapsto \xi$ and $\alpha \mapsto \xi^p$.

Theorem 3.20. *The integral cohomology ring of the real quadric $Q_{p,q}$ has the following form:*

$$\mathrm{H}^\bullet(Q_{p,q}, \mathbb{Z} \oplus \mathbb{Z}(1)) \cong \begin{cases} \mathbb{Z}[\xi, \alpha, \beta]/(2\xi, \xi^{p+1}, \xi\alpha, \alpha^2, \beta^2 - \alpha\beta) & \text{if } p = q \text{ and } p \text{ is even} \\ \mathbb{Z}[\xi, \alpha, \beta]/(2\xi, \xi^{p+1}, \xi\alpha, \alpha^2, \beta^2) & \text{in all other cases} \end{cases}$$

with generators of degrees $|\xi| = (1, \bar{1})$, $|\alpha| = (p, \overline{p+1})$, $|\beta| = (q, \overline{q+1})$. The generators ξ and α are pullbacks under π_1 (cf. Figure 2) of the generators of $\mathrm{H}^\bullet(\mathbb{RP}^p, \mathbb{Z} \oplus \mathbb{Z}(1))$ that have the same names in (3.19). Under mod-2-reduction, $\xi \mapsto \xi$, $\alpha \mapsto \xi^p$ and $\beta \mapsto \zeta$. See Table 1 for some examples.

Proof of Theorem 3.20, Part I (integral products and reduction formulas). It is easy to verify that the quotient rings $\mathbb{Z}[\xi, \alpha, \beta]/(\dots)$ displayed in the theorem have the correct additive structure. To verify that the ring structure is correct, we begin by computing the products of the non-torsion classes (α and β).

We first consider the cases with $p < q$. In these cases, non-torsion classes α and β of the degrees specified in Theorem 3.20 are unique up to signs, and we already know from the additive structure that $\mathrm{H}^i(Q_{p,q}, \mathbb{Z}(s)) = 0$ in the bidegrees $(i, s) = (2p, \bar{0})$ and $(i, s) = (2q, \bar{0})$ in which α^2 and β^2 reside. So $\alpha^2 = \beta^2 = 0$. It now follows from (the twisted version of) Poincaré duality (see [**DK01**, Theorem 5.7 and the following remarks]) that the product $\alpha\beta$ is a generator.

The case $p = q$ requires more thought. We need to understand the product

$$\mathrm{H}^p(Q_{p,p}, \mathbb{Z}(p+1)) \times \mathrm{H}^p(Q_{p,p}, \mathbb{Z}(p+1)) \to \mathrm{H}^{2p}(Q_{p,p}, \mathbb{Z}),$$

where $\mathrm{H}^p(Q_{p,p}, \mathbb{Z}(p+1)) = \mathbb{Z} \oplus \mathbb{Z}$. Recall from Figure 2 our notation π for the two-fold cover $\pi\colon S^p \times S^p \to Q_{p,p}$, the notation $\pi_1, \pi_2\colon Q_{p,p} \rightrightarrows \mathbb{RP}^p$ for the compositions of the two-fold cover $Q_{p,p} \to \mathbb{RP}^p \times \mathbb{RP}^p$ with the projections onto the two factors, and $\pi_1^S, \pi_2^S\colon S^p \times S^p \rightrightarrows S^p$ for the projections onto the two sphere factors. Consider the two-fold cover $S^p \to \mathbb{RP}^p$. The induced map on the top cohomology group, $\mathrm{H}^p(S^p, \mathbb{Z}) \leftarrow \mathrm{H}^p(\mathbb{RP}^p, \mathbb{Z}(p+1))$, is multiplication by ± 2, as can be seen either

	0	1	2	3	4	5	6	7	8	9	10	11	
$\mathrm{H}^\bullet(Q_{4,4},\mathbb{Z})$	$\mathbb{Z}1$	0	$\mathbb{z}\xi^2$	0	$\mathbb{z}\xi^4$	$\mathbb{z}\xi\beta$	0	$\mathbb{z}\xi^3\beta$	$\mathbb{Z}\alpha\beta$				$(\beta^2 = \alpha\beta)$
$\mathrm{H}^\bullet(Q_{4,4},\mathbb{Z}(1))$	0	$\mathbb{z}\xi$	0	$\mathbb{z}\xi^3$	$\mathbb{Z}\alpha \oplus \mathbb{Z}\beta$	0	$\mathbb{z}\xi^2\beta$	0	$\mathbb{z}\xi^4\beta$				
$\mathrm{h}^\bullet(Q_{4,4})$	$\mathbb{z}1$	$\mathbb{z}\xi$	$\mathbb{z}\xi^2$	$\mathbb{z}\xi^3$	$\mathbb{z}\xi^4 \oplus \mathbb{z}\zeta$	$\mathbb{z}\xi\zeta$	$\mathbb{z}\xi^2\zeta$	$\mathbb{z}\xi^3\zeta$	$\mathbb{z}\xi^4\zeta$				$(\zeta^2 = \xi^4\zeta)$
$\mathrm{H}^\bullet(Q_{4,5},\mathbb{Z})$	$\mathbb{Z}1$	0	$\mathbb{z}\xi^2$	0	$\mathbb{z}\xi^4$	$\mathbb{Z}\beta$	0	$\mathbb{z}\xi^2\beta$	0	$\mathbb{z}\xi^4\beta$			
$\mathrm{H}^\bullet(Q_{4,5},\mathbb{Z}(1))$	0	$\mathbb{z}\xi$	0	$\mathbb{z}\xi^3$	$\mathbb{Z}\alpha$	0	$\mathbb{z}\xi\beta$	0	$\mathbb{z}\xi^3\beta$	$\mathbb{Z}\alpha\beta$			
$\mathrm{h}^\bullet(Q_{4,5})$	$\mathbb{z}1$	$\mathbb{z}\xi$	$\mathbb{z}\xi^2$	$\mathbb{z}\xi^3$	$\mathbb{z}\xi^4$	$\mathbb{z}\zeta$	$\mathbb{z}\xi\zeta$	$\mathbb{z}\xi^2\zeta$	$\mathbb{z}\xi^3\zeta$	$\mathbb{z}\xi^4\zeta$			
$\mathrm{H}^\bullet(Q_{5,5},\mathbb{Z})$	$\mathbb{Z}1$	0	$\mathbb{z}\xi^2$	0	$\mathbb{z}\xi^4$	$\mathbb{Z}\alpha \oplus \mathbb{Z}\beta$	0	$\mathbb{z}\xi^2\beta$	0	$\mathbb{z}\xi^4\beta$	$\mathbb{Z}\alpha\beta$		$(\beta^2 = 0)$
$\mathrm{H}^\bullet(Q_{5,5},\mathbb{Z}(1))$	0	$\mathbb{z}\xi$	0	$\mathbb{z}\xi^3$	0	$\mathbb{z}\xi^5$	$\mathbb{z}\xi\beta$	0	$\mathbb{z}\xi^3\beta$	0	$\mathbb{z}\xi^5\beta$		
$\mathrm{h}^\bullet(Q_{5,5})$	$\mathbb{z}1$	$\mathbb{z}\xi$	$\mathbb{z}\xi^2$	$\mathbb{z}\xi^3$	$\mathbb{z}\xi^4$	$\mathbb{z}\xi^5 \oplus \mathbb{z}\zeta$	$\mathbb{z}\xi\zeta$	$\mathbb{z}\xi^2\zeta$	$\mathbb{z}\xi^3\zeta$	$\mathbb{z}\xi^4\zeta$	$\mathbb{z}\xi^5\zeta$		$(\zeta^2 = 0)$
$\mathrm{H}^\bullet(Q_{5,6},\mathbb{Z})$	$\mathbb{Z}1$	0	$\mathbb{z}\xi^2$	0	$\mathbb{z}\xi^4$	$\mathbb{Z}\alpha$	0	$\mathbb{z}\xi\beta$	0	$\mathbb{z}\xi^3\beta$	0	$\mathbb{z}\xi^4\beta$	
$\mathrm{H}^\bullet(Q_{5,6},\mathbb{Z}(1))$	0	$\mathbb{z}\xi$	0	$\mathbb{z}\xi^3$	0	$\mathbb{z}\xi^5$	$\mathbb{Z}\beta$	0	$\mathbb{z}\xi^2\beta$	0	$\mathbb{z}\xi^4\beta$	$\mathbb{Z}\alpha\beta$	
$\mathrm{h}^\bullet(Q_{5,6})$	$\mathbb{z}1$	$\mathbb{z}\xi$	$\mathbb{z}\xi^2$	$\mathbb{z}\xi^3$	$\mathbb{z}\xi^4$	$\mathbb{z}\xi^5$	$\mathbb{z}\zeta$	$\mathbb{z}\xi\zeta$	$\mathbb{z}\xi^2\zeta$	$\mathbb{z}\xi^3\zeta$	$\mathbb{z}\xi^4\zeta$	$\mathbb{z}\xi^5$	

TABLE 1. The integral cohomology and the mod-2-cohomology $\mathrm{h}^\bullet := \mathrm{H}^\bullet(-,\mathbb{Z}/2)$ of some real split quadrics, with $\mathbb{z}$ denoting $\mathbb{Z}/2$.

explicitly from the cellular computations or from degree considerations. We may therefore choose generators $\sigma \in \mathrm{H}^p(S^p,\mathbb{Z})$ and $\alpha \in \mathrm{H}^p(\mathbb{RP}^p,\mathbb{Z}(p+1))$ such that α maps to 2σ under this pullback map. Define $\alpha_i := \pi_i^*(\alpha)$ and $\sigma_i := (\pi_i^S)^*\sigma$, so that

$$\pi^*(\alpha_i) = 2\sigma_i.$$

Choose a generator β of $\ker(\Delta^*) \subset \mathrm{H}^p(Q_{p,p},\mathbb{Z}(p+1))$. As π_1^* is split by Δ^*, it follows that α_1, β form a $\mathbb{Z}$-basis of $\mathrm{H}^p(Q_{p,p},\mathbb{Z}(p+1))$. The pullback of β lives in the kernel of $(\Delta^S)^*$, hence can be written as

$$\pi^*\beta = a(\sigma_1 - \sigma_2)$$

for some $a \in \mathbb{Z}$.

We now analyse what happens in cohomological degree $2p$. As the generators α_i are pulled back from $\mathbb{RP}^p$, we find that $\alpha_1^2 = \alpha_2^2 = 0$. In $\mathrm{H}^\bullet(S^p \times S^p,\mathbb{Z})$, we know that similarly $\sigma_1^2 = \sigma_2^2 = 0$, while $\sigma_1\sigma_2$ is a generator in degree $2p$. The pullback $\pi^*\colon \mathrm{H}^{2p}(S^p \times S^p,\mathbb{Z}) \leftarrow \mathrm{H}^{2p}(Q_{p,p},\mathbb{Z})$ is multiplication by ± 2. (Indeed, degree considerations show that π^* is either ± 2 or zero, and by composing with the transfer we find that π^* is either ± 2 or ± 1.) We may therefore choose a generator $\gamma \in \mathrm{H}^{2p}(Q_{p,p},\mathbb{Z})$ such that $\pi^*\gamma = 2\sigma_1\sigma_2$. For this generator, we find $\alpha_1\alpha_2 = 2\gamma$. For the remaining products of α_1 and β, we obtain:

$$\pi^*(\beta^2) = a^2(-\sigma_1\sigma_2 - (-1)^{p^2}\sigma_1\sigma_2) = -a^2\tfrac{1}{2}(1+(-1)^p)\pi^*\gamma,$$
$$\text{so} \quad \beta^2 = -a^2\tfrac{1}{2}(1+(-1)^p)\gamma,$$
$$\pi^*(\alpha_1\beta) = 2a\sigma_1(\sigma_1 - \sigma_2) = -2a\sigma_1\sigma_2 = -a\pi^*\gamma, \qquad \text{so} \quad \alpha_1\beta = -a\gamma$$

Thus, in the basis of $\mathrm{H}^p(Q_{p,p},\mathbb{Z}(p+1))$ given by α_1,β, the product is described by the following matrix:

$$\begin{pmatrix} 0 & -(-1)^p a \\ -a & -a^2\frac{1}{2}(1+(-1)^p) \end{pmatrix}$$

By Poincaré duality, this matrix must define a perfect pairing on $\mathbb{Z}^2$. So $a=\pm 1$, and by changing the sign of β if necessary, we may as well assume that $a=1$. It follows that $\alpha_1\beta$ is a generator of $\mathrm{H}^{2p}(Q_{p,p},\mathbb{Z})$. For even p, we moreover find that $\beta^2=\alpha_1\beta$; for odd p, we find that $\beta^2=0$. Altogether, this is precisely the result displayed above, with α_1 written as α. (We also find that $2\beta=\alpha_1-\alpha_2$.)

We now determine the images of the various generators under mod-2-reduction. As the elements ξ and α are pulled back from cohomology classes of $\mathbb{RP}^p$ under π_1, the formulas for these elements follow from the corresponding formulas for $\mathbb{RP}^p$. The elements $\beta\in\mathrm{H}^p(Q_{p,p},\mathbb{Z}(p+1))$ and $\zeta\in\mathrm{H}^p(Q_{p,p},\mathbb{Z}/2)$ were both chosen as generators of the kernel of the map induced by $\Delta\colon\mathbb{RP}^p\hookrightarrow Q_{p,p}$. As Δ is split by π_1, it follows that $\beta\mapsto\zeta$, as claimed. □

End of proof of Proposition 3.18. We have just observed that β reduces to ζ. The formula for ζ^2 displayed in Proposition 3.18 is therefore immediate from the formula for β^2 that we have already verified. □

Proof of Theorem 3.20, Part II (torsion products). As observed in Corollary 3.16, each cohomology group $\mathrm{H}^i(Q_{p,q},\mathbb{Z}(s))$ is either isomorphic to $\mathbb{Z}/2$ or free abelian. For the proof of Theorem 3.20, it remains to compute the products of two-torsion cohomology classes with arbitrary cohomology classes. So take a two-torsion class $\mu\in\mathrm{H}^i(Q_{p,q},\mathbb{Z}(s))$ and an arbitrary class $\mu'\in\mathrm{H}^{i'}(Q_{p,q},\mathbb{Z}(s'))$. Then the product $\mu\mu'\in\mathrm{H}^{i+i'}(Q_{p,q},\mathbb{Z}(s+s'))$ is again a two-torsion class. If $\mathrm{H}^{i+i'}(Q_{p,q},\mathbb{Z}(s+s'))$ is free abelian, we deduce $\mu\mu'=0$. If, on the other hand, $\mathrm{H}^{i+i'}(Q_{p,q},\mathbb{Z}(s+s'))\cong\mathbb{Z}/2$, then by Corollary 3.16 the reduction map $\mathrm{H}^{i+i'}(Q_{p,q},\mathbb{Z}(s+s'))\to\mathrm{H}^{i+i'}(Q_{p,q},\mathbb{Z}/2)$ is injective. We can therefore compute the product $\mu\mu'$ by passing to the mod-2-reductions $\bar\mu$ and $\bar\mu'$ and computing the product $\bar\mu\bar\mu'$ in $\mathrm{H}^\bullet(Q_{p,q},\mathbb{Z}/2)$, whose ring structure we already know. □

4. The blow-up setup of Balmer-Calmès

We now turn to algebraic geometry. Recall from Section 2 that Q_n denotes the split n-dimensional quadric over a smooth scheme S over a field F, where $n\geq 3$. We begin by summarizing some material from **[Nen09]**.

Lemma 4.1. *Let $\mathcal{N}_{\mathbb{P}^p_x}$ denote the normal bundle of $\mathbb{P}^p_x$ in Q_n. Then* $\det(\mathcal{N}_{\mathbb{P}^p_x})\cong\mathcal{O}_{\mathbb{P}^p_x}(n-p-1)$.

Proof. See **[Nen09**, Lemma 6.1 and above Theorem 6.4]. □

Definition 4.2. An affine space bundle of rank r is a Zariski locally trivial fibre bundle $f\colon X\to Y$ with fibres isomorphic to $\mathbb{A}^r$. This means that $Y=\cup_{i\in I}U_i$ can be covered by open subschemes U_i such that $f^{-1}(U_i)\cong U_i\times\mathbb{A}^r$ over U_i for every $i\in I$.

Lemma 4.3. *The morphism $\rho\colon Q_n-\mathbb{P}^p_x\to\mathbb{P}^p_y$ described in Section 2 is an affine space bundle of rank $p+1$ with $i_y\colon\mathbb{P}^p_y\hookrightarrow Q_n-\mathbb{P}^p_x$ as a global section. When n is even, ρ is even a vector bundle, and i_y is its zero section.*

Proof. This seems to be well-known (see [**Nen09**, Section 6]). As we are not aware of a suitable reference, we provide here a proof for the case of odd n for the reader's convenience. The case of even n is similar. Let $U \cong \mathbb{A}^p$ be the open subscheme of $\mathbb{P}^p_y$ defined by $y_i \neq 0$. We will check that $\rho^{-1}(U) \cong U \times \mathbb{A}^{p+1}$. We may assume that $S = \mathrm{Spec}(F)$, as all schemes are already defined over F. Let us moreover assume for ease of notation that $i = 0$. Then $\rho^{-1}(U)$ is an affine quadric in $\mathbb{A}^{2p+2} \cong \mathbb{P}^{2p+2} - V(y_0)$, defined in terms of the coordinates $[x_0 : \ldots : x_p : 1 : y_1 : \ldots : y_p : z]$ by the equation $x_0 + x_1y_1 + \cdots + x_py_p + z^2 = 0$. This affine quadric is isomorphic to $U \times \mathbb{A}^{p+1}$ via an isomorphism of coordinate rings as follows:

$$\begin{aligned}\frac{F[x_0, x_1, \ldots, x_p, y_1, \ldots, y_p, z]}{(x_0 + x_1y_1 + \cdots + x_py_p + z^2)} &\to F[x_1, \ldots, x_p, y_1, \ldots, y_p, z]\\ x_0 &\mapsto -(x_1y_1 + \cdots + x_py_p + z^2)\\ x_i &\mapsto x_i \quad \text{for } i \neq 0\\ y_i &\mapsto y_i \quad \text{for all } i\end{aligned}$$

This isomorphism is clearly compatible with the projections to U. □

Lemma 4.4. *Let $n \geq 3$. Let $\iota_w\colon \mathbb{P}^p_w \to Q_n$ denote one of the closed embeddings ι_x or ι_y (or $\iota_{x'}$ or $\iota_{y'}$ if n is even). This morphism induces an isomorphism $\iota_w^*\colon \mathrm{Pic}(Q_n) \xrightarrow{\cong} \mathrm{Pic}(\mathbb{P}^p_w)$ under which $\iota_w^*\mathcal{O}_Q(k) \cong \mathcal{O}_{\mathbb{P}^p_w}(k)$. In particular, each of the mentioned embeddings induces the* same *isomorphism on Picard groups.*

Proof. We only deal with $\mathbb{P}^p_y$ here, as the other cases are analogous. Recall that $\mathrm{Pic}(X) \cong \mathrm{CH}^1(X)$ when X is smooth over a field. Consider the pullback map induced by the morphism along the top row of Figure 1:

$$\mathbb{Z}\oplus\mathrm{Pic}(S) \cong \mathrm{CH}^1(\mathbb{P}^p_y) \xleftarrow{i_y^*} \mathrm{CH}^1(Q_n - \mathbb{P}^p_x) \xleftarrow{j^*} \mathrm{CH}^1(Q_n) \xleftarrow{i^*} \mathrm{CH}^1(\mathbb{P}^{n+1}) \cong \mathbb{Z}\oplus\mathrm{Pic}(S)$$

Using Lemma 4.3 and the homotopy invariance of Chow groups, we see that i_y^* is an isomorphism. Using the assumption $n \geq 3$ and the localization sequence of Chow groups, we see that j^* is an isomorphism. As the composition $i \circ j \circ i_y$ is a linear embedding of $\mathbb{P}^p_y$ into $\mathbb{P}^{n+1}$, the composition of all these pullback maps is likewise an isomorphism, sending $\mathcal{O}_{\mathbb{P}^{n+1}}(1)$ to $\mathcal{O}_{\mathbb{P}^p_y}(1)$. The claim follows. □

The closed immersion $\iota\colon \mathbb{P}^p_x \hookrightarrow Q_n$ is a regular immersion of codimension p (if $n = 2p$) or $p+1$ (if $n = 2p+1$), since $\mathbb{P}^p_x$ and Q_n are both smooth (cf. [**EGA4**, Ch. 4, 17.12.1]). Let Bl denote the blow-up of Q_n along $\mathbb{P}^p_x$ and let E denote the exceptional fibre. The following proposition shows that this setup satisfies [**BC09**, Hypothesis 1.2].

Proposition 4.5. *There exists a morphism $\tilde{\rho}\colon Bl \to \mathbb{P}^p_y$ making the following diagram commutative:*

(4.6)

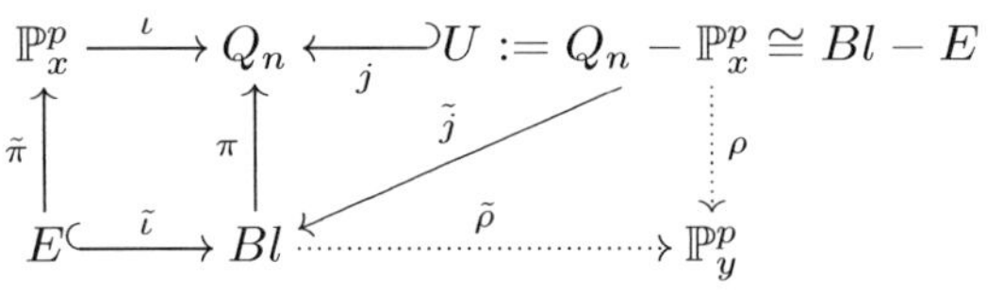

Here, ι and $\tilde{\iota}$ are closed immersions, π and $\tilde{\pi}$ are projections, j and $\tilde{j}$ are open immersions, and ρ is the affine space bundle from Lemma 4.3.

PROOF. We concentrate on the case $n = 2p$; the case of odd n is similar. Then Bl is contained in (in fact equal to) the closed subscheme of

$$Q_n \times \mathbb{P}^p = \operatorname{Proj}(\mathcal{O}_S[x_0, \ldots, x_p, y_0, \ldots, y_p]/(x_0y_0 + \cdots + x_py_p)) \times \operatorname{Proj}(\mathcal{O}_S[T_0, \ldots, T_p])$$

defined by the homogeneous polynomials $y_kT_j - y_jT_k$ (for $k, j = 0, 1, \ldots, p$) and $\sum_{i=0}^{p} x_iT_i = 0$. This follows from the universal property of a blow-up, or from the geometric description of the blow-up along $\mathbb{P}^p_x$ as the closure of the graph of the rational map $Q_n \to \mathbb{P}^d, [x_0 : \ldots : x_p : y_0 : \ldots : y_p] \mapsto [y_0 : \ldots : y_p]$, see [**Har92**, paragraph above Exercise 7.19]. Define $\tilde{\rho}$ to be the composition $Bl \hookrightarrow Q_n \times \mathbb{P}^p \to \mathbb{P}^p_y$, where the first map is the inclusion and the last map is the projection onto the second factor. The relations $y_kT_j = y_jT_k$ guarantee the commutativity of diagram (4.6). □

Diagram (4.6) descends to the following diagram of Picard groups:

$$(4.7)\qquad \begin{array}{ccc} \operatorname{Pic}(Q_n) & \xrightarrow{\quad j^* \simeq \quad} & \operatorname{Pic}(U) \\ {\scriptstyle \binom{1}{0}} \big\downarrow {\scriptstyle \pi^*} & \nearrow^{\tilde{j}^*} & \big\downarrow {\scriptstyle (\rho^*)^{-1} \simeq} \\ \operatorname{Pic}(Q_n) \oplus \mathbb{Z}[E] \cong \operatorname{Pic}(Bl) & \xleftarrow[\tilde{\rho}^*]{\binom{1}{\lambda}} & \operatorname{Pic}(\mathbb{P}^p_y) \end{array}$$

The map $\lambda\colon \operatorname{Pic}(\mathbb{P}^p_y) \to \mathbb{Z}[E]$ does not vanish in general. Hence, the square does *not* generally commute (cf. [**BC09**, Remark 2.2]), but both triangles in this diagram commute. For even n, we now compute the value of $\lambda(\mathcal{O}(1))$ as -1:

PROPOSITION 4.8. *Assume $n = 2p$ with $p \geq 2$. The pullback homomorphism $\tilde{\rho} = \binom{1}{\lambda}\colon \operatorname{Pic}(\mathbb{P}^p_y) \to \operatorname{Pic}(Q_n) \oplus \mathbb{Z}[E]$ sends $\mathcal{O}_{\mathbb{P}^p}(1)$ to $(\mathcal{O}_Q(1), -1)$.*

PROOF. All schemes in sight are smooth over a field, so we can identify all Picard groups with codimension one Chow groups as above. The dimensions of our schemes are $\dim(Bl) = \dim(Q_n) = 2p$ and $\dim(E) = 2p - 1$. The identification of $\mathrm{CH}^1(Bl)$ with $\mathrm{CH}^1(Q_n) \oplus \mathbb{Z}[E]$ is given by $\pi^*\colon \mathrm{CH}^1(Q_n) \to \mathrm{CH}^1(Bl)$ and by $\tilde{\iota}_*\colon \mathrm{CH}^0(E) = \mathbb{Z}[E] \to \mathrm{CH}^1(Bl)$. This identification is obtained by noting that the usual localization sequence of Chow groups is split exact:

$$(4.9)\qquad 0 \to \mathrm{CH}^0(E) \xrightarrow{\tilde{\iota}_*} \mathrm{CH}^1(Bl) \xrightarrow{\tilde{j}^*} \mathrm{CH}^1(U) \to 0$$

To see this, recall that for smooth schemes this exact sequence fits into a long exact sequence of motivic cohomology (cf. [**DI07**, § 3]):

$$\cdots \to \mathrm{H}^{1,1}(Bl) \xrightarrow{\tilde{j}^*} \mathrm{H}^{1,1}(U) \xrightarrow{\partial} \mathrm{CH}^0(E) \xrightarrow{\tilde{\iota}_*} \mathrm{CH}^1(Bl) \xrightarrow{\tilde{j}^*} \mathrm{CH}^1(U) \to 0$$

The sequences breaks down into short split exact sequences because $\tilde{j}^*$ is split surjective via $\tilde{\rho}^* \circ (\rho^*)^{-1}$ for all degrees. The map $\tilde{j}^*\colon \mathrm{CH}^1(Bl) \to \mathrm{CH}^1(U)$ in degree one can also be split by $\pi^* \circ (j^*)^{-1}$.

Next, we describe the maps in diagram (4.7) explicitly in terms of generators. Let us write $[y_0] \in \mathrm{CH}^1(Q_n)$ for the cycle on Q_n corresponding to the subscheme defined by $y_0 = 0$, and analogously for cycles on other schemes. In this notation, each of the groups $\mathrm{CH}^1(\mathbb{P}^p_y)$, $\mathrm{CH}^1(U)$ and $\mathrm{CH}^1(Q_n)$ is generated by the cycle $[y_0]$ corresponding to the line bundle $\mathcal{O}(1)$ in each Picard group. The pullback maps are determined by $\rho^*[y_0] = [y_0] = j^*[y_0]$, $\tilde{\rho}^*[y_0] = [T_0]$ and $\pi^*[y_0] = [y_0]$. By the commutativity of both triangles in diagram (4.7), we see $\tilde{j}^*[T_0] = [y_0] = \tilde{j}^*[y_0]$ in $\mathrm{CH}^1(U)$. So the exact sequence 4.9 shows that

$$[T_0] - [y_0] = \lambda[E] \tag{4.10}$$

in $\mathrm{CH}^1(Bl)$ for some integer λ. This is the integer that we need to compute.

To compute λ, first note that for the closed subschemes $V(y_0)$ and $V(T_0)$ of Bl we have an equality of sets $V(y_0) = V(T_0) \cup E$ with neither of E or $V(T_0)$ contained in one another. Here, the exceptional divisor E is the smooth subscheme defined by $\{([x_0 : \ldots : x_p], [T_0 : \ldots : T_p]) \in \mathbb{P}^p \times \mathbb{P}^p : \sum x_i T_i = 0\}$. In particular, E is integral, hence an irreducible component of $V(y_0)$. Using [**Ful78**, Section 1.5], we conclude

$$[y_0] = [T_0] + \ell[E], \tag{4.11}$$

where ℓ is the length of $\mathcal{O}_{V(y_0),E}$ as a module over itself. Now note that $\mathcal{O}_{V(y_0),E}$ is a field: $\mathcal{O}_{Bl,E}$ is a discrete valuation ring with maximal ideal given by $(y_0, y_1, \cdots, y_p)$; as $y_i = y_0 \frac{T_p}{T_0}$, this ideal is principal, generated by y_0. The relation $y_0 = 0$ in $\mathcal{O}_{V(y_0),E}$ kills this maximal ideal of $\mathcal{O}_{Bl,E}$, so $\mathcal{O}_{V(y_0),E}$ is a field as claimed. It follows that $\ell = 1$.

By comparing (4.10) and (4.11), we conclude that $\lambda = -1$. □

5. I-cohomology: additive structure

We now embark on our computations of **I**-cohomology of split quadrics, keeping the notation established in Section 2 and in the previous section. For the definition and basic properties of **I**- and $\bar{\mathbf{I}}$-cohomology, we refer to [**Fas08**], [**Mor12**], [**AF16**] and the survey lectures of Jean Fasel in these Proceedings [**Fas19**]. In particular, these groups may be described using a variant of the Gersten complex, in which the entries are powers of the fundamental ideal in the Witt ring of the appropriate residue fields. We also refer to [**Fas08**] for twisted coefficients and how they appear when studying pushforwards, and to Lemmas 4.1 and 4.4 for possible twists that appear for split quadrics. Finally, we note that similarly to Chow groups, **I**-cohomology groups satisfy homotopy invariance for affine space bundles over smooth bases:

THEOREM 5.1. *Let $f\colon X \to Y$ be an affine space bundle (recall Definition 4.2) over a smooth variety Y over a base field F of characteristic $\neq 2$. Then*

$$f^*\colon \mathrm{H}^i(Y, \mathbf{I}^j, \mathcal{L}) \to \mathrm{H}^i(X, \mathbf{I}^j, f^*\mathcal{L})$$

is an isomorphism for any line bundle $\mathcal{L}$ on Y.

PROOF. Since Y is quasicompact, we may assume that the open cover in Definition 4.2 is a finite cover by open affines. The pullback along the restriction of f to any affine subset is an isomorphism by [**Fas08**, Corollaire 11.2.8]. Arguing by induction, we are reduced to the following commutative ladder diagram with U affine, in which the rows are exact Mayer-Vietoris sequences:

$$\begin{array}{ccccccc}
\cdots \longrightarrow & \mathrm{H}^i(Y,\mathbf{I}^j,\mathcal{L}) & \longrightarrow & \mathrm{H}^i(U,\mathbf{I}^j,\mathcal{L}|_U)\oplus\mathrm{H}^i(V,\mathbf{I}^j,\mathcal{L}|_V) & \longrightarrow & \mathrm{H}^i(U\cap V,\mathbf{I}^j,\mathcal{L}|_{U\cap V}) & \longrightarrow \cdots \\
& \downarrow{\scriptstyle f^*} & & \downarrow{\scriptstyle (f_U^*,f_V^*)} & & \downarrow{\scriptstyle f_{U\cap V}^*} & \\
\cdots \longrightarrow & \mathrm{H}^i(X,\mathbf{I}^j,f^*\mathcal{L}) & \longrightarrow & \mathrm{H}^i(X_U,\mathbf{I}^j,f^*(\mathcal{L}|_U))\oplus\mathrm{H}^i(X_V,\mathbf{I}^j,f^*(\mathcal{L}|_V)) & \longrightarrow & \mathrm{H}^i(X_{U\cap V},\mathbf{I}^j,f^*(\mathcal{L}|_{U\cap V})) & \longrightarrow \cdots
\end{array}$$

The claim now follows by the five lemma.

Note that the required Mayer-Vietoris sequence can be deduced from the localization sequence [**Fas08**, Théorème 9.3.4] and excision along open embeddings for **I**-cohomology. (More generally, we have excision along flat morphisms, cf. [**CF17**, Lemma 3.7].) □

THEOREM 5.2 (Base change formula [**AF16**, Theorem 2.12]). *Suppose that $f\colon X \to Y$ is a regular codimension c embedding of smooth schemes that fits into a cartesian diagram of smooth schemes as follows:*

$$\begin{array}{ccc} X' & \xrightarrow{\;v\;} & X \\ \Big\downarrow{\scriptstyle g} & & \Big\downarrow{\scriptstyle f} \\ Y' & \xrightarrow{\;u\;} & Y \end{array}$$

Suppose that the natural map on normal bundles $\Omega\colon \mathcal{N}_{X'}Y' \to v^*\mathcal{N}_XY$ induced by u and v is an isomorphism. Then $u^* \circ f_* = g_* \circ \det\Omega^\vee \circ v^*$, i.e. the following square commutes for any line bundle $\mathcal{L}$ over Y:

$$\begin{array}{ccc} \mathrm{H}^i(X', \mathbf{I}^j, v^*f^*\mathcal{L} \otimes v^*\det(\mathcal{N}_XY)^\vee) & \xleftarrow{\;v^*\;} & \mathrm{H}^i(X, \mathbf{I}^j, f^*\mathcal{L} \otimes \det(\mathcal{N}_XY)^\vee) \\ {\scriptstyle \det\Omega^\vee}\Big\downarrow{\scriptstyle\cong} & & \\ \mathrm{H}^i(X', \mathbf{I}^j, g^*u^*\mathcal{L} \otimes \det(\mathcal{N}_{X'}Y')^\vee) & & \Big\downarrow{\scriptstyle f_*} \\ {\scriptstyle g_*}\Big\downarrow & & \\ \mathrm{H}^{i+c}(Y', \mathbf{I}^{j+c}, u^*\mathcal{L}) & \xleftarrow{\;u^*\;} & \mathrm{H}^{i+c}(Y, \mathbf{I}^{j+c}, \mathcal{L}) \end{array}$$

Fasel's computations for projective spaces. Let S be a smooth scheme over a field F of characteristic $\neq 2$. The following isomorphisms of graded abelian groups are proved by Fasel in [**Fas13**]: see Corollary 5.8, Definition 5.9, and Theorems 9.1, 9.2 and 9.4 of loc. cit. (Note that the sequence in 9.4 of loc. cit. splits as the bundle E Fasel considers is trivial in our case.)

THEOREM 5.3 (Fasel).

$$\mathrm{H}^i(\mathbb{P}^p, \mathbf{I}^j, \mathcal{O}(l)) \cong \begin{cases} \Big(\bigoplus\limits_{\substack{m \text{ even} \\ 2\le m\le p}} \mathrm{H}^{i-m}(S, \bar{\mathbf{I}}^{j-m})\Big) \oplus \mathrm{H}^i(S, \mathbf{I}^j) & \text{if } l \text{ is even and } p \text{ is even} \\ \Big(\bigoplus\limits_{\substack{m \text{ even} \\ 2\le m\le p}} \mathrm{H}^{i-m}(S, \bar{\mathbf{I}}^{j-m})\Big) \oplus \mathrm{H}^i(S, \mathbf{I}^j) \oplus \mathrm{H}^{i-p}(S, \mathbf{I}^{j-p}) & \text{if } l \text{ is even and } p \text{ is odd} \\ \Big(\bigoplus\limits_{\substack{m \text{ odd} \\ 1\le m\le p}} \mathrm{H}^{i-m}(S, \bar{\mathbf{I}}^{j-m})\Big) \oplus \mathrm{H}^{i-p}(S, \mathbf{I}^{j-p}) & \text{if } l \text{ is odd and } p \text{ is even} \\ \bigoplus\limits_{\substack{m \text{ odd} \\ 1\le m\le p}} \mathrm{H}^{i-m}(S, \bar{\mathbf{I}}^{j-m}) & \text{if } l \text{ is odd and } p \text{ is odd} \end{cases}$$

Let $p\colon \mathbb{P}^p \to S$ be the projection map and $s\colon S \to \mathbb{P}^p$ a rational point. The isomorphism above on the component $\mathrm{H}^{i-m}(S, \mathbf{I}^{j-m}) \to \mathrm{H}^i(\mathbb{P}^p, \mathbf{I}^j, \mathcal{O}(l))$ is given by the pullback p^* if $l = m = 0$ and by the pushforward s_* if $m = p$ and $l = p - 1$. The map

$$\mu^m_{\mathcal{L}}\colon \mathrm{H}^{i-m}(S, \bar{\mathbf{I}}^{j-m}) \to \mathrm{H}^i(\mathbb{P}^p, \mathbf{I}^j, \mathcal{L}(-m)) \tag{5.4}$$

in the isomorphism above is defined in [**Fas13**, Definition 5.1]; explicitly it is the composition

$$\mathrm{H}^{i-m}(S,\bar{\mathbf{I}}^{j-m}) \xrightarrow{p^*} \mathrm{H}^{i-m}(\mathbb{P}^p,\bar{\mathbf{I}}^{j-m}) \xrightarrow{\partial_{\mathcal{L}(-1)}} \mathrm{H}^{i-m+1}(\mathbb{P}^p,\mathbf{I}^{j-m+1},\mathcal{L}(-1)) \xrightarrow{c(\mathcal{O}(1))^{m-1}} \mathrm{H}^{i}(\mathbb{P}^p,\mathbf{I}^{j},\mathcal{L}(-m))$$

where $\partial_{\mathcal{L}(-1)}$ is the connecting homomorphism (or Bockstein homomorphism) [**Fas13**, § 2.1] and $c(\mathcal{O}(1))$ is the Euler class homomorphism [**Fas13**, § 3].

I-cohomology of completely split quadrics. The aim of this section is to obtain corresponding results for split quadrics, i. e. to determine the **I**-cohomology of split quadrics in all bidegrees, with all twists. Over a field, we will later also compute the ring structure; see Section 6.

We will always assume $n \geq 3$. Recall that $Q_1 \cong \mathbb{P}^1$ (for which the previous computation applies) and that $Q_2 \cong \mathbb{P}^1 \times \mathbb{P}^1$ (using the Segre embedding in $\mathbb{P}^3$).

THEOREM 5.5. *For the split quadric Q_n of dimension $n \geq 3$, we have isomorphisms of groups as follows:*

$$\mathrm{H}^i(Q_n,\mathbf{I}^j) \cong \begin{cases} \Big(\bigoplus_{m\in T_n} \mathrm{H}^{i-m}(S,\bar{\mathbf{I}}^{j-m})\Big) \oplus \mathrm{H}^i(S,\mathbf{I}^j) \oplus \mathrm{H}^{i-n}(S,\mathbf{I}^{j-n}) & \text{if } n=2p \text{ and } p \text{ is even} \\ \Big(\bigoplus_{m\in T_n} \mathrm{H}^{i-m}(S,\bar{\mathbf{I}}^{j-m})\Big) \oplus \mathrm{H}^i(S,\mathbf{I}^j) \oplus \mathrm{H}^{i-p}(S,\mathbf{I}^{j-p})^{\oplus 2} \oplus \mathrm{H}^{i-n}(S,\mathbf{I}^{j-n}) & \text{if } n=2p \text{ and } p \text{ is odd} \\ \Big(\bigoplus_{m\in T_n} \mathrm{H}^{i-m}(S,\bar{\mathbf{I}}^{j-m})\Big) \oplus \mathrm{H}^i(S,\mathbf{I}^j) \oplus \mathrm{H}^{i-p-1}(S,\mathbf{I}^{j-p-1}) & \text{if } n=2p+1 \text{ and } p \text{ is even} \\ \Big(\bigoplus_{m\in T_n} \mathrm{H}^{i-m}(S,\bar{\mathbf{I}}^{j-m})\Big) \oplus \mathrm{H}^i(S,\mathbf{I}^j) \oplus \mathrm{H}^{i-p}(S,\mathbf{I}^{j-p}) & \text{if } n=2p+1 \text{ and } p \text{ is odd} \end{cases}$$

where $T_n := \big\{1 \leq m \leq n : m \text{ even if } 1 \leq m \leq \lfloor \tfrac{n}{2} \rfloor \text{ and } m \text{ odd if } \lceil \tfrac{n}{2} \rceil + 1 \leq m \leq n\big\}$.

$$\mathrm{H}^i(Q_n,\mathbf{I}^j,\mathcal{O}(1)) \cong \begin{cases} \Big(\bigoplus_{m\in U_n} \mathrm{H}^{i-m}(S,\bar{\mathbf{I}}^{j-m})\Big) \oplus \mathrm{H}^{i-p}(S,\mathbf{I}^{j-p})^{\oplus 2} & \text{if } n=2p \text{ and } p \text{ is even} \\ \Big(\bigoplus_{m\in U_n} \mathrm{H}^{i-m}(S,\bar{\mathbf{I}}^{j-m})\Big) & \text{if } n=2p \text{ and } p \text{ is odd} \\ \Big(\bigoplus_{m\in U_n} \mathrm{H}^{i-m}(S,\bar{\mathbf{I}}^{j-m})\Big) \oplus \mathrm{H}^{i-p}(S,\mathbf{I}^{j-p}) \oplus \mathrm{H}^{i-n}(S,\mathbf{I}^{j-n}) & \text{if } n=2p+1 \text{ and } p \text{ is even} \\ \Big(\bigoplus_{m\in U_n} \mathrm{H}^{i-m}(S,\bar{\mathbf{I}}^{j-m})\Big) \oplus \mathrm{H}^{i-p-1}(S,\mathbf{I}^{j-p-1}) \oplus \mathrm{H}^{i-n}(S,\mathbf{I}^{j-n}) & \text{if } n=2p+1 \text{ and } p \text{ is odd} \end{cases}$$

where $U_n := \big\{1 \leq m \leq n : m \text{ odd if } 1 \leq m \leq \lfloor \tfrac{n}{2} \rfloor \text{ and } m \text{ even if } \lceil \tfrac{n}{2} \rceil + 1 \leq m \leq n\big\}$.

The remainder of this section constitutes a proof of this theorem. The proof relies on homotopy invariance (Theorem 5.1), localization and dévissage [**Fas08**, Théorème 9.3.4 and Remarque 9.3.5] for **I**-cohomology. We need to distinguish cases based on the parities of n, p and the twist, so there are eight different cases to consider.

Twisting homomorphisms for quadrics.

DEFINITION 5.6. Let $\lambda_{\mathcal{L}}^m\colon \mathrm{H}^{i-m}(S,\bar{\mathbf{I}}^{j-m}) \to \mathrm{H}^i(Q_n,\mathbf{I}^j,\mathcal{O}(-m))$ denote the following composition:

$$\mathrm{H}^{i-m}(S,\bar{\mathbf{I}}^{j-m}) \xrightarrow{q^*} \mathrm{H}^{i-m}(Q_n,\bar{\mathbf{I}}^{j-m}) \xrightarrow{\partial_{\mathcal{L}(-1)}} \mathrm{H}^{i-m+1}(Q_n,\mathbf{I}^{j-m+1},\mathcal{L}(-1)) \xrightarrow{c(\mathcal{O}(1))^{m-1}} \mathrm{H}^i(Q_n,\mathbf{I}^j,\mathcal{L}(-m))$$

where $\partial_{\mathcal{L}(-1)}$ is the connecting homomorphism (or Bockstein homomorphism) [**Fas13**, § 2.1] and $c(\mathcal{O}(1))$ is the Euler class homomorphism [**Fas13**, § 3].

This homomorphism is analogous to the map $\mu_{\mathcal{L}}^m$ defined by Fasel on projective spaces (5.4).

LEMMA 5.7. *The following diagram commutes.*

$$\begin{array}{ccc}
\mathrm{H}^i(Q_n,\mathbf{I}^j,\mathcal{L}(-m)) & \xrightarrow{j^*} & \mathrm{H}^i(Q_n-\mathbb{P}^p_x,\mathbf{I}^j,\mathcal{L}(-m)) \\
\uparrow{\scriptstyle \lambda_{\mathcal{L}}^m} & & \uparrow{\scriptstyle \rho^*} \\
\mathrm{H}^{i-m}(S,\bar{\mathbf{I}}^{j-m}) & \xrightarrow{\mu_{\mathcal{L}}^m} & \mathrm{H}^i(\mathbb{P}^p_y,\mathbf{I}^j,\mathcal{L}(-m))
\end{array}$$

PROOF. Similarly to $\lambda_{\mathcal{L}}^m$ for Q_n and $\mu_{\mathcal{L}}^m$ for $\mathbb{P}^p$, we can define twist homomorphisms $\delta_{\mathcal{L}}^m\colon \mathrm{H}^{i-m}(S,\bar{\mathbf{I}}^{j-m}) \to \mathrm{H}^i(Q_n-\mathbb{P}^p_x,\mathbf{I}^j,\mathcal{O}(-m))$ for $Q-\mathbb{P}^p_x$. The result then follows as the Bockstein homomorphisms and the Euler class homomorphisms commute with the pullback homomorphisms j^* and ρ^*, cf. [**Fas13**, Proposition 2.1 and § 3]. $\square$

The key short split exact sequence. If $n=2p$ (resp. $n=2p+1$), we defined $q:=p$ (resp. $q:=p+1$). By dévissage, homotopy invariance and localization, we obtain a long exact sequence

$$\cdots \to \mathrm{H}^{i-q}(\mathbb{P}^p_x,\mathbf{I}^{j-q},\omega_x\otimes\mathcal{O}(l)) \xrightarrow{(\iota_x)_*} \mathrm{H}^i(Q_n,\mathbf{I}^j,\mathcal{O}(l)) \xrightarrow{\iota_y^*} \mathrm{H}^i(\mathbb{P}^p_y,\mathbf{I}^j,\mathcal{O}(l)) \xrightarrow{\partial} \mathrm{H}^{i-q+1}(\mathbb{P}^p_x,\mathbf{I}^{j-q},\omega_x\otimes\mathcal{O}(l)) \to \cdots$$

with $\omega_x \cong \mathcal{O}(1-q)$ (see Lemma 4.1) and q the codimension of $\mathbb{P}^p_x$ in Q_n. In this section, we prove the following result:

THEOREM 5.8. *The sequence*

$$0 \longrightarrow \mathrm{H}^{i-q}(\mathbb{P}^p_x,\mathbf{I}^{j-q},\omega_x\otimes\mathcal{O}(l)) \xrightarrow{(\iota_x)_*} \mathrm{H}^i(Q_n,\mathbf{I}^j,\mathcal{O}(l)) \xrightarrow{\iota_y^*} \mathrm{H}^i(\mathbb{P}^p_y,\mathbf{I}^j,\mathcal{O}(l)) \longrightarrow 0$$

is split exact.

PROOF. **Case I.** We consider first the case $n=\dim(Q_n)=2p$ with p even and l even. We will show that the map $\iota_y^*\colon \mathrm{H}^i(Q_n,\mathbf{I}^j) \to \mathrm{H}^i(\mathbb{P}^p_y,\mathbf{I}^j)$ is split surjective. By homotopy invariance, the vector bundle projection ρ of Lemma 4.3 induces an isomorphism $\rho^*\colon \mathrm{H}^i(\mathbb{P}^p_y,\mathbf{I}^j) \to \mathrm{H}^i(Q_n-\mathbb{P}^p_x,\mathbf{I}^j)$. Consider the following commutative diagram:

$$\begin{array}{ccc}
\mathrm{H}^i(Q_n,\mathbf{I}^j) & \xrightarrow{\qquad j^* \qquad} & \mathrm{H}^i(Q_n-\mathbb{P}^p_x,\mathbf{I}^j) \\
\uparrow{\scriptstyle \alpha} & \searrow{\scriptstyle \iota_y^*} & \uparrow{\scriptstyle \rho^*} \\
\Big(\bigoplus\limits_{\substack{2\le m\le p\\ m \text{ even}}} \mathrm{H}^{i-m}(S,\bar{\mathbf{I}}^{j-m})\Big) \oplus \mathrm{H}^i(S,\mathbf{I}^j) & \xrightarrow{\qquad \beta \qquad} & \mathrm{H}^i(\mathbb{P}^p_y,\mathbf{I}^j)
\end{array}$$

The commutativity of the upper right triangle follows from the commutative diagram in Figure 1 and $i_y^* = (\rho^*)^{-1}$. Here, β is the isomorphism $\sum_m \mu^m + p^*$ of [**Fas13**, Theorem 9.1], cf. Definitions 5.1 and 5.9 and Corollary 5.8 of loc. cit. The map α is defined similarly as β, namely as $\alpha := \sum_m \lambda^m + q^*$. The diagram commutes by Lemma 5.7. Since β and ρ^* are both isomorphisms, we obtain a splitting $\alpha \circ \beta^{-1}$ of ι_y^*.

The cases

- $n = 2p+1$, p even, l even
- $n = 2p$, p odd, l odd
- $n = 2p+1$, p odd, l odd

are proved similarly.

Case II. Next, we consider the case $n = \dim(Q_n) = 2p$ with p odd and l even. Let Bl be the blow-up of Q_n along $\mathbb{P}_x^p$ and let E be the exceptional fibre. Recall from Proposition 4.5 above that this setup satisfies Hypothesis 1.2 of [**BC09**]. We use the same notation as in Proposition 4.5. As in the previous case, we want to show that $j^*\colon \mathrm{H}^i(Q_n, \mathbf{I}^j) \to \mathrm{H}^i(Q_n - \mathbb{P}_x^p, \mathbf{I}^j)$ is split surjective. We draw the following diagram:

$$\begin{array}{ccccccc}
\mathrm{H}^i_{\mathbb{P}_x^p}(Q_n,\mathbf{I}^j) & \longrightarrow & \mathrm{H}^i(Q_n,\mathbf{I}^j) & \xrightarrow{j^*} & \mathrm{H}^i(Q_n-\mathbb{P}_x^p,\mathbf{I}^j) & \xrightarrow{\partial} & \mathrm{H}^{i+1}_{\mathbb{P}_x^p}(Q_n,\mathbf{I}^j) \\
 & & \uparrow{\scriptstyle \pi_*} & & \uparrow{\scriptstyle \rho^*\ \cong} & & \\
 & & \mathrm{H}^i(Bl,\mathbf{I}^j,\omega_\pi) & & \mathrm{H}^i(\mathbb{P}_y^p,\mathbf{I}^j) & & \\
 & & \downarrow{\scriptstyle \cong} & \swarrow{\scriptstyle \tilde{\rho}^*} & & & \\
 & & \mathrm{H}^i(Bl,\mathbf{I}^j) & & & &
\end{array}$$

Using [**BC09**, Proposition A.11.(iii)], we find that ω_π is isomorphic to $\mathcal{O}(p-1)$. As $\mathbf{I}$-cohomology is two-periodic in the twist, and as $p-1$ is even, we have an isomorphism as indicated by the lower vertical arrow. We can push forward along π by an analogue of [**BC09**, Proposition 2.1(A)]. Then, arguing as in the proof of [**BC09**, Theorem 2.3], we see that the middle diagram is commutative: as $\lambda(\mathcal{O}) = 0$ and as p is odd, $\lambda(\mathcal{O}) \equiv p-1 \mod 2$, as required.

The case $n = 2p$, p even and l odd is obtained similarly. In this case, we need to use Proposition 4.8 to compute $\lambda(\mathcal{O}(1)) = -1$.

Case III. Now, we consider the case $n = 2p+1 = \dim(Q_n)$ is odd and p is odd and l is even. We show that $\iota_*\colon \mathrm{H}^{i-p-1}(\mathbb{P}_x^p, \mathbf{I}^{j-p-1}, \mathcal{O}(-p)) \to \mathrm{H}^i(Q_n, \mathbf{I}^j)$ is split injective. Consider the following diagram:

$$\begin{array}{ccc}
\mathrm{H}^{i-p-1}(\mathbb{P}_x^p,\mathbf{I}^{j-p-1},\mathcal{O}(-p)) & \xrightarrow{\iota_*} & \mathrm{H}^i(Q_n,\mathbf{I}^j) \\
\downarrow{\scriptstyle \mathrm{red}_{\mathcal{O}(-p)}} & & \downarrow{\scriptstyle \mathrm{red}} \\
\mathrm{H}^{i-p-1}(\mathbb{P}_x^p,\bar{\mathbf{I}}^{j-p-1}) & \xrightarrow{\iota_*} & \mathrm{H}^i(Q_n,\bar{\mathbf{I}}^j)
\end{array}$$

The diagram is commutative since pushforward commutes with mod-2-reduction. The lower horizontal map $\iota_*\colon \mathrm{H}^{i-p-1}(\mathbb{P}_x^p, \bar{\mathbf{I}}^{j-p-1}) \to \mathrm{H}^i(Q_n, \bar{\mathbf{I}}^j)$ is split injective because $\bar{\mathbf{I}}$-cohomology is oriented and because ι_* is part of a localization sequence. Let $s_1\colon \mathrm{H}^i(Q_n, \bar{\mathbf{I}}^j) \to \mathrm{H}^{i-p-1}(\mathbb{P}_x^p, \bar{\mathbf{I}}^{j-p-1})$ denote a splitting. The map $\mathrm{red}_{\mathcal{O}(-p)}$ is also split injective: a splitting is given by the commutative diagram

$$\begin{array}{ccc}
\mathrm{H}^{i-p-1}(\mathbb{P}^p_x, \bar{\mathbf{I}}^{j-p-1}) & \xleftarrow{\mathrm{red}_{\mathcal{O}(-p)}} & \mathrm{H}^{i-p-1}(\mathbb{P}^p_x, \mathbf{I}^{j-p-1}, \mathcal{O}(-p)) \\
\cong \Big\uparrow \sum \bar{\mu}^m & & \sum \mu^m \Big\uparrow \cong \\
\bigoplus_{0 \le m \le p} \mathrm{H}^{i-p-1-m}(S, \bar{\mathbf{I}}^{j-p-1-m}) & \xrightarrow{\mathrm{pr}} & \bigoplus_{m \text{ odd}}^{0 \le m \le p} \mathrm{H}^{i-p-1-m}(S, \bar{\mathbf{I}}^{j-p-1-m})
\end{array}$$

The commutativity and vertical isomorphisms are all established in [**Fas13**]. Let this splitting be denoted by s_2. Now, the composition

$$\mathrm{H}^i(Q_n, \mathbf{I}^j) \xrightarrow{\mathrm{red}} \mathrm{H}^i(Q_n, \bar{\mathbf{I}}^j) \xrightarrow{s_1} \mathrm{H}^{i-p-1}(\mathbb{P}^p_x, \bar{\mathbf{I}}^{j-p-1}) \xrightarrow{s_2} \mathrm{H}^{i-p-1}(\mathbb{P}^p_x, \mathbf{I}^{j-p-1}, \mathcal{O}(-p))$$

provides a splitting for $\iota_*\colon \mathrm{H}^{i-p-1}(\mathbb{P}^p_x, \mathbf{I}^{j-p-1}, \mathcal{O}(-p)) \to \mathrm{H}^i(Q_n, \mathbf{I}^j)$.

The case $n = 2p+1$, p even, l odd is obtained similarly. □

PROOF OF THEOREM 5.5. By Theorem 5.8, we can compute the **I**-cohomology of quadrics from the **I**-cohomology of projective spaces (Theorem 5.3). □

6. I-cohomology: multiplicative structure

In this section, our base S is always the spectrum of a field F of characteristic $\neq 2$. Given a smooth scheme X over F, we define[1]

$$\mathrm{H}^{\mathrm{tot}}(X, \mathbf{I}) := \bigoplus_{\substack{i,j \in \mathbb{Z} \\ \mathcal{L} \in \mathrm{Pic}(X)/2}} \mathrm{H}^i(X, \mathbf{I}^j, \mathcal{L})$$

$$\mathrm{H}^{\star,\bullet}(X, \bar{\mathbf{I}}) := \bigoplus_{i,j \in \mathbb{Z}} \mathrm{H}^i(X, \bar{\mathbf{I}}^j)$$

The $\bar{\mathbf{I}}$-cohomology ring is a commutative $\mathbb{Z}\oplus\mathbb{Z}$-graded ring. The **I**-cohomology ring is $\mathbb{Z} \oplus \mathbb{Z} \oplus (\mathrm{Pic}(X)/2)$-graded, and it is graded commutative in the sense that

$$aa' = (-1)^{ii'} a'a$$

for homogeneous elements of degrees $|a| = (i, j, \bar{l})$ and $|a'| = (i', j', \bar{l}')$, respectively. For $X = \mathbb{P}^n$ or $X = Q_n$, we have $\mathrm{Pic}(X) \cong \mathbb{Z}$, so that the **I**-cohomology ring is a $\mathbb{Z} \oplus \mathbb{Z} \oplus \mathbb{Z}/2$-graded algebra over the $\mathbb{Z} \oplus \mathbb{Z} \oplus \mathbb{Z}/2$-graded coefficient ring

$$\mathbb{I} := \bigoplus_{\substack{i,j \in \mathbb{Z} \\ \bar{l} \in \mathbb{Z}/2}} \mathrm{H}^i(F, \mathbf{I}^j, \bar{l})$$

concentrated in degrees $(0, *, \bar{0})$:

$$\mathrm{H}^i(F, \mathbf{I}^j, l) := \begin{cases} \mathbf{I}^j(F) & \text{if } i = 0 \text{ and } l \equiv 0 \mod 2 \\ 0 & \text{otherwise} \end{cases}$$

Note that we have a graded ideal $\mathbb{I}' := (\mathrm{H}^0(F, \mathbf{I}^1, 0)) \cdot \mathbb{I} \subset \mathbb{I}$.

---

[1]A reader worried about the indexing set $\mathrm{Pic}(X)/2$ may consult [**BC12**].

The ring $\mathrm{H}^{\mathrm{tot}}(\mathbb{P}^p, \mathbf{I})$.

THEOREM 6.1. *Consider $\mathbb{P}^p = \mathbb{P}^p_F$, where F is a field of characteristic $\neq 2$. For any $p \geq 1$, we have a $\mathbb{Z} \oplus \mathbb{Z} \oplus \mathbb{Z}/2$-graded ring isomorphism*

$$\mathrm{H}^{\mathrm{tot}}(\mathbb{P}^p, \mathbf{I}) \cong \mathbb{I}[\xi, \alpha]/(\mathbb{I}'\xi, \xi^{p+1}, \xi\alpha, \alpha^2) \quad \textit{with} \begin{cases} |\xi| = (1, 1, \bar{1}) \\ |\alpha| = (p, p, \overline{p+1}) \end{cases}$$

PROOF. From the additive result of Fasel quoted as Theorem 5.3 above, we conclude:

$$\mathrm{H}^i(\mathbb{P}^p, \mathbf{I}^j, \mathcal{O}(l)) \cong \begin{cases} \bar{\mathrm{I}}^{j-i}(F) & \text{if } 1 \leq i \leq p \text{ and } l \equiv i \mod 2 \\ \mathrm{I}^{j-i}(F) & \text{if } i = 0 \text{ and } l \text{ is even} \\ \mathrm{I}^{j-i}(F) & \text{if } i = p \text{ and } l \equiv -p-1 \mod 2 \\ 0 & \text{otherwise} \end{cases} \tag{6.2}$$

Let ξ be the generator of $\mathrm{H}^1(\mathbb{P}^p, \mathbf{I}^1, \mathcal{O}(-1)) \cong \mathbb{Z}/2$, and let more generally $\xi_m \in \mathrm{H}^m(\mathbb{P}^p, \mathbf{I}^m, \mathcal{O}(-m))$ be the additive generator for $1 \leq m \leq p$. This generator is the image of 1 under the map μ^m (cf. (5.4)).

We first check that the product $\phi\xi_m = p^*\phi \cup \xi_m$ vanishes for an element $\phi \in \mathbf{I}^j(F)$ if and only if $\bar{\phi} = 0 \in \bar{\mathbf{I}}^j(F)$, i.e. if and only if $\phi \in \mathbf{I}^{j+1}(F)$. To see this, consider the following diagram, in which the horizontal maps in the right square are the mod-2-reductions:

$$\begin{array}{ccccc}
\mathrm{H}^0(F, \mathbf{I}^j) \times \mathrm{H}^0(F, \bar{\mathbf{I}}^0) & \xrightarrow{(p^*, \mu^m)} & \mathrm{H}^0(\mathbb{P}^p, \mathbf{I}^j) \times \mathrm{H}^m(\mathbb{P}^p, \mathbf{I}^m, \mathcal{O}(-m)) & \xrightarrow{\mathrm{red} \times \mathrm{red}} & \mathrm{H}^0(\mathbb{P}^p, \bar{\mathbf{I}}^j) \times \mathrm{H}^m(\mathbb{P}^p, \bar{\mathbf{I}}^m) \\
\downarrow \cup & & \downarrow \cup & & \downarrow \cup \\
\mathrm{H}^0(F, \bar{\mathbf{I}}^j) & \xrightarrow{\mu^m} & \mathrm{H}^m(\mathbb{P}^p, \mathbf{I}^{j+m}, \mathcal{O}(-m)) & \xrightarrow{\mathrm{red}} & \mathrm{H}^m(\mathbb{P}^p, \bar{\mathbf{I}}^{j+m})
\end{array}$$

The whole diagram commutes. Indeed, Lemma 5.2 of **[Fas13]** identifies the compositions $\mathrm{red} \circ \mu^m$ with multiplication with ξ^m. (Note that we are applying the lemma in the case when $i = 0$ and Fasel's base scheme X is $\mathrm{Spec}(F)$, the spectrum of a field. The second summand on the right side of the formula given in the lemma is therefore zero since $\partial_{\mathcal{L}}(\alpha)$ lies in the trivial group $\mathrm{H}^1(F, \mathbf{I}^{j+1})$.) The commutativity of the outer square of the diagram therefore follows from the ring structure on $\mathrm{H}^{\star,\bullet}(\mathbb{P}^p, \bar{\mathbf{I}})$. The commutativity of the square in the right half of the diagram is clear. Moreover, we know from the additive computations that both factors in the lower horizontal composition are isomorphisms. In particular, the commutativity of the square on the left follows from the commutativity of the other two squares.

Now suppose $p^*\phi \cup \xi_m = 0$. As $\xi_m = \mu^m(1)$, the commutativity of the left square in the above diagram shows that this is equivalent to the condition $\mu^m(\bar{\phi}) = 0$. As μ^m is an isomorphism, this in turn is equivalent to $\bar{\phi} = 0 \in \bar{\mathbf{I}}^j(F)$, as claimed.

A similar argument as above also shows the commutativity of the following diagram:

$$\begin{array}{ccc}
\mathrm{H}^0(F, \bar{\mathbf{I}}^0) \times \mathrm{H}^0(F, \bar{\mathbf{I}}^0) & \xrightarrow{(\mu^l, \mu^m)} & \mathrm{H}^l(\mathbb{P}^p, \mathbf{I}^l, \mathcal{O}(-l)) \times \mathrm{H}^m(\mathbb{P}^p, \mathbf{I}^m, \mathcal{O}(-m)) \\
\downarrow \cup & & \downarrow \cup \\
\mathrm{H}^0(F, \bar{\mathbf{I}}^0) & \xrightarrow{\mu^{l+m}} & \mathrm{H}^{l+m}(\mathbb{P}^p, \mathbf{I}^{l+m}, \mathcal{O}(-l-m))
\end{array}$$

This shows that $\xi_l\xi_m = \xi_{l+m}$ in the appropriate degrees, and hence that $\xi_m = \xi_1^m$ is an additive generator for $1 \leq m \leq p$. Note that $\xi^{p+1} = 0$ for degree reasons.

Finally, let $s\colon S \to \mathbb{P}^p$ be a rational point, and let α be the image of $1 \in \mathrm{W}(F) = \mathrm{H}^0(F, \mathbf{I}^0)$ under the isomorphism $s_*\colon \mathrm{H}^0(F,\mathbf{I}^0) \to \mathrm{H}^p(\mathbb{P}^p, \mathbf{I}^p, \mathcal{O}(-p-1))$. Again for degree reasons, $\xi\alpha = 0$ and $\alpha^2 = 0$. It remains to check that $\alpha\phi = \alpha \cup p^*\phi$ vanishes for an element $\phi \in \mathrm{H}^i(F, \mathbf{I}^j, k)$ if and only if $\phi = 0$. This is immediate from the identity $\alpha \cup p^*\phi = s_*(\phi)$, which in turn follows from the projection formula of **[CF17]**: for any $\phi \in \mathrm{H}^0(F, \mathbf{I}^j)$, this formula gives $s_*\psi \cup p^*\phi = s_*(\psi \cup s^*p^*\phi)$, and we conclude by taking $\psi = 1$ and noting that $p \circ s = \mathrm{id}$. □

Milnor cohomology and $\bar{\mathbf{I}}$-cohomology.

THEOREM 6.3 (Karpenko-Merkurjev, Dugger-Isaksen). *The Chow rings of the split quadrics Q_n over a field F can be described as follows:*

$$\mathrm{CH}^\bullet(Q_n) \cong \begin{cases} \mathbb{Z}[x,y]/(x^{p+1} - 2xy, y^2 - x^p y) & \text{if } n = 2p \text{ with } p \text{ even} \\ \mathbb{Z}[x,y]/(x^{p+1} - 2xy, y^2) & \text{if } n = 2p \text{ with } p \text{ odd} \\ \mathbb{Z}[x,y]/(x^{p+1} - 2y, y^2) & \text{if } n = 2p+1 \end{cases}$$

where $|x| = 1$ and $|y| = p$ if n is even, $|y| = p+1$ if n is odd.

PROOF. See [**DI07**, Theorems A.4 and A.10]. □

COROLLARY 6.4. *The bigraded Milnor sheaf cohomology rings $\mathrm{H}^{\star,\bullet}(Q_n, \mathbf{K}^M) := \bigoplus_{i,j\in\mathbb{Z}} \mathrm{H}^i(Q_n, \mathbf{K}^M_j)$ of the split quadrics Q_n over a field F can be described as follows:*

$$\mathrm{H}^{\star,\bullet}(Q_n, \mathbf{K}^M) \cong \begin{cases} \mathrm{H}^{\star,\bullet}(F, \mathbf{K}^M)[x,y]/(x^{p+1} - 2xy, y^2 - x^p y) & \text{if } n{=}2p \text{ with } p \text{ even} \\ \mathrm{H}^{\star,\bullet}(F, \mathbf{K}^M)[x,y]/(x^{p+1} - 2xy, y^2) & \text{if } n{=}2p \text{ with } p \text{ odd} \\ \mathrm{H}^{\star,\bullet}(F, \mathbf{K}^M)[x,y]/(x^{p+1} - 2y, y^2) & \text{if } n{=}2p+1 \end{cases}$$

where $|x| = (1,1)$ and $|y| = (p,p)$ if n is even, $|y| = (p+1, p+1)$ if n is odd.

PROOF. In the computation of the bigraded motivic cohomology ring of quadrics [**DI07**, Proposition 4.3] we can replace motivic cohomology with Milnor sheaf cohomology. To pass from the geometric bidegrees to the full bigraded ring, note that $\mathrm{H}^{\star,\bullet}(Q_n, \mathbf{K}^M)$ is a free module over $\mathrm{H}^{\star,\bullet}(F, \mathbf{K}^M)$ with generators in geometric bidegrees $(0,0)$, $(1,1)$, …, (n,n) when n is odd, and in geometric bidegrees $(0,0)$, $(1,1)$, …, (n,n) plus an extra generator in bidegree $(\frac{n}{2}, \frac{n}{2})$ when n is even. This is proved by an analogous argument as in [**DI07**, Proposition 4.1], using localization and homotopy invariance of bigraded Milnor cohomology. □

COROLLARY 6.5. *We have isomorphisms of graded commutative rings as follows:*

$$\mathrm{H}^{\star,\bullet}(Q_n, \bar{\mathbf{I}}) \cong \begin{cases} \mathrm{H}^{\star,\bullet}(F, \bar{\mathbf{I}})[\bar{\xi}, \bar{\beta}]/(\bar{\xi}^{p+1}, \bar{\beta}^2 - \bar{\xi}^p \bar{\beta}) & \text{if } n = 2p \text{ with } p \text{ even} \\ \mathrm{H}^{\star,\bullet}(F, \bar{\mathbf{I}})[\bar{\xi}, \bar{\beta}]/(\bar{\xi}^{p+1}, \bar{\beta}^2) & \text{if } n = 2p+1 \text{ or } (n = 2p \text{ with } p \text{ odd}) \end{cases}$$

Here, $|\bar{\xi}| = (1,1)$ and $|\bar{\beta}| = (q,q)$ where $q = p$ if $n = 2p$ is even and $q = p+1$ if $n = 2p+1$ is odd.

PROOF. By the Milnor conjecture, we can identify $K^M_\bullet(F)/2$ with $\bar{\mathbf{I}}^\bullet(F)$ via the Pfister norm map, and we can further identify the ring $\mathrm{H}^{\star,\bullet}(Q_n, \bar{\mathbf{I}})$ with $\mathrm{H}^{\star,\bullet}(Q_n, \mathbf{K}^M/2)$, cf. [**Fas07**, Proposition 4.11]. The result then follows from Corollary 6.4. □

REMARK 6.6. If n is odd, the generators $\bar{\xi}$ and $\bar{\beta}$ are uniquely defined by the explicit computation of $\bar{\mathbf{I}}$-cohomology above, so there is no ambiguity in choosing the generators. If n is even, the generator $\bar{\xi}$ is still uniquely defined, but the generator $\bar{\beta}$ has an ambiguity because $\mathrm{H}^p(Q_{2p}, \bar{\mathbf{I}}^p) \cong \mathbb{Z}/2 \oplus \mathbb{Z}/2$. Consider the following short split exact sequence:

$$\begin{array}{ccccccccc} 0 & \longrightarrow & \mathrm{H}^0(\mathbb{P}^p_x, \bar{\mathbf{I}}^0) & \xrightarrow{(\iota_x)_*} & \mathrm{H}^p(Q_{2p}, \bar{\mathbf{I}}^p) & \xrightarrow{(\iota_y)^*} & \mathrm{H}^p(\mathbb{P}^p_y, \bar{\mathbf{I}}^p) & \longrightarrow & 0 \\ & & & & \uparrow{\scriptstyle (\iota_{x'})_*} & & \uparrow{\scriptstyle (s_y)_*}\ \cong & & \\ & & & & \mathrm{H}^0(\mathbb{P}^p_{x'}, \bar{\mathbf{I}}^0) & \xrightarrow[\cong]{(s_{x'})^*} & \mathrm{H}^0(F, \bar{\mathbf{I}}^0) & & \end{array}$$

The subschemes $\mathbb{P}^p_y$ and $\mathbb{P}^p_{x'}$ of Q_{2p} intersect transversally in a point, so we have a cartesian square as in diagram (6.10) below. The induced pullback and pushforward morphisms fit into a square as in the diagram above, and the base change formula shows that this square commutes. We choose $\bar{\beta}$ and $\bar{\alpha}$ to be $(\iota_x)_*(1)$ and $(\iota_{x'})_*(1)$, respectively. Note that the pushforward maps of $\bar{\mathbf{I}}$-cohomology and $\mathbf{K}^M/2$-cohomology are compatible via the norm map [**Fas08**, Lemma 10.4.4]. It follows that our generators $\bar{\beta}$ and $\bar{\alpha}$ map to the mod-2-reductions of Dugger-Isaksen's generators α and β in [**DI07**, Section A.5], respectively. Therefore, we have also the relations $\bar{\alpha} + \bar{\beta} = \bar{\xi}^p$, $\bar{\beta}^2 = \bar{\xi}^p\bar{\beta}$ if p is even, and $\bar{\beta}^2 = 0$ if p is odd in $\bar{\mathbf{I}}$-cohomology, as in the proof of [**DI07**, Theorem A.10].

The ring $\mathrm{H}^{\mathrm{tot}}(Q_n, \mathbf{I})$. In the following computations, we use the base change theorem for $\mathbf{I}$-cohomology (Theorem 5.2) and the projection formula [**Fas13**, § 2.2] several times without explicitly mentioning them. This requires even more care than in the additive case when dealing with twists by line bundles, so we will be even more explicit about these.

Let $q = p$ if $n = 2p$ is even and $q = p + 1$ if $n = 2p + 1$ is odd. Recall the split exact sequence

(6.7)
$$0 \longrightarrow \mathrm{H}^{i-q}(\mathbb{P}^p_x, \mathbf{I}^{j-q}, \omega_x \otimes \mathcal{O}(l)) \xrightarrow{(\iota_x)_*} \mathrm{H}^i(Q_n, \mathbf{I}^j, \mathcal{O}(l)) \xrightarrow{\iota_y^*} \mathrm{H}^i(\mathbb{P}^p_y, \mathbf{I}^j, \mathcal{O}(l)) \longrightarrow 0$$

with $\omega_x \cong \mathcal{O}(1-q)$ of Theorem 5.8. Note that, strictly speaking, we have identified $\mathcal{O}(l)$ with $(\iota_y)^*\mathcal{O}(l)$ in this exact sequence. This should not cause any confusion: for any closed immersion $\iota\colon X \to Y$ of smooth closed subschemes of a fixed projective space, we have a *canonical* isomorphism $\iota^*\mathcal{O}_Y(l) \cong \mathcal{O}_X(l)$. We will make this simplification of notation throughout this section whenever possible, without further comment.

THEOREM 6.8. *Let F be a field of characteristic $\neq 2$. For $p \geq 1$, we have a $\mathbb{Z} \oplus \mathbb{Z} \oplus \mathbb{Z}/2$-graded ring (and even $\mathbb{I}$-algebra) isomorphism*

$$\mathrm{H}^{\mathrm{tot}}(Q_{2p+1}, \mathbf{I}) \cong \mathbb{I}[\xi, \alpha, \beta]/(\mathbb{I}'\xi, \xi^{p+1}, \xi\alpha, \alpha^2, \beta^2)$$

Here, the generators are of degrees $|\xi| = (1, 1, \bar{1})$, $|\alpha| = (p, p, \overline{p-1})$, and $|\beta| = (p+1, p+1, \bar{p})$. The reduction map $\mathrm{H}^{\mathrm{tot}}(Q_{2p+1}, \mathbf{I}) \to \mathrm{H}^{\star,\bullet}(Q_{2p+1}, \bar{\mathbf{I}})$, which collapses the $\mathbb{Z}/2$-grading, is given by $\xi \mapsto \bar{\xi}$, $\alpha \mapsto \bar{\xi}^p$ and $\beta \mapsto \bar{\beta}$.

PROOF. From the split exact sequence (6.7) for $n = 2p+1$, or by specializing Theorem 5.5 to the case $S = \mathrm{Spec}(F)$, we obtain:

$$\mathrm{H}^i(Q_{2p+1}, \mathbf{I}^j, \mathcal{O}(l)) = \begin{cases} \bar{\mathrm{I}}^{j-i}(F) & \text{if } 1 \leq i \leq p \text{ and } l \equiv i \mod 2 \\ \bar{\mathrm{I}}^{j-i}(F) & \text{if } p+2 \leq i \leq 2p+1 \text{ and } l \not\equiv i \mod 2 \\ \mathrm{I}^{j-i}(F) & \text{if } (i = p \text{ or } i = p+1) \text{ and } l \not\equiv i \mod 2 \\ \mathrm{I}^{j-i}(F) & \text{if } (i = 0 \text{ or } i = 2p+1) \text{ and } l \equiv i \mod 2 \\ 0 & \text{otherwise} \end{cases}$$

We claim that the reduction map $\mathrm{H}^i(Q_n, \mathbf{I}^j, \mathcal{O}(l)) \to \mathrm{H}^i(Q_n, \bar{\mathbf{I}}^j) \cong \bar{\mathrm{I}}^{j-i}(F)$ is an isomorphism for $1 \leq i \leq p$ and $l \equiv i \mod 2$, and also for $p+2 \leq i \leq 2p+1$ and $l \not\equiv i \mod 2$. Indeed, in the first case, it follows from degree considerations and (6.2) that the map ι_y^* in sequence (6.7) is an isomorphism; in the second case, the map $(\iota_x)_*$ in this sequence is an isomorphism for degree reasons again. Consider the following commutative diagrams:

$$\begin{array}{ccc} \mathrm{H}^i(Q_n, \mathbf{I}^j, \mathcal{O}(l)) & \to & \mathrm{H}^i(Q_n, \bar{\mathbf{I}}^j) \\ \downarrow \iota_x^* & & \downarrow \iota_x^* \\ \mathrm{H}^i(\mathbb{P}^p_x, \mathbf{I}^j, \mathcal{O}(l)) & \to & \mathrm{H}^i(\mathbb{P}^p_x, \bar{\mathbf{I}}^j) \end{array} \qquad \begin{array}{ccc} \mathrm{H}^{i-p-1}(\mathbb{P}^p_y, \mathbf{I}^{j-p-1}, \omega_y \otimes \mathcal{O}(l)) & \to & \mathrm{H}^{i-p-1}(\mathbb{P}^p_y, \bar{\mathbf{I}}^{j-p-1}) \\ \downarrow (\iota_y)_* & & \downarrow (\iota_y)_* \\ \mathrm{H}^i(Q_n, \mathbf{I}^j, \mathcal{O}(l)) & \longrightarrow & \mathrm{H}^i(Q_n, \bar{\mathbf{I}}^j) \end{array}$$

Using Fasel's additive computations for projective spaces (6.2) once again, we see that the lower arrow of the left square is an isomorphism for $1 \leq i \leq p$ and $l \equiv i \mod 2$, and that the upper arrow of the right square is an isomorphism for $p+2 \leq i \leq 2p+1$ and $l \not\equiv i \mod 2$. The claim follows.

Let $\xi \in \mathrm{H}^1(Q_n, \mathbf{I}^1, \mathcal{O}(1)) \cong \mathbb{Z}/2$ be the unique generator, and choose generators α and β of $\mathrm{H}^p(Q_n, \mathbf{I}^p, \mathcal{O}(p-1)) \cong \mathrm{W}(F)$ and $\mathrm{H}^{p+1}(Q_n, \mathbf{I}^{p+1}, \mathcal{O}(p)) \cong \mathrm{W}(F)$, respectively. We see that the reduction map sends ξ to $\bar{\xi}$, α to $\bar{\xi}^p$ and β to $\bar{\beta}$. The relations $\mathbb{I}'\xi$, ξ^{p+1} and $\xi\alpha$ in the ring structure follow from the additive results and by considering reduction to $\mathrm{H}^{\star,\bullet}(Q_n, \bar{\mathbf{I}})$. The relation $\beta^2 = 0$ is clear for degree reasons ($\mathrm{H}^{2p+2}(Q_{2p+1}, \mathbf{I}^{2p+2}, \mathcal{O}(l)) = 0$ since $2p+2 > \dim Q$), and similarly $\alpha^2 = 0$ since $\mathrm{H}^{2p}(Q_{2p+1}, \mathbf{I}^{2p}, \mathcal{O}(2p-2)) = 0$. □

We now consider the even-dimensional split quadrics over F, i.e. Q_n with $n = 2p$. In this case $p = q$ in sequence (6.7). We choose the following additive generators:

- Let $\xi \in \mathrm{H}^1(Q_{2p}, \mathbf{I}^1, \mathcal{O}(-1)) \cong \bar{\mathrm{I}}^0(F) \cong \mathbb{Z}/2$ be the unique generator.
- Let $\beta \in \mathrm{H}^p(Q_{2p}, \mathbf{I}^p, \mathcal{O}(p-1))$ be the image of $1 = \langle 1 \rangle \in \mathrm{W}(F)$ under the following morphism of $\mathrm{W}(F)$-modules:

$$\mathrm{W}(F) \xrightarrow{p^*} \mathrm{H}^0(\mathbb{P}^p_x, \mathbf{I}^0) \xrightarrow{t_x} \mathrm{H}^0(\mathbb{P}^p_x, \mathbf{I}^0, \omega_x \otimes \mathcal{O}(p-1)) \xrightarrow{(\iota_x)_*} \mathrm{H}^p(Q_{2p}, \mathbf{I}^p, \mathcal{O}(p-1))$$

  The second map is defined by an isomorphism $\mathcal{O} \to \omega_x \otimes \mathcal{O}(p-1)$. We will write 1_x for the element $1_x := t_x p^* 1 \in \mathrm{H}^0(\mathbb{P}^p_x, \mathbf{I}^0, \omega_x \otimes \mathcal{O}(p-1))$, and similarly for the other embeddings $\iota_{x'}$, ι_y, etc.
- Let $\alpha \in \mathrm{H}^p(Q_{2p}, \mathbf{I}^p, \mathcal{O}(p-1))$ similarly be defined as $\alpha := (\iota_{x'})_*(1_{x'}) = (\iota_{x'})_* t_{x'} p^*_{x'}(1)$:

$$\mathrm{W}(F) \xrightarrow{p^*} \mathrm{H}^0(\mathbb{P}^p_{x'}, \mathbf{I}^0) \xrightarrow{t_{x'}} \mathrm{H}^0(\mathbb{P}^p_{x'}, \mathbf{I}^0, \omega_{x'} \otimes \mathcal{O}(p-1)) \xrightarrow{(\iota_{x'})_*} \mathrm{H}^p(Q_{2p}, \mathbf{I}^p, \mathcal{O}(p-1)),$$

  where $t_{x'}$ is defined by an isomorphism $\mathcal{O} \to \omega_{x'} \otimes \mathcal{O}(p-1)$.

We will choose the isomorphisms defining t_x and t'_x more carefully in the proof of the following theorem. However, these choices have no effect on the final result.

THEOREM 6.9. *Let F be a field of characteristic $\neq 2$. For any $p \geq 2$, we have a $\mathbb{Z} \oplus \mathbb{Z} \oplus \mathbb{Z}/2$-graded ring (and even $\mathbb{I}$-algebra) isomorphism*

$$\mathrm{H}^{\mathrm{tot}}(Q_{2p}, \mathbf{I}) \cong \begin{cases} \mathbb{I}[\xi, \alpha, \beta]/(\mathbb{I}'\xi, \xi^{p+1}, \xi\alpha + \xi\beta, \alpha^2 + \beta^2, \alpha\beta) & \text{if } p \text{ is even} \\ \mathbb{I}[\xi, \alpha, \beta]/(\mathbb{I}'\xi, \xi^{p+1}, \xi\alpha + \xi\beta, \alpha^2, \beta^2) & \text{if } p \text{ is odd} \end{cases}$$

As can be seen from the explicit definitions above, the generators here have degrees $|\xi| = (1, 1, \overline{1})$ and $|\alpha| = |\beta| = (p, p, \overline{p-1})$. The reduction map $\mathrm{H}^{\mathrm{tot}}(Q_{2p}, \mathbf{I}) \to \mathrm{H}^{\star,\bullet}(Q_{2p}, \bar{\mathbf{I}})$, which collapses the $\mathbb{Z}/2$-grading, is given by $\xi \mapsto \bar{\xi}$, $\beta \mapsto \bar{\beta}$ and $\alpha \mapsto \bar{\xi}^p - \bar{\beta}$.

PROOF. From the split exact sequence (6.7) or by specializing Theorem 5.5, we obtain:

$$\mathrm{H}^i(Q_{2p}, \mathbf{I}^j, \mathcal{O}(l)) = \begin{cases} \bar{\mathrm{I}}^{j-i}(F) & \text{if } 1 \leq i \leq p \text{ and } i \equiv l \mod 2 \\ \bar{\mathrm{I}}^{j-i}(F) & \text{if } p+1 \leq i \leq 2p \text{ and } i \not\equiv l \mod 2 \\ \mathrm{I}^{j-i}(F) \oplus \mathrm{I}^{j-i}(F) & \text{if } i = p \text{ and } i \not\equiv l \mod 2 \\ \mathrm{I}^{j-i}(F) & \text{if } (i = 0 \text{ or } i = 2p) \text{ and } l \equiv i \\ 0 & \text{otherwise} \end{cases}$$

Analogously to the case when $n = 2p+1$ is odd, we note that the reduction map $\mathrm{H}^i(Q_{2p}, \mathbf{I}^j, \mathcal{O}(l)) \to \mathrm{H}^i(Q_{2p}, \bar{\mathbf{I}}^j) \cong \bar{\mathrm{I}}^{j-i}(F)$ is an isomorphism for $1 \leq i \leq p$ and $l \equiv i$, and also for $p+1 \leq i \leq 2p$ and $l \not\equiv i$. The reduction map sends ξ to $\bar{\xi}$, α to $\bar{\xi}^p - \bar{\beta}$ and β to $\bar{\beta}$.

Let us now compute the ring structure of $\mathrm{H}^{\mathrm{tot}}(Q_{2p}, \mathbf{I})$. The relations $\mathbb{I}'\xi, \xi^{p+1}$ and $\xi(\alpha + \beta)$ follow by using the reduction map as before.

Let s_x, s_y, $s_{x'}$ and $s_{y'}$ denote the inclusions of a point into $\mathbb{P}^p$ determined by the following cartesian diagrams:

$$(6.10) \qquad \begin{array}{ccc} \mathrm{Spec}(F) & \xrightarrow{s_y} & \mathbb{P}^p_y \\ \downarrow{\scriptstyle s_{x'}} & & \downarrow{\scriptstyle \iota_y} \\ \mathbb{P}^p_{x'} & \xrightarrow{\iota_{x'}} & Q_{2p} \end{array} \qquad\qquad \begin{array}{ccc} \mathrm{Spec}(F) & \xrightarrow{s_x} & \mathbb{P}^p_x \\ \downarrow{\scriptstyle s_{y'}} & & \downarrow{\scriptstyle \iota_x} \\ \mathbb{P}^p_{y'} & \xrightarrow{\iota_{y'}} & Q_{2p} \end{array}$$

Let s and $\tilde{s}$ denote the rational points $\mathrm{Spec}(F) \to Q_{2p}$ defined by the diagonal of the left and right diagram, respectively. For $i = p$ and $l = 1 - p$, we can use the maps in the left cartesian diagram to make the splitting in Theorem 5.8 explicit:

$$\begin{array}{ccccccccc} 0 & \longrightarrow & \mathrm{H}^0(\mathbb{P}^p_x, \mathbf{I}^{j-p}) & \xrightarrow{(\iota_x)_*} & \mathrm{H}^p(Q_{2p}, \mathbf{I}^j, \mathcal{O}(p-1)) & \xrightarrow{(\iota_y)^*} & \mathrm{H}^p(\mathbb{P}^p_y, \mathbf{I}^j, \mathcal{O}(p-1)) & \longrightarrow & 0 \\ & & & & \uparrow{\scriptstyle (\iota_{x'})_*} & & \uparrow{\scriptstyle (s_y)_*}\ \cong & & \\ & & & & \mathrm{H}^0(\mathbb{P}^p_{x'}, \mathbf{I}^{j-p}) & \xrightarrow[\cong]{(s_{x'})^*} & \mathrm{H}^0(F, \mathbf{I}^{j-p}) & & \end{array}$$

It follows that α and β form a basis of the degree p part $\mathrm{H}^p(Q_{2p}, \mathbf{I}^p, \mathcal{O}(p-1))$ considered as a rank two free $\mathrm{W}(F)$-module. In order to compute the various products of the generators α and β, let us consider a general element

$$(a,b) \in \mathrm{H}^0(\mathbb{P}^p_{x'}, \mathbf{I}^0, \omega_{x'} \otimes \mathcal{O}(p-1)) \times \mathrm{H}^0(\mathbb{P}^p_{y}, \mathbf{I}^0, \omega_{y} \otimes \mathcal{O}(p-1)).$$

Using the base change formula 5.2 for the left cartesian diagram in (6.10) and the projection formula, we find:

$$\begin{aligned}(\iota_{x'})_* a \cup (\iota_y)_* b &= (\iota_{x'})_*\big(a \cup \iota_{x'}^*(\iota_y)_* b\big)\\ &= (\iota_{x'})_*\big(a \cup (s_{x'})_* s_y^* b\big)\\ &= (\iota_{x'})_*(s_{x'})_*\big(s_{x'}^* a \cup s_y^* b\big)\\ &= s_*(s_{x'}^* a \cup s_y^* b)\end{aligned} \tag{6.11}$$

Similarly, using the right diagram in (6.10), we find

$$(\iota_x)_* a \cup (\iota_{y'})_* b = \tilde{s}_*(s_x^* a \cup s_{y'}^* b).$$

When $n = 2p$ with odd p, we consider the $\mathbb{A}^1$-homotopy $h\colon \mathbb{P}^p \times \mathbb{A}^1 \to Q_{2p} \times \mathbb{A}^1$ given by

$$\begin{aligned}&h([a_0, a_1, \dots, a_p], t)\\ &\quad := ([ta_1, -ta_0, \dots, ta_p, -ta_{p-1}; (1-t)a_0, (1-t)a_1, \dots, (1-t)a_{p-1}, (1-t)a_p], t).\end{aligned}$$

This homotopy fits into the following commutative diagram consisting of two cartesian squares:

$$\begin{array}{ccc} \mathbb{P}^p & \xrightarrow{\;h_0\;} & Q_{2p} \\ {\scriptstyle \mathrm{id}\times 0}\big\downarrow & & \big\downarrow{\scriptstyle \mathrm{id}\times 0} \\ \mathbb{P}^p \times \mathbb{A}^1 & \xrightarrow{\;h\;} & Q_{2p} \times \mathbb{A}^1 \\ {\scriptstyle \mathrm{id}\times 1}\big\uparrow & & \big\uparrow{\scriptstyle \mathrm{id}\times 1} \\ \mathbb{P}^p & \xrightarrow[\;h_1\;]{} & Q_{2p} \end{array} \tag{6.12}$$

Note that $h_0 = \iota_y$ and $h_1 = \iota_x \phi$ for some linear change of coordinates $\phi\colon \mathbb{P}^p \to \mathbb{P}^p$. Using the base change formula and homotopy invariance, we find that the pushforward maps along h_0 and h_1 coincide up to an identification of the respective normal bundles. Thus, up to an identification of the respective normal bundles, $(\iota_x)_*$ and $(\iota_y)_*$ also agree up to some unit in $\mathrm{H}^0(\mathbb{P}^p, \mathbf{I}^0) \cong \mathrm{W}(F)$ depending on ϕ. To be more precise, choose an isomorphism $t_h\colon \mathcal{O} \to \omega_h \otimes \mathcal{O}(1-p)$. Define $t_i := (\mathrm{id} \times i)^* t_h$ for $i \in \{0,1\}$, and choose the isomorphisms t_x and t_y in the definitions of α and β above Theorem 6.9 such that $t_0 = t_y$ and $t_1 = \phi^* t_x$. Then $(\iota_x)_* t_x$ and $(\iota_y)_* t_y$ agree up to multiplication with a unit in $\mathrm{W}(F)$.

By combining the identification of $(\iota_x)_* t_x$ with $(\iota_y)_* t_y$ and formula (6.11), we find that $\alpha\beta$ is a generator in degree $2p$. (Note that s_* in (6.11) is an isomorphism by the previous additive computations.) The following identities moreover prove the relation $\alpha^2 = 0$. We write "$\sim$" for "equal up to multiplication by a unit":

(6.13)
$$\begin{array}{rcll}(\iota_x)_*1_x \cup (\iota_x)_*1_x & \sim & (\iota_x)_*1_x \cup (\iota_y)_*1_y & \text{by the } \mathbb{A}^1\text{-homotopy} \\ & = & (\iota_x)_*(1_x \cup \iota_x^*(\iota_y)_*1_y) & \text{by the projection formula} \\ & = & 0 & \text{since } \iota_x^*(\iota_y)_*1_y = 0 \text{ by sequence (6.7)}\end{array}$$

Similarly, we obtain the relation $\beta^2 = 0$.

When $n = 2p$ with even p, we consider the following two $\mathbb{A}^1$-homotopies $\mathbb{P}^p \times \mathbb{A}^1 \to Q_{2p} \times \mathbb{A}^1$:

$$\begin{aligned}&h([a_0, a_1, \ldots, a_p], t) \\ &\qquad := ([a_0, ta_1, ta_2, \ldots, ta_p; 0,\ (1-t)a_2, (t-1)a_1, \ldots, (1-t)a_p, (t-1)a_{p-1}], t) \\ &\tilde{h}([a_0, a_1, \ldots, a_p], t) \\ &\qquad := ([0,\ ta_1, ta_2, \ldots, ta_p; a_0, (1-t)a_2, (t-1)a_1, \ldots, (1-t)a_p, (t-1)a_{p-1}], t)\end{aligned}$$

Each homotopy fits into a commutative diagram of the form (6.12). Note that $h_1 = \iota_x$, $h_0 = \iota_{y'}\phi$, $\tilde{h}_1 = \iota_{x'}$ and $\tilde{h}_0 = \iota_y\tilde{\phi}$ for some some linear changes of coordinates $\phi, \tilde{\phi}\colon \mathbb{P}^p \to \mathbb{P}^p$.

As in the previous case, we choose isomorphisms $t\colon \mathcal{O} \to \omega_h \otimes \mathcal{O}(1-p)$ and $\tilde{t}\colon \mathcal{O} \to \omega_{\tilde{h}} \otimes \mathcal{O}(1-p)$ and we define $t_i := (\mathrm{id} \times i)^* t_h$ and $\tilde{t}_i := (\mathrm{id} \times i)^*\tilde{t}$ for $i \in \{0, 1\}$. Then $(h_1)_*t_1 = (h_0)_*t_0$ and $(\tilde{h}_1)_*\tilde{t}_1 = (\tilde{h}_0)_*\tilde{t}_0$ as maps $\mathrm{H}^0(\mathbb{P}^p, \mathbf{I}^0) \to \mathrm{H}^p(Q_{2p}, \mathbf{I}^p, \mathcal{O}(p-1))$. We choose t_x, t_y, $t_{x'}$ and $t_{y'}$ such that $t_1 = t_x$, $t_0 = \phi^* t_{y'}$, $\tilde{t}_1 = t_{x'}$ and $\tilde{t}_0 = \tilde{\phi}^* t_y$. Then we find that $(\iota_x)_*t_x$ and $(\iota_{y'})_*t_{y'}$ agree up to multiplication by a unit in $\mathrm{W}(F)$, and likewise $(\iota_{x'})_*$ and $(\iota_y)_*$ agree up to multiplication by a unit.

From a computation analogous to (6.13) and the above $\mathbb{A}^1$-homotopies, we may now deduce the relation $\alpha\beta = 0$. Similarly, using the homotopy h and diagram (6.11), we find that α^2 is the additive generator $s_*(t_0 1 \cup t_1 1)$ in degree $2p$. Using the homotopy $\tilde{h}$ and (6.11) with x' and y replaced by x and y', respectively, we find that β^2 is the additive generator $\tilde{s}_*(\tilde{t}_0 1 \cup \tilde{t}_1 1)$ in degree $2p$. (To be precise, in these expressions for α^2 and β^2 the isomorphisms "t_i" and "$\tilde{t}_i$" really denote the pullbacks of the respective trivializations t_i from $\mathbb{P}^p$ to the base. For example, for $t_1 = t_x$ we use the following commutative diagram:

$$\begin{array}{ccc}\mathrm{W}(F) & \xrightarrow{\quad p^* \quad} & \mathrm{H}^0(\mathbb{P}^p, \mathbf{I}^0) \\ \Big\downarrow{\scriptstyle s_x^* t_x} & & \Big\downarrow{\scriptstyle t_x} \\ \mathrm{W}(F, s_x^*\omega_x(p-1)) & \xrightarrow{\ p^*\ } & \mathrm{H}^0(\mathbb{P}^p, \omega_x \otimes \mathcal{O}(p-1))\end{array}$$

However, we drop s_x^* from our notation for simplicity.)

To compare $\alpha^2 = s_*(t_0 1 \cup t_1 1)$ and $\beta^2 = \tilde{s}_*(\tilde{t}_0 1 \cup \tilde{t}_1 1)$, we consider the following involution $\tau\colon \mathbb{P}^{2p+1} \to \mathbb{P}^{2p+1}$:

$$\tau([x_0 : x_1 : \cdots : x_p; y_0 : y_1 \cdots : y_p]) := [y_0 : x_1 : \cdots : x_p; x_0 : y_1 : \cdots : y_p]$$

This involution can be restricted to an involution of Q_{2p}, which we still denote by τ. It fits into the following commutative diagram of cartesian squares:

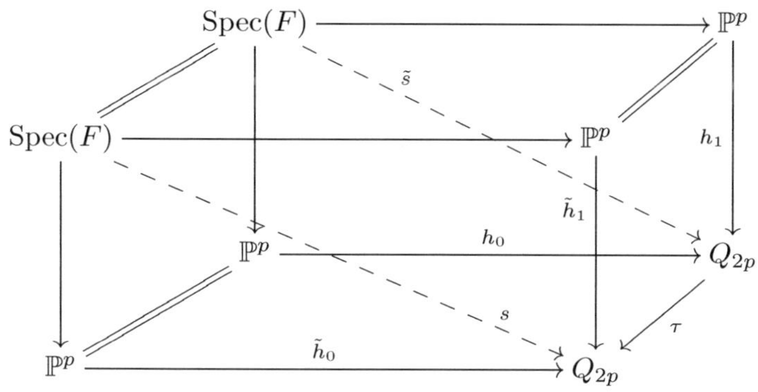

As an automorphism of $\mathbb{P}^{2p+1}$, τ induces a canonical isomorphism of line bundles $\tau^*\mathcal{O}(1) \cong \mathcal{O}(1)$. This isomorphism of line bundles restricts to a canonical isomorphism between the respective line bundles $\mathcal{O}(1)$ and $\tau^*\mathcal{O}(1)$ on Q_{2p}, and we implicitly use this canonical identification in the following.

The product $\tau \times \mathrm{id}$ relates the homotopies h and $\tilde{h}$, in the sense that we also have the following commutative diagram:

$$\begin{array}{ccc} \mathbb{P}^p \times \mathbb{A}^1 & \xrightarrow{h} & Q_{2p} \times \mathbb{A}^1 \\ \downarrow{\scriptstyle \tilde{h}} & \swarrow{\scriptstyle \tau\times\mathrm{id}} & \\ Q_{2p} \times \mathbb{A}^1 & & \end{array}$$

In particular, $\tau \times \mathrm{id}$ induces a canonical isomorphism of the relative canonical line bundles of h and $\tilde{h}$, $\Omega_{\tau\times\mathrm{id}} \colon \omega_h \to \omega_{\tilde{h}}$. Changing our (previously arbitrary) choices of t and $\tilde{t}$ if necessary, we may assume that the following diagram of bundles over $\mathbb{P}^p \times \mathbb{A}^1$ is commutative:

$$\begin{array}{ccc} \mathcal{O} & \xrightarrow{t} & \omega_h \otimes \mathcal{O}(1-p) \\ \downarrow{\scriptstyle \tilde{t}} & \swarrow{\scriptstyle \Omega_\tau} & \\ \omega_{\tilde{h}} \otimes \mathcal{O}(1-p) & & \end{array}$$

Consider the map $\tau^* \colon \mathrm{H}^{2p}(Q_{2p}, \mathbf{I}^{2p}, \mathcal{O}(2p-2)) \to \mathrm{H}^{2p}(Q_{2p}, \mathbf{I}^{2p}, \mathcal{O}(2p-2))$ induced by τ and the canonical identification $\tau^*\mathcal{O}(1) \cong \mathcal{O}(1)$. By the base change formula, we find that

$$\tau^* s_*(t_0 1 \cup t_1 1) = \tilde{s}_* \mathrm{id}^*(\tilde{t}_0 1 \cup \tilde{t}_1 1)$$

in $\mathrm{H}^{2p}(Q_{2p}, \mathbf{I}^{2p}, \mathcal{O}(2p-2))$, i.e. $\tau^*\alpha^2 = \beta^2$. We claim that the map τ^* is precisely $-\mathrm{id}$, so that we obtain $\alpha^2 = -\beta^2$.

To prove the claim, we first note that the pushforward map along the inclusion $i \colon Q_{2p} \hookrightarrow \mathbb{P}^{2p+1}$ induces an isomorphism

$$i_* \colon \mathrm{H}^{2p}(Q_{2p}, \mathbf{I}^{2p}, \omega_i \otimes i^*\mathcal{L}) \to \mathrm{H}^{2p+1}(\mathbb{P}^{2p+1}, \mathbf{I}^{2p+1}, \mathcal{L})$$

for any even twist $\mathcal{L}$. This can be seen from the additive computation above and the identification of the relative canonical line bundle ω_i of the inclusion as $\omega_i \cong \mathcal{O}(-2)$.

The study of τ^* on Q_{2p} can therefore be translated to the study of τ^* on $\mathbb{P}^{2p+1}$. Consider a fixed point of τ on $\mathbb{P}^{2p+1}$:

$$\begin{array}{ccc} \mathrm{Spec}(F) & \xrightarrow{\mathrm{pt}} & \mathbb{P}^{2p+1} \\ \downarrow{=} & & \downarrow{\tau} \\ \mathrm{Spec}(F) & \xrightarrow{\mathrm{pt}} & \mathbb{P}^{2p+1} \end{array}$$

The base change formula 5.2 implies that $\tau^*\mathrm{pt}_*$ differs from pt_* by multiplication with the determinant of the map on normal bundles $\mathcal{N}_{\mathrm{pt}} \to \mathcal{N}_{\mathrm{pt}}$ induced by τ. It is not hard to see that this determinant is just the determinant of the matrix of the linear change of coordinates τ, which is -1. □

7. Geometric bidegrees and real realization

We now compare Theorems 6.8 and 6.9 with Theorem 3.20. They both apply to the split quadrics Q_n with $n = (p,p)$ or $n = (p+1,p)$ and $n \geq 3$, while the latter theorem also holds more generally. The $\mathbb{Z}\oplus\mathbb{Z}\oplus\mathbb{Z}/2$-graded ring $\mathrm{H}^{\mathrm{tot}}(X,\mathbf{I})$ contains the $\mathbb{Z}\oplus\mathbb{Z}/2$-graded subring (even a $\mathrm{W}(F)$-algebra) given by the condition that the two $\mathbb{Z}$-degrees coincide. This corresponds to motivic bidegrees $(2i,i)$ and hence to Chow and Chow-Witt groups. Following **[HW19]**, **[HWXZ]** and others, we denote this important sub-$\mathrm{W}(F)$-algebra by $\mathrm{H}^\bullet(X,\mathbf{I}^\bullet,\mathcal{O}\oplus\mathcal{O}(1))$. The computations of Sections 5 and 6 easily restrict to these $\mathrm{W}(F)$-submodules/subalgebras. For example, restricting Theorem 6.8 to $\mathrm{H}^\bullet(Q_n,\mathbf{I}^\bullet,\mathcal{O}\oplus\mathcal{O}(1))$ means replacing $\mathbb{I}$ by $\mathrm{W}(F)$ and $\mathbb{I}'$ by $\mathrm{I}(F)$.

Now if $\mathrm{W}(F) \cong \mathbb{Z}$, e.g. if $F = \mathbb{R}$, then there is an explicit ring isomorphism from this graded commutative ring $\mathrm{H}^\bullet(Q_n,\mathbf{I}^\bullet,\mathcal{O}\oplus\mathcal{O}(1))$ in algebraic geometry to the graded commutative ring $\mathrm{H}^\bullet(X(\mathbb{R})^{an},\mathbb{Z}\oplus\mathbb{Z}(1))$ in topology, given by mapping α to $\alpha-\beta$ if $n = 2p$, and the identity on all other generators. Moreover, there is another obvious isomorphism between the corresponding rings with $\mathbb{Z}/2$-coefficients, and both isomorphisms are compatible via the reduction maps.

None of this is a coincidence, of course. As observed in Remark 3.17, the strategy of our computation of the $\mathbf{I}$-cohomology of the algebro-geometric Q_n also lends itself to a computation of the integral singular cohomology $\mathrm{H}^\bullet(Q_n,\mathbb{Z}\oplus\mathbb{Z}(1))$ of the topological real quadric. Even more is true. We recall from **[Jac17]** that there is a real realization functor

$$\mathrm{H}^\bullet(X,\mathbf{I}^\bullet) \to \mathrm{H}^\bullet(X(\mathbb{R})^{an},\mathbb{Z})$$

for a smooth variety X over $\mathbb{R}$, which is known to be an isomorphism in some degrees in special cases, see loc. cit. and previous work by Fasel and others. By recent work of the authors and Matthias Wendt **[HWXZ]**, we have a stronger result for smooth cellular varieties.

THEOREM 7.1. *For any smooth variety X over $\mathbb{R}$, the real realization morphism*

$$\mathrm{H}^\bullet(X,\mathbf{I}^\bullet) \to \mathrm{H}^\bullet(X(\mathbb{R})^{an},\mathbb{Z})$$

is a ring homomorphism. If X is cellular, then this is an isomorphism.

PROOF. See **[HWXZ]**. □

This applies in particular to the split quadrics Q_n. The preprint [**HWXZ**] also studies twisted coefficients and establishes a generalization of Theorem 7.1, replacing $\mathbf{I}^\bullet$ and $\mathbb{Z}$ by $(\mathbf{I}^\bullet, \mathcal{O} \oplus \mathcal{O}(1))$ and $\mathbb{Z} \oplus \mathbb{Z}(1)$, respectively. We deduce that over $F = \mathbb{R}$ the description of Theorems 6.8 and 6.9 extends to all quadrics covered by Theorem 3.20, in particular to $\mathbb{P}^1 \times \mathbb{P}^1$. However, the Picard group is larger in this case, so we have not computed cohomology with respect to all possible twists for $\mathbb{P}^1 \times \mathbb{P}^1$. If more generally F is a subfield of $\mathbb{R}$, then the base change composed with real realization is still a ring homomorphism.

8. Chow-Witt rings

This section contains a short self-contained presentation and computation of the Chow-Witt ring of Q_n, arising from the fiber product description first used in [**HW19**] that emphasizes the relationship with Chow groups. This description relies on the isomorphism $\mathrm{Ch}^\bullet(X) \cong \mathrm{H}^\bullet(X, \bar{\mathbf{I}}^\bullet)$, and hence on the proof of the Milnor conjecture of Voevodsky et al.

Recall that the $\mathbb{Z} \oplus \mathbb{Z} \oplus \mathbb{Z}/2$-graded ring $\mathrm{H}^{\mathrm{tot}}(Q_n, \mathbf{I})$ restricts to Chow-Witt groups in bidegrees (i, i), corresponding to motivic bidegrees $(2i, i)$. By Theorem 6.3, we know that $\mathrm{CH}^\bullet(Q_{p,p})$ and $\mathrm{CH}^\bullet(Q_{p,p+1})$ have no two-torsion. Hence to compute $\widetilde{\mathrm{CH}}^\bullet(Q_{p,p})$ and $\widetilde{\mathrm{CH}}^\bullet(Q_{p,p+1})$, we may apply part (1) of Proposition 2.11 of [**HW19**], which we briefly recall.

PROPOSITION 8.1. *Let X be a smooth scheme over a field F of characteristic $\neq 2$. The canonical ring homomorphism*

$$\widetilde{\mathrm{CH}}^\bullet(X) \to \mathrm{H}^\bullet(X, \mathbf{I}^\bullet) \times_{\mathrm{Ch}^\bullet(X)} \ker \partial$$

induced from the cartesian square defining Milnor-Witt K-theory is always surjective. It is injective if $\mathrm{CH}^\bullet(X)$ has no non-trivial two-torsion.

Here, $\partial\colon \mathrm{CH}^\bullet(X) \to \mathrm{H}^{\bullet+1}(X, \mathbf{I}^{\bullet+1})$ is defined as in loc. cit. The kernel of ∂ is a subring of $\mathrm{CH}^\bullet(X)$ because the Bockstein homomorphism satisfies a Leibniz formula [**Fas07**, proof of Proposition 4.7].

Note that the general assumption of [**HW19**] that the base field is perfect is not necessary in Proposition 8.1: the proof of [**HW19**, Proposition 2.11] applies verbatim if we use Zariski sheaf cohomology in the "key diagram" considered in loc. cit., and Zariski sheaf cohomology can be computed from the Gersten complex [**Fas07**, Theorem 3.26]. Moreover, Proposition 8.1 also applies to twisted coefficients: the proof of loc. cit. goes through when twisting with a line bundle $\mathcal{L}$.

LEMMA 8.2. *In terms of our chosen generators, the isomorphism of commutative $\mathbb{Z}/2$-algebras $\mathrm{Ch}^\bullet(Q_n) \xrightarrow{\cong} \mathrm{H}^\bullet(Q_n, \bar{\mathbf{I}}^\bullet)$ has the form*

$$\bar{x} \mapsto \bar{\xi}$$
$$\bar{y} \mapsto \begin{cases} \overline{\beta} + \delta\overline{\xi}^p & \text{if } n \text{ is even, where } \delta \in \{0,1\} \\ \overline{\beta} & \text{if } n \text{ is odd} \end{cases}$$

PROOF. This is obvious, given the degree constraints and that we are working over $\mathbb{Z}/2$. □

Putting everything together, we obtain the following result:

THEOREM 8.3. *Let Q_n be the split quadric over a field F of characteristic $\neq 2$ as above, and $\mathcal{L}$ a line bundle over Q_n. Then the graded* $\mathrm{GW}(F)$*-module* $\widetilde{\mathrm{CH}}^\bullet(Q_n, \mathcal{L})$ *is given by*

$$\widetilde{\mathrm{CH}}^\bullet(Q_n, \mathcal{L}) \xrightarrow{\cong} \mathrm{H}^\bullet(Q_n, \mathbf{I}^\bullet, \mathcal{L}) \times_{\mathrm{Ch}^\bullet(Q_n)} \ker \partial_{\mathcal{L}}$$

with $\ker \partial_{\mathcal{L}}$ *given as follows:*

$$\text{for } \mathcal{L} = \mathcal{O} : \begin{cases} \mathbb{Z}\langle 2x, 2y, xy, x^2, y^2 \rangle & \text{when } n = 2p, \text{ and } p \text{ is even} \\ \mathbb{Z}\langle 2x, x^2, x^p, y \rangle & \text{when } n = 2p, \text{ and } p \text{ is odd} \\ \mathbb{Z}\langle 2x, x^2, y \rangle & \text{when } n = 2p+1, \text{ and } p \text{ is even} \\ \mathbb{Z}\langle 2x, 2y, xy, x^2, x^p \rangle & \text{when } n = 2p+1, \text{ and } p \text{ is odd} \end{cases}$$

where $\mathbb{Z}\langle \textit{elements} \rangle$ *denotes the subring of* $\mathrm{CH}^\bullet(Q_n)$ *generated by the specified elements (see Theorem 6.3). The following four twisted cases are described as submodules of* $\mathrm{CH}^\bullet(Q_n)$ *over the respective four rings above:*

$$\text{for } \mathcal{L} = \mathcal{O}(1) : \begin{cases} x \cdot \mathbb{Z}\langle x^2, xy \rangle & \\ \qquad + y \cdot \mathbb{Z} + \ker(\operatorname{mod} 2) & \text{when } n = 2p, \text{ and } p \text{ is even} \\ x \cdot \mathbb{Z}\langle x^2, x^p, y \rangle & \\ \qquad + \ker(\operatorname{mod} 2) & \text{when } n = 2p, \text{ and } p \text{ is odd} \\ x \cdot \mathbb{Z}\langle x^2, y, x^{p-1}y \rangle & \\ \qquad + \ker(\operatorname{mod} 2) & \text{when } n = 2p+1, \text{ and } p \text{ is even} \\ x \cdot \mathbb{Z}\langle x^2, x^p, xy, x^{p-1}y \rangle & \\ \qquad + y \cdot \mathbb{Z} + \ker(\operatorname{mod} 2) & \text{when } n = 2p+1, \text{ and } p \text{ is odd} \end{cases}$$

In this description, $\operatorname{mod} 2$ *denotes the mod-2-reduction map* $\mathrm{CH}^\bullet(Q_n) \to \mathrm{Ch}^\bullet(Q_n)$.

PROOF. Everything except the computation of $\ker \partial_{\mathcal{L}}$ has been established already. For the latter, one uses that in the notations of the diagram in [**HW19**, Section 2.4] we have $\partial_{\mathcal{L}} = \beta_{\mathcal{L}} \circ \operatorname{mod} 2$ and $\ker \beta_{\mathcal{L}} = \operatorname{Im} \rho_{\mathcal{L}}$. (Strictly speaking, the diagram in [**HW19**, Section 2.4] does include twists by line bundles, but line bundles can be fitted into the diagram without any difficulty.) Our computation of the $\mathrm{W}(F)$-algebra $\mathrm{H}^\bullet(Q_{p,q}, \mathbf{I}^\bullet)$ and the reduction map $\rho_{\mathcal{L}}$ (using Lemma 8.2) allow us to completely compute the image of the latter. Note that for $n = 2p$ and p even this is independent of the value of δ, using the relation $\bar{\xi}^{p+1} = 0$ both in the untwisted and twisted case and moreover the relation $y^2 - x^p y$ in the twisted case. From this we easily deduce the kernels of $\partial_{\mathcal{L}}$ and $\beta_{\mathcal{L}}$ both for $\mathcal{L} = \mathcal{O}$ and $\mathcal{L} = \mathcal{O}(1)$. □

For a more concise description of the $\mathbb{Z} \oplus \mathbb{Z}/2$-graded $\mathrm{GW}(F)$-algebra $\widetilde{\mathrm{CH}}^\bullet(Q_n, \mathcal{O} \oplus \mathcal{O}(1))$, we introduce an artificial $\mathbb{Z}/2$-grading on $\mathrm{CH}^\bullet(Q_n)$ and $\mathrm{Ch}^\bullet(Q_n)$ as follows:

$$\mathrm{CH}^\bullet(Q_n, \mathcal{O} \oplus \mathcal{O}(1)) := \mathrm{CH}^\bullet(Q_n)[\tau]/(\tau^2 - 1)$$
$$\mathrm{Ch}^\bullet(Q_n, \mathcal{O} \oplus \mathcal{O}(1)) := \mathrm{Ch}^\bullet(Q_n)[\tau]/(\tau^2 - 1)$$

In this notation, the same arguments as in the proof above yield the following description of the Chow-Witt ring:

$$\begin{aligned}&\widetilde{\mathrm{CH}}^\bullet(Q_n,\mathcal{O}\oplus\mathcal{O}(1))\\&\quad\cong \mathrm{H}^\bullet(Q_n,\mathbf{I}^\bullet,\mathcal{O}\oplus\mathcal{O}(1))\times_{\mathrm{Ch}^\bullet(Q_n,\mathcal{O}\oplus\mathcal{O}(1))}\mathbb{Z}\langle 2\tau,2x,2y,x\tau,x^p\tau^{p+1},y\tau^{q+1}\rangle,\end{aligned}$$

where $\mathbb{Z}\langle\text{elements}\rangle$ denotes the subring of $\mathrm{CH}^\bullet(Q_n,\mathcal{O}\oplus\mathcal{O}(1))$ generated by the specified elements.

9. Milnor-Witt cohomology

The additive computations of **I**-cohomology can be transferred to Milnor-Witt cohomology. Again, we begin by recalling the corresponding result for projective spaces:

THEOREM 9.1 ([**Fas13**, Theorem 11.7]). *Let $\mathbb{P}^n=\mathbb{P}^n_F$, where F is a field of characteristic $\neq 2$. For degrees $i\in\{0,\dots,n\}$, the Milnor-Witt cohomology groups of $\mathbb{P}^n$ are as follows:*

$$\mathrm{H}^i(\mathbb{P}^n,\mathbf{K}^{MW}_j,\mathcal{O}(l))\cong\begin{cases}\mathrm{K}^{MW}_{j-i}(F) & \text{if } (i=0 \text{ and } l\equiv i \mod 2)\\ & \text{or } (i=n \text{ and } l\not\equiv i \mod 2)\\ \mathrm{K}^M_{j-i}(F) & \text{if } i\neq 0 \text{ and } l\equiv i \mod 2\\ 2\mathrm{K}^M_{j-i}(F) & \text{if } i\neq n \text{ and } l\not\equiv i \mod 2\end{cases}$$

In degrees $i\notin\{0,\dots,n\}$, these groups vanish.

Now consider once again our split quadric Q_n over a field. Write $n=p+q$ with $q=p$ if n is even and $q=p+1$ if n is odd, so that q is the codimension of $\iota_x\colon\mathbb{P}^p_x\hookrightarrow Q_n$. Recall from Lemma 4.1 that $\det\mathcal{N}\cong\mathcal{O}(q-1)$ for the normal bundle $\mathcal{N}$ of this inclusion. As in **I**-cohomology, localization and dévissage yield a long exact sequence as follows [**Fas08**, Corollaire 10.4.10]:

$$\cdots\to\mathrm{H}^{i-q}(\mathbb{P}^p_x,\mathbf{K}^{MW}_{j-q},\mathcal{L}(1-q))\xrightarrow{(\iota_x)_*}\mathrm{H}^i(Q_n,\mathbf{K}^{MW}_j,\mathcal{L})\to\mathrm{H}^i(Q_n-\mathbb{P}^p_x,\mathbf{K}^{MW}_j,\mathcal{L})\to\cdots$$

Homotopy invariance of **I**-cohomology (Theorem 5.1) together with the corresponding result for Milnor cohomology and the usual five lemma argument implies homotopy invariance for Milnor-Witt cohomology. Applying this to the affine space bundle $\rho\colon Q_n-\mathbb{P}^p_x\to\mathbb{P}^p_y$, we obtain the following exact sequence:

(9.2)

$$\cdots\to\mathrm{H}^{i-q}(\mathbb{P}^p_x,\mathbf{K}^{MW}_{j-q},\mathcal{L}(1-q))\xrightarrow{(\iota_x)_*}\mathrm{H}^i(Q_n,\mathbf{K}^{MW}_j,\mathcal{L})\xrightarrow{\iota_y^*}\mathrm{H}^i(\mathbb{P}^p_y,\mathbf{K}^{MW}_j,\mathcal{L})\to\cdots$$

THEOREM 9.3. *Let Q_n be the n-dimensional split quadric over a field F of characteristic $\neq 2$, with $n\geq 3$. Let $\mathcal{L}$ be a line bundle over Q_n. The exact sequence (9.2) above splits, so that we have isomorphisms*

$$\mathrm{H}^i(Q_n,\mathbf{K}^{MW}_j,\mathcal{L})\cong\mathrm{H}^{i-q}(\mathbb{P}^p_x,\mathbf{K}^{MW}_{j-q},\mathcal{L}(1-q))\oplus\mathrm{H}^i(\mathbb{P}^p_y,\mathbf{K}^{MW}_j,\mathcal{L})$$

PROOF. Suppose $n=2p+1$. Then for degree reasons the map ι_y^* in sequence (9.2) is an isomorphism in the range $0\leq i<p$, and the map $(\iota_x)_*$ is an isomorphism in the range $p+1<i\leq n$. In degrees $i=p$ and $i=p+1$, we have

an exact sequence

$$(9.4)\qquad \begin{aligned} 0 \to \mathrm{H}^p(Q_n, \mathbf{K}^{MW}_j, \mathcal{L}) &\xrightarrow{\iota_y^*} \mathrm{H}^p(\mathbb{P}^p_y, \mathbf{K}^{MW}_j, \mathcal{L}) \\ &\xrightarrow{\partial} \mathrm{H}^0(\mathbb{P}^p_x, \mathbf{K}^{MW}_{j-q}, \mathcal{L}\otimes\omega_x) \xrightarrow{(\iota_x)_*} \mathrm{H}^{p+1}(Q_n, \mathbf{K}^{MW}_j, \mathcal{L}) \to 0 \end{aligned}$$

Consider first the case $\mathcal{L} = \mathcal{O}(p+1)$. Consider the following commutative diagram induced by the short exact sequences of sheaves $0 \to \mathbf{I}^{j+1} \to \mathbf{K}^{MW}_j \to \mathbf{K}^M_j \to 0$ (cf. [**Fas13**, above Lemma 11.3]), keeping in mind that the twists with respect to any $\mathcal{L}$ are invisible to $\mathbf{K}^M_j$:

$$\begin{array}{ccccccc}
\mathrm{H}^p(Q_n, \mathbf{I}^{j+1}, \mathcal{L}) & \longrightarrow & \mathrm{H}^p(Q_n, \mathbf{K}^{MW}_j, \mathcal{L}) & \longrightarrow & \mathrm{H}^p(Q_n, \mathbf{K}^M_j) & \longrightarrow & \mathrm{H}^{p+1}(Q_n, \mathbf{I}^{j+1}, \mathcal{L}) \\
\cong\downarrow \iota_y^* & & \downarrow \iota_y^* & & \downarrow\!\!\downarrow \iota_y^* & & \downarrow \\
\mathrm{H}^p(\mathbb{P}^p_y, \mathbf{I}^{j+1}, \mathcal{L}) & \longrightarrow & \mathrm{H}^p(\mathbb{P}^p_y, \mathbf{K}^{MW}_j, \mathcal{L}) & \longrightarrow & \mathrm{H}^p(\mathbb{P}^p_y, \mathbf{K}^M_j) & \longrightarrow & 0
\end{array}$$

From Theorem 5.5, we know that the group $\mathrm{H}^{p+1}(Q_n, \mathbf{I}^{j+1}, \mathcal{O}(p+1))$ in the top right corner vanishes. Our proof of Theorem 5.5 moreover shows that ι_y induces an isomorphism on $\mathbf{I}$-cohomology in degree p, so the vertical map on the far left is an isomorphism. Recall from [**Fas13**, Theorem 11.1] that $\mathrm{H}^p(\mathbb{P}^p_y, \mathbf{K}^M_j) = \mathrm{K}^M_{j-p}(F) \cdot e(\mathcal{O}(1))^p$. By considering the Euler class of the bundle $\mathcal{O}(1)$ over Q_n (and using Lemma 4.4), we find that ι_y^* is surjective on $\mathbf{K}^M$-cohomology in degree p. Altogether, it follows from a version of the five lemma that the second vertical map in the diagram is also surjective. So in sequence (9.4), $\partial = 0$ and both ι_y^* and $(\iota_x)_*$ are isomorphisms.

Now consider the case $\mathcal{L} = \mathcal{O}(p)$. Recall from Lemma 4.1 that $\det(\mathcal{N})^\vee = \mathcal{O}(-p)$. Consider the following commutative diagram induced by the same short exact sequence of sheaves as in the previous case:

$$\begin{array}{ccccccc}
0 & \longrightarrow & \mathrm{H}^0(\mathbb{P}^p_x, \mathbf{I}^{j-q+1}) & \longrightarrow & \mathrm{H}^0(\mathbb{P}^p_x, \mathbf{K}^{MW}_{j-q}) & \longrightarrow & \mathrm{H}^0(\mathbb{P}^p_x, \mathbf{K}^M_{j-q}) \\
\downarrow & & \cong\downarrow (\iota_x)_* & & \downarrow (\iota_x)_* & & \cong\downarrow (\iota_x)_* \\
\mathrm{H}^p(Q_n, \mathbf{K}^M_j) & \xrightarrow{\delta} & \mathrm{H}^{p+1}(Q_n, \mathbf{I}^{j+1}, \mathcal{O}(p)) & \longrightarrow & \mathrm{H}^{p+1}(Q_n, \mathbf{K}^{MW}_j, \mathcal{O}(p)) & \longrightarrow & \mathrm{H}^{p+1}(Q_n, \mathbf{K}^M_j)
\end{array}$$

We have isomorphisms on the left and right in this diagram by the computation of $\mathbf{I}$-cohomology and Milnor cohomology of quadrics above. (One sees this directly from the localization sequence and degree considerations.) We claim that the boundary map δ vanishes, and that therefore $(\iota_x)_*$ is injective on $\mathbf{K}^{MW}$ cohomology. To see this, we compare δ with the boundary map in the long exact sequences associated with the coefficient sequence $0 \to \mathbf{I}^{j+1} \to \mathbf{I}^j \to \bar{\mathbf{I}}^j \to 0$, via the following commutative diagram (see [**Fas13**, Lemma 11.3]):

$$\begin{array}{ccccccc}
\ldots & \longrightarrow & \mathrm{H}^p(Q_n, \mathbf{K}^M_j) & \xrightarrow{\delta} & \mathrm{H}^{p+1}(Q_n, \mathbf{I}^{j+1}, \mathcal{O}(p)) & \longrightarrow & \ldots \\
 & & \downarrow & & \| & & \\
\mathrm{H}^p(Q_n, \mathbf{I}^j, \mathcal{O}(p)) & \longrightarrow & \mathrm{H}^p(Q_n, \bar{\mathbf{I}}^j) & \xrightarrow{\partial} & \mathrm{H}^{p+1}(Q_n, \mathbf{I}^{j+1}, \mathcal{O}(p)) & \longrightarrow & \ldots
\end{array}$$

Our computations of $\mathbf{I}$- and $\bar{\mathbf{I}}$-cohomology show that the reduction map on the left of the lower sequence is surjective for odd-dimensional quadrics. So the boundary map ∂ is zero, and hence so is δ.

Suppose now that $n = 2p$. In this case, the map ι_y^* in sequence (9.2) is an isomorphism for degree reasons in the range $0 \leq i < p-1$, and the map $(\iota_x)_*$ is an isomorphism for degree reasons in the range $p+1 < i \leq n$. It remains to show that ι_y^* is also an isomorphism in degree $p-1$, that $(\iota_x)_*$ is an isomorphism in degree $p+1$, and that the following sequence in degree p is split exact:

$$0 \longrightarrow \mathrm{H}^0(\mathbb{P}_x^p, \mathbf{K}_{j-p}^{MW}, \omega_x \otimes \mathcal{L}) \xrightarrow{(\iota_x)_*} \mathrm{H}^p(Q_n, \mathbf{K}_j^{MW}, \mathcal{L}) \xrightarrow{\iota_y^*} \mathrm{H}^p(\mathbb{P}_y^p, \mathbf{K}_j^{MW}, \mathcal{L}) \longrightarrow 0$$

To verify that this sequence is indeed split exact, one can use the cartesian diagram (6.10) and the base change formula. The other two claims, concerning degrees $p-1$ and $p+1$, then follow. □

COROLLARY 9.5. *The Milnor-Witt cohomology groups of a split quadric Q_n over a field F of characteristic $\neq 2$ in degrees $i \in \{0, \ldots, n\}$ are as follows:*
Case $n = 2p$:

$$\mathrm{H}^i(Q_{2p}, \mathbf{K}_j^{MW}, \mathcal{O}(l)) \cong \begin{cases} \mathrm{K}_{j-i}^{MW}(F) & \text{if } (i=0 \text{ and } l \equiv i \mod 2) \\ & \text{or } (i=n \text{ and } l \equiv i \mod 2) \\ \mathrm{K}_{j-i}^{MW}(F) \oplus \mathrm{K}_{j-i}^{MW}(F) & \text{if } i=p \text{ and } l \not\equiv i \mod 2 \\ \mathrm{K}_{j-i}^{M}(F) \oplus 2\mathrm{K}_{j-i}^{M}(F) & \text{if } i=p \text{ and } l \equiv i \mod 2 \\ \mathrm{K}_{j-i}^{M}(F) & \text{if } (0<i<p \text{ and } l \equiv i \mod 2) \\ & \text{or } (p<i\leq n \text{ and } l \not\equiv i \mod 2) \\ 2\mathrm{K}_{j-i}^{M}(F) & \text{if } (0 \leq i<p \text{ and } l \not\equiv i \mod 2) \\ & \text{or } (p<i<n \text{ and } l \equiv i \mod 2) \end{cases}$$

Case $n = 2p+1$:

$$\mathrm{H}^i(Q_{2p+1}, \mathbf{K}_j^{MW}, \mathcal{O}(l)) \cong \begin{cases} \mathrm{K}_{j-i}^{MW}(F) & \text{if } (i=0 \text{ and } l \equiv i \mod 2) \\ & \text{or } (i=p \text{ and } l \not\equiv i \mod 2) \\ & \text{or } (i=p+1 \text{ and } l \not\equiv i \mod 2) \\ & \text{or } (i=n \text{ and } l \equiv i \mod 2) \\ \mathrm{K}_{j-i}^{M}(F) & \text{if } (0<i\leq p \text{ and } l \equiv i \mod 2) \\ & \text{or } (p+1<i\leq n \text{ and } l \not\equiv i \mod 2) \\ 2\mathrm{K}_{j-i}^{M}(F) & \text{if } (0\leq i<p \text{ and } l \not\equiv i \mod 2) \\ & \text{or } (p+1\leq i<n \text{ and } l \equiv i \mod 2) \end{cases}$$

In degrees $i \notin \{0, \ldots, n\}$, all groups are zero.

References

[AF16] A. Asok and J. Fasel, *Comparing Euler classes*, Q. J. Math. **67** (2016), no. 4, 603–635. MR3609848

[Ba01] Paul Balmer, *Witt cohomology, Mayer-Vietoris, homotopy invariance and the Gersten conjecture*, K-Theory **23** (2001), no. 1, 15–30, DOI 10.1023/A:1017594924542. MR1852452

[BC09] Paul Balmer and Baptiste Calmès, *Geometric description of the connecting homomorphism for Witt groups*, Doc. Math. **14** (2009), 525–550. MR2565903

[BC12] Paul Balmer and Baptiste Calmès, *Bases of total Witt groups and lax-similitude*, J. Algebra Appl. **11** (2012), no. 3, 1250045, 24, DOI 10.1142/S0219498811005725. MR2928113

[BM00] Jean Barge and Fabien Morel, *Groupe de Chow des cycles orientés et classe d'Euler des fibrés vectoriels* (French, with English and French summaries), C. R. Acad. Sci. Paris Sér. I Math. **330** (2000), no. 4, 287–290, DOI 10.1016/S0764-4442(00)00158-0. MR1753295

[CGT] Michel Cahen, Simone Gutt, and Andrzej Trautman, *Spin structures on real projective quadrics*, J. Geom. Phys. **10** (1993), no. 2, 127–154, DOI 10.1016/0393-0440(93)90025-A. MR1204145

[Cal08] B. Calmès. A talk on Witt groups of split quadrics in Nottingham, 2008.

[CF17] B. Calmès and J. Fasel. *The category of finite MV-correspondences.* Preprint 2017, arxiv 1412.2989.

[DK01] James F. Davis and Paul Kirk, *Lecture notes in algebraic topology*, Graduate Studies in Mathematics, vol. 35, American Mathematical Society, Providence, RI, 2001. MR1841974

[DF17] F. Déglise and J. Fasel. *MW-motivic complexes.* Preprint 2017, arxiv 1708.06095.

[Die82] J. Dieudonné, *Éléments d'analyse. Tome IX. Chapitre XXIV* (French), Cahiers Scientifiques [Scientific Reports], XL11, Gauthier-Villars, Paris, 1982. MR658305

[DI07] Daniel Dugger and Daniel C. Isaksen, *The Hopf condition for bilinear forms over arbitrary fields*, Ann. of Math. (2) **165** (2007), no. 3, 943–964, DOI 10.4007/annals.2007.165.943. MR2335798

[EGA4] A. Grothendieck, *Éléments de géométrie algébrique. IV. Étude locale des schémas et des morphismes de schémas. II* (French), Inst. Hautes Études Sci. Publ. Math. **24** (1965), 231. MR0199181

[Fas07] Jean Fasel, *The Chow-Witt ring*, Doc. Math. **12** (2007), 275–312. MR2350291

[Fas08] Jean Fasel, *Groupes de Chow-Witt* (French, with English and French summaries), Mém. Soc. Math. Fr. (N.S.) **113** (2008), viii+197, DOI 10.24033/msmf.425. MR2542148

[Fas13] Jean Fasel, *The projective bundle theorem for $\mathbf{I}^j$-cohomology*, J. K-Theory **11** (2013), no. 2, 413–464, DOI 10.1017/is013002015jkt217. MR3061003

[Fas19] J. Fasel. *Lectures on Chow-Witt groups.* This volume.

[FS09] J. Fasel and V. Srinivas, *Chow-Witt groups and Grothendieck-Witt groups of regular schemes*, Adv. Math. **221** (2009), no. 1, 302–329, DOI 10.1016/j.aim.2008.12.005. MR2509328

[Ful78] William Fulton, *Intersection theory*, 2nd ed., Ergebnisse der Mathematik und ihrer Grenzgebiete. 3. Folge. A Series of Modern Surveys in Mathematics [Results in Mathematics and Related Areas. 3rd Series. A Series of Modern Surveys in Mathematics], vol. 2, Springer-Verlag, Berlin, 1998. MR1644323

[Har92] Joe Harris, *Algebraic geometry*, Graduate Texts in Mathematics, vol. 133, Springer-Verlag, New York, 1995. A first course; Corrected reprint of the 1992 original. MR1416564

[Hat02] Allen Hatcher, *Algebraic topology*, Cambridge University Press, Cambridge, 2002. MR1867354

[HW19] J. Hornbostel and M. Wendt. *Chow–Witt rings of classifying spaces for symplectic and special linear groups.* J. Topol. 12 (2019), 915–965.

[HWXZ] J. Hornbostel, M. Wendt, H. Xie, and M. Zibrowius. *The real cycle class map.* Preprint 2019, arXiv:1911.04150.

[Jac17] Jeremy A. Jacobson, *Real cohomology and the powers of the fundamental ideal in the Witt ring*, Ann. K-Theory **2** (2017), no. 3, 357–385, DOI 10.2140/akt.2017.2.357. MR3658988

[Ker09] Moritz Kerz, *The Gersten conjecture for Milnor K-theory*, Invent. Math. **175** (2009), no. 1, 1–33, DOI 10.1007/s00222-008-0144-8. MR2461425

[KM91] N. A. Karpenko and A. S. Merkur′ev, *Chow groups of projective quadrics* (Russian), Algebra i Analiz **2** (1990), no. 3, 218–235; English transl., Leningrad Math. J. **2** (1991), no. 3, 655–671. MR1073215

[Mor12] Fabien Morel, *$\mathbb{A}^1$-algebraic topology over a field*, Lecture Notes in Mathematics, vol. 2052, Springer, Heidelberg, 2012. MR2934577

[Mur94] M. Pavaman Murthy, *Zero cycles and projective modules*, Ann. of Math. (2) **140** (1994), no. 2, 405–434, DOI 10.2307/2118605. MR1298718

[Nen09] Alexander Nenashev, *On the Witt groups of projective bundles and split quadrics: geometric reasoning*, J. K-Theory **3** (2009), no. 3, 533–546, DOI 10.1017/is008007001jkt050. MR2507729

[RS18] Kay Rülling and Shuji Saito, *Higher Chow groups with modulus and relative Milnor K-theory*, Trans. Amer. Math. Soc. **370** (2018), no. 2, 987–1043, DOI 10.1090/tran/7018. MR3729494

[Spa81] Edwin H. Spanier, *Algebraic topology*, Springer-Verlag, New York-Berlin, 1981. Corrected reprint. MR666554

[Wen18] M. Wendt. *Chow-Witt rings of Grassmannians*. Preprint 2018, arxiv 1805.06142.

[WL62] Wen-juan Wu and Pei-xen Li, *The characteristic classes of certain real quadrics*, Chinese Math. **3** (1962), 218–231. MR0154298

[Xie19] Heng Xie, *Witt groups of smooth projective quadrics*, Adv. Math. **346** (2019), 70–123, DOI 10.1016/j.aim.2019.01.038. MR3905194

[Zib11] Marcus Zibrowius, *Witt groups of complex cellular varieties*, Doc. Math. **16** (2011), 465–511. MR2823367

Jens Hornbostel, Bergische Universität Wuppertal, Gaussstrasse 20, 42119 Wuppertal, Germany

Email address: hornbostel@math.uni-wuppertal.de

Heng Xie, Bergische Universität Wuppertal, Gaussstrasse 20, 42119 Wuppertal, Germany

Email address: sysuxieheng@gmail.com

Marcus Zibrowius, Heinrich-Heine-Universität Düsseldorf, Universitätsstr. 1, 40225 Düsseldorf, Germany

Email address: marcus.zibrowius@cantab.net

Contemporary Mathematics
Volume **745**, 2020
https://doi.org/10.1090/conm/745/15025

Lectures on quadratic enumerative geometry

Marc Levine

ABSTRACT. We give a survey of some recent developments in quadratic enumerative geometry. This includes discussions of Euler characteristics, Euler classes, Riemann-Hurwitz formulas, Becker-Gottlieb transfers, and computations of Pontryagin and Euler classes in Witt cohomology.

CONTENTS

Introduction

These notes are based on four lectures I gave at the conference "Motivic homotopy theory and enumerative geometry" held at the University of Duisburg-Essen in May 2018. They give an introduction to an emerging area of study, in which one uses methods and results from motivic homotopy theory to enlarge the scope of enumerative geometry to allow for a more general type of invariant than the classical integer ones. We have in mind the invariants in the Grothendieck-Witt ring that arise via Morel's theorem identifying the endomorphism ring of the motivic sphere spectrum with the Grothendieck-Witt ring, or other "motivic" approaches, such as replacing the intersection theory given by the Chow ring with that formed by the Chow-Witt ring, or replacing algebraic K-theory with hermitian K-theory. These lectures explore some of the techniques that have been developed recently to

2010 *Mathematics Subject Classification.* 14F42, 18G55, 55P42.

The author thanks the DFG for support through the SFB Transregio 45 and the SPP 1786.

give such a quadratic enumerative geometry a suitable toolkit to enable an effective exploration of classical enumerative problems, such as curve counting, in the quadratic setting. We have not discussed the many concrete examples that have been worked out already using these and other techniques, such as the quadratic count of lines on a cubic surface by Kass-Wickelgren [**10**] or our quadratic version of Welschinger invariants [**13**], leaving a survey of these results to another day.

This is mainly a survey of results already available in the literature or posted on the arXiv, but we hope that in its current form these notes will provide the reader with an accessible and enjoyable introduction to this area.

Finally, I would like to thank Federico Binda, Manh Toan Nguyen and Oliver Röndigs for their work in organizing this volume and the referee for carefully going through an earlier version of this paper and making a number of corrections and useful suggestions.

Lecture 1. Euler characteristics

In this first lecture, we discuss one of the most basic numerical invariants, the Euler characteristic, from a number of different points of view. We first look at the Euler characteristic of a dualizable object in a symmetric monoidal category, then specialize to the case of a perfect field k and the suspension spectrum of a smooth (projective in the case of positive characteristic) k-scheme, viewed as an object in the motivic stable homotopy category $\mathrm{SH}(k)$. Via Morel's theorem computing the endomorphism ring of the motivic sphere spectrum as the Grothendieck-Witt ring, this gives us the Euler characteristic $\chi(X/k) \in \mathrm{GW}(k)$. We formulate a version of a "motivic Gauß-Bonnet formula" which relates this rather abstractly constructed invariant to more concrete versions, and use this to give "realizations" of the motivic Euler characteristics with respect to several different cohomology theories. Two particular cases, the cohomology of the Milnor-Witt sheaves, and hermitian K-theory, give two complementary methods for computing $\chi(X/k)$. The second realizes a suggestion of Serre, that $\chi(X/k)$ should be the quadratic form on $\oplus_{p,q} H^q(X, \Omega^p_{X/k})[p-q]$ (in the derived category) given by cup product and the trace map $H^{\dim X}(X, \Omega^{\dim X}_{X/k}) \to k$.

A general version of a motivic Gauß-Bonnet formula, which we do not use here, appears in the paper of Déglise-Jin-Khan [**6**]; many of the details of the ideas sketched here are worked out in a paper of ours with Arpon Raksit [**12**]. For a general discussion of the $\mathrm{SH}(k)$ Euler characteristic and its basic properties, we refer the reader to [**11**].

1.1. Categorical Euler characteristics.

There is a general notion of a *trace* in a symmetric monoidal category, and the Euler characteristic of a dualizable object x is defined as the trace of the identity map on x.

DEFINITION 1.1.1. Let $(\mathcal{C}, \otimes, \tau, \mathbb{1})$ be a symmetric monoidal category.

1. A *dual* of an object $x \in \mathcal{C}$ is a triple $(x^\vee, \delta_x, ev_x)$ with

$$\delta_x : \mathbb{1} \to x \otimes x^\vee, \; ev_x : x^\vee \otimes x \to \mathbb{1}$$

morphisms in $\mathcal{C}$ such that the compositions

$$x = \mathbb{1} \otimes x \xrightarrow{\delta_x \otimes \mathrm{Id}_x} x \otimes x^\vee \otimes x \xrightarrow{\mathrm{Id}_x \otimes ev_x} x \otimes \mathbb{1} = x$$

and

$$x^\vee = x^\vee \otimes 1 \xrightarrow{\mathrm{Id}_{x^\vee}\otimes\delta_x} x^\vee \otimes x \otimes x^\vee \xrightarrow{ev_x\otimes \mathrm{Id}_{x^\vee}} 1 \otimes x^\vee = x^\vee$$

are identity morphisms. The triple $(x^\vee, \delta_x, ev_x)$ is unique up to unique isomorphism; we call x *dualizable* if x has a dual.

2. Let $f : x \to y$ be a morphism of dualizable objects in $\mathcal{C}$. We have the dual morphism $f^\vee : y^\vee \to x^\vee$ defined as the composition

$$y^\vee = y^\vee \otimes 1 \xrightarrow{\mathrm{Id}_{y^\vee}\otimes\delta_x} y^\vee \otimes x \otimes x^\vee \xrightarrow{\mathrm{Id}_{y^\vee}\otimes f\otimes \mathrm{Id}_x} y^\vee \otimes y \otimes x^\vee \xrightarrow{ev_y\otimes \mathrm{Id}_x} 1 \otimes x^\vee = x^\vee$$

Then $(x, f) \mapsto (x^\vee, f^\vee)$ defines an involution on the full subcategory of dualizable objects $\mathcal{C}^\times$ in $\mathcal{C}$.

3. For $f : x \to x$ a morphism in $\mathcal{C}^\times$, we have $\mathrm{Tr}(f) \in \mathrm{End}_{\mathcal{C}}(1)$ defined as the composition

$$1 \xrightarrow{\delta_x} x \otimes x^\vee \xrightarrow{f\otimes \mathrm{Id}_{x^\vee}} x \otimes x^\vee \xrightarrow{\tau_{x,x^\vee}} x^\vee \otimes x \xrightarrow{ev_x} 1$$

The *Euler characteristic* $\chi_{\mathcal{C}}(x) \in \mathrm{End}_{\mathcal{C}}(1)$ is defined to be $\mathrm{Tr}(\mathrm{Id}_x)$.

Clearly $\mathrm{Tr}_{\mathcal{C}}(f)$ and $\chi_{\mathcal{C}}(x)$ are functorial: if $F : \mathcal{C} \to \mathcal{D}$ is a symmetric monoidal functor and $x \in \mathcal{C}^\times$, then $F(x) \in \mathcal{D}^\times$ and $F(\mathrm{Tr}_{\mathcal{C}}(f)) = \mathrm{Tr}_{\mathcal{D}}(F(f))$.

EXAMPLE 1.1.2. For X a finite simplicial set or finite CW complex, we have the chain complex $C_*(X;\mathbb{Z}) \in D(\mathbf{Ab})$, a dualizable object, and $\chi_{D(\mathbf{Ab})}(C_*(X;\mathbb{Z}))$ is the usual topological Euler characteristic $\chi^{top}(X)$. We have $\mathrm{End}_{D(\mathbf{Ab})}(\mathbb{Z}) = \mathbb{Z}$; as $\mathrm{End}_{D(\mathbb{Q})}(\mathbb{Q}) = \mathbb{Q}$, we can pass to $D(\mathbb{Q})$ and then note that $C_*(X;\mathbb{Q}) \cong \oplus_{i=0}^{\dim X} H_i(X,\mathbb{Q})[i]$; the dual is $C_*(X;\mathbb{Q})^\vee = \oplus_{i=0}^{\dim X} H_i(X,\mathbb{Q})[-i]$ and the trace of the identity gives the trace of the identity map on $H_i(X,\mathbb{Q})$, times the factor $(-1)^i$ coming from the symmetry $\tau(a\otimes b) = (-1)^{\deg a \deg b} b \otimes a$.

2. The same computation says that

$$\chi_{D(\mathrm{Sh}_{\text{ét}}(k))}(R\Gamma(X_{\text{ét}}, \mathbb{Q}_\ell)) = \chi^{\text{ét}}(X) = \sum_{i=0}^{2\dim X} (-1)^i \dim_{\mathbb{Q}_\ell} H^i_{\text{ét}}(X, \mathbb{Q}_\ell)$$

for $k = \bar{k}$ and X a finite type k-scheme.

3. For $\mathcal{C} = \mathrm{DM}(k)$, we have $\mathrm{End}(\mathbb{Z}(0)) = \mathbb{Z}$, and we can see by either an étale realization or the Betti realization for $k \subset \mathbb{C}$, that for X smooth over k

$$\chi_{\mathrm{DM}(k)}(M(X)) = \chi^{\text{ét}}(X_{\bar{k}}) = \chi^{top}(X(\mathbb{C})).$$

Similarly, for $\mathcal{C} = \mathrm{SH}$, we have $\mathrm{End}_{\mathrm{SH}}(\mathbb{S}) = \mathbb{Z}$, so we can compute in $D(\mathbf{Ab})$ and we get

$$\chi_{\mathrm{SH}}(\Sigma^\infty X_+) = \chi_{D(\mathbf{Ab})}(C_*(X;\mathbb{Z})) = \chi^{top}(X).$$

for X a finite simplicial set/CW complex.

Our main object of study is the "motivic" Euler characteristic $\chi_{\mathrm{SH}(k)}(\Sigma^\infty_T X_+)$. For k a field, we have the category of *pointed spaces over* k, $\mathbf{Spc}_\bullet(k)$, this being the category of presheaves on $\mathbf{Sm}/k$ with values in pointed simplicial sets. There is a good notion of "$\mathbb{A}^1$-weak equivalence" replacing the classical notion of weak homotopy equivalence, giving us the motivic unstable pointed homotopy category over k, $\mathcal{H}_\bullet(k) := \mathbf{Spc}_\bullet(k)[WE^{-1}_{\mathbb{A}^1}]$ [**18**]. The motivic analog of classical stabilization through S^1-suspension spectra is the so-called T-stabilization, $T := \mathbb{A}^1/(\mathbb{A}^1 - \{0\})$, giving rise to the symmetric monoidal triangulated category $\mathrm{SH}(k)$, the *motivic stable homotopy over* k [**9, 23**]. Each $X \in \mathbf{Sm}/k$ gives an object $X_+ \in \mathbf{Spc}_\bullet(k)$

defined as the pointed representable sheaf of sets $Y \mapsto \mathrm{Hom}_{\mathbf{Sm}/k}(Y, X) \cup \{*\}$. There is an infinite suspension functor

$$\Sigma_T^\infty : \mathcal{H}_\bullet(k) \to \mathrm{SH}(k)$$

and for $X \in \mathbf{Sch}/k$, smooth and proper over k, the infinite suspension $\Sigma_T^\infty X_+$ is dualizable (see e.g. [**22**, §2], [**9**, §5]). We write $\chi(X/k)$ for $\chi_{\mathrm{SH}(k)}(\Sigma_T^\infty X_+)$.

For k of characteristic zero, $\Sigma_T^\infty X_+$ is dualizable for all $X \in \mathbf{Sm}/k$; for simplicity, we will either assume k has characteristic zero, or that "all relevant X" have $\Sigma_T^\infty X_+$ dualizable. We will assume as needed that $\mathrm{char}\, k \neq 2$.

The first point is the theorem of Morel:

THEOREM 1.1.3 (Morel [**15**, §6] [**16**, §6]). *For k a perfect field there is a canonical isomorphism*

$$\pi_0(\mathbb{S}_k)_n \cong \mathcal{K}_n^{MW}.$$

To explain: $\mathcal{K}_*^{MW} = \oplus_{n\in\mathbb{Z}}\mathcal{K}_n^{MW}$ is a Nisnevich sheaf of graded rings on $\mathbf{Sm}/k$ with value $K_*^{MW}(F)$ on fields, defined by Hopkins-Morel [**15**, §6], [**16**, §6]. Here $K_*^{MW}(F)$ is defined by generators and relations.

Generators: for $u \in F^\times$, the generator $[u] \in K_1^{MW}(F)$ and an additional generator $\eta \in K_{-1}^{MW}(F)$.

Relations:
(0) $\eta[u] = [u]\eta$
(1) $[u][1-u] = 0$ for $u \in F - \{0, 1\}$
(2) $[uv] = [u] + [v] + \eta[u][v]$
(3) Let $h = 2 + \eta[-1]$. Then $\eta \cdot h = 0$.

There are "tame symbol maps" $\partial_t : K_n^{MW}(F) \to K_{n-1}^{MW}(f)$ for each DVR $\mathcal{O}$ with quotient field F, residue field f and generator t for the maximal ideal, and $\mathcal{K}_n^{MW}$ is the associated unramified sheaf on $\mathbf{Sm}/k$: For $X \in \mathbf{Sm}/k$ irreducible we have

$$\mathcal{K}_n^{MW}(X) = \{\alpha \in K_n^{MW}(k(X)) \mid \partial_{t_x}(\alpha) = 0\ \forall x \in X^{(1)}, (t_x) = \mathfrak{m}_x \subset \mathcal{O}_{X,x}\}$$

where $X^{(n)} \subset X$ is the set of codimension n points of X. The fact that this does indeed form a (Nisnevich) sheaf on $\mathbf{Sm}/k$ is not obvious, but is proven in [**14**].

For $\mathcal{E} \in \mathrm{SH}(k)$, we have the bi-graded sheaf of homotopy groups, with $\pi_a(\mathcal{E})_b$ associated to the presheaf

$$U \mapsto [\Sigma_{S^1}^a \Sigma_T^\infty U_+, \Sigma_{\mathbb{G}_m}^b \mathcal{E}]_{\mathrm{SH}(k)}.$$

Morel's connectedness theorem [**17**] tells us that $\pi_a(\mathbb{S})_b = 0$ for $a < 0$ and all b, so the $\pi_0(\mathbb{S}_k)_*$ are the "first" non-vanishing sheaves; these should be thought of as the analog of $\pi_0(\mathbb{S}) = \mathbb{Z}$ in SH.

In addition, Morel computes $K_n^{MW}(F)$ as follows (for simplicity, we work away from characteristic 2):

THEOREM 1.1.4 (Morel [**15**, §6], [**16**, §6]). *Let* $\mathrm{GW}(F)$ *be the Grothendieck-Witt ring of non-degenerate quadratic forms over F, $W(F)$ the Witt ring* $\mathrm{GW}(F)/(H)$, *H the hyperbolic form $x^2 - y^2$ and $I(F) \subset \mathrm{GW}(F)$ the ideal of forms of virtual rank zero. Sending the quadratic form $q_u(x) = ux^2$, $u \in F^\times$, to $\langle u\rangle := 1 + \eta[u]$ defines an isomorphism*

$$\mathrm{GW}(F) \to K_0^{MW}(F)$$

For $n > 0$, sending q_u to $\eta^n(1+\eta[u])$ defines an isomorphism

$$W(F) \to K^{MW}_{-n}(F)$$

Sending $[u_1]\cdot \ldots \cdot [u_n]$ to the symbol $\{u_1, \ldots, u_n\}$ and sending η to zero defines a surjection $\pi : K^{MW}_*(F) \to K^M(F)$. For $n > 0$, there is an exact sequence

$$0 \to I(F)^{n+1} \to K^{MW}_n(F) \xrightarrow{\pi} K^M_n(F) \to 0$$

As consequence, we can identify $\mathrm{End}_{\mathrm{SH}(k)}(\mathbb{S}) = \pi_0(\mathbb{S})_0(k)$ with $\mathrm{GW}(k)$, and thus we have the Euler characteristic $\chi(X/k) \in \mathrm{GW}(k)$, or more generally, $\chi(\mathcal{X}/k) := \chi_{\mathrm{SH}(k)}(\Sigma^\infty \mathcal{X}) \in \mathrm{GW}(k)$ for a dualizable pointed space $\mathcal{X} \in \mathbf{Spc}_\bullet(k)$.

Elementary properties of the Euler characteristic (see [**11**, §§1.1, 1.2])

1. Let $S^{p,q}_k = S^{p-q} \wedge \mathbb{G}_m^{\wedge q}$. Then $\chi(S^{p,q}_k/k) = (-1)^p \cdot \langle -1\rangle^q$.

2. Let F, X and Y be in $\mathbf{Sm}/k$ and let $p : Y \to X$ be a Nisnevich (or Zariski) locally trivial fiber bundle with fiber F. Then

$$\chi(Y/k) = \chi(X/k) \cdot \chi(F/k).$$

3. Let X be in $\mathbf{Sm}/k$ and let $p : V \to X$ be a rank r vector bundle. Then $\mathrm{Th}(V)$ is dualizable and

$$\chi(\mathrm{Th}(V)/k) = \langle -1\rangle^r \cdot \chi(X/k).$$

4. Let X be in $\mathbf{Sm}/k$ and let $p : V \to X$ be a rank r vector bundle. Let $q : \mathbb{P}(V) \to X$ be the associated projective space bundle $\mathrm{Proj}_X(\mathrm{Sym}^* V^\vee)$. Then $\mathbb{P}(V)$ is dualizable and

$$\chi(\mathbb{P}(V)/k) = (\sum_{i=0}^{r-1} \langle -1\rangle^i) \cdot \chi(X/k).$$

5. Let $i : Z \to X$ be a codimension c closed immersion in $\mathbf{Sm}/k$. Let $\tilde{X}$ be the blow up of X along Z and let $p : N_Z \to Z$ be the normal bundle of i. . Then

$$\chi(\tilde{X}/k) = \chi(X/k) + (\sum_{i=1}^{c-1} \langle -1\rangle^i) \cdot \chi(Z/k).$$

6. Let $\sigma : k \to F$ be an extension of fields, inducing the homomorphism $\sigma_* : \mathrm{GW}(k) \to \mathrm{GW}(F)$. Then

$$\chi(X_F/F) = \sigma_*(\chi(X/k)).$$

7. For $\sigma : k \to \mathbb{C}$, and for $X \in \mathbf{Sm}/k$, $\mathrm{rank}\chi(X/k) = \chi^{top}(X^\sigma(\mathbb{C}))$.

8. For $\sigma : k \to \mathbb{R}$, $\mathrm{sig}(\sigma_*\chi(X/k)) = \chi^{top}(X^\sigma(\mathbb{R}))$. In consequence $\chi^{top}(X^\sigma(\mathbb{R})) \equiv \chi^{top}(X^\sigma_{\mathbb{C}}(\mathbb{C}))$ mod 2.

1.2. Motivic Gauss-Bonnet. One can define Euler classes of vector bundles in a variety of situations. Déglise-Jin-Khan [**6**] have given a definition with values in the motivic cohomotopy of the Thom space of $-V$; we discuss here a version with values in an SL-oriented theory, following [**12**]. For notions concerning SL-orientations and the basic properties of SL-oriented theories, we refer the reader to [**2**].

DEFINITION 1.2.1. An SL-*orientation* on a motivic commutative ring spectrum $\mathcal{E} \in \mathrm{SH}(k)$ is the functorial assignment of a *Thom class* $\theta_{V,\phi} \in \mathcal{E}^{2r,r}(\mathrm{Th}(V))$ for each rank r vector bundle $V \to X$, $X \in \mathbf{Sm}/k$ endowed with a trivialization of the determinant $\phi : \det V \xrightarrow{\sim} \mathcal{O}_X$, satisfying

1. For $V = \oplus_{i=1}^r \mathcal{O}_X \cdot e_i$, and ϕ the isomorphism $e_1 \wedge \ldots \wedge e_r \mapsto 1$, $\theta_{V,\phi} \in \mathcal{E}^{2r,r}(\mathrm{Th}(V))$ is the image of the unit map $\mathbb{S} \to \mathcal{E}$ via the canonical isomorphisms
$$[\mathbb{S}, \mathcal{E}]_{\mathrm{SH}(k)} \xrightarrow{\sim} [\Sigma_T^r \mathbb{S}, \Sigma_T^r \mathcal{E}]_{\mathrm{SH}(k)} \xrightarrow{\sim} [\mathrm{Th}(\mathcal{O}_X^r), \Sigma_T^r \mathcal{E}]_{\mathrm{SH}(k)} = \mathcal{E}^{2r,r}(\mathrm{Th}(V)).$$
2. $\theta_{V_1 \oplus V_2, \phi_1 \wedge \phi_2} = p_1^* \theta_{V_1,\phi_1} \cup p_2^* \theta_{V_2,\phi_2}$.

For $\mathcal{E}$ an SL-oriented theory, and $V \to X$ a rank r vector bundle with trivialized determinant, the map
$$\vartheta_{V,\phi} := (- \cup \theta_{V,\phi}) \circ \pi^* : \mathcal{E}^{a,b}(X) \to \mathcal{E}^{a+2r,b+r}(\mathrm{Th}(V))$$
is an isomorphism. For $L \to X$ a line bundle, one defines
$$\mathcal{E}^{a,b}(X; L) := \mathcal{E}^{a+2,b+1}(\mathrm{Th}(L)).$$
For $p : V \to X$ a vector bundle let $\pi : L \to X$ be the line bundle $\det V \to X$. Then we have canonical trivializations ϕ_1, ϕ_2 of the determinant bundles of $p_1 : \pi^*(V \oplus L^{-1}) \to L$ and $p_2 : p^*(L^{-1} \oplus L) \to V$. The diagram
$$\begin{array}{ccccc} & & \mathcal{E}^{a+2r+4,b+r+2}(\mathrm{Th}(V \oplus L \oplus L^{-1})) & & \\ & \nearrow^{\vartheta_{\pi^*(V \oplus L^{-1}),\phi_1}} & & \nwarrow^{\vartheta_{p^*(L^{-1} \oplus L),\phi_2}} & \\ \mathcal{E}^{a+2,b+1}(\mathrm{Th}(L)) & & & & \mathcal{E}^{a+2r,b+r}(\mathrm{Th}(V)) \end{array}$$
gives an isomorphism
$$\vartheta_V : \mathcal{E}^{a,b}(X; \det V) \to \mathcal{E}^{a+2r,b+r}(\mathrm{Th}(V))$$
If we have $\phi : \det V \xrightarrow{\sim} \mathcal{O}_X$, the suspension isomorphism
$$\mathcal{E}^{a,b}(X) \cong \mathcal{E}^{a+2,b+1}(\mathrm{Th}(\mathcal{O}_X))$$
gives the isomorphism $\phi_* : \mathcal{E}^{a,b}(X) \to \mathcal{E}^{a,b}(X; \det^{-1} V)$ and one checks that the composition
$$\mathcal{E}^{a,b}(X) \xrightarrow{\phi_*} \mathcal{E}^{a,b}(X; \det{}^{-1} V) \xrightarrow{\vartheta_V} \mathcal{E}^{2r,r}(\mathrm{Th}(V))$$
is $\vartheta_{V,\phi}$.

Let $\theta_V \in \mathcal{E}^{2r,r}(\mathrm{Th}(V); \det^{-1} V)$ be the element corresponding to
$$\theta_{V \oplus \det^{-1} V, can} \in \mathcal{E}^{2r+2,r+1}(\mathrm{Th}(V \oplus \det{}^{-1} V));$$
we call θ_V the *canonical Thom class* for V. Again, if $\phi : \det V \to \mathcal{O}_X$ is a trivialization of $\det V$, then via the isomorphism
$$\mathcal{E}^{2r,r}(\mathrm{Th}(V); \det{}^{-1} V) \xrightarrow{\phi_*^{-1}} \mathcal{E}^{2r,r}(\mathrm{Th}(V))$$
as above, θ_V maps to $\theta_{V,\phi}$, so using the "orientation bundle" $\det^{-1} V$ we have canonical Thom classes without choice of a trivialization. See [**2**, §4] for details on these constructions.

DEFINITION 1.2.2. Let $\mathcal{E}$ be an SL-oriented theory and let $V \to X$ be a rank r vector bundle on $X \in \mathbf{Sm}/k$ with 0-section $s : X \to V$. Define the *Euler class* $e(V) \in \mathcal{E}^{2r,r}(X; \det^{-1} V)$ by
$$e(V) := s^* \theta_V.$$

For $\mathcal{E}$ an SL-oriented theory, one has functorial push-forward maps

$$p_* : \mathcal{E}^{a,b}(Y, \omega_{Y/k} \otimes f^*L) \to \mathcal{E}^{a-2d,b-d}(X, \omega_{X/k} \otimes L)$$

for each proper map $p : Y \to X$ of relative dimension d in $\mathbf{Sm}/k$. To see this, we introduce a part of the Grothendieck six-functor formalism, available for the motivic stable homotopy category through work of Ayoub [**1**], Cisinski-Déglise [**5**] and Hoyois [**9**]. This involves two sets of adjoint functors for each morphism of finite type B-schemes $f : Y \to X$ over some fixed base-scheme B:

$$f^* : \mathrm{SH}(X) \rightleftarrows \mathrm{SH}(Y) : f_*$$

$$f_! : \mathrm{SH}(Y) \rightleftarrows \mathrm{SH}(X) : f^!.$$

The left adjoints on the left, the ones with lower decorations are covariant in f, those with upper decorations are contravariant. One has the respective units and co-units of adjunction, and a natural transformation $\eta_!^f : f_! \to f_*$, which is an isomorphism if f is proper.

Back to our base-scheme $B = \mathrm{Spec}\, k$ and a proper map $p : Y \to X$ in $\mathbf{Sm}/k$. Let $\pi_X : X \to \mathrm{Spec}\, k$, $\pi_Y : Y \to \mathrm{Spec}\, k$ be the structure maps. We have the natural transformation (*proper pullback*)

$$p^* : \pi_{X!} \to \pi_{Y!} \circ p^* : \mathrm{SH}(X) \to \mathrm{SH}(k) \tag{1.2.1}$$

defined as the composition

$$\pi_{X!} \xrightarrow{u_p} \pi_{X!}p_*p^* \xrightarrow{(\eta^p_{!*})^{-1}} \pi_{X!}p_!p^* \cong \pi_{Y!} \circ p^*.$$

Applying p^* to 1_X and noting that $\pi_{X!}(1_X) = \mathrm{Th}(-T_{X/k})$, $\pi_{Y!}(1_Y) = \mathrm{Th}(-T_{Y/k})$ gives the morphism

$$p^* : \mathrm{Th}(-T_{X/k}) \to \mathrm{Th}(-T_{Y/k})$$

in $\mathrm{SH}(k)$. One checks easily that $(pq)^* = q^*p^*$ for composable proper morphisms p and q.

Since $\det(-T_{X/k}) = \det^{-1}(T_{X/k}) = \omega_{X/k}$, and similarly for Y, we have

$$\mathcal{E}^{a,b}(Y, \omega_{Y/k} \otimes f^*L) = \mathcal{E}^{a+2\mathrm{dim}Y, b+\mathrm{dim}Y}(\mathrm{Th}(-T_{Y/k}); f^*L)$$

and

$$\mathcal{E}^{a',b'}(X, \omega_{X/k} \otimes L) = \mathcal{E}^{a'+2\mathrm{dim}X, b'+\mathrm{dim}X}(\mathrm{Th}(-T_{X/k}); L)$$

so the map p^* induces the map

$$p_* : \mathcal{E}^{a,b}(Y, \omega_{Y/k} \otimes f^*L) \to \mathcal{E}^{a-2d,b-d}(X, \omega_{X/k} \otimes L)$$

on $\mathcal{E}$-cohomology.

Via these constructions, we have another description of $\theta_V \in \mathcal{E}^{2r,r}(X; \det^{-1} V)$ as

$$\theta_V = s_*(1_X^{\mathcal{E}})$$

where $1_X^{\mathcal{E}} \in \mathcal{E}^{0,0}(X)$ is the unit and $s : X \to V$ is the 0-section.

The diagonal $\Delta_X : X \to X \times_k X$ induces the map

$$\Delta_* : X \to X \times_k X/(X \times_k X - \Delta_X(X)) \cong \mathrm{Th}(N_{\Delta_X})$$

The composition

$$T_{X/k} \xrightarrow{p_2^*} \Delta_X^* p_2^* T_{X/k} \xrightarrow{i_2} \Delta_X^*(p_1^*T_{X/k} \oplus p_2^*T_{X/k}) \cong \Delta_X^* T_{X \times_k X/k} \xrightarrow{\pi} N_{\Delta_X}$$

is an isomorphism.

Letting $j : X \times_k X - \Delta_X(X) \to X \times_k X$ be the inclusion, the map Δ_* induces the map $\beta_X : \mathrm{Th}(-T_{X/k}) \to \Sigma^\infty_T X_+$ as the composition

$$\begin{aligned}\mathrm{Th}(-T_{X/k}) \xrightarrow{\mathrm{Th}(\Delta_X)} \mathrm{Th}(-p_1^*T_{X/k}) &\xrightarrow{q} \mathrm{Th}(-p_1^*T_{X/k})/\mathrm{Th}(-j^*p_1^*T_{X/k}) \\ &\cong \mathrm{Th}(-T_{X/k} \oplus N_{\Delta_X}) \cong \mathrm{Th}(-T_{X/k} \oplus T_{X/k}) \cong \Sigma^\infty_T X_+\end{aligned}$$

LEMMA 1.2.3. *For X smooth and proper over k, $\chi(X/k)$ is equal to the composition*

$$\mathbb{S}_k \xrightarrow{\pi_X^*} \mathrm{Th}(-T_{X/k}) \xrightarrow{\beta_X} \Sigma^\infty_T X_+ \xrightarrow{\pi_X} \mathbb{S}_k$$

PROOF. Let $p_i : X \times_k X \to X$, $i = 1, 2$, be the projections. The map

$$ev_X : (\Sigma^\infty_T X_+)^\vee \wedge \Sigma^\infty_T X_+ \to \mathbb{S}_k$$

is the composition

$$\begin{aligned}(\Sigma^\infty_T X_+)^\vee \wedge \Sigma^\infty_T X_+ &= \mathrm{Th}(-T_{X/k}) \wedge \Sigma^\infty_T X_+ \\ &= \mathrm{Th}(-p_1^*T_{X/k}) \xrightarrow{q} \mathrm{Th}(-p_1^*T_{X/k})/\mathrm{Th}(-j^*p_1^*T_{X/k}) \cong \mathrm{Th}(-T_{X/k} \oplus N_{\Delta_X}) \\ &\cong \mathrm{Th}(-T_{X/k} \oplus T_{X/k}) \cong \Sigma^\infty_T X_+ \xrightarrow{\pi_X} \mathbb{S}_k\end{aligned}$$

Thus $\pi_X \circ \beta_X = ev_X \circ \mathrm{Th}(\Delta_X)$. Also, $\pi_X^* = \pi_X^\vee$ (see [**9**, Corollary 6.13]) and $\pi_X^\vee$ is given by

$$\mathbb{S}_k \xrightarrow{\delta_X} \Sigma^\infty_T X_+ \wedge (\Sigma^\infty_T X_+)^\vee = \mathrm{Th}(-p_2^*T_{X/k}) \xrightarrow{p_2} \mathrm{Th}(-T_{X/k})$$

It follows from the construction of the map δ_X (see for example [**9**, §5.3, Corollary 6.13]) that

$$\delta_X = \mathrm{Th}(\Delta_X) \circ \pi_X^*.$$

This gives us the commutative diagram

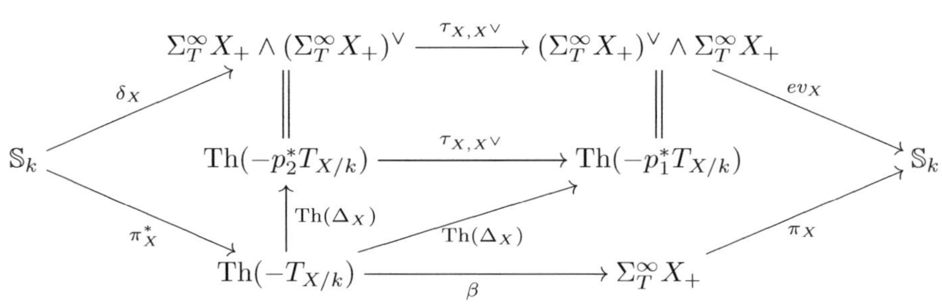

and the result follows from the definition of $\chi(X/k)$ as $ev_X \circ \tau_{X,X^\vee} \circ \delta_X$. □

THEOREM 1.2.4 (Motivic Gauss-Bonnet [**12**, Theorem 5.2]). *Let $\mathcal{E}$ be an SL-oriented theory, $\pi_X \to \mathrm{Spec}\, k$ a smooth and projective k-scheme, let $u_\mathcal{E} : \mathbb{S} \to \mathcal{E}$ be the unit map. Then*

$$\pi_{X*}(e^\mathcal{E}(T_{X/k})) = u_{\mathcal{E}*}(\chi(X/k)) \in \mathcal{E}^{0,0}(k).$$

PROOF. We use a Jouanolou cover $p : \tilde{X} \to X$, that is, p is an affine space bundle and $\tilde{X}$ is affine. There is then a vector bundle $p_\nu : \nu_{\tilde{X}} \to \tilde{X}$ and an isomorphism

$$p^*T_{X/k} \oplus \nu_{\tilde{X}} \cong \mathcal{O}^N_{\tilde{X}},$$

which gives the isomorphism of virtual bundles (i.e. in $K_0(\tilde{X})$) $\nu_{\tilde{X}} \cong -p^*T_{X/k} \oplus \mathcal{O}_{\tilde{X}}^N$, giving in turn an isomorphism in $\mathrm{SH}(k)$

$$\Sigma_T^N \mathrm{Th}(-T_{X/k}) \cong \Sigma_T^\infty \mathrm{Th}(\nu_{\tilde{X}}).$$

Let $i_1 : \nu_{\tilde{X}} \to \nu_{\tilde{X}} \oplus p^*T_{X/k}$ be the inclusion. Via the projection $\pi_1 : \nu_{\tilde{X}} \oplus p^*T_{X/k} \to \nu_{\tilde{X}}$, we can view $\nu_{\tilde{X}} \oplus p^*T_{X/k}$ as the vector bundle $p_\nu^* p^* T_{X/k}$ on $\nu_{\tilde{X}}$ with 0-section i_1.

We have the maps $\Delta_X^1 : \tilde{X} \to \tilde{X} \times_k X$, $\Delta_X^2 : \tilde{X} \to X \times_k \tilde{X}$ induced by p. The map $\Sigma_T^N \beta_X$ is the stabilization of the map

$$\begin{aligned}\mathrm{Th}(\nu_{\tilde{X}}) &\xrightarrow{\mathrm{Th}(\Delta_X^1)} \mathrm{Th}(p_1^*\nu_{\tilde{X}}) \xrightarrow{q} \mathrm{Th}(p_1^*\nu_{\tilde{X}})/\mathrm{Th}(j^*p_1^*\nu_{\tilde{X}}) \\ &\cong \mathrm{Th}(\nu_{\tilde{X}} \oplus N_{\Delta_X^1}) \cong \mathrm{Th}(\nu_{\tilde{X}} \oplus p^*T_{X/k}) \cong \mathrm{Th}(\mathcal{O}_{\tilde{X}}^N) = \Sigma_T^N \tilde{X}_+ \cong \Sigma_T^N X_+\end{aligned}$$

Thus the map on $\mathcal{E}^{a,b}$ induced by β_X is the stabilization of $\mathrm{Th}(i_1)^* \circ susp$, with

$$\mathrm{Th}(i_1)^* : \mathcal{E}^{a+2N,b+N}(\mathrm{Th}(\nu_{\tilde{X}} \oplus p^*T_{X/k})) \to \mathcal{E}^{a+2N,b+N}(\mathrm{Th}(\nu_{\tilde{X}})) = \mathcal{E}^{a,b}(\mathrm{Th}(-T_{X/k}))$$

and $susp$ the suspension isomorphism

$$\mathcal{E}^{a,b}(\tilde{X}) \cong \mathcal{E}^{a+2N,b+N}(\mathrm{Th}(\mathcal{O}_{\tilde{X}}^N)) \cong \mathcal{E}^{a+2N,b+N}(\mathrm{Th}(\nu_{\tilde{X}} \oplus p^*T_{X/k})).$$

Identifying $\mathcal{E}^{a+2N,b+N}(\mathrm{Th}(\nu_{\tilde{X}} \oplus p^*T_{X/k}))$ with the $\mathcal{E}$-cohomology with support $\mathcal{E}_0^{a+2N,b+N}(\nu_{\tilde{X}} \oplus p^*T_{X/k})$ and $\mathcal{E}^{a+2N,b+N}(\nu_{\tilde{X}})$ with $\mathcal{E}_0^{a+2N,b+N}(\nu_{\tilde{X}})$ ($0 =$ respective 0-section) identifies $\mathrm{Th}(i_1)^*$ with the map

$$i_1^* : \mathcal{E}_0^{a+2N,b+N}(\nu_{\tilde{X}} \oplus p^*T_{X/k}) \to \mathcal{E}_0^{a+2N,b+N}(\nu_{\tilde{X}})$$

Via the isomorphism $\mathcal{E}_0^{a+2N,b+N}(\nu_{\tilde{X}}) \cong \mathcal{E}^{a,b}(\mathrm{Th}(-T_{X/k}))$, this gives the description of $\mathcal{E}^{a,b}(\beta_X)^* : \mathcal{E}^{a,b}(X) \to \mathcal{E}^{a,b}(\mathrm{Th}(-T_{X/k}))$ as $i_1^* \circ susp$.

The Thom classes $\theta_{p^*T_{X/k}} = p^*\theta_{T_{X/k}}$ and $\theta_{\nu_{\tilde{X}}}$ give the Thom isomorphisms

$$\vartheta_{p^*T_{X/k}} : \mathcal{E}^{0,0}(\tilde{X}) \to \mathcal{E}_0^{2d_X,d_X}(p^*T_{X/k}, \det{}^{-1}(p^*T_{X/k}))$$

$$\vartheta_{p_\nu^*p^*T_{X/k}} : \mathcal{E}^{2(N-d_X),N-d_X}(\nu_{\tilde{X}}) \to \mathcal{E}_0^{2N,N}(\nu_{\tilde{X}} \oplus p^*T_{X/k}, \det{}^{-1}(\nu_{\tilde{X}} \oplus p^*T_{X/k}))$$

and

$$\vartheta_{\nu_{\tilde{X}}} : \mathcal{E}^{0,0}(\tilde{X}) \to \mathcal{E}_0^{2(N-d_X)+,N-d_X}(\nu_{\tilde{X}}, \det{}^{-1}(\nu_{\tilde{X}})).$$

Moreover, we have

$$\pi_1^*\theta_{\nu_{\tilde{X}}} \cup \pi_2^*\theta_{p^*T_{X/k}} = \theta_{\nu_{\tilde{X}} \oplus p^*T_{X/k}} = \theta_{\mathcal{O}_{\tilde{X}}^N},$$

giving us the commutative diagram

$$\begin{array}{ccc} \mathcal{E}^{0,0}(\tilde{X}) & \xrightarrow{\vartheta_{\nu_{\tilde{X}}}} & \mathcal{E}_0^{2(N-d_X),N-d_X}(\nu_{\tilde{X}}, \det{}^{-1}\nu_{\tilde{X}}) \\ {\scriptstyle \vartheta_{p^*T_{X/k}}}\downarrow & \searrow^{susp} & \downarrow{\scriptstyle \vartheta_{p_\nu^*p^*T_{X/k}}} \\ \mathcal{E}_0^{2d_X,d_X}(p^*T_{X/k}, \det{}^{-1}p^*T_{X/k}) & \xrightarrow{\vartheta_{\nu_{\tilde{X}}}} & \mathcal{E}_0^{2N,N}(\nu_{\tilde{X}} \oplus p^*T_{X/k}, \det{}^{-1}p^*T_{X/k} \otimes \det{}^{-1}\nu_{\tilde{X}}). \end{array}$$

With $s : \tilde{X} \to p^*T_{X/k}$ the 0-section, this gives us the commutative diagram

$$
\begin{array}{ccc}
\mathcal{E}^{0,0}(\tilde{X}) & \xrightarrow{\vartheta_{\nu_{\tilde{X}}}} & \mathcal{E}_0^{2(N-d_X),N-d_X}(\nu_{\tilde{X}}, \det^{-1}\nu_{\tilde{X}}) \\
\downarrow{\scriptstyle \vartheta_{p^*T_{X/k}}} & \searrow{\scriptstyle susp} & \downarrow{\scriptstyle \vartheta_{p_\nu^* p^* T_{X/k}}} \\
\mathcal{E}_0^{2d_X,d_X}(p^*T_{X/k}, \det^{-1} p^*T_{X/k}) & \xrightarrow{\vartheta_{\nu_{\tilde{X}}}} & \mathcal{E}_0^{2N,N}(\nu_{\tilde{X}} \oplus p^*T_{X/k}, \det^{-1}\nu_{\tilde{X}} \otimes \det^{-1} p^*T_{X/k}) \\
\downarrow{\scriptstyle s^*} & & \downarrow{\scriptstyle i_1^*} \\
\mathcal{E}^{2d_X,d_X}(\tilde{X}, \det^{-1} p^*T_{X/k}) & \xrightarrow{\vartheta_{\nu_{\tilde{X}}}} & \mathcal{E}_0^{2N,N}(\nu_{\tilde{X}}, \det^{-1} p^*T_{X/k} \otimes \det^{-1}\nu_{\tilde{X}}) \\
 & \searrow{\scriptstyle \mathrm{Id}} & \uparrow{\scriptstyle \vartheta_{\nu_{\tilde{X}}}} \\
 & & \mathcal{E}^{2d_X,d_X}(\tilde{X}, \det^{-1} p^*T_{X/k})
\end{array}
$$

(with the curved arrow $\times e^{\mathcal{E}}(p^*T_{X/k})$ from $\mathcal{E}^{0,0}(\tilde{X})$ to $\mathcal{E}^{2d_X,d_X}(\tilde{X}, \det^{-1} p^*T_{X/k})$)

After the identifications

$$\mathcal{E}_0^{2N,N}(\nu_{\tilde{X}}, \det^{-1}\nu_{\tilde{X}} \otimes \det^{-1}p^*T_{X/k}) \cong \mathcal{E}^{2N,N}(\mathrm{Th}(\nu_{\tilde{X}})) \cong \mathcal{E}^{0,0}(\mathrm{Th}(-T_{X/k}))$$

$$p^* : \mathcal{E}^{0,0}(X) \xrightarrow{\sim} \mathcal{E}^{0,0}(\tilde{X}),\ p^* : \mathcal{E}^{2d_X,d_X}(X, \omega_{X/k}) \xrightarrow{\sim} \mathcal{E}^{2d_X,d_X}(\tilde{X}, \det^{-1}p^*T_{X/k})$$

and the identity $\mathcal{E}(\beta_X)^* = i_1^* \circ susp$, this gives

$$\times e^{\mathcal{E}}(T_{X/k}) = \vartheta_{-T_{X/k}}^{-1} \circ \mathcal{E}(\beta_X)^* : \mathcal{E}^{0,0}(X) \to \mathcal{E}^{2d_X,d_X}(X, \omega_{X/k})$$

Moreover, the identification $\mathrm{Th}(-T_{X/k}) = \pi_{X!}(1_X) = (\Sigma_T^\infty X_+)^\vee$ gives the commutative diagram

$$
\begin{array}{ccc}
\mathcal{E}^{2d_X,d_X}(X, \omega_{X/k}) & \xrightarrow{\pi_{X*}} & \mathcal{E}^{0,0}(k) \\
\downarrow{\scriptstyle \vartheta_{-T_{X/k}}} & \nearrow{\scriptstyle \mathcal{E}(\pi_X^*)^*} & \uparrow{\scriptstyle \mathcal{E}(\pi_X^\vee)^*} \\
\mathcal{E}^{0,0}(\mathrm{Th}(-T_{X/k})) & = & \mathcal{E}^{0,0}((\Sigma_T^\infty X_+)^\vee)
\end{array}
$$

Via Lemma 1.2.3, this gives

$$
\begin{aligned}
\pi_{X*}(e^{\mathcal{E}}(T_{X/k})) &= \pi_{X*}((\times e^{\mathcal{E}}(T_{X/k})) \circ (\pi_X^*(u_{\mathcal{E}} : \mathbb{S} \to \mathcal{E}))) \\
&= \mathcal{E}(\pi_X^*)^* \circ (\vartheta_{-T_{X/k}} \circ \times e^{\mathcal{E}}(T_{X/k})) \circ \mathcal{E}(\pi_X)^*(u_{\mathcal{E}} : \mathbb{S} \to \mathcal{E}) \\
&= \mathcal{E}(\pi_X^*)^* \circ \mathcal{E}(\beta_X)^* \circ \mathcal{E}(\pi_X)^*(u_{\mathcal{E}} : \mathbb{S} \to \mathcal{E}) \\
&= \mathcal{E}(\pi_X \circ \beta_X \circ \pi_X^*)^*(u_{\mathcal{E}} : \mathbb{S} \to \mathcal{E}) \\
&= \mathcal{E}(\chi(X/k))^*(u_{\mathcal{E}} : \mathbb{S} \to \mathcal{E}) \\
&= u_{\mathcal{E}*}(\chi(X/k))
\end{aligned}
$$

□

1.3. Chow-Witt groups. The twisted Chow-Witt groups are defined using the cohomology of the twisted Milnor-Witt sheaves (see [**3, 7**] for a somewhat different definition; see also Jean Fasel's lectures [**8**] for many details concerning the Chow-Witt groups). Sending $u \in F^\times$ to $\langle u \rangle := 1 + \eta[u] \in K_0^{MF}(F)^\times$ extends to a map of sheaves of abelian groups $\mathbb{G}_m \to (\mathcal{K}_0^{MW})^\times$. If $L \to X$ is a line bundle on some $X \in \mathbf{Sm}/k$, we have the principal $\mathbb{G}_m$-bundle $L^\times = L - 0 \to X$ and we can define the sheaf $\mathcal{K}_n^{MW}(L)$ on X by

$$\mathcal{K}_n^{MW}(L) := \mathcal{K}_n^{MW} \times^{\mathbb{G}_m} L^\times$$

Explicitly, if λ is a generating section of L over an open subset U of X, a section s of $\mathcal{K}_n^{MW}(L)$ over U may be written as $s = x \otimes \lambda$, with the relation $x \otimes u \cdot \lambda = \langle u \rangle \cdot x \otimes \lambda$ for $x \in H^0(U, \mathcal{K}_n^{MW})$, $u \in H^0(U, \mathcal{O}_X^\times)$.

The isomorphism $\mathcal{K}_0^{MW} \to \mathcal{GW}$ extends to an isomorphism $\mathcal{K}_0^{MW}(L) \to \mathcal{GW}(L)$, where $\mathcal{GW}(L)$ is the sheaf of L-valued quadratic forms; we have a similar isomorphism $\mathcal{K}_n^{MW}(L) \cong \mathcal{W}(L)$ for all $n < 0$ and an exact sequence

$$0 \to \mathcal{I}(L)^{n+1} \to \mathcal{K}_n^{MW}(L) \to \mathcal{K}_n^M \to 0$$

for $n > 0$, where $\mathcal{I}(L) \subset \mathcal{GW}(L)$ is the sheaf of twisted augmentation ideals, that is, the kernel of the rank map $\mathcal{GW}(L) \to \mathbb{Z}$.

DEFINITION 1.3.1. Let $L \to X$ be a line bundle, $X \in \mathbf{Sm}/k$. The nth twisted Chow-Witt group $\widetilde{\mathrm{CH}}^n(X; L)$ is defined as $H^n(X, \mathcal{K}_n^{MW}(L))$.

The product structure on $\mathcal{K}_*^{MW}$ gives rise to products

$$\tilde{\mathrm{CH}}^n(X; L) \times \tilde{\mathrm{CH}}^m(X, L') \to \tilde{\mathrm{CH}}^{n+m}(X; L \otimes L')$$

There are canonical isomorphisms

$$\tilde{\mathrm{CH}}^n(X; L \otimes M^{\otimes 2}) \cong \tilde{\mathrm{CH}}^n(X; L)$$

arising from the identity $\langle vu^2 \rangle = \langle v \rangle$ for u, v units.

For $f : Y \to X$ in $\mathbf{Sm}/k$, the fact that $\mathcal{K}_n^{MW}$ is a sheaf gives pull-back maps

$$f^* : \tilde{\mathrm{CH}}^n(X, L) \to \tilde{\mathrm{CH}}^n(Y, f^*L)$$

Moreover, for $f : Y \to X$ proper of relative dimensions d, there are push-forward maps

$$f_* : \tilde{\mathrm{CH}}^n(Y, \omega_{Y/k} \otimes f^*L) \to \tilde{\mathrm{CH}}^{n-d}(X, \omega_{X/k} \otimes L)$$

The push-forward maps may be described using the *Rost-Schmid complexes* (see [**14**, §5])

$$\mathcal{C}^*_{X,n,L} := \oplus_{x \in X^{(0)}} i_{x*} K_n^{MW}(k(x); L) \to \oplus_{x \in X^{(1)}} K_{n-1}^{MW}(k(x); \det^{-1}\mathfrak{m}_x/\mathfrak{m}_x^2 \otimes L) \to \\ \ldots \to \oplus \oplus_{x \in X^{(q)}} K_{n-q}^{MW}(k(x); \det^{-1}\mathfrak{m}_x/\mathfrak{m}_x^2 \otimes L) \to \ldots$$

The complex $\mathcal{C}^*_{X,n,L}$ gives a flasque resolution of $\mathcal{K}_n^{MW}(L)$ on X and taking global section gives

$$H^n(X, \mathcal{K}_n^{MW}(L)) \cong H^n(C^*_{X,L,n})$$

where $C^*_{X,L,n} = \Gamma(X, \mathcal{C}^*_{X,L,n})$. For k perfect, this gives generators for $\tilde{\mathrm{CH}}^n(Y; \omega_{Y/k} \otimes f^*L)$ as a subgroup of $\oplus_{y \in Y^{(n)}} \mathrm{GW}(k(y); f^*L \otimes k(y))$. For $x = f(y)$ with $\mathrm{codim}_X\, x = \mathrm{codim}_Y\, y - d = n - d$ and $k(x) \to k(y)$ a separable extension, the trace maps

$$\mathrm{Tr}_{k(y)/k(x)} : \mathrm{GW}(k(y); f^*L \otimes k(y)) \to \mathrm{GW}(k(x); L \otimes k(x))$$

induces the map $f_* : \tilde{\mathrm{CH}}^n(Y, \omega_{Y/k} \otimes f^*L) \to \tilde{\mathrm{CH}}^{n-d}(X, \omega_{X/k} \otimes L)$. In general, one needs to handle inseparability issues, which is done by Fasel [**7**, §5, 6]. For $n = \dim_k Y$, one can simply pass to the perfect closure of k, noting that we have assumed the characteristic is odd and that the map on Grothendieck-Witt groups associated to an odd degree extension is always injective.

Similarly, one has a description of the cohomology with supports $H^p_Z(X, \mathcal{K}_n^{MW}(L))$ as the pth cohomology of the subcomplex of $C^*_{X,L,n}$ constructed by replacing $\oplus_{x \in X^{(q)}}$ with $\oplus_{x \in X^{(q)} \cap Z}$.

In particular, for $p : V \to X$ a vector bundle of rank r with 0-section s and for $x \in X^{(0)}$ the generic point (assuming X irreducible) taking $1 \in \mathrm{GW}(k(s(x)))$ gives a well-defined class $\theta_V^{CW} \in H^r(V, \mathcal{K}_r^{MW}(\det^{-1} V))$ which in fact gives a twisted Thom class. This gives the cohomology theory

$$X \mapsto H^*(X, \mathcal{K}_*^{MW})$$

the properties one would expect from an SL-orientation.

For instance, we have the Euler class $e^{CW}(V) \in H^r(X, \mathcal{K}_r^{MW}(\det^{-1} V)) = \tilde{\mathrm{CH}}^r(X; \det^{-1} V)$ for $V \to X$ a rank r vector bundle, defined as

$$e^{CW}(V) := s^* \theta_V^{CW}$$

for $s : X \to V$ the 0-section.

DEFINITION 1.3.2. Let $\pi_X : X \to \mathrm{Spec}\, k$ be a smooth projective k-scheme. The Chow-Witt Euler characteristic $\chi^{CW}(X/k) \in \tilde{\mathrm{CH}}^0(\mathrm{Spec}\, k) = \mathrm{GW}(k)$ is defined as

$$\chi^{CW}(X/k) := \pi_{X*} e^{CW}(T_{X/k}).$$

In fact, this cohomology theory is represented by a motivic commutative ring spectrum with an SL-orientation. Let $\mathrm{EM}(\mathcal{K}_n^{MW}, n)$ be the presheaf of spectra on $\mathbf{Sm}/k$ associated to the presheaf $\mathcal{K}_n^{MW}[n]$. There is a canonical isomorphism of presheaves $X \mapsto \mathcal{K}_{n+1}^{MW}(X_+ \wedge \mathbb{G}_m)$ with $\mathcal{K}_n^{MW}$, giving the canonical weak equivalence

$$\epsilon_n : \mathrm{EM}(\mathcal{K}_n^{MW}, n) \to \Omega_T \, \mathrm{EM}(\mathcal{K}_{n+1}^{MW}, n+1)$$

Thus the sequence $(\mathrm{EM}(\mathcal{K}_0^{MW}, 0), \mathrm{EM}(\mathcal{K}_1^{MW}, 1), \ldots)$ with bonding maps ϵ_n gives rise to a T-spectrum $\mathrm{EM}(\mathcal{K}_*^{MW})$ with

$$\mathrm{EM}(\mathcal{K}_*^{MW})^{a+b,b}(X) \cong H^a(X, \mathcal{K}_b^{MW})$$

The Thom classes in Milnor-Witt cohomology described above thus give an SL-orientation for $\mathrm{EM}(\mathcal{K}_*^{MW})$ and one has a canonical isomorphism of $\mathrm{EM}(\mathcal{K}_*^{MW})^{a+b,b}(X; L)$ with the group $H^a(X, \mathcal{K}_b^{MW}(L))$ defined via the twisted Milnor-Witt cohomology. One shows that the resulting push-forward maps in $\mathrm{EM}(\mathcal{K}_*^{MW})$-cohomology for projective morphisms agree with those already defined for twisted Chow-Witt groups by Fasel (see [**11**, Theorem 5.17]). This gives us

THEOREM 1.3.3. *Let X be a smooth projective k scheme. Then*

$$\chi(X/k) = \chi^{CW}(X/k)$$

in $\mathrm{GW}(k)$

PROOF. By the motivic Gauss-Bonnet theorem, we have

$$\chi^{CW}(X/k) = u_*^{\mathrm{EM}(\mathcal{K}_*^{MW})}(\chi(X/k))$$

But on $\pi_0(-)_0$, the unit map $\mathbb{S} \to \mathrm{EM}(\mathcal{K}_*^{MW})$ is just Morel's isomorphism

$$\pi_0(\mathbb{S}_k)_0(k) = \mathbb{S}_k^{0,0}(k) \to \mathrm{EM}(\mathcal{K}_*^{MW})^{0,0}(k) = \mathrm{GW}(k)$$

giving us the desired identity. □

REMARK 1.3.4. A similar argument, using the classical Chow groups instead of the Chow-Witt groups, show that

$$\mathrm{rnk}\chi(X/k) = \chi^{CH}(X/k) := \pi_{X*}(c_{\dim_k X}(T_{X/k})) \in \mathrm{CH}^0(\mathrm{Spec}\, k) = \mathbb{Z}.$$

Here we represent the Chow groups as the cohomology of the Milnor K-sheaves $\mathcal{K}^M$, and use the fact that the unit map $\mathbb{S} \to \mathrm{EM}(\mathcal{K}^M_*)$ induces the rank homomorphism $\mathrm{GW}(k) \to \mathbb{Z} = H^0(\mathrm{Spec}\, k, \mathcal{K}^M_0)$. The $H^*(-, \mathcal{K}^M_*)$ Euler class of a rank r vector bundle $V \to X$ is the usual Chern class $c_r(V) \in H^r(X, \mathcal{K}^M_r) \cong \mathrm{CH}^r(X)$.

1.4. Hodge cohomology. Another SL-oriented theory is algebraic K-theory, represented by the T-spectrum KGL. KGL is oriented, i.e. has Thom classes $\theta_V \in \mathrm{KGL}^{2r,r}(\mathrm{Th}(V))$ for all vector bundles $p : V \to X$, $r = \mathrm{rnk}\, V$. The Thom class is given by the class of the standard Koszul complex

$$\mathrm{Kos}(p^*V^\vee, s^\vee_{can}) : \ \Lambda^* p^* V^\vee : 0 \to \Lambda^r p^* V^\vee \to \ldots \to p^* V^\vee \xrightarrow{s^\vee_{can}} \mathcal{O}_V \to 0$$

associated to the tautological section $s_{can} : V \to p^*V$, and giving a resolution of $s_* \mathcal{O}_X$. Here $\mathrm{Kos}(p^*V^\vee, s^\vee_{can})$ is to be considered as a perfect complex on V with $j^*\mathrm{Kos}(p^*V^\vee, s^\vee_{can})$ acyclic, where $j : V - 0_V \to V$ is the inclusion, i.e., $\mathrm{Kos}(p^*V^\vee, s^\vee_{can})$ defines an element of the relative K_0, $K_0(\mathrm{Perf}(V, V - 0_V))$. Algebraic Bott periodicity gives a canonical isomorphism

$$\mathrm{KGL}^{2r,r}(\mathrm{Th}(V)) \cong K_0(\mathrm{Perf}(V, V - 0_V))$$

and $\theta^{\mathrm{KGL}}_V = [\Lambda^* p^* V^\vee]$. Then

$$e^{\mathrm{KGL}}(V) = s^*[\Lambda^* p^* V^\vee] = \sum_{j=0}^{\mathrm{rnk}V} [\Lambda^j V^\vee [j]] = \sum_{j=0}^{\mathrm{rnk}V} (-1)^j [\Lambda^j V^\vee].$$

The pushforward associated to a proper map $f : Y \to X$ of relative dimension d,

$$\pi_{X*} : \mathrm{KGL}^{2r,r}(Y) \to \mathrm{KGL}^{2(r-d),r-d}(X),$$

is the usual push-forward

$$\pi_{X*}([\mathcal{E}]) = \sum_{i=0}^{\dim Y} [R^i f_* \mathcal{E}] \in K_0(X) = \mathrm{KGL}^{2(r-d),r-d}(X),$$

for $[\mathcal{E}] \in \mathrm{KGL}^{2r,r}(Y) = K_0(Y)$.

We apply this to $e(T_{X/k})$ for $\pi_X : X \to \mathrm{Spec}\, k$ smooth and projective.

$$e(T_{X/k}) = \sum_{j=0}^{\dim X} (-1)^j [\Omega^j_{X/k}]$$

and

$$\pi_{X*}(e(T_{X/k})) = \sum_{i,j=0}^{\dim X} (-1)^{i+j} [H^i(X, \Omega^j_{X/k})]$$

The unit map $u^{\mathrm{KGL}} : \mathrm{GW}(k) \to K_0(k) = \mathbb{Z}$ is the rank map, so the motivic Gauss-Bonnet theorem gives

THEOREM 1.4.1. *For $X \in \mathbf{Sm}/k$ projective, we have*

$$\chi^{\acute{e}t}(X) = \sum_{i,j=0}^{\dim X} (-1)^{i+j} \dim_k H^i(X, \Omega^j_{X/k}).$$

Always assuming $\mathrm{char}\, k \neq 2$, this can be refined to give a formula for $\chi(X/k)$ in terms of Hodge cohomology by using hermitian K-theory. By work of Panin-Walter [**19**] and Schlichting-Tripathi [**21**], hermitian K-theory is represented by a motivic commutative ring spectrum $\mathrm{KO} \in \mathrm{SH}(k)$. This is an SL-oriented theory with a direct connection with Schlichting's Grothendieck-Witt groups [**20**]. More precisely, there are functorial isomorphisms

$$\mathrm{KO}^{2r,r}(X;L) \cong \mathrm{GW}(D_{perf}(X), L[r], can)$$

where $L \to X$ is a line bundle and $\mathrm{GW}(D_{perf}(X), L[r], can)$ is the Grothendieck-Witt group of *$L[r]$-valued quadratic forms on $D_{perf}(X)$*.

DEFINITION 1.4.2. Let $L \to X$ is a line bundle be a line bundle. An $L[n]$-valued quadratic form on $C \in D_{perf}(X)$ is a map

$$\phi : C \otimes_{\mathcal{O}_X} C \to L[n]$$

in $D_{perf}(X)$ which is

i. *non-degenerate*: the induced map $C \to \mathcal{RH}om(C, L[n])$ is an isomorphism in $D_{perf}(X)$.
ii. *symmetric*: $\phi \circ \tau = \phi$, where $\tau : C \otimes_{\mathcal{O}_X} C \to C \otimes_{\mathcal{O}_X} C$ is the symmetry isomorphism.

For a rank r vector bundle $p : V \to X$, the Thom class

$$\theta_V^{\mathrm{KO}} \in \mathrm{KO}^{2r,r}(V; p^*\det^{-1} V)$$

is given by the Koszul complex $\mathrm{Kos}(p^*V^\vee, s_{can}^\vee)$, with the quadratic form

$$\phi : \mathrm{Kos}(p^*V^\vee, s_{can}^\vee) \otimes \mathrm{Kos}(p^*V^\vee, s_{can}^\vee) \to p^*\det^{-1} V[r] = \Lambda^r V^\vee[r]$$

given by the usual exterior product

$$- \wedge - : \Lambda^i V^\vee \otimes \Lambda^{r-i} V^\vee \to \Lambda^r V^\vee.$$

For $f : Y \to X$ a proper map of relative dimension d_f in $\mathbf{Sm}/k$, the resulting push-forward map

$$f_* : \mathrm{KO}^{2r,r}(Y, \omega_{Y/k} \otimes f^*L) \to \mathrm{KO}^{2r-2d_f, r-d_f}(X, \omega_{X/k} \otimes L)$$

is induced by Grothendieck-Serre duality (see [**4**], where this is worked out for Witt-theory; the same construction works for the Grothendieck-Witt groups) as follows: Given a quadratic form $\phi : C \otimes C \to \omega_{Y/k} \otimes f^*L[r]$ we have the corresponding isomorphism

$$\tilde{\phi} : C \to \mathcal{RH}om(C, \omega_{Y/k} \otimes f^*L[r]) = \mathcal{RH}om(C, \omega_{Y/X} \otimes f^*(\omega_{X/k} \otimes L[r]))$$

Grothendieck-Serre duality gives the isomorphism

$$Rf_*\mathcal{RH}om(C, \omega_{Y/X} \otimes f^*(\omega_{X/k} \otimes L[r])) \xrightarrow[\sim]{\psi} \mathcal{RH}om(Rf_*C, \omega_{X/k} \otimes L[r - d_f])$$

which gives the isomorphism

$$\psi \circ \tilde{\phi} : Rf_*C \to \mathcal{RH}om(Rf_*C, \omega_{X/k} \otimes L[r - d_f])$$

or

$$Rf_*(\phi) : Rf_*C \otimes Rf_*C \to \omega_{X/k} \otimes L[r - d_f]$$

which one shows is symmetric; explicitly, one has $f_*(C, \phi) = (Rf_*C, Rf_*(\phi))$.

If we apply this to $V = T_{X/k}$, $f = \pi_X : X \to \operatorname{Spec} k$ a smooth and proper k-scheme, we have the following formula for $\pi_{X*}(e^{\mathrm{KO}}(T_{X/k}))$

$$\pi_{X*}(e^{\mathrm{KO}}(T_{X/k})) = (\oplus_{i,j=0}^{\dim_k X} H^i(X, \Omega^j_{X/k})[j-i], \mathrm{Tr})$$

where

$$\mathrm{Tr} : (\oplus_{i,j=0}^{\dim_k X} H^i(X, \Omega^j_{X/k})[j-i]) \otimes (\oplus_{i,j=0}^{\dim_k X} H^i(X, \Omega^j_{X/k})[j-i]) \to k$$

is the quadratic form in $D^b(k)$ given by the compositions

$$H^i(X, \Omega^j_{X/k}) \otimes H^{d_X - i}(X, \Omega^{d_X - j}_{X/k}) \xrightarrow{\cup} H^{d_X}(X, \Omega^{d_X}_{X/k}) \xrightarrow{\mathrm{Tr}} k,\ d_X = \dim_k X.$$

Indeed, for $s : X \to T_{X/k}$ the 0-section, we have

$$e^{\mathrm{KO}}(T_{X/k}) = s^*(\mathrm{Kos}(T_{X/k}), \phi) = (\oplus_{j=0}^{d_X} \Omega^j_{X/k}[j], s^*\phi)$$

with $s^*\phi$ given by the products

$$\Omega^j_{X/k}[j] \otimes \Omega^{d_X - j}_{X/k}[d_X - j] \to \omega_{X/k}[d_X]$$

and thus $\pi_*^{\mathrm{KO}}(e^{\mathrm{KO}}(T_{X/k}))$ is $\oplus_{i,j=0}^{d_X} H^i(X, \Omega^j_{X/k})[j-i])$ with the quadratic form Tr as described above.

Since the unit map $u^{\mathrm{KO}} : \mathbb{S} \to \mathrm{KO}$ induces the identity map

$$\mathrm{GW}(k) \cong \mathbb{S}^{0,0}(k) \xrightarrow{u_*^{\mathrm{KO}}} \mathrm{KO}^{0,0}(k) \cong \mathrm{GW}(k),$$

applying the motivic Gauss-Bonnet theorem gives the following formula proposed by Serre:

THEOREM 1.4.3 ([**12**, Theorem 8.6]). *Let X be a smooth projective k-scheme. Then*

$$\chi(X/k) = (\oplus_{i,j=0}^{\dim_k X} H^i(X, \Omega^j_{X/k})[j-i], \mathit{Tr}).$$

COROLLARY 1.4.4. *For X a smooth and projective k-scheme of odd dimension $2n-1$, $\chi(X/k) = m \cdot H$, H the hyperbolic form $x^2 - y^2$, with*

$$m = \sum_{i+j<2n-1} \dim_k H^i(X, \Omega^j_{X/k}) + \sum_{\substack{0 \le i < j \\ i+j = 2n-1}} \dim_k H^i(X, \Omega^j_{X/k})$$

PROOF. The quadratic form Tr reduces to perfect pairings

$$H^i(X, \Omega^j_{X/k}) \otimes H^{2n-1-i}(X, \Omega^{2n-1-j}_{X/k}) \to k$$

for $i + j < 2n - 1$, or $0 \le i < j$ and $i + j = 2n - 1$, which identifies Tr with the canonical (hyperbolic) form on $V \oplus V^\vee$, where

$$V = \bigoplus_{i+j<2n-1} H^i(X, \Omega^j_{X/k}) \oplus \bigoplus_{\substack{0 \le i < j \\ i+j = 2n-1}} H^i(X, \Omega^j_{X/k})$$

□

References: Lecture 1

[1] Joseph Ayoub, *Les six opérations de Grothendieck et le formalisme des cycles évanescents dans le monde motivique. I* (French, with English and French summaries), Astérisque **314** (2007), x+466 pp. (2008). MR2423375

[2] A. Ananyevskiy, SL-*oriented cohomology theories*, Motivic Homotopy Theory and Refined Enumerative Geometry, Contemporary Mathematics **745**, 2020.

[3] Jean Barge and Fabien Morel, *Groupe de Chow des cycles orientés et classe d'Euler des fibrés vectoriels* (French, with English and French summaries), C. R. Acad. Sci. Paris Sér. I Math. **330** (2000), no. 4, 287–290, DOI 10.1016/S0764-4442(00)00158-0. MR1753295

[4] Baptiste Calmès and Jens Hornbostel, *Push-forwards for Witt groups of schemes*, Comment. Math. Helv. **86** (2011), no. 2, 437–468, DOI 10.4171/CMH/230. MR2775136

[5] Denis-Charles Cisinski and Frédéric Déglise, *Triangulated Categories of Mixed Motives* Springer Monographs in Mathematics, 1st ed. 2019, XLII, 406 p. ISBN 978-3-030-33241-9

[6] Frédéric Déglise, Fangzhou Jin, Adeel A. Khan, *Fundamental classes in motivic homotopy theory.* preprint (2018) arXiv:1805.05920 [math.AG].

[7] Jean Fasel, *Groupes de Chow-Witt* (French, with English and French summaries), Mém. Soc. Math. Fr. (N.S.) **113** (2008), viii+197, DOI 10.24033/msmf.425. MR2542148

[8] Jean Fasel, *Lectures on Chow-Witt*, Motivic Homotopy Theory and Refined Enumerative Geometry, Contemporary Mathematics **745**, 2020.

[9] Marc Hoyois, *The six operations in equivariant motivic homotopy theory*, Adv. Math. **305** (2017), 197–279, DOI 10.1016/j.aim.2016.09.031. MR3570135

[10] Kirsten Wickelgren, *An arithmetic count of the lines on a smooth cubic surface*, Notices Amer. Math. Soc. **65** (2018), no. 4, 404–405. MR3752384

[11] Marc Levine, *Toward an enumerative geometry with quadratic forms* preprint (2017) arXiv:1703.03049 [math.AG]

[12] Marc Levine, Arpon Raksit, *Motivic Gauß-Bonnet formulas*, preprint (2018) arXiv:1808.08385 [math.AG].

[13] Marc Levine, *Toward an algebraic theory of Welschinger invariants.* preprint (2018) arXiv:1808.02238 [math.AG].

[14] Fabien Morel, *$\mathbb{A}^1$-algebraic topology over a field*, Lecture Notes in Mathematics, vol. 2052, Springer, Heidelberg, 2012. MR2934577

[15] Fabien Morel, *An introduction to $\mathbb{A}^1$-homotopy theory*, Contemporary developments in algebraic K-theory, ICTP Lect. Notes, XV, Abdus Salam Int. Cent. Theoret. Phys., Trieste, 2004, pp. 357–441. MR2175638

[16] Fabien Morel, *On the motivic π_0 of the sphere spectrum*, Axiomatic, enriched and motivic homotopy theory, NATO Sci. Ser. II Math. Phys. Chem., vol. 131, Kluwer Acad. Publ., Dordrecht, 2004, pp. 219–260, DOI 10.1007/978-94-007-0948-5_7. MR2061856

[17] Fabien Morel, *The stable $\mathbb{A}^1$-connectivity theorems*, K-Theory **35** (2005), no. 1-2, 1–68, DOI 10.1007/s10977-005-1562-7. MR2240215

[18] Fabien Morel and Vladimir Voevodsky, *$\mathbf{A}^1$-homotopy theory of schemes*, Inst. Hautes Études Sci. Publ. Math. **90** (1999), 45–143 (2001). MR1813224

[19] I. Panin and C. Walter, *On the motivic commutative ring spectrum* **BO**, Algebra i Analiz **30** (2018), no. 6, 43–96. MR3882540

[20] Marco Schlichting, *Hermitian K-theory of exact categories*, J. K-Theory **5** (2010), no. 1, 105–165, DOI 10.1017/is009010017jkt075. MR2600285

[21] Marco Schlichting and Girja S. Tripathi, *Geometric models for higher Grothendieck-Witt groups in $\mathbb{A}^1$-homotopy theory*, Math. Ann. **362** (2015), no. 3-4, 1143–1167, DOI 10.1007/s00208-014-1154-z. MR3368095

[22] Vladimir Voevodsky, *Motivic cohomology with $\mathbf{Z}/2$-coefficients*, Publ. Math. Inst. Hautes Études Sci. **98** (2003), 59–104, DOI 10.1007/s10240-003-0010-6. MR2031199

[23] Vladimir Voevodsky, *$\mathbf{A}^1$-homotopy theory*, Proceedings of the International Congress of Mathematicians, Vol. I (Berlin, 1998), Doc. Math. **Extra Vol. I** (1998), 579–604. MR1648048

Lecture 2. Riemann-Hurwitz formulas

We describe methods for computing the Euler characteristics and Euler classes in Chow-Witt theory based on the classical approach of fibering by a Lefschetz pencil. This gives an algebraic version of classical Morse theory, and recovers the calculation of Euler characteristics via Morse theory when looking at an $\mathbb{R}$-realization. Most of these results are taken from [**4**, §§8-13]

Throughout this lecture we will be working with the Euler classes and Euler characteristics in Chow-Witt theory.

2.1. Euler class of a dual bundle and tensor product with a line bundle. For a morphism $f : X \to Y$ one has the associated global section df of $\Omega_{X/k} \otimes f^*\Omega^\vee_{Y/k}$; for Y a curve, this is a twist of $T^\vee_{X/k}$ by a line bundle. The classical Riemann-Hurwitz formula (for Y a curve) is simply a computation of the top Chern class of $T^\vee_{X/k} \otimes f^*\omega_{Y/k}$ in terms of the Chern classes $c_{\dim X}(\Omega_{X/k})$, $c_{\dim X-1}(\Omega_{X/k})$ and $c_1(\omega_{Y/k})$, together with the identity $c_{\dim X}(\Omega_{X/k}) = (-1)^{\dim X} c_{\dim X}(T_{X/k})$. This latter identity in the case of the Chow-Witt valued Euler class becomes

THEOREM 2.1.1. *Let $V \to X$ be a rank r vector bundle over $X \in \mathbf{Sm}/k$. Then*

$$e(V) = (-\langle -1\rangle)^r e(V^\vee) \in \widetilde{\mathrm{CH}}(X; \det{}^{-1}V) = \widetilde{\mathrm{CH}}(X; \det{}^{-1}V^\vee)$$

In case $\det V \cong \mathcal{O}_X$, this is proven by Asok-Fasel [**1**] using the an interpretation of the Euler class as an obstruction class. Following their approach, we have extended their proof to cover the case of a general V in [**4**, §11].

This settles the relation of $e(T_{X/k})$ and $e(\Omega_{X/k})$, we now turn to the problem of twisting with a line bundle. The main problem here is the lack of "lower Euler classes" comparable with $c_i(V)$, $2 \le i < \mathrm{rank}(V)$; we can always use $e(\det V)$ as a refinement for $c_1(V)$. In spite of this, we have the following result

THEOREM 2.1.2 ([**4**, Corollary 10.9]). *Let $V \to X$ be a rank r vector bundle on $X \in \mathbf{Sm}/k$ and $L \to X$ a line bundle. Assume that $L \cong M^{\otimes 2}$ for some line bundle M on X. Then*

$$e(V \otimes L) = e(V) + \bar{h} \cdot c_1(M) \cdot (\sum_{i=1}^{r} c_{r-i}(V) c_1(L)^{i-1}) \in \widetilde{\mathrm{CH}}^r(X, \det{}^{-1}V).$$

Note that $\widetilde{\mathrm{CH}}^r(X, \det^{-1} V) \cong \widetilde{\mathrm{CH}}^r(X, \det^{-1}(V \otimes L))$, since $\det^{-1}(V \otimes L)$ and $\det^{-1} V$ differ by the square of some line bundle

We need to explain the notation

$$\bar{h} \cdot c_1(M) \cdot (\sum_{i=1}^{r} c_{r-i}(V) c_1(L)^{i-1}) \in \widetilde{\mathrm{CH}}^r(X, \det{}^{-1}V).$$

For this, the relation (4) defining the Milnor-Witt ring $K_*^{MW}(F)$

$$0 = \eta \cdot h = \eta \cdot (\langle 1\rangle + \langle -1\rangle)$$

tells us that the map

$$h\times : \mathcal{K}_*^{MW}(L) \to \mathcal{K}_*^{MW}(L)$$

factors through the surjection $\pi : \mathcal{K}_*^{MW}(L) \to \mathcal{K}_*^{MW}(L)/(\eta) = \mathcal{K}_*^M$. We denote the resulting "hyperbolic map" by

$$\bar{h}\cdot : \mathcal{K}_*^M \to \mathcal{K}_*^{MW}(L)$$

and use the same notation for the map in cohomology

$$\bar{h}\cdot : H^a(X, \mathcal{K}_b^M) \to H^a(X, \mathcal{K}_b^{MW}(L)).$$

One easily shows that

$$\bar{h} \cdot c_r(V) = h \cdot e(V) \in \widetilde{\mathrm{CH}}^r(X; \det^{-1} V).$$

for $V \to X$ a rank r vector bundle.

To apply Theorem 2.1.2, we need the notion of local contributions to the Euler class. For this, we recall that for $V \to X$ a rank r vector bundle with 0-section $s_0 : X \to V$, we have

$$e(V) = s_0^*(\theta_V)$$

where $\theta_V \in H^r_{0_V}(V; \mathcal{K}_r^{MW}(\det^{-1} V))$ is the Thom class. Letting $s : X \to V$ be an arbitrary section, it follows from $\mathbb{A}^1$-homotopy invariance that

$$s^*\theta_V = s_0^*\theta_V \in H^r(X; \mathcal{K}_r^{MW}(\det^{-1} V))$$

However, we can remember the supports, giving us the class

$$e_Z(V, s) := s^*\theta_V \in H^r_Z(X; \mathcal{K}_r^{MW}(\det^{-1} V))$$

for any closed subset $Z \supset s^{-1}(0_V)$, i.e. Z containing the 0-locus of s.

For $\{s = 0\}$ a union of closed points $\{x_1, \ldots, x_m\}$ we have

$$H^r_{\{x_1,\ldots,x_m\}}(X; \mathcal{K}_r^{MW}(\det^{-1} V)) = \oplus_{i=1}^m H^r_{\{x_i\}}(X; \mathcal{K}_r^{MW}(\det^{-1} V))$$

giving the local contribution $e_{x_i}(V; s) \in H^r_{\{x_i\}}(X; \mathcal{K}_r^{MW}(\det^{-1} V))$ to $e(V)$.

Kass and Wickelgren [**3**] have given a formula for these local contributions in terms of an explicit quadratic form, originally developed by Eisenbud-Levine and Khimshiashvili in the setting of real algebraic geometry. In the case of a non-degenerate zero of s at x with $r = \dim X$, one has a simple description of $e_x(V; s)$. Let $\lambda_1, \ldots, \lambda_r$ be a basis of sections for V near x, giving the local basis $\lambda_1^\vee, \ldots, \lambda_r^\vee$ for $V^\vee$ and let $t_1, \ldots, t_r \in \mathcal{O}_{X,x}$ be local parameters, giving the basis $\partial/\partial t_1, \ldots, \partial/\partial t_r$ for $(\mathfrak{m}/\mathfrak{m}^2)^\vee$. Then s is given by

$$s = \sum_{i=1}^r s_i \cdot \lambda_i$$

with $s_i \in \mathfrak{m}_x = (t_1, \ldots, t_r)$, so

$$s_i = \sum_{j=1}^r a_{ij} t_j, \ a_{ij} \in \mathcal{O}_{X,x}$$

and s has x as a non-degenerate zero exactly when $\det(a_{ij})$ is a unit in $\mathcal{O}_{X,x}$. This gives us the rank one quadratic form $\langle \det(a_{ij}) \rangle \in \mathrm{GW}(k(x))$. The local contribution $e_x(V, s)$ is given by the element

$$\langle \det(a_{ij}) \rangle \otimes \partial/\partial t_1 \wedge \ldots \partial/\partial t_r \otimes \lambda_1^\vee \wedge \ldots \wedge \lambda_r^\vee \in \mathrm{GW}(k(x); \det^{-1} \mathfrak{m}_x/\mathfrak{m}_x^2 \otimes \det^{-1} V),$$

where we use the purity isomorphism

$$H^r_x(X, \mathcal{K}_r^{MW}(\det^{-1} V)) \cong \mathrm{GW}(k(x); \det^{-1}(\mathfrak{m}_x/\mathfrak{m}_x^2) \otimes \det^{-1} V).$$

to identify the local cohomology.

One can also apply the Kass-Wickelgren formula to the case of a "diagonal" section with a higher order zero at x. For $s = \sum_i u_i t_i^{n_i}\lambda_i$ with $u_i \in \mathcal{O}_{X,x}^\times$ $n_i > 0$, one has

$$e_x(V,s) = \prod_{i=1}^{r}(\langle \bar{u}_i\rangle \cdot n_{i\epsilon}) \otimes \partial/\partial t_1 \wedge \ldots \partial/\partial t_r \otimes \lambda_1^\vee \wedge \ldots \wedge \lambda_r^\vee$$

in $\mathrm{GW}(k(x); \det^{-1}\mathfrak{m}_x/\mathfrak{m}_x^2 \otimes \det^{-1} V)$, where $\bar{u}_i \in k(x)^\times$ is the reduction of u_i and for $n > 0$, $n \in \mathbb{Z}$, we define $n_\epsilon = \sum_{i=0}^{n-1}\langle -1\rangle^i \in \mathrm{GW}(k)$. One has the same formula for $e_x(V,s)$ if one only assumes that

$$s \equiv \sum_{i=1}^{r} u_i t_i^{n_i}\lambda_i \mod \mathfrak{m}_x \cdot (t_1^{n_1}, \ldots, t_r^{n_r}).$$

COROLLARY 2.1.3 (Riemann-Hurwitz formula [**4**, Corollary 12.3]). *Let $f : X \to C$ be a flat morphism, with X and C smooth and projective over k and C a curve. Assume that $\omega_{C/k} \cong M^{\otimes 2}$ for some line bundle M on C and that df has only isolated zeros $x_1, \ldots, x_m$. Then*

$$(-\langle -1\rangle)^{d_X}\chi(X/k) = \sum_{i=1}^{m} Tr_{k(x_i)/k} e_{x_i}(\Omega_{X/k} \otimes f^*\omega_{C/k}^{-1}, df) \\ + \frac{1}{2}\deg_k(c_{d_X - 1}(\Omega_{X/k}) \cdot f^*c_1(\omega_{C/k})) \cdot h \in \mathrm{GW}(k).$$

This follows from Theorem 2.1.2 applied to $V = \Omega_{X/k}$, $L = f^*\omega_{C/k}^{-1}$ and Theorem 2.1.1 applied to $V = T_{X/k}$. To make sense of the term $Tr_{k(x_i)/k}e_{x_i}(\Omega_{X/k} \otimes f^*\omega_{C/k}, df)$, we note that, if k is perfect, $e_{x_i}(\Omega_{X/k} \otimes f^*\omega_{C/k}, df)$ lives in

$$\mathrm{GW}(k(x_i); (\det^{-1}\mathfrak{m}_{x_i}/\mathfrak{m}_{x_i}^2 \otimes \omega_{X/k}^{-1} \otimes f^*\omega_{C/k}^{\otimes d_X}) \\ = \mathrm{GW}(k(x_i); \omega_{X/k}^{-1} \otimes k(x_i) \otimes (\omega_{X/k}^{-1} \otimes f^*M^{-\otimes 2d_X})) = \mathrm{GW}(k(x_i)).$$

The trace is the usual trace map $\mathrm{Tr}_{k(x_i)/k} : \mathrm{GW}(k(x_i)) \to \mathrm{GW}(k)$, as $k(x_i)/k$ is separable. In general, one uses Fasel's pushforward map, as described in [**2**, §5,6].

Before we apply the Riemann-Hurwitz formula, we give some ideas toward the proof of the twisting theorem 2.1.2.

The first thing to note is that to verify this identity, we are free to pull back to any $Y \in \mathbf{Sm}/k$ such that the pullback map on $\widetilde{\mathrm{CH}}^r$ is injective. We can for instance use a Jouanolou cover and assume that X is affine. We can also use homotopy invariance and replace $e(V)$ and $e(V \otimes L)$ with the pullbacks of the respective Thom classes by any section we chose, not just the 0-section.

We fix a general section s of L with divisor $D \subset X$. Multiplication by s gives the map

$$\phi := \times s : V \to V \otimes L$$

so we compute $e(V \otimes L)$ by $v^*(\phi^*\theta_{V\otimes L})$ for $v : X \to V$ some section. We note that $\phi^{-1}(0_{V\otimes L})$ splits into two pieces

$$\phi^{-1}(0_{V\otimes L}) = \pi_V^{-1}(D) \cup 0_V$$

$\pi_V : V \to X$ the projection, and that $F := \pi_V^{-1}(D) \cap 0_V$ has codimension $r+1$ on V. If $v : X \to V$ is general, then $v^{-1}(F)$ will have codimension $r+1$ on X, so the restriction map $\widetilde{\mathrm{CH}}^r(X,-)\to\widetilde{\mathrm{CH}}^r(X-v^{-1}(F),-)$ is injective and we can safely remove F.

Letting $U=V-F$, $\phi^{-1}(0_{V\otimes L})\cap U$ splits into a disjoint union of two pieces

$$\phi^{-1}(0_{V\otimes L})\cap U=\pi_V^{-1}(D)\cap U\amalg 0_V\cap U$$

and thus $\phi^*(\theta_{V\otimes L})$ restricted to U decomposes accordingly

$$j_U^*\phi^*(\theta_{V\otimes L})=\phi_{0_V}^*\theta_{V\otimes L}+\phi_D^*\theta_{V\otimes L}$$

with

$$\begin{aligned}\phi_{0_V}^*\theta_{V\otimes L}&\in H^r_{0_V\cap U}(U,\mathcal{K}_r^{MW}(\det{}^{-1}V\otimes L)),\\ \phi_D^*\theta_{V\otimes L}&\in H^r_{\pi_V^{-1}(D)\cap U}(U,\mathcal{K}_r^{MW}(\det{}^{-1}(V\otimes L))).\end{aligned}$$

Using $\times s$ to identify $\det V$ with $\det V\otimes L$ over $U-\pi^{-1}(D)$ gives

$$\phi_{0_V}^*\theta_{V\otimes L}=j_U^*\theta_V$$

For the other term, we restrict to $U'=U-0_V$, noting that $\pi_V^{-1}(D)\cap U=\pi_V^{-1}(D)\cap U'$. Over U' the tautological section $s':\mathcal{O}_V\to V$ of V gives the rank $r-1$ vector bundle $W:=j_{U'}^*(V)/s'(\mathcal{O}_{U'})$ and the exact sequence

$$0\to j_{U'}^*L\xrightarrow{s'}j_{U'}^*V\otimes L\to W\otimes L\to 0$$

Pulling back again to a Jouanolou cover of U' to split this sequence, and using the multiplicativity of the Thom classes gives the identity

$$\phi_D^*\theta_{V\otimes L}=j_{U'}^*e_D(L,s)\cup e(W\otimes L)$$

Now we use the assumption that $L\cong M^{\otimes 2}$. We can take the section s to be of the form $s=t^{\otimes 2}$ for t a suitably general section of M. Using the relation in K_1^{MW}:

$$[x^2]=(\langle 1\rangle+\langle -1\rangle)[x]=h\cdot[x]$$

gives

$$e_D(L,s)=e_D(L,t^{\otimes 2})=h\cdot e_D(M)=\bar{h}\cdot c_{1D}(M)$$

and thus

$$\begin{aligned}j_{U'}^*e_D(L,s)\cup e(W\otimes L)&=h\cdot(j_{U'}^*(e_D(M))\cup e(W\otimes L))\\ &=\bar{h}\cdot(j_{U'}^*c_{1D}(M)\cup c_{r-1}(W\otimes L))\\ &=\bar{h}\cdot((j_{U'}^*c_{1D}(M)\cup\sum_{i=0}^{r-1}c_{r-1-i}(W)\cup j_{U'}^*c_1(L)^i)\end{aligned}$$

As the exact sequence $0\to\mathcal{O}_{U'}\to j_{U'}^*V\to W\to 0$ gives $c_i(W)=j_{U'}^*c_i(V)$ for $i=0,\dots,r-1$, we have

$$\phi_D^*\theta_{V\otimes L}=j_{U'}^*(\bar{h}\cdot c_{1D}(M)\cup\sum_{i=1}^{r}c_{r-i}(V)\cup c_1(L)^{i-1})$$

Noting that $\pi_V^{-1}(D)\cap U'=\pi_V^{-1}(D)\cap U$, pulling back

$$j_U^*\phi^*\theta_{V\otimes L}=\phi_{0_V}^*\theta_{V\otimes L}+\phi_D^*\theta_{V\otimes L}$$

by a general section $v:X\to V$ gives the desired identity in

$$H^r(X-v^{-1}(F),\mathcal{K}_r^{MW}(\det{}^{-1}V\otimes L)),$$

which gives the identity over X since $v^{-1}(F)$ has codimension $\geq r+1$ on X.

2.2. Applications.

THEOREM 2.2.1 ([**4**, Theorem 12.7]). *Let $f : X \to C$ be a separable morphism of smooth projective curves $X, C \in \mathbf{Sm}/k$. Suppose that $\omega_{C/k} \cong M^{\otimes 2}$ for some $M \in \mathrm{Pic}(C)$ and let $x_1, \dots, x_m \in X$ be the zeros of df. Then*

$$\sum_{i=1}^{m} Tr_{k(x_i)/k} e_{x_i}(\omega_{X/C}, df) \in \mathrm{GW}(k)$$

is hyperbolic.

PROOF. This follows from the fact that $\chi(X/k)$ is hyperbolic, as X has odd dimension 1, and the Riemann-Hurwitz formula 2.1.3. □

REMARK 2.2.2. We can make this statement more concrete, for example, if $C = \mathbb{P}^1$. We assume that $y_i := f(x_i) \in \mathbb{A}^1 = \mathrm{Spec}\, k[t] = \mathbb{P}^1 - \{\infty\}$ for all $i = 1, \dots, m$ and that f is tamely ramified. Let $g_i \in k[t]$ be the normalized irreducible polynomial of y_i over k and let $s_i := g_i(t)/g_i'(t) \in \mathfrak{m}_{y_i}$. Then $ds_i = dt$. We have an isomorphism $\omega_{\mathbb{P}^1/k} \cong \mathcal{O}_{\mathbb{P}^1}(-2)$, and we may chose this isomorphism so that dt is the image $\lambda^{\otimes 2}$ for λ a generator of $\mathcal{O}_{\mathbb{P}^1}(-1)$ over $\mathbb{A}^1$. This trivializes $f^*\omega_{\mathbb{P}^1/k}$ at all the points x_i, giving the isomorphism

$$\mathrm{GW}(k(x_i), \omega_{X/k}^{-1} \otimes f^*\omega_{\mathbb{P}^1/k} \otimes \omega_{X/k}^{-1}) \cong \mathrm{GW}(k(x_i)).$$

explicitly, we write $f^*(s_i) = u_i t_i^{n_i}$, then via this trivialization we have $df = d(f^*(s_i)) = n_i u_i t_i^{n_i - 1} dt_i$, giving the local contribution

$$e_{x_i}(\omega_{X/k} \otimes f^*\omega_{\mathbb{P}^1/k}^{-1}, df) = \langle n_i \bar{u}_i \rangle \cdot (n_i - 1)_\epsilon$$

As

$$\langle n_i \bar{u}_i \rangle \cdot (n_i - 1)_\epsilon = \begin{cases} [(n_i - 1)/2] \cdot h & \text{for } n_i \text{ odd,} \\ \langle n_i \bar{u}_i \rangle + [(n_i - 2)/2] \cdot h & \text{for } n_i \text{ even,} \end{cases}$$

the Riemann-Hurwitz identity gives an identity on the "leading terms" u_i:

$$\sum_{n_i \text{ even}} \mathrm{Tr}_{k(x_i)/k} \langle n_i \bar{u}_i \rangle \text{ is hyperbolic in } \mathrm{GW}(k).$$

As another application, we have

THEOREM 2.2.3 ([**4**, Theorem 13.1]). *Let $X \subset \mathbb{P}_k^{2n+1}$ be a smooth hypersurface defined by an equation of the form $\sum_{i=0}^{2n+1} a_i T_i^m$. Then there are integers $A_{n,m}$ depending only on n and m such that*

$$\chi(X/k) = \begin{cases} A_{n,m} \cdot h + \langle m \rangle & \textit{for } m \textit{ odd,} \\ A_{n,m} \cdot h + \langle m \rangle + \langle -m \prod_{i=0}^{2n+1} a_i \rangle & \textit{for } m \textit{ even.} \end{cases}$$

In fact, as $\mathrm{rank}(\chi(X/k)) = \deg_k c_{2n}(T_{X/k})$ and

$$c_{2n}(T_{X/k}) = i_X^* \left[\frac{(1+H)^{2n+2}}{1+mH} \right]_{2n},$$

where $H = c_1(\mathcal{O}(1))$ is the hyperplane class and $[-]_{2n}$ means the degree $2n$ component, we have the following formula for $A_{n,m}$:

$$A_{n,m} = \begin{cases} (\deg \left[\frac{mH \cdot (1+H)^{2n+2}}{1+mH} \right]_{2n+1} - 1)/2 & \text{for } m \text{ odd,} \\ \\ (\deg \left[\frac{mH \cdot (1+H)^{2n+2}}{1+mH} \right]_{2n+1} - 2)/2 & \text{for } m \text{ even.} \end{cases}$$

The idea of the proof is to use the Lefschetz pencil with base-locus $T_{2n} = T_{2n+1} = 0$. The intersection with X is the hypersurface $Z \subset \mathbb{P}^{2n-1}$ defined by $\sum_{i=0}^{2n-1} a_i T_i^m$. Letting $\tilde{X} \to X$ be the blow-up along Z and $f : \tilde{X} \to \mathbb{P}^1$ the morphism arising from the pencil, we use the blow-up formula to compute

$$\chi(\tilde{X}/k) = \chi(X/k) + \langle -1 \rangle \chi(Z/k)$$

The map f is ramified at the closed subscheme x of X defined by $T_0 = \ldots = T_{2n-1} = 0$. We compute the local contribution $e_x(\Omega_{\tilde{X}/k} \otimes f^*\omega^{-1}_{\mathbb{P}^1/k}, df)$ and then use the Riemann-Hurwitz formula to compute $\chi(\tilde{X}/k)$. We then use induction to compute $\chi(Z/k)$, giving the formula for $\chi(X/k)$.

COROLLARY 2.2.4. *Let $Q \subset \mathbb{P}^{2n+1}$ be a smooth quadric hypersurface of even dimension $2n$ defined by the quadratic polynomial $q \in k[T_0, \ldots, T_{2n+1}]$. Then*

$$\chi(Q/k) = n \cdot h + \langle 2 \rangle + \langle -2\delta_q \rangle$$

where δ_q is the discriminant of q; here we use the convention that $\delta_q = \det Q$, where Q is a matrix of the symmetric bilinear form associated to q.

This follows from the well-know topological Euler characteristic $2n+2$ and the fact that we can diagonalize q.

References: Lecture 2

[1] A. Asok and J. Fasel, *Comparing Euler classes*, Q. J. Math. **67** (2016), no. 4, 603–635. MR3609848

[2] Jean Fasel, *Groupes de Chow-Witt* (French, with English and French summaries), Mém. Soc. Math. Fr. (N.S.) **113** (2008), viii+197, DOI 10.24033/msmf.425. MR2542148

[3] Jesse Leo Kass and Kirsten Wickelgren, *The class of Eisenbud-Khimshiashvili-Levine is the local $\mathbf{A}^1$-Brouwer degree*, Duke Math. J. **168** (2019), no. 3, 429–469, DOI 10.1215/00127094-2018-0046. MR3909901

[4] Marc Levine, *Toward an enumerative geometry with quadratic forms* preprint (2017) arXiv:1703.03049 [math.AG]

Lecture 3. Pontryagin classes, splitting principles and Becker-Gottlieb transfers

We have already mentioned that the lack of Chow-Witt versions of the "middle" Chern classes make the theory of the Euler class somewhat unwieldy. This can be remedied by passing from the Chow-Witt groups to a simpler theory, namely, (twisted) Witt cohomology $H^*(-, \mathcal{W}(L))$. This is simply the localization of Milnor-Witt cohomology with respect to η: $\mathcal{W}(L) = \mathcal{K}_*^{MW}(L)[\eta^{-1}]$. Thus, passing to Witt cohomology we have an SL-oriented and η-invertible theory.

3.1. Borel classes, Pontryagin classes and Ananyevskiy's SL_2 splitting principle.

DEFINITION 3.1.1. Let $\mathcal{E} \in \mathrm{SH}(k)$ be a motivic commutative ring spectrum. We call $\mathcal{E}$ an *η-invertible* if $\times\eta : \mathcal{E}^{a,b}(k) \to \mathcal{E}^{a-1,b-1}(k)$ is an isomorphism for all a, b.

For an η-invertible theory, $\times\eta : \mathcal{E}^{a,b}(\mathcal{X}) \to \mathcal{E}^{a-1,b-1}(\mathcal{X})$ is an isomorphism for all $\mathcal{X} \in \mathbf{Spc}_\bullet(k)$ or even $\mathcal{X} \in \mathrm{SH}(k)$.

Ananyevskiy [**1**] has studied theories $\mathcal{E}$ which are both SL-oriented and η-invertible. For such theories, the *Pontryagin classes* of vector bundles give a good theory of characteristic classes; we review briefly how these are defined.

Analogous to the notion of an SL-orientation [**1**, Definition 1] is that of a *symplectic orientation.* A symplectic bundle is a vector bundle $V \to X$ equipped with a non-degenerate alternating form $\omega : V \wedge V \to \mathcal{O}_X$. Then V has even rank, say $2m$ and then $\wedge^m\omega$ trivializes $\det V$, so V is also an SL_{2m}-bundle.

The analog of the projective space $\mathbb{P}^n = \mathrm{Gr}(1, n+1)$ and the tautological line bundle $O_{\mathbb{P}^n}(1)$ in the setting of symplectic bundles is the symplectic projective space $H\mathbb{P}^n \subset \mathrm{Gr}(2, 2n+2)$. Give $V = \mathbb{A}^{2n+2}$ the standard symplectic form

$$\omega_n := \begin{pmatrix} 0 & I \\ -I & 0 \end{pmatrix} \in M_{2n+2}$$

and let $H\mathbb{P}^n \subset \mathrm{Gr}(2, 2n+2)$ be the open subscheme parametrizing two-planes $\pi \subset \mathbb{A}^{2n+2}$ such that ω_n restricts to a non-degenerate form on π. The tautological rank two subbundle $\mathcal{E}_2 \to \mathrm{Gr}(2, 2n+2)$ thus restricts to a rank two symplectic bundle $(O_{H\mathbb{P}^n}(-1) \to H\mathbb{P}^n, \omega)$.

A symplectic orientation for a commutative ring spectrum $\mathcal{E}$ is a functorial choice of classes $\theta^{\mathcal{E}}_{V,\omega} \in \mathcal{E}^{4r,2r}(X)$ for each rank $2r$ symplectic bundle $(V \to X, \omega)$, such that for the "trivial" bundle $(V = \mathcal{O}_X^r \oplus \mathcal{O}_X^r, \omega_r)$, $\theta_{V,\omega}$ corresponds to the unit of $\mathcal{E}$ under the suspension isomorphism. In addition, one has a product formula for direct sums [**5**, §7].

DEFINITION 3.1.2. For $\mathcal{E}$ a symplectically oriented motivic ring spectrum, the 1st Borel class $b_1(E_2, \omega) \in \mathcal{E}^{4,2}(X)$ of a rank two symplectic bundle $(E_2 \to X, \omega)$ is the Euler class $s_0^*\theta^{\mathcal{E}}_{E_2,\omega}$.

An SL-oriented $\mathcal{E}$ is also symplectically oriented, and $b_1(E_2, \omega)$ is just the usual Euler class in $\mathcal{E}^{4,2}(X, \det^{-1}(E_2))$, where we use the isomorphism $\omega : E_2 \wedge E_2 \to \mathcal{O}_X$ to trivialize $\det E_2$, giving the identification $\mathcal{E}^{4,2}(X, \det^{-1}(E_2)) \cong \mathcal{E}^{4,2}(X)$.

THEOREM 3.1.3 (Panin-Walter [**5**, Theorem 8.1]). *Let $\mathcal{E}$ be a symplectically oriented commutative motivic ring spectrum. Then $\mathcal{E}^{*,*}(H\mathbb{P}^n)$ is a truncated polynomial ring over $\mathcal{E}^{*,*}(k)$, with generator $b_1(O_{H\mathbb{P}^n}(-1),\omega)$ and relation $b_1^{n+1}=0$:*

$$\mathcal{E}^{*,*}(H\mathbb{P}^n)=\mathcal{E}^{*,*}(k)[b_1]/(b_1^{n+1})$$

Panin-Walter follow the Grothendieck playbook to construct higher *Borel classes*[1] $b_1(E,\omega),\ldots,b_r(E,\omega)$ for (E,ω) a rank $2r$ symplectic bundle,

$$b_i(E,\omega)\in\mathcal{E}^{4i,2i}(X)$$

For $V\to X$ a rank r vector bundle, we have the rank $2r$ vector bundle $V\oplus V^\vee$ with canonical symplectic form ω_V:

$$\omega_V((s,s^\vee),(t,t^\vee)):=s^\vee(t)-t^\vee(s).$$

DEFINITION 3.1.4 ([**1**, Definition 19]). Let $\mathcal{E}$ be an η-invertible, SL-oriented theory and let $V\to X$ be a rank r vector bundle. Define the ith Pontryagin class $p_i(V)\in\mathcal{E}^{8i,4i}(X)$, $i=1,\ldots,[r/2]$ by

$$p_i(V):=(-1)^ib_{2i}(V\oplus V^\vee,\omega_V)$$

We set $p_0(V)=1_X$ and $p(V)=\sum_{i=0}^{[r/2]}p_i(V)$.

It turns out that the odd Borel classes of $(V\oplus V^\vee,\omega_V)$ are zero. This gives the Whitney formula: for

$$0\to V'\to V\to V''\to 0$$

an exact sequence of vector bundles on some $X\in\mathbf{Sm}/k$ we have

$$p(V)=p(V')\cdot p(V'').$$

In addition, for V of even rank $r=2m$, one has

$$p_m(V)=e(V)^2\in\mathcal{E}^{8m,4m}(\mathcal{E},\det{}^{-2}V)=\mathcal{E}^{4m,2m}(\mathcal{E}).$$

We recall that for an odd rank bundle V, one has $\eta\cdot e(V)=0$, so $e(V)=0$ in an SL-oriented, η-invertible theory.

Ananyevskiy proves a number of fundamental results about the Pontryagin classes in an η-invertible, SL-oriented theory. We mention two of these.

We have the classifying space BSL_n with universal rank n bundle $\tilde{E}_n\to\mathrm{BSL}_n$.

THEOREM 3.1.5 (Ananyevskiy [**1**, Theorem 10]). *Let $p_i=p_i(\tilde{E}_n)$, $i=1,\ldots,[n/2]$, and for $n=2m$, let $e=e(\tilde{E}_n)$. Let $\mathcal{E}$ be an SL-oriented, η-invertible commutative motivic ring spectrum. Then*

$$\mathcal{E}^{**}(\mathrm{BSL}_n)=\begin{cases}\mathcal{E}^{**}(k)[p_1,\ldots,p_m] & \text{for } n=2m+1\\ \mathcal{E}^{**}(k)[p_1,\ldots,p_m,e]/(p_m-e^2) & \text{for } n=2m\end{cases}$$

THEOREM 3.1.6 (Ananyevskiy [**1**, Theorem 6, Theorem 10]). *Let $i_m:\mathrm{BSL}_2^m\to\mathrm{BSL}_{2m}$ be the map induced by the diagonal inclusion $\mathrm{SL}_2^m\to\mathrm{SL}_{2m}$ and let $\mathcal{E}$ be an SL-oriented, η-invertible commutative motivic ring spectrum. Then*

1. *$\mathcal{E}^{**}(\mathrm{BSL}_2^m)=\mathcal{E}^{**}(\mathrm{BSL}_2)^{\otimes_{\mathcal{E}^{**}}m}=\mathcal{E}^{**}(k)[e_1,\ldots,e_m]$, where e_i is the pullback $p_i^*e(\tilde{E}_2)$ via the projection onto the ith factor $p_i:\mathrm{BSL}_2^m\to\mathrm{BSL}_2$.*

[1] Unfortunately, these classes are called "Pontryagin classes" in [**5**], but fortunately, this terminology has been replaced with the more appropriate term "Borel classes" in the literature.

2. The pull-back

$$i_m^* : \mathcal{E}^{**}(\mathrm{BSL}_{2m}) \to \mathcal{E}^{**}(\mathrm{BSL}_2^m)$$

is injective, and one has

$$i_m^* p(\tilde{E}_{2m}) = \prod_{i=1}^m (1+e_i^2),\ i_m^* e(\tilde{E}_{2m}) = \prod_{i=1}^m e_i.$$

3.2. Becker-Gottlieb transfers. We are interested in the following problem: For $\rho : \mathrm{GL}_n \to \mathrm{GL}_m$ a representation, one has the associated functor S_ρ from rank n vector bundles to rank m vector bundles. How can one compute $p(S_\rho E)$ and $e(S_\rho E)$ in terms of $p(E)$ and $e(E)$? A related question: Given bundles E_1, E_2 what is $p(E_1 \otimes E_2)$ and $e(E_1 \otimes E_2)$ in terms of the Pontryagin and Euler classes of E_1 and E_2?

For the Chern classes, that is, for characteristic classes in an oriented theory, the answer is given by the classical splitting principle, together with the fact that ρ is determined by its restriction to the maximal torus $T \cong \mathbb{G}_m^n$ of GL_n. This reduces the question of computing these characteristic classes to a problem in symmetric functions for the Weyl group S_n of GL_n. Ananyevskiy's two theorems described above reduce this problem, at least for bundles with trivialized determinant, to the case of rank two bundles. However, as SL_2 has irreducible representations of arbitrary rank (namely, the kth symmetric power representation) the SL_2-splitting principle on its own does not give a complete answer.

A first approach to the problem of finding a good splitting principle is furnished by a motivic version of Becker-Gottlieb transfers. In classical homotopy theory, this is the construction of a stable "wrong way" map $p^* : \Sigma^\infty B_+ \to \Sigma^\infty E_+$ for a fiber bundle $p : E \to B$ with fiber F having the homotopy type of a finite CW complex. This transfer map has the property that, at least for oriented theories, $p \circ p^*$ induces multiplication by the Euler characteristic $\chi^{top}(F)$.

In the setting of motivic homotopy theory, Hoyois has constructed a transfer map for $p : E \to B$ smooth and projective. Here we extend this construction to a wider class of maps, using the Grothendieck six-functor formalism on $\mathrm{SH}(-) : \mathbf{Sch}/S \to \mathbf{Tr}$ introduced in § 1.2.

For $f : X \to Y$ in $\mathbf{Sch}/S$ we have the adjoint pairs $f^* \dashv f_*$, $f_! \dashv f^!$. For f smooth, f^* has the left adjoint $f_\#$ and the infinite suspension spectrum $\Sigma_T^\infty X_+ \in \mathrm{SH}(Y)$ is just $f_\#(1_X)$. We write X/Y for $f_\#(1_X)$; the map f induces the map $f/Y : X/Y \to Y/Y = 1_Y$.

Suppose we have $f : E \to B \in \mathbf{Sm}/B$ such that E/B is dualizable with dual $(E/B^\vee, \delta_{E/B}, ev_{E/B})$. We define the transfer map

$$\mathrm{Tr}_{f/B} : 1_B \to E/B$$

in $\mathrm{SH}(B)$ as follows: Upon applying $f_\#$, the diagonal $\Delta_E : E \to E \times_B E$ gives the map

$$\Delta_E/B : E/B \to E \times_B E/B \cong E/B \wedge_B E/B$$

Duality gives the adjunction

$$(-)^{\mathrm{tr}} : \mathrm{Hom}_{\mathrm{SH}(B)}(E/B, E/B \wedge_B E/B) \cong \mathrm{Hom}_{\mathrm{SH}(B)}(E/B^\vee \wedge_B E/B, E/B)$$

so we have the morphism

$$\Delta_E/B^{\mathrm{tr}} : E/B^\vee \wedge_B E/B \to E/B$$

Define $\mathrm{Tr}_{f/B}$ as the composition

$$1_B \xrightarrow{\delta_{E/B}} E/B \wedge_B E/B^\vee \xrightarrow{\tau_{E/B,E/B^\vee}} E/B^\vee \wedge_B E/B \xrightarrow{\Delta_{E/B}^{\mathrm{tr}}} E/B$$

Applying $f_\#$ gives the maps

$$\mathrm{Tr}_{f/S} : B/S \to E/S,\ f/S : E/S \to B/S$$

Since E/B is dualizable in $\mathrm{SH}(B)$, we have the Euler characteristic $\chi(E/B) \in \mathrm{End}_{\mathrm{SH}(B)}(1_B)$.

THEOREM 3.2.1 (Motivic Becker-Gottlieb transfers [**3**, Lemma 1.7, Theorem 1.10]). *1. For $f : E \to B$ in **Sm**$/B$ with E/B dualizable, we have*

$$f/B \circ Tr(f/B) = \chi(E/B) : 1_B \to 1_B$$

*2. Take F and B in **Sm**$/S$ with F/S dualizable in $\mathrm{SH}(S)$. Let $f : E \to B$ be a Nisnevich locally trivial fiber bundle with fiber F. Then E/B is dualizable in $\mathrm{SH}(B)$.*

3. Take $\mathcal{E} \in \mathrm{SH}(S)$ so that $\chi(F/S) : 1_S \to 1_S$ induces an isomorphism on $\mathcal{E}^{0,0}(S)$. Then the map

$$f/S^* : \mathcal{E}^{**}(B/S) \to \mathcal{E}^{**}(E/S)$$

is split injective.

The assertion (1) is a rather easy diagram chase from the definitions. For (2), one uses May's theorem [**4**] that the Euler characteristic for $\mathrm{SH}(B)$ is additive in distinguished triangles to reduce to the case of a product, for which the assertion is easy. For (3), it suffices to show that the map $\mathrm{Tr}(E/S) \circ f/S^*$ is an isomorphism. In the case of a product, this is just (1), noting that $\chi(F \times_S B/B) = \pi_B^*\chi(F/S)$. One then makes a Nisnevich descent argument to reduce to this case.

3.3. Splitting principles. We would like to apply this to the $\mathcal{E}$-cohomology of BG for G a reductive group. Take for example $G = \mathrm{GL}_n$ and let $i : T \to \mathrm{GL}_n$ be the inclusion of the maximal torus of diagonal matrices. We have the homotopy fiber sequence

$$T\backslash \mathrm{GL}_n \to BT \xrightarrow{Bi} \mathrm{BGL}_n$$

We note that $T\backslash \mathrm{GL}_n$ is $\mathbb{A}^1$ homotopy equivalent to $B\backslash \mathrm{GL}_n$, where $B \supset T$ is a Borel subgroup, giving the full flag manifold $\mathcal{F}l_n = B\backslash \mathrm{GL}_n$. Factoring $\mathcal{F}l_n$ as a tower of projective space bundles and using the projective bundle theorem, this says that for $\mathcal{E}$ an orientable theory, the map

$$Bi^* : \mathcal{E}^{**}(\mathrm{BGL}_n) \to \mathcal{E}^{**}(BT)$$

is injective and in fact identifies $\mathcal{E}^{**}(\mathrm{BGL}_n)$ with the invariants in $\mathcal{E}^{**}(BT)$ for the action of the Weyl group $W(\mathrm{GL}_n) = S_n$.

However, this does not hold for non-oriented theories, even for SL-oriented theories. In fact, if we have an SL-oriented and η-invertible theory, the map Bi^* maps all the non-trivial elements in $\mathcal{E}^{**}(\mathrm{BGL}_n)$ to zero.

One can still recover a splitting principle that holds for every theory $\mathcal{E}$ by replacing the $W(\mathrm{GL}_n)$-invariants in $\mathcal{E}^{**}(BT)$ with $\mathcal{E}^{**}(BN_T)$, where N_T is the normalizer of T in GL_n. In fact $N_T \cong T \rtimes S_n$, with $S_n \subset \mathrm{GL}_n$ the subgroup of permutation matrices.

We have the following conjecture (formulated by Fabien Morel)

CONJECTURE 3.3.1. *Let G be a split reductive group scheme over a perfect field k with maximal split torus T and N_T the normalizer of T. Then*

Strong form $\chi(N_T\backslash G) = 1$

Weak form $\chi(N_T\backslash G)$ *is invertible in* $\mathrm{GW}(k)$.

It follows using the motivic Becker-Gottlieb transfers that for G special (i.e., $G = \mathrm{GL}_n, \mathrm{SL}_n, \mathrm{Sp}_n$) the above conjecture implies the splitting principle for the normalizer: that $Bi^*_{N_T} : BG \to BN_T$ is injective. Indeed, for G special the bundle

$$N_T\backslash G \to BN_T \to BG$$

is Zariski locally trivial (over the smooth finite type schemes that approximate BG) and then one can apply Theorem 3.2.1, assuming the weak form of the conjecture.

We sketch a proof below of:

THEOREM 3.3.2. *The weak form of the conjecture holds for $G = \mathrm{GL}_n, \mathrm{SL}_n$ over a characteristic zero field, and the strong form holds for $G = \mathrm{GL}_2, \mathrm{SL}_2$ over a perfect field.*

In order to prove the weak form of the conjecture for a given group scheme G over a field of characteristic zero, it suffices to take G over $\mathbb{Q}$. We need to show that $\chi(N_T\backslash G)$ is invertible in $\mathrm{GW}(\mathbb{Q})$; the known structure of $\mathrm{GW}(\mathbb{Q})$ implies that it suffices to show that $\chi(N_T\backslash G)$ has both rank and signature equal to one, in other words, that

$$\chi^{top}((N_T\backslash G)(\mathbb{C})) = 1 = \chi^{top}((N_T\backslash G)(\mathbb{R})).$$

The identity $\chi^{top}((N_T\backslash G)(\mathbb{C}))$ is well-known: Since $\mathbb{C}$ is algebraically closed, we have

$$(N_T\backslash G)(\mathbb{C}) \cong N_T(\mathbb{C})\backslash G(\mathbb{C}) \cong W(G(\mathbb{C}))\backslash(T(\mathbb{C})\backslash G(\mathbb{C}))$$

Now $T(\mathbb{C})\backslash G(\mathbb{C})$ is homotopy equivalent to $B(\mathbb{C})\backslash G(\mathbb{C})$ and the Bruhat decomposition

$$B(\mathbb{C})\backslash G(\mathbb{C}) = \amalg_{w\in W(G(\mathbb{C}))} B(\mathbb{C})\backslash B(\mathbb{C})wB(\mathbb{C})$$

gives a cell decomposition of $B(\mathbb{C})\backslash G(\mathbb{C})$ with $\#W(G(\mathbb{C}))$ even dimensional cells. Thus $\chi(T(\mathbb{C})\backslash G(\mathbb{C})) = \#W(G(\mathbb{C}))$ and since

$$T(\mathbb{C})\backslash G(\mathbb{C}) \to N_T(\mathbb{C})\backslash G(\mathbb{C})$$

is a covering space of degree $\#W(G(\mathbb{C}))$, it follows that

$$\chi^{top}((N_T\backslash G)(\mathbb{C})) = 1.$$

For the computation of $\chi^{top}((N_T\backslash G)(\mathbb{R}))$, we restrict to the case $G = \mathrm{GL}_n$. $N_T = \mathbb{G}_m^n \rtimes S_n$, so we have

$$N_T\backslash \mathrm{GL}_n = S_n\backslash(\mathbb{G}_m^n\backslash \mathrm{GL}_n)$$

and $\mathbb{G}_m^n\backslash \mathrm{GL}_n$ is the open subscheme of $(\mathbb{P}^{n-1})^n$ parametrizing n-tuples of lines through 0 in $\mathbb{A}^n$ which span $\mathbb{A}^n$. This realizes $N_T\backslash \mathrm{GL}_n$ as an open subscheme of $\mathrm{Sym}^n\mathbb{P}^{n-1}$, and gives a decomposition of the $\mathbb{R}$-points $x \in (N_T\backslash \mathrm{GL}_n)(\mathbb{R})$ in terms of the residue fields of the corresponding closed points $y_1, \dots, y_s$ of $(\mathbb{P}^{n-1})^n$ lying over x. This decomposes $(N_T\backslash \mathrm{GL}_n)(\mathbb{R})$ as a disjoint union of open submanifolds

$$(N_T\backslash \mathrm{GL}_n)(\mathbb{R}) = \amalg_{i=0}^n U_i$$

where U_i parametrizes the spanning collections of lines with exactly i real lines. Of course U_i is empty if $n-i$ is odd.

We claim that $\chi^{top}(U_i)=0$ if $i\geq 2$. Indeed, we can form a finite covering space $\tilde{U}_i\to U_i$ by killing the permutation action on the real lines and choosing an orientation on each real line. Then putting a metric on $\mathbb{C}^n$, we see that $\tilde{U}_i$ is homotopy equivalent to an $SO(i)$-bundle over the space parametrizing the (unordered) collection of the remaining lines. Since $\chi^{top}(SO(i))=0$ for $i\geq 2$, we have $\chi^{top}(\tilde{U}_i)=0$ and hence $\chi^{top}(U_i)=0$ as well.

Now assume $n=2m$ is even. U_0 parametrizes the unordered collections of m pairs of complex conjugate lines that span $\mathbb{C}^n$, so we may form the degree $m!$ covering space $\tilde{U}_0\to U_0$ of ordered m-tuples of such conjugate pairs. Each pair $\ell,\bar{\ell}$ spans a $\mathbb{A}^2\subset\mathbb{A}^n$ defined over $\mathbb{R}$, so we may map $\tilde{U}_0$ to the real points of the quotient $(\mathrm{GL}_2)^m\backslash\mathrm{GL}_n$, which is homotopy equivalent to the real points of the partial flag manifold $Fl_m:=P_m\backslash\mathrm{GL}_n$, where P_m is the parabolic with Levi subgroup $(\mathrm{GL}_2(\mathbb{C}))^m$. In fact, the map $\tilde{U}_0\to Fl_m(\mathbb{R})$ is also a homotopy equivalence: given a two-plane $\pi\subset\mathbb{A}^n$ defined over $\mathbb{R}$, the pairs of conjugate lines in π are parametrized by the upper half-plane in the Riemann sphere $\mathbb{P}(\pi)$, which is contractible.

We claim that $\chi^{top}(Fl_m(\mathbb{R}))=m!$ for all $m\geq 2$. To compute the Euler characteristic, we use the following fact:

If $X\in\mathbf{Sm}/\mathbb{R}$ is *cellular*, that is, X admits a finite stratification by locally closed subschemes $X_i\cong\mathbb{A}^{n_i}$, then $\chi^{top}(X(\mathbb{R}))=\mathrm{rnk}\ \mathrm{CH}^{even}(X)-\mathrm{rnk}\ \mathrm{CH}^{odd}(X)$.

This follows from [**2**, Remark 1.11, Proposition 1.14], using the GW($\mathbb{R}$)-valued Euler characteristic $\chi(X/\mathbb{R})$:

(3.3.1)

i. For $X\in\mathbf{Sm}/\mathbb{R}$, $\chi^{top}(X(\mathbb{R}))$ is equal to the signature of $\chi(X/\mathbb{R})$ [**2**, Remark 1.11].

ii. For $X\in\mathbf{Sm}/k$ cellular, $\chi(X/k)=\mathrm{rnk}\ \mathrm{CH}^{even}\cdot\langle 1\rangle+\mathrm{rnk}\ \mathrm{CH}^{odd}\cdot\langle -1\rangle$ [**2**, Proposition 1.14].

Alternatively, one can apply the Lefschetz trace formula to complex conjugation c acting on $H^*(X(\mathbb{C}),\mathbb{Q})$, and use the fact that for X cellular, $H^{odd}(X(\mathbb{C}),\mathbb{Q})=0$, $H^{2n}(X(\mathbb{C}),\mathbb{Q})$ has dimension $\mathrm{rnk}\ \mathrm{CH}^n(X)$ and c acts by $(-1)^n$, since the cycle class map cl^n is a c-equivariant isomorphism from the c-invariant $\mathbb{Q}$ vector space $\mathrm{CH}^n(X)\otimes\mathbb{Q}$ to $H^{2n}(X(\mathbb{C}),\mathbb{Q}(2\pi i)^n)=\mathbb{Q}\cdot\mathrm{cl}^n(\mathrm{CH}^n(X))$.

For $m=2$, $Fl_2=\mathrm{Gr}(2,4)$, which has 4 Schubert cells of even dimension and 2 of odd dimension. Thus $\chi(Fl_2/\mathbb{R})=4\langle 1\rangle+2\langle -1\rangle$, which has signature 2, and hence $\chi^{top}(Fl_2(\mathbb{R}))=2$.

In general, we have the fibration $Fl_m\to Fl_{m-1}$ with fiber $\mathrm{Gr}(2,2m)$. We can compare the Schubert cells for $\mathrm{Gr}(2,2m)$ with those for $\mathrm{Gr}(2,2m-2)$: the "new" ones come from the partitions $(2m-3,i)$, $i=0,\ldots,2m-3$ and $(2m-2,i)$, $i=0,\ldots,2m-2$. This adds the quadratic form $(2m-1)\langle 1\rangle+(2m-2)\langle -1\rangle$ of signature $+1$ to $\chi(\mathrm{Gr}(2,2m-2))$, which implies that $\chi^{top}(\mathrm{Gr}(2,2m)(\mathbb{R}))=\chi^{top}(\mathrm{Gr}(2,2m-2))+1$. By induction, this says that $\chi^{top}(\mathrm{Gr}(2,2m)(\mathbb{R}))=m$, and similarly by induction we find $\chi^{top}(Fl_m)=m\cdot\chi^{top}(Fl_{m-1})=m!$. Thus $\chi^{top}(\tilde{U}_0)=m!$ and hence $\chi^{top}(U_0)=1$.

In case $n = 2m + 1$ is odd, we can again choose a metric and fiber U_1 over the real projective space $\mathbb{RP}^{2m}$, with fiber over $[\ell] \in \mathbb{RP}^{2m}$ homotopy equivalent to the U_0 for the hyperplane perpendicular to the $\mathbb{C}$-span of ℓ. As $\chi^{top}(\mathbb{RP}^{2m}) = 1$, this yields $\chi^{top}(U_1) = 1$.

We note that $N_T(\mathrm{GL}_n)\backslash \mathrm{GL}_n \cong N_T(\mathrm{SL}_n)\backslash \mathrm{SL}_n$ so the weak form of the conjecture also holds for $G = \mathrm{SL}_n$.

We conclude with a proof of the strong form of the conjecture for GL_2 (or SL_2). We argue as above: $(\mathbb{G}_m)^2\backslash \mathrm{GL}_2 = \mathbb{P}^1 \times \mathbb{P}^1 - \Delta$, where $\Delta \cong \mathbb{P}^1$ is the diagonal, and thus $N_T\backslash \mathrm{GL}_2 = S_2\backslash(\mathbb{P}^1 \times \mathbb{P}^1 - \Delta)$. We have $S_2\backslash\mathbb{P}^1 \times \mathbb{P}^1 = \mathrm{Sym}^2\mathbb{P}^1 \cong \mathbb{P}^2$. Via this isomorphism the quotient map

$$q : \mathbb{P}^1 \times \mathbb{P}^1 \to \mathbb{P}^2 = \mathrm{Sym}^2\mathbb{P}^1$$

is given by symmetric functions,

$$q((x_0 : x_1), (y_0 : y_1)) = (x_0y_0 : x_0y_1 + x_1y_0 : x_1y_1)$$

Thus the restriction of q to Δ is the map $\bar{q}(x_0 : x_1) = (x_0^2 : 2x_0x_1 : x_1^2)$ with image the conic $C \subset \mathbb{P}^2$ defined by $Q := T_1^2 - 4T_0T_2$. Furthermore, this identifies the double cover $\mathbb{P}^1 \times \mathbb{P}^1 \to \mathbb{P}^2$ with Spec $\mathcal{O}_{\mathbb{P}^2}(\sqrt{Q})$, that is, the closed subscheme of $O_{\mathbb{P}^2}(1)$ defined by pulling back the section of $O_{\mathbb{P}^2}(2)$ defined by Q via the squaring map $O_{\mathbb{P}^2}(1) \to O_{\mathbb{P}^2}(2)$.

For k of characteristic $\neq 2$, the map $\bar{q} : \mathbb{P}^1 \to C$ is an isomorphism. If $\mathrm{char}\, k = 2$, then C is just the line $T_1 = 0$ and $N_T\backslash \mathrm{GL}_2 \cong \mathbb{A}^2$.

In any case, applying (3.3.1)(ii), the identification of $N_T\backslash \mathrm{GL}_2$ with $\mathbb{P}^2 - C \cong \mathbb{P}^2 - \mathbb{P}^1$ gives

$$\chi((N_T\backslash \mathrm{GL}_2)/k) = \chi(\mathbb{P}^2/k) - \chi(\mathbb{P}^1/k) = 2\langle 1\rangle + \langle -1\rangle - (\langle 1\rangle + \langle -1\rangle) = \langle 1\rangle$$

verifying the strong form of the conjecture in this case.

PROPOSITION 3.3.3. *For k a perfect field, $\chi((N_T\backslash \mathrm{GL}_2)/k) = 1$.*

References: Lecture 3

[1] Alexey Ananyevskiy, *The special linear version of the projective bundle theorem*, Compos. Math. **151** (2015), no. 3, 461–501, DOI 10.1112/S0010437X14007702. MR3320569

[2] Marc Levine, *Toward an enumerative geometry with quadratic forms* preprint (2017) arXiv:1703.03049 [math.AG]

[3] Marc Levine, *Motivic Euler characteristics and Witt-valued characteristic classes.* preprint (2018) arXiv:1806.10108 [math.AG]. Published online, Nagoya J. Math. https://doi.org/10.1017/nmj.2019.6

[4] J. P. May, *The additivity of traces in triangulated categories*, Adv. Math. **163** (2001), no. 1, 34–73, DOI 10.1006/aima.2001.1995. MR1867203

[5] Ivan Panin, Charles Walter, *Quaternionic Grassmannians and Pontryagin classes in algebraic geometry.* preprint (2010) arXiv.1011.0649 [math.AG].

Lecture 4. Reduction to the normalizer

In this final lecture, we consider the normalizer $N = N_T \subset \mathrm{SL}_2$ and compute the characteristic classes for all representations $\rho : N \to \mathrm{GL}_n$. In fact, $N = \mathbb{G}_m \rtimes \mathbb{Z}/2$ and all irreducible representations are either one- or two-dimensional. This reduces us to computing the Euler classes of the bundles associated to the two-dimensional representations. The N_T-splitting principle reduces the computation of the characteristic classes of bundles formed from a rank 2 SL_2-bundle by applying a representation $\rho : \mathrm{SL}_2 \to \mathrm{GL}_n$ to the case of a bundle arising from an irreducible N_T-representation, so knowing the Euler classes of these combined with Ananyevskiy's SL_2-splitting principle gives a complete calculus of characteristic classes of SL (or even GL) vector bundles. Here we give the computation of the Euler classes in Witt cohomology for N_T-bundles.

The representation theory of N is closely related to the real representation theory of $O(2)$. The Euler classes for the irreducible representations of $SO(2)$ have been computed, for example by Okonek-Teleman [**5**], by noting that an $SO(2)$-bundle has a canonical $\mathbb{C}$-structure, and in the rank two case, the Euler class is just the 1st Chern class of the associated $\mathbb{C}$-line bundle. Our method is to mimic this approach; most of this lecture is taken from [**3**].

4.1. BGL_n and BSL_n. We mention a result that reduces the study of characteristic classes for arbitrary bundles to the case of bundles with trivial determinant. For $\mathcal{E} \in \mathrm{SH}(k)$, we have a canonical action of the bi-graded presheaf $U \mapsto \mathbb{S}^{*,*}(U)$ on the bi-graded presheaf $U \mapsto \mathcal{E}^{*,*}(U)$; passing from sheaf to presheaf and inverting η gives a canonical map of presheaves $[U \mapsto \mathbb{S}^{*,*}(U)] \to [U \mapsto \mathcal{W}(U)]$. We say that $\mathcal{E}^{**}$ is a $\mathcal{W}$-module if the $U \mapsto \mathbb{S}^{*,*}(U)$-action descends to a $U \mapsto \mathcal{W}(U)$-action.

THEOREM 4.1.1 ([**3**, Theorem 4.1]). *Let $\mathcal{E}$ be an* SL*-oriented and η-invertible theory. We assume in addition that $\mathcal{E}^{**}$ is a $\mathcal{W}$-module. Let $O(1) \to \mathrm{BGL}_n$ be the determinant of the universal bundle $E_n \to \mathrm{BGL}_n$. Then the inclusion $i_n : \mathrm{SL}_n \to \mathrm{GL}_n$ induces an isomorphism*

$$\mathcal{E}^{**}(\mathrm{BGL}_n) \oplus \mathcal{E}^{**}(\mathrm{BGL}_n; O(1)) \xrightarrow[\sim]{Bi_n^*} \mathcal{E}^{**}(\mathrm{BSL}_n)$$

More precisely, via Ananyevskiy's isomorphism

$$\mathcal{E}^{**}(\mathrm{BSL}_n) \cong \begin{cases} \mathcal{E}^{**}(k)[p_1, \ldots, p_m] & \text{for } n = 2m+1 \\ \mathcal{E}^{**}(k)[p_1, \ldots, p_m, e]/p_m - e^2 & \text{for } n = 2m \end{cases}$$

we have

$$\mathcal{E}^{**}(\mathrm{BGL}_n) \xrightarrow[\sim]{Bi_n^*} \mathcal{E}^{**}(\mathrm{BSL}_n); \quad \mathcal{E}^{**}(\mathrm{BGL}_n; O(1)) = 0$$

for $n = 2m+1$ odd,

$$Bi_n^*(\mathcal{E}^{**}(\mathrm{BGL}_n)) = \mathcal{E}^{**}(k)[p_1, \ldots, p_m]$$
$$Bi_n^*(\mathcal{E}^{**}(\mathrm{BGL}_n, O(1))) = \mathcal{E}^{**}(k)[p_1, \ldots, p_m] \cdot e$$

for $n = 2m$ even.

The result is proven by realizing $\mathrm{BSL}_n \to \mathrm{BGL}_n$ as the $\mathbb{G}_m$-bundle $O(1) - 0_{O(1)} \to \mathrm{BGL}_n$ and analyzing the localization sequence for the diagram

$$\begin{array}{ccccc} \mathrm{BSL}_n = O(1) - 0_{O(1)} & \xhookrightarrow[j]{} & O(1) & \xhookleftarrow[s_0]{} & \mathrm{BGL}_n \\ & & \Big\uparrow{\scriptstyle \wr}\, p^* & & \\ & & \mathrm{BGL}_n & & \end{array}$$

4.2. $\mathcal{W}$-cohomology of BN_T. Via Becker-Gottlieb transfers, we know that the pull-back map for $BN_T \to \mathrm{BSL}_2$ is injective; we give here an explicit computation of $H^*(BN_T, \mathcal{W})$ in terms of $H^*(\mathrm{BSL}_2, \mathcal{W}) = W(k)[e]$.

Recall the description of $N_T\backslash \mathrm{SL}_2$ as $\mathbb{P}^2 - C$ from § 3.3.We use the model $E\,\mathrm{SL}_2 = E\,\mathrm{GL}_2 =$ the colimit of $2 \times N$ matrices of rank two. Then

$$BN_T = N_T\backslash E\,\mathrm{GL}_2 = (N_T\backslash \mathrm{SL}_2) \times^{\mathrm{SL}_2} E\,\mathrm{GL}_2$$

i.e., $BN_T \to \mathrm{BSL}_2 = \mathrm{SL}_2\backslash E\,\mathrm{GL}_2$ is a Zariski locally trivial bundle with fiber $N_T\backslash \mathrm{SL}_2 \cong \mathbb{P}^2 - C$.

Letting $Q = T_1^2 - 4T_0T_2$, then Q is the defining equation for C and the double cover $T\backslash \mathrm{SL}_2 = \mathbb{P}^1 \times \mathbb{P}^1 - \Delta \to \mathbb{P}^2 - C$ is identified with $\mathrm{Spec}_{\mathcal{O}_{\mathbb{P}^2 - C}}(\sqrt{Q})$. One checks that Q is SL_2-invariant and thus the quadratic form $\langle q \rangle$ (locally $\langle Q/L^2 \rangle$) gives an SL_2-invariant global section of $\mathcal{W}$ over $\mathbb{P}^2 - C$, which then gives

$$\langle q \rangle \in H^0(BN_T, \mathcal{W})$$

PROPOSITION 4.2.1. *Let $p : BN_T \to \mathrm{BSL}_2$ be the projection. Then*

$$p^* : H^n(\mathrm{BSL}_2, \mathcal{W}) \to H^n(BN_T, \mathcal{W})$$

is an isomorphism for $n \geq 1$. For $n = 0$, we have an exact sequence

$$0 \to H^0(\mathrm{BSL}_2, \mathcal{W}) = W(k) \xrightarrow{p^*} H^0(BN_T, \mathcal{W}) \to W(k) \to 0$$

split by the class $\langle q \rangle$. Explicitly

$$H^*(BN_T, \mathcal{W}) = W(k)[p^*e] \oplus W(k) \cdot \langle q \rangle$$

PROOF. In invariant terms $N_T\backslash \mathrm{SL}_2 = \mathbb{P}(\mathrm{Sym}^2 F) - \mathbb{P}(F)$, where $F = \mathbb{A}^2$ is the tautological right representation of SL_2. We analyze $H^*(BN_T, \mathcal{W})$ via the localization sequence for

$$(N_T\backslash \mathrm{SL}_2) \times^{\mathrm{SL}_2} E\,\mathrm{GL}_2 \hookrightarrow \mathbb{P}(\mathrm{Sym}^2 F) \times^{\mathrm{SL}_2} E\,\mathrm{GL}_2 \hookleftarrow \mathbb{P}(F) \times^{\mathrm{SL}_2} E\,\mathrm{GL}_2$$

Using η-invertibility and $\mathbb{A}^1$-homotopy invariance gives

$$\begin{aligned} H^*(\mathbb{P}(F) \times^{\mathrm{SL}_2} E\,\mathrm{GL}_2, \mathcal{W}) &= H^*(\mathbb{A}^2 - \{0\} \times^{\mathrm{SL}_2} E\,\mathrm{GL}_2, \mathcal{W}) \\ &= H^*(\mathrm{SL}_2 \times^{\mathrm{SL}_2} E\,\mathrm{GL}_2, \mathcal{W}) = H^*(\mathrm{Spec}\, k, \mathcal{W}) \end{aligned}$$

so

$$H^n(\mathbb{P}(F) \times^{\mathrm{SL}_2} E\,\mathrm{GL}_2, \mathcal{W}) = \begin{cases} 0 & \text{for } n > 0 \\ W(k) & \text{for } n = 0. \end{cases}$$

Fasel [**2**, Theorem 11.7] has computed $H^*(\mathbb{P}^2_L, \mathcal{W})$ for a field L

$$H^n(\mathbb{P}^2_L, \mathcal{W}) = \begin{cases} 0 & \text{for } n > 0 \\ W(L) & \text{for } n = 0. \end{cases}$$

and then a Leray spectral sequence gives

$$H^*(\mathrm{BSL}_2, \mathcal{W}) \cong H^*(\mathbb{P}(\mathrm{Sym}^2 F) \times^{\mathrm{SL}_2} E\,\mathrm{GL}_2, \mathcal{W})$$

The localization sequence completes the proof. □

4.3. The bundles $\tilde{O}(m)$. Let $\sigma \in N_T(k)$ be the matrix

$$\sigma = \begin{pmatrix} 0 & 1 \\ -1 & 0 \end{pmatrix}$$

For each $m = 1, 2, \ldots$, we have the representation

$$\rho_m = \rho_m^+ : N_T \to \mathrm{GL}_2$$

defined by

$$\rho_m \begin{pmatrix} t & 0 \\ 0 & t^{-1} \end{pmatrix} = \begin{pmatrix} t^m & 0 \\ 0 & t^{-m} \end{pmatrix}; \quad \rho_m(\sigma) = \begin{pmatrix} 0 & 1 \\ (-1)^m & 0 \end{pmatrix}$$

We have as well the representation ρ_m^- with the same values on T and with $\rho_m^-(\sigma) = -\rho_m^+(\sigma)$. We note that for m odd, $\rho_m^\pm$ has values in SL_2, but this is not the case for m even. We have as well the one-dimensional representations $\rho_0 = \mathrm{Id}$ and ρ_0^- with $\rho_0^-(T) = \mathrm{Id}$ and $\rho_0^-(\sigma) = -1$.

This gives us the rank two bundles $\tilde{O}^\pm(m) \to BN_T$, $m > 0$ and the rank one bundle $\gamma = \tilde{O}^-(0) \to BN_T$. We note that $\mathrm{Pic}(BN_T) = \mathbb{Z}/2$ with generator γ. Our main object is to compute the Euler class of $\tilde{O}^\pm(m)$ for $m > 0$. We note that $e(\tilde{O}^-(m)) = -e(\tilde{O}(m))$ since we have the orientation reversing isomorphism $(x, y) \to (-x, y)$ of ρ_m with ρ_m^-. Also $\tilde{O}(1)$ is the tautological bundle, so has Euler class p^*e.

Our approach is the following: Let $s^{(m)} : BN_T \to \tilde{O}(m)$ be the zero-section. The map

$$\psi_m : \mathbb{A}^2 \to \mathbb{A}^2; \quad \psi_m(x, y) = (x^m, y^m)$$

is N_T-equivariant, if we let N_T act by ρ_1 on the domain and by ρ_m on the target. This gives us the map over BN_T

$$\psi_m : \tilde{O}(1) \to \tilde{O}(m)$$

and as $\psi_m \circ s^{(1)} = s^{(m)}$. Thus

$$e(\tilde{O}(m)) = s^{(1)*}(\psi_m^*(\theta_{\tilde{O}(m)})$$

so we reduce to calculating

$$\psi_m^*(\theta_{\tilde{O}(m)}) \in H^2_{0_{\tilde{O}(1)}}(\tilde{O}(1), \mathcal{W}) \cong H^0(BN_T, \mathcal{W}) = W(k) \oplus W(k) \cdot \langle q \rangle$$

In fact, it suffices to calculate the image of $\psi_m^*(\theta_{\tilde{O}(m)})$ in $H^2(\tilde{O}(1), \mathcal{W}) = W(k) \cdot p^*e$, so we need only compute $\psi_m^*(\theta_{\tilde{O}(m)})$ modulo the kernel of the surjection

$$H^2_{0_{\tilde{O}(1)}}(\tilde{O}(1), \mathcal{W}) \to H^2(\tilde{O}(1), \mathcal{W})$$

This information is furnished by an explicit computation

LEMMA 4.3.1 ([**3**, Lemma 5.4]). *The kernel of* $H^2_{0_{\tilde{O}(1)}}(\tilde{O}(1), \mathcal{W}) \to H^2(\tilde{O}(1), \mathcal{W})$ *is* $W(k) \cdot (1 + \langle q \rangle)$.

Thus, rather than make the computation of $\psi_m^*(\theta_{\tilde{O}(m)})$ in $H^2_{0_{\tilde{O}(1)}}(\tilde{O}(1),\mathcal{W})$, we may instead restrict to the fiber of $\tilde{O}(1)$ over any point $x \in BN_T$ with $\langle q(x)\rangle = \langle -1\rangle = -1$ in $W(k)$, because we have the commutative diagram

$$\begin{array}{ccc} W(k)\oplus W(k)\langle q\rangle & \xrightarrow{\quad i_x^* \quad} & W(k) \\ \| & & \| \\ H^2_{0_{\tilde{O}(1)}}(\tilde{O}(1),\mathcal{W}) & \xrightarrow{i_x^*} & H^2_{0_{\tilde{O}(1)_x}}(\tilde{O}(1)_x,\mathcal{W}) \end{array}$$

and $i_x^*(a+b\cdot\langle q\rangle) = a+b\cdot\langle q(x)\rangle = a-b$.

The map $\psi_m(x):\tilde{O}(1)_x\to\tilde{O}(m)_x$ induces the map of projective spaces

$$\mathbb{P}(\psi_m(x)):\mathbb{P}(\tilde{O}(1)_x)\to\mathbb{P}(\tilde{O}(m)_x)$$

For m odd, the bundle $\tilde{O}(m)$ carries a canonical trivialization of $\det\tilde{O}(m)$ so we may replace $\mathbb{P}(\tilde{O}(1)_x)$ and $\mathbb{P}(\tilde{O}(m)_x)$ with any choice of bases $v_{1,1}, v_{2,1}$, $v_{1,m}, v_{2,m}$ such that

$$\det(v_{1,1},v_{2,1}) = \det(v_{1,m},v_{2,m})$$

We then have the map

$$\mathbb{P}(\psi_m(x)):\mathbb{P}^1\to\mathbb{P}^1$$

and the induced map

$$\psi_m(x)^*: W(k) = H^2_{0_{\tilde{O}(m)_x}}(\tilde{O}(m)_x,\mathcal{W})\to H^2_{0_{\tilde{O}(1)_x}}(\tilde{O}(1)_x,\mathcal{W}) = W(k)$$

is multiplication by the $\mathbb{A}^1$-degree of $\mathbb{P}(\psi_m(x))$.

We take $k=\mathbb{Q}$ and $x\in BN_T(\mathbb{Q})$ to be the image of

$$\tilde{x} = \begin{pmatrix}1 & i\\ 1 & -i\end{pmatrix}$$

For $((x_0:x_1),(y_0:y_1))\in\mathbb{P}^1\times\mathbb{P}^1-\Delta$, the image in $\mathbb{P}^2-C$ is given by $(x_0y_0 : x_0y_1+x_1y_0 : x_1y_1)$, in other words

$$x = (1:0:1),\ Q(x) = -4\equiv -1 \mod \mathbb{Q}^{\times 2},$$

so $\langle q(x)\rangle = \langle -1\rangle = -1\in W(\mathbb{Q})$.

The basis $v_{1,m}, v_{2,m}$ is chosen to be

$$\begin{pmatrix}v_{1,m}\\ v_{2,m}\end{pmatrix} := \begin{cases}\begin{pmatrix}1 & i\\ i & 1\end{pmatrix} & \text{for } m\equiv 1 \mod 4\\ \begin{pmatrix}1 & -i\\ -i & 1\end{pmatrix} & \text{for } m\equiv 3 \mod 4\end{cases}$$

Then $v_{1,m}\wedge v_{2,m} = 2\cdot e_1\wedge e_2$, so

$$\left\langle\frac{v_{1,1}\wedge v_{2,1}}{v_{1,m}\wedge v_{2,m}}\right\rangle = 1\in W(k)$$

One computes

$$\psi_m(x_0v_{1,1}+x_1v_{2,1}) = \begin{cases}\Re(x_0+ix_1)^m v_{1,m} + \Im(x_0+ix_1)^m v_{2,m} & \text{for } m\equiv 1 \mod 4\\ \Re(x_0-ix_1)^m v_{1,m} + \Im(x_0-ix_1)^m v_{2,m} & \text{for } m\equiv 3 \mod 4\end{cases}$$

in other words,

$$\psi_m(x_0 : x_1) = \begin{cases} (\Re(x_0 + ix_1)^m : \Im(x_0 + ix_1)^m) \text{ for } m \equiv 1 \mod 4 \\ (\Re(x_0 - ix_1)^m : \Im(x_0 - ix_1)^m) \text{ for } m \equiv 3 \mod 4 \end{cases}$$

To complete the story (for m odd) we have

PROPOSITION 4.3.2. *Let $p > 2$ be prime. For $p \equiv 1 \mod 4$, we have $\deg_{\mathbb{A}^1} \psi_m = p$. For $p \equiv 3 \mod 4$ we have $\deg_{\mathbb{A}^1} \psi_m = -p$.*

PROOF. Since $(x_0 : x_1) \mapsto (x_0 : -x_1)$ intertwines the formulas for ψ_m with $m \equiv 1 \mod 4$ and ψ_m with $m \equiv 3 \mod 4$, we reduce to showing that the map

$$(x_0 : x_1) \mapsto (\Re(x_0 + ix_1) : \Im(x_0 + ix_1)^p)$$

has $\mathbb{A}^1$-degree $p \in \mathbb{W}(k)$ for all primes $p > 2$. We actually show that the $\mathbb{A}^1$-degree in $\mathrm{GW}(k)$ is p.

For this, we compute the $\mathbb{A}^1$-degree as the *Brouwer degree* (see [**4**, Chapter 2]), that is, we take a regular k-rational value $s \in \mathbb{A}^1 = \mathbb{P}^1 - \{(0 : 1)\}$ so that $\psi_m^{-1}(s) \subset \mathbb{A}^1$. We use the parameter $t = x_0/x_1$ and let $g : \mathbb{A}^1 \to \mathbb{P}^1$ be the map

$$g(t) = \frac{\Re(t + i)^p}{\Im(t + i)^p}.$$

Then for each $t \in g^{-1}(s)$ we have the one-dimensional quadratic form $\langle g'(t)\rangle$ and the Brouwer degree is

$$\sum_{t \in g^{-1}(s)} \mathrm{Tr}_{k(t)/k}\langle g'(t)\rangle \in \mathrm{GW}(k).$$

We take $s = (0 : 1)$, giving

$$g^{-1}(s) = \{\cot(\ell\pi/2p) \mid \ell = 1, \ldots, p\}$$

At the point $t_0 = (0 : 1)$ ($\ell = p$) we have $g'(t_0) = p$. The other points are conjugate over $\mathbb{Q}$, and generate the real cyclotomic field $F_{4p} := \mathbb{Q}(\zeta_{4p} + \zeta_{4p}^{-1})$, giving the closed point $t_1 \in \mathbb{A}^1$. One computes that $g'(t_1)$ is a square in F_{4p}, so the Brouwer degree is

$$\langle p\rangle + \mathrm{Tr}_{F_{4p}/\mathbb{Q}}(\langle 1\rangle)$$

i.e., $\langle p\rangle$ plus the trace form for F_{4p} over $\mathbb{Q}$.

One can compute this trace form, or rather its invariants: rank, signature, discriminant and Hasse-Witt invariant, to show that $\langle p\rangle + \mathrm{Tr}_{F_{4p}/\mathbb{Q}}(\langle 1\rangle)$ has rank p, signature p, and trivial discriminant and Hasse-Witt invariant, and therefore in $\mathrm{GW}(\mathbb{Q})$ we have

$$[\langle p\rangle + \mathrm{Tr}_{F_{4p}/\mathbb{Q}}(\langle 1\rangle)] = p \cdot \langle 1\rangle.$$

Alternatively, an integral model for the trace form $\mathrm{Tr}_{F_{4p}/\mathbb{Q}}(\langle 1\rangle)$ has been computed by Bayer and Suarez [**1**, pg. 222-"The root lattice $\mathbb{A}_{p-1}$"], who show that $\mathrm{Tr}_{F_{4p}/\mathbb{Q}}(\langle 1\rangle)$ is equivalent to the lattice $\mathbb{A}_{p-1} \subset \mathbb{Z}^p$, i.e., the hyperplane $\sum_{i=1}^p x_i = 0$, with quadratic form induced by vector dot product. But then $\langle p\rangle + \mathrm{Tr}_{F_{4p}/\mathbb{Q}}(\langle 1\rangle)$ is the form associated to the lattice $(1, \ldots, 1) + A_{p-1} \subset \mathbb{Z}^p$, and as

$$\mathbb{Q}((1, \ldots, 1) + A_{p-1}) = \mathbb{Q}^p$$

we see that over $\mathbb{Q}$, $\langle p\rangle + \mathrm{Tr}_{F_{4p}/\mathbb{Q}}(\langle 1\rangle)$ is equivalent to a sum of p squares. □

The case $p=2$ and $m=4$ are handled much the same way, with the difference being that, $\tilde{O}(2m)$ having non-trivial determinant γ, $e(\tilde{O}(2m))$ is in $H^2(BN_T,\gamma)$ and hence is not comparable with p^e. However, an analogous computation shows that the map $\psi_2:\tilde{O}(2)\to\tilde{O}(4)$ has degree -2, so

$$e(\tilde{O}(4)) = -2e(\tilde{O}(2))$$

and also, that

$$p_1(\tilde{O}(2)) = e(\tilde{O}(2))^2 = 4p^*e.$$

The upshot is the following. For an odd integer $m>0$, define $\epsilon(m)\in\{\pm1\}$ by $\epsilon(m)=+1$ if $m\equiv 1 \mod 4$ and $\epsilon(m)=-1$ if $m\equiv 3 \mod 4$. Let $\tilde{e}=e(\tilde{O}(2))\in H^2(BN_T,\mathcal{W}(\gamma))$.

THEOREM 4.3.3 ([**3**, Theorem 7.1]). *Let $m>0$ be an integer. Let k be a field of characteristic zero or of characteristic q prime to m. Write $m=2^rn$ with n odd. If $r=0$, then $e(\tilde{O}(m))=\epsilon(m)\cdot m\cdot p^*e$. For $r=1$, then $e(\tilde{O}(m))=(m/2)\cdot\tilde{e}$ and if $r>1$, then $e(\tilde{O}(m))=-(m/2)\cdot\tilde{e}$. Moreover $\tilde{e}^2=4p^*e$ and $e(\tilde{O}^-(m))=-e(\tilde{O}(m))$.*

COROLLARY 4.3.4. *Let $k\geq 1$ be an integer and let $E\to X$ be a rank two bundle over $X\in\mathbf{Sm}/k$. Then for $k=2m-1$, we have*

$$e(\mathrm{Sym}^kE) = k!!e(E)^m \in H^{2m}(X,\mathcal{W})$$

where $k!! := \prod_{i=0}^{[k/2]} k-2i$. For $k=2m$, $e(\mathrm{Sym}^kE)=0$. Moreover, for all $k\geq 1$,

$$p(\mathrm{Sym}^kE) = \prod_{i=0}^{[k/2]}(1+(k-2i)^2e(E)^2)$$

This follows directly from the theorem, together with the identity

$$\mathrm{Sym}^k\rho_1 \cong \oplus_{i=0}^{[k/2]}\rho_{k-2i}^{(-1)^i}.$$

and the N_T splitting principle.

One can similarly show

THEOREM 4.3.5. *Let $E_1\to X$, $E_2\to Y$ be rank two bundle. Then the bundle $\pi_1^*E_1\otimes\pi_2^*E_2$ on $X\times Y$ has characteristic classes*

$$e(\pi_1^*E_1\otimes\pi_2^*E_2) = \pi_1^*e(E_1)^2 - \pi_2^*e(E_2)^2$$
$$p_1(\pi_1^*E_1\otimes\pi_2^*E_2) = 2(\pi_1^*e(E_1)^2 + \pi_2^*e(E_2)^2)$$

and

$$p_2(\pi_1^*E_1\otimes\pi_2^*E_2) = (\pi_1^*e(E_1)^2 - \pi_2^*e(E_2)^2)^2.$$

References: Lecture 4

[1] Eva Bayer-Fluckiger and Ivan Suarez, *Ideal lattices over totally real number fields and Euclidean minima*, Arch. Math. (Basel) **86** (2006), no. 3, 217–225, DOI 10.1007/s00013-005-1469-9. MR2215310

[2] Jean Fasel, *The projective bundle theorem for* $\mathbf{I}^j$*-cohomology*, J. K-Theory **11** (2013), no. 2, 413–464, DOI 10.1017/is013002015jkt217. MR3061003

[3] Marc Levine, *Motivic Euler characteristics and Witt-valued characteristic classes.* preprint (2018) arXiv:1806.10108 [math.AG]. Published online, Nagoya J. Math. https://doi.org/10.1017/nmj.2019.6

[4] Fabien Morel, *On the motivic* π_0 *of the sphere spectrum*, Axiomatic, enriched and motivic homotopy theory, NATO Sci. Ser. II Math. Phys. Chem., vol. 131, Kluwer Acad. Publ., Dordrecht, 2004, pp. 219–260, DOI 10.1007/978-94-007-0948-5_7. MR2061856

[5] Christian Okonek and Andrei Teleman, *Intrinsic signs and lower bounds in real algebraic geometry*, J. Reine Angew. Math. **688** (2014), 219–241, DOI 10.1515/crelle-2012-0055. MR3176620

Universität Duisburg-Essen, Fakultät Mathematik, Thea-Leymann-Strasse 9, 45127 Essen, Germany

Email address: marc.levine@uni-due.de

Contemporary Mathematics
Volume **745**, 2020
https://doi.org/10.1090/conm/745/15026

Remarks on motivic Moore spectra

Oliver Röndigs

ABSTRACT. The term "motivic Moore spectrum" refers to a cone of an element $\alpha\colon \Sigma^{s,w}\mathbf{1} \to \mathbf{1}$ in the motivic stable homotopy groups of spheres. Homotopy groups, multiplicative structures, and Voevodsky's slice spectral sequence are discussed for motivic Moore spectra.

1. Introduction

Let R be a ring and $a \in R$ an element, generating a two-sided ideal $(a) \subset R$. The projection onto the quotient $R \to R/a := R/(a)$ is then a ring homomorphism. If R is commutative, then so is R/a. In homotopy theory, the situation is more subtle. The topological sphere spectrum $\mathbb{S}$ is the unit in a closed symmetric monoidal category modeling the stable homotopy category, and in particular a commutative monoid. Given any endomorphism $a\colon \mathbb{S} \to \mathbb{S}$, the homotopy-theoretic quotient is almost never a commutative monoid. The first instance occurs when $a = 2$: The homotopy-theoretic quotient $\mathbb{S}/2$, also known as the Moore spectrum for the group $\mathbb{Z}/2$, satisfies $\pi_2\mathbb{S}/2 \cong \mathbb{Z}/4$ by [**3**], and in particular admits no unital multiplicative structure. Even if a unital multiplicative structure exists on $\mathbb{S}/n$ for $n > 2$, its associativity or commutativity in the stable homotopy category is not automatic. See [**16**] and the references therein for details.

Within motivic or $\mathbf{A}^1$-homotopy theory, Moore spectra have appeared for example in [**18**], [**6**], and [**11**]. As the structure of the endomorphisms of the motivic sphere spectrum is much richer, also the notion of Moore spectra should be more sophisticated. The degree zero part of these endomorphisms over a field is the Milnor-Witt K-theory graded by weight [**14**], and the weight zero part of that is the Grothendieck-Witt ring of the field. Since neither is a principal ideal domain in general, and usually far from Noetherian, Moore spectra with respect to ideals instead of single elements are more sensible. See for example [**11**, Remark 1.4]. Nevertheless, an elementary approach is chosen here, which still suffices to illustrate a few interesting phenomena. More precisely, multiplicative structures on motivic Moore spectra – whose existence may depend on the ground field – are discussed in Section 5, based to some extent on results concerning Toda brackets listed in Section 4. These in turn rely on some preliminaries on the few first stable stems of

2010 *Mathematics Subject Classification.* Primary 14F42.

This work was supported through DFG grants within the SPP 1786 "Homotopy theory and algebraic geometry", and a guest professorship at the University of Oslo. Many thanks to the anonymous referee for helpful comments!

the motivic sphere spectrum, to be discussed in Section 2, which follows and partly expands [**21**]. The article closes with some results on slices and slice differentials for special motivic Moore spectra in Section 6. These results may be used for slice spectral sequence computations of homotopy groups of motivic Moore spectra. A noteworthy feature in comparison with corresponding slice spectral sequence computations for the motivic sphere spectrum is the absence of motivic cohomology groups with integral coefficients; motivic cohomology groups with finite coefficients are understood much better.

2. Preliminaries on π_0, π_1, and π_2

Determining the existence of multiplications or pairings on Moore spectra requires information about stable homotopy groups of motivic spheres. Let $\pi_{s,w}\mathsf{E}$ denote the abelian group $[\Sigma^{s,w}\mathbf{1},\mathsf{E}]$, where E is a motivic spectrum and $\mathbf{1}$ is the motivic sphere spectrum. Set $\pi_{s+(w)}\mathsf{E} := \pi_{s+w,w}\mathsf{E}$, and let

$$\pi_{s+(\star)}\mathsf{E} = \bigoplus_{w\in\mathbb{Z}} \pi_{s+w,w}\mathsf{E}$$

denote the direct sum, considered as a $\mathbb{Z}$-graded module over the $\mathbb{Z}$-graded ring $\pi_{0+(\star)}\mathbf{1}$. The notation $\pi_{s-(\star)}\mathsf{E} := \pi_{s+(-\star)}\mathsf{E}$ will be used frequently. The strictly $\mathbf{A}^1$-invariant sheaf obtained as the associated Nisnevich sheaf of $U \mapsto \pi_{s,w}\mathsf{E}_U$ for $U \in \mathbf{Sm}_F$ is denoted $\underline{\pi}_{s,w}\mathsf{E}$, which gives rise to $\underline{\pi}_{s+(\star)}\mathsf{E}$. See [**14**] for the following statement.

THEOREM 2.1 (Morel). *Let F be a field. Then $\pi_{0-(\star)}\mathbf{1}$ is the Milnor-Witt K-theory of F.*

The Milnor-Witt K-theory of F is denoted $\mathbf{K}^{\mathsf{MW}}(F)$, or simply $\mathbf{K}^{\mathsf{MW}}$, following the convention that the base field or scheme may be ignored in the notation. Its generators are denoted $\eta \in \mathbf{K}^{\mathsf{MW}}_{-1} = \pi_{1,1}\mathbf{1}$ and $[u] \in \mathbf{K}^{\mathsf{MW}}_1(F) = \pi_{-1,-1}\mathbf{1}_F$ for every unit $u \in F^\times$. The abbreviations

$$\begin{aligned}\langle u\rangle &:= 1+\eta[u] \in \mathbf{K}^{\mathsf{MW}}_0(F)\\ \langle u_1,\dots,u_m\rangle &:= \langle u_1\rangle + \cdots + \langle u_m\rangle \in \mathbf{K}^{\mathsf{MW}}_0(F)\\ \varepsilon &:= -\langle -1\rangle\\ \mathsf{h} &:= \langle 1,-1\rangle = 1-\varepsilon\end{aligned}$$

for units $u,u_1,\dots,u_m \in F^\times$ will be convenient. Under the identification of $\mathbf{K}^{\mathsf{MW}}_0(F)$ with the Grothendieck-Witt ring $\mathbf{GW}(F)$ of F, the element $\langle u_1,\dots,u_m\rangle$ corresponds to the quadratic form given by the appropriate diagonal matrix. Milnor K-theory [**12**] is expressed as the quotient $\mathbf{K}^{\mathsf{M}}_\star \cong \mathbf{K}^{\mathsf{MW}}_\star/(\eta)$. Set $\mathbf{k}^{\mathsf{M}}_\star := \mathbf{K}^{\mathsf{M}}_\star/2$. Theorem 2.1 implies that for every motivic spectrum E and for every integer s, $\pi_{s+(\star)}\mathsf{E}$ is a graded $\mathbf{K}^{\mathsf{MW}}$-module. As a first instance besides the motivic sphere spectrum $\mathbf{1}$, consider the very effective cover $\mathsf{kq} \to \mathsf{KQ}$ of the motivic spectrum representing hermitian K-theory [**1**], [**2**]. Using kq instead of the effective cover $\mathsf{f}_0\mathsf{KQ} \to \mathsf{KQ}$ leads to a slight improvement on the computation [**21**, Theorem 5.5].

THEOREM 2.2 (Röndigs-Spitzweck-Østvær). *Let F be a field of exponential characteristic $e \neq 2$. The unit map $\mathbf{1} \to \mathsf{kq}$ induces an isomorphism $\pi_{0+(\star)}\mathbf{1} \to \pi_{0+(\star)}\mathsf{kq}$, and a surjection $\pi_{1+(\star)}\mathbf{1} \to \pi_{1+(\star)}\mathsf{kq}$ whose kernel coincides with $\mathbf{K}^{\mathsf{M}}_{2-\star}/24$ after inverting e.*

Voevodsky's slice filtration $\{\mathsf{f}_q\mathsf{E} \to \mathsf{E}\}_{q\in\mathbb{Z}}$ allows to be more precise, and in particular to describe the $\mathbf{K}^{\mathsf{MW}}$-module structure. Let $\mathbf{M}\mathbb{Z}$ be Voevdosky's integral motivic Eilenberg-MacLane spectrum representing motivic cohomology, and let $\mathbf{M}\mathbb{Z}/2$ be the version with coefficients in $\mathbb{Z}/2$. Moreover, set for k a natural number $H^{\star-k,\star} := \pi_{k-(\star)}\mathbf{M}\mathbb{Z}$ and $h^{\star-k,\star} := \pi_{k-(\star)}\mathbf{M}\mathbb{Z}/2$. Note the $\mathbf{K}^{\mathsf{MW}}$-module isomorphism $h^{\star-k,\star} \cong \mathbf{k}^{\mathsf{M}}_{\star-k}$ given by multiplication with τ^k, where

$$\tau = -1 \in h^{0,1} = \{1,-1\} = \ker\big(H^{1,1} \xrightarrow{x\mapsto x^2} H^{1,1}\big)$$

is the unique nontrivial element. The $\mathbf{K}^{\mathsf{MW}}$-module $\pi_{1-(\star)}\mathsf{kq}$ is an extension of the $\mathbf{K}^{\mathsf{MW}}$-module $H^{\star-1,\star} = \pi_{1-(\star)}\mathsf{s}_0\mathsf{kq}$ (on which η operates trivially) and the $\mathbf{K}^{\mathsf{MW}}$-module given by the image of $\pi_{1-(\star)}\mathsf{f}_1\mathsf{kq}$ in $\pi_{1-(\star)}\mathsf{kq}$.

LEMMA 2.3. *The $\mathbf{K}^{\mathsf{MW}}$-module $\pi_{1-(\star)}\mathsf{f}_1\mathsf{kq}$ is generated by the image of η_{top} under the unit map* $\mathrm{u}\colon \mathbf{1} \to \mathsf{kq}$, *and has the presentation*

$$\mathbf{K}^{\mathsf{MW}}/(2,\eta^2) \cong \pi_{1-(\star)}\mathsf{f}_1\mathsf{kq}.$$

PROOF. The determination of the relevant part of the slice spectral sequence for kq given in [**1**, Proposition 27] implies that the $\mathbf{K}^{\mathsf{MW}}$-module $\pi_{1-(\star)}\mathsf{f}_1\mathsf{kq}$ is an extension of the $\mathbf{K}^{\mathsf{MW}}$-module $h^{\star,1+\star} = \pi_{1-(\star)}\mathsf{s}_1\mathsf{kq}$ (on which η operates trivially) and the $\mathbf{K}^{\mathsf{MW}}$-module $h^{1+\star,2+\star}/\mathsf{Sq}^2h^{\star-1,1+\star}$ (on which η operates trivially as well). The $\mathbf{K}^{\mathsf{M}}$-module $h^{\star,1+\star} = \pi_{1-(\star)}\mathsf{s}_1\mathsf{kq}$ is generated by the image of $\eta_{\mathsf{top}} \in \pi_{1,0}\mathbf{1}$, and the $\mathbf{K}^{\mathsf{M}}$-module $h^{1+\star,2+\star}/\mathsf{Sq}^2h^{\star-1,1+\star}$ is generated by the image of $\eta\eta_{\mathsf{top}} \in \pi_{2,1}\mathbf{1}$. Since $2\eta_{\mathsf{top}} = 0$, the extension describing $\pi_{1-(\star)}\mathsf{f}_1\mathsf{kq}$ splits in every degree as a short exact sequence of abelian groups. In other words, the surjection $\mathbf{K}^{\mathsf{MW}} \to \pi_{1-(\star)}\mathsf{kq}$ factors through a surjection $\mathbf{K}^{\mathsf{MW}}/2 \to \pi_{1-(\star)}\mathsf{f}_1\mathsf{kq}$. The relation $0 = \eta^2\mathrm{u}(\eta_{\mathsf{top}}) \in \pi_{3,2}\mathsf{f}_1\mathsf{kq}$ follows from the slice spectral sequence computation (which even gives $\pi_{3,2}\mathsf{kq} = 0$).

In order to show that the obtained surjection $\mathbf{K}^{\mathsf{MW}}/(2,\eta^2) \to \pi_{1-(\star)}\mathsf{f}_1\mathsf{kq}$ is an isomorphism, observe that it fits into a natural transformation

$$\begin{array}{ccccc} \eta\mathbf{K}^{\mathsf{MW}}/(2,\eta^2) & \longrightarrow & \mathbf{K}^{\mathsf{MW}}/(2,\eta^2) & \longrightarrow & \mathbf{K}^{\mathsf{MW}}/(2,\eta) \\ \downarrow & & \downarrow & & \downarrow \\ h^{1+\star,2+\star}/\mathsf{Sq}^2h^{\star-1,1+\star} & \longrightarrow & \pi_{1-(\star)}\mathsf{f}_1\mathsf{kq} & \longrightarrow & \pi_{1-(\star)}\mathsf{s}_1\mathsf{kq} \end{array} \tag{2.1}$$

of short exact sequences, where the outer vertical morphisms are isomorphisms. This implies the result. $\square$

The difference between $\pi_{1-(\star)}\mathsf{f}_1\mathsf{kq}$ and its image $\mathsf{f}_1\pi_{1-(\star)}\mathsf{kq}$ in $\pi_{1-(\star)}\mathsf{kq}$ is given by the image of $\pi_{2-(\star)}\mathsf{s}_0\mathsf{kq} \cong H^{\star-2,\star} \to \pi_{1-(\star)}\mathsf{f}_1\mathsf{kq}$. Since $2\eta_{\mathsf{top}} = 0$, this map factors over $H^{\star-2,\star}/2$. The short exact sequence displayed in (2.1) then induces a long exact sequence

$$\begin{aligned} \mathrm{Hom}_{\mathbf{k}^{\mathsf{M}}}(H^{\star-2,\star}/2,\mathbf{k}^{\mathsf{M}}_{\star+1}/\rho^2) &\to \mathrm{Hom}_{\mathbf{K}^{\mathsf{MW}}}(H^{\star-2,\star}/2,\pi_{1-(\star)}\mathsf{f}_1\mathsf{kq}) \\ &\to \mathrm{Hom}_{\mathbf{k}^{\mathsf{M}}}(H^{\star-2,\star}/2,\mathbf{k}^{\mathsf{M}}) \to \cdots \end{aligned}$$

in which the first group is zero for any finite field or any number field. Here ρ is the class of -1 in $h^{1,1}$. The homomorphism in question maps to the restriction of the motivic Steenrod square Sq^2 in the abelian group $\mathrm{Hom}_{\mathbf{k}^{\mathsf{M}}}(H^{\star-2,\star}/2, h^{\star,\star+1} \cong \mathbf{k}^{\mathsf{M}}_{\star})$, and is thereby uniquely determined for prime fields.

Following a specific request, a probably well-known identification, in which $\mathcal{O}^\times$ denotes the sheaf of units, can be derived from Lemma 2.3. In principle, any sequence of strictly $\mathbf{A}^1$-invariant sheaves which is exact when evaluated on fields is already exact, by a theorem of Morel. This applies in particular to Theorem 2.2, Lemma 2.3, and Theorem 2.5. However, the following case can be proved directly instead.

PROPOSITION 2.4. *Let F be a field of characteristic not two. The sheaf $\underline{\pi}_{1,0}\mathsf{KQ}$ is isomorphic to the sheaf $\mathcal{O}^\times/2 \times \mathbb{Z}/2$.*

PROOF. By construction, the canonical maps induce isomorphisms $\underline{\pi}_{1,0}\mathsf{f}_1\mathsf{kq} \cong \underline{\pi}_{1,0}\mathsf{kq} \cong \underline{\pi}_{1,0}\mathsf{KQ}$. This uses the vanishing $\underline{\pi}_{1,0}\mathsf{s}_0\mathsf{kq} = \underline{\pi}_{2,0}\mathsf{s}_0\mathsf{kq} = 0$, since $\mathsf{s}_0\mathsf{kq} = \mathbf{M}\mathbb{Z}$ and motivic cohomology of smooth schemes in weight zero is concentrated in degree zero. The homotopy cofiber sequence

$$\mathsf{f}_2\mathsf{kq} \to \mathsf{f}_1\mathsf{kq} \to \mathsf{s}_1\mathsf{kq}$$

induces, for every smooth connected F-scheme U, a short exact sequence

$$0 = \pi_{2,0}\mathsf{s}_1\mathsf{kq}(U) \to \pi_{1,0}\mathsf{f}_2\mathsf{kq}(U) \to \pi_{1,0}\mathsf{f}_1\mathsf{kq}(U) \to \pi_{1,0}\mathsf{s}_1\mathsf{kq}(U) = h^{0,1}(U) = \mathbb{Z}/2 \to 0$$

which splits naturally via the map $\mathbb{Z}/2 = \pi_1\mathbb{S} \to \pi_{1,0}\mathsf{f}_1\mathsf{kq}$. In case U is (the spectrum of) a field extension, this sequence appears in diagram (2.1). Already if U is an essentially smooth local F-scheme, then $\pi_{1,0}\mathsf{f}_2\mathsf{kq}(U) \cong \pi_{1,0}\mathsf{s}_2\mathsf{kq}(U) \cong h^{1,2}(U)$ which supplies the result on the level of Nisnevich sheaves. □

THEOREM 2.5. *Let $\nu \in \pi_{3,2}\mathbf{1}$ be the second algebraic Hopf map, obtained by the Hopf construction on $\mathbf{SL}_2$, and let $\eta_{\mathsf{top}} \in \pi_{1,0}\mathbf{1}$ be the first topological Hopf map. The $\mathbf{K}^{\mathsf{MW}}$-module map*

$$\mathbf{K}^{\mathsf{MW}}_{2+\star} \oplus \mathbf{K}^{\mathsf{MW}} \to \pi_{1-(\star)}\mathsf{f}_1\mathbf{1}$$

sending (a,b) to $a\cdot\nu + b\cdot\eta_{\mathsf{top}}$ induces an isomorphism

$$\mathbf{K}^{\mathsf{MW}}_{2+\star}\{\nu\} \oplus \mathbf{K}^{\mathsf{MW}}\{\eta_{\mathsf{top}}\}/(\eta\nu, 2\eta_{\mathsf{top}}, \eta^2\eta_{\mathsf{top}} - 12\nu) \cong \pi_{1-(\star)}\mathsf{f}_1\mathbf{1}$$

after inverting the exponential characteristic.

PROOF. Observe first that ν naturally lifts to $\pi_{3,2}\mathsf{f}_2\mathbf{1}$, hence defines also an element $\nu \in \pi_{3,2}\mathsf{f}_1\mathbf{1}$. The element η_{top} has image $0 \in \pi_{1,0}\mathsf{s}_0\mathbf{1} = \pi_{1,0}\mathbf{M}\mathbb{Z} = H^{-1,0}$, and hence also lifts to $\pi_{1,0}\mathsf{f}_1\mathbf{1}$ – even uniquely, since $\pi_{2,0}\mathsf{s}_0\mathbf{1} = \pi_{2,0}\mathbf{M}\mathbb{Z} = H^{-2,0} = 0$. Thus the $\mathbf{K}^{\mathsf{MW}}$-module map

$$\mathbf{K}^{\mathsf{MW}}_{2+\star} \oplus \mathbf{K}^{\mathsf{MW}} \to \pi_{1-(\star)}\mathsf{f}_1\mathbf{1}$$

sending (a,b) to $a\cdot\nu + b\cdot\eta_{\mathsf{top}}$ is well-defined over any base scheme in which motivic cohomology in weight zero vanishes in degree -1. Certain relations hold before inverting the exponential characteristic. The relation $\eta\nu = 0$ holds by [**4**] over Spec($\mathbb{Z}$). Also $2\eta_{\mathsf{top}} = 0$, again over any base scheme. If 2 is invertible in the base scheme, $\mathrm{u}(\nu) = 0$, where $\mathrm{u}\colon \mathbf{1} \to \mathsf{kq}$ is the unit map figuring in Lemma 2.3.

Assuming that the exponential characteristic is now implicitly inverted, the exact sequence from Theorem 2.2 implies that $\pi_{3,2}\mathbf{1} = \mathbb{Z}/24$, generated by ν. Either by reference to complex realization or by the multiplicative structure of the slice spectral sequence (more precisely, only the effect of multiplying with the first algebraic Hopf map η), the element $\eta^2\eta_{\mathsf{top}} \in \pi_{3,2}\mathbf{1}$ is the unique nontrivial element of order two, hence $\eta^2\eta_{\mathsf{top}} = 12\nu = 6\mathsf{h}\nu$. Thus there is an induced $\mathbf{K}^{\mathsf{MW}}$-module homomorphism

$$\mathbf{K}^{\mathsf{MW}}_{2+\star}\{\nu\} \oplus \mathbf{K}^{\mathsf{MW}}\{\eta_{\mathsf{top}}\}/(\eta\nu, 2\eta_{\mathsf{top}}, \eta^2\eta_{\mathsf{top}} - 12\nu) \to \pi_{1-(\star)}\mathsf{f}_1\mathbf{1}.$$

The $\mathbf{K}^{\mathsf{MW}}$-module homomorphism $\pi_{1-(\star)}\mathsf{f}_1\mathbf{1} \to \pi_{1-(\star)}\mathsf{f}_1\mathsf{kq}$ induced by the unit map $\mathsf{u}\colon \mathbf{1} \to \mathsf{kq}$ then factors over the isomorphism

$$\mathbf{K}^{\mathsf{MW}}/(2, \eta^2) \cong \pi_{1-(\star)}\mathsf{f}_1\mathsf{kq}$$

from Lemma 2.3. There results a commutative diagram

$$\begin{array}{ccccc}
\frac{\mathbf{K}^{\mathsf{MW}}_{2+\star}\{\nu\}}{(24\nu,\eta\nu)} & \longrightarrow & \frac{\mathbf{K}^{\mathsf{MW}}_{2+\star}\{\nu\}\oplus\mathbf{K}^{\mathsf{MW}}\{\eta_{\mathsf{top}}\}}{(\eta\nu,2\eta_{\mathsf{top}},\eta^2\eta_{\mathsf{top}}-12\nu)} & \longrightarrow & \frac{\mathbf{K}^{\mathsf{MW}}\{\eta_{\mathsf{top}}\}}{(2\eta_{\mathsf{top}},\eta^2\eta_{\mathsf{top}})} \\
\downarrow & & \downarrow & & \downarrow \\
\mathbf{K}^{\mathsf{M}}_{2+\star}/24 & \longrightarrow & \pi_{1-(\star)}\mathsf{f}_1\mathbf{1} & \longrightarrow & \pi_{1-(\star)}\mathsf{f}_1\mathsf{kq}
\end{array} \tag{2.2}$$

of short exact sequences, where the upper horizontal row is short exact by direct computation. Note that the relations $2\eta_{\mathsf{top}} = 0$ and $\eta^2\eta_{\mathsf{top}} = 12\nu$ imply $24\nu = 0$. The exactness of the lower horizontal row follows from the exactness of the sequence in Theorem 2.2 and the fact that the unit $\mathsf{u}\colon \mathbf{1} \to \mathsf{kq}$ induces an isomorphism on zero slices. The right vertical map in diagram (2.2) is an isomorphism by Lemma 2.3, and the left vertical map in diagram (2.2) is an isomorphism by direct inspection, whence the result. □

As in the case of kq, the difference between $\pi_{1-(\star)}\mathsf{f}_1\mathbf{1}$ and its image $\mathsf{f}_1\pi_{1-(\star)}\mathbf{1}$ in $\pi_{1-(\star)}\mathbf{1}$ is given by the image of $\pi_{2-(\star)}\mathsf{s}_0\mathbf{1} \cong H^{\star-2,\star} \to \pi_{1-(\star)}\mathsf{f}_1\mathbf{1}$. The short exact sequence

$$0 \to \mathsf{f}_1\pi_{1-(\star)}\mathbf{1} \to \pi_{1-(\star)}\mathbf{1} \to \pi_{1-(\star)}\mathsf{s}_0\mathbf{1} = H^{\star-1,\star} \to 0$$

then describes the difference between the first two $\mathbf{K}^{\mathsf{MW}}$-modules. In particular, the canonical map $\pi_{1+w,w}\mathsf{f}_1\mathbf{1} \to \pi_{1+w,w}\mathbf{1}$ is an isomorphism for $w > -2$. For the applications to motivic Moore spectra, an important weight is $w = 0$, where a short exact sequence

$$0 \to \mathbf{K}^{\mathsf{M}}_2/24 \to \pi_{1,0}\mathbf{1} \to \mathbb{Z}/2 \oplus \mathbf{K}^{\mathsf{M}}_1/2 \to 0 \tag{2.3}$$

results. One result of Theorem 2.5 is that the sequence (2.3) splits as a sequence of abelian groups, but not as a sequence of $\mathbf{K}^{\mathsf{MW}}_0$-modules. In particular, $24\cdot\pi_{1,0}\mathbf{1} = 0$, in contrast with $24\cdot\pi_{-1,-2}\mathbf{1}_{\mathbb{Q}} \neq 0$, which is similar to $24\cdot\pi_{-1,-2}\mathsf{KGL}_{\mathbb{Q}} \cong 24\cdot\mathbb{Z}/48 \neq 0$.

LEMMA 2.6. *The action of* $\mathbf{GW}$ *on* $\pi_{1+(\star)}\mathbf{1}$ *is determined by the following equations for* $u \in F^\times$.

$$\langle u\rangle \cdot \nu = \nu \in \pi_{3,2}\mathbf{1}$$
$$\langle u\rangle \cdot \eta\eta_{\mathsf{top}} = \eta\eta_{\mathsf{top}} + [u^{12}]\nu \in \pi_{2,1}\mathbf{1}$$
$$\langle u\rangle \cdot \eta_{\mathsf{top}} = \eta_{\mathsf{top}} + [u]\eta\eta_{\mathsf{top}} \in \pi_{1,0}\mathbf{1}$$

In particular, $\mathsf{h}\cdot\eta_{\mathsf{top}} = \rho\eta\eta_{\mathsf{top}}$, $\mathsf{h}\cdot\eta\eta_{\mathsf{top}} = 0$, *and* $\mathsf{h}\cdot\nu = 2\nu$. *If* n *is even, then* $n\mathsf{h}$ *acts on* $\pi_{1+(\star)}\mathsf{f}_1\mathbf{1}$ *as multiplication by* $2n$.

PROOF. This follows from Theorem 2.5, once the identification $\pi_{1+(w)}\mathsf{f}_1\mathbf{1} = \pi_{1+(w)}\mathbf{1}$ for $w > -2$ is deduced from vanishing results for integral motivic cohomology of fields in weight < 2. □

The situation for $\pi_{2+(\star)}\mathbf{1}$ is a bit more delicate. Nevertheless, the following can be read off from the slice spectral sequence.

THEOREM 2.7. *The element* $\nu^2\colon \Sigma^{6,4}\mathbf{1} \to \mathbf{1}$ *induces an inclusion*

$$\mathbf{K}^{\mathsf{MW}}_{4+\star}\{\nu^2\}/(\eta\nu^2, 2\nu^2) \to \pi_{2-(\star)}\mathbf{1}$$

of $\mathbf{K}^{\mathsf{MW}}$*-modules, which is an isomorphism for all* $\star < -1$*. In particular, for all* $w > 4$*, the group* $\pi_{2+w,w}\mathbf{1} \cong 0$*.*

PROOF. Since $\eta\nu = 0$ [**4**], also $\eta\nu^2 = 0$. Moreover, the ε-graded commutativity of $\pi_{*+(\star)}\mathbf{1}$ implies that $\nu^2 = -\nu^2$, whence a map $\mathbf{K}^{\mathsf{MW}}_{4+\star}\{\nu^2\}/(\eta\nu^2, 2\nu^2) \to \pi_{2-(\star)}\mathbf{1}$ of $\mathbf{K}^{\mathsf{MW}}$-modules exists. The slice spectral sequence for $\pi_{2+(\star)}\mathbf{1}$ shows its injectivity, as well as the isomorphism statement, using results from [**21**] and [**17**, Theorem 8.3]; details are to be given in [**20**]. □

3. Elementary properties of motivic Moore spectra

DEFINITION 3.1. Let $s \geq w \in \mathbb{Z}$, and let $\alpha\colon \Sigma^{s,w}\mathbf{1} \to \mathbf{1}$ be an endomorphism. Choosing a homotopy cofiber sequence

$$\Sigma^{s,w}\mathbf{1} \xrightarrow{\alpha} \mathbf{1} \xrightarrow{c} \mathsf{C}_\alpha \xrightarrow{d} \Sigma^{s+1,w}\mathbf{1}$$

defines the motivic Moore spectrum C_α.

By definition, a motivic Moore spectrum for a is unique up to equivalence. The restriction $s \geq w$ is reasonable at least over a field by Morel's connectivity theorem [**15**].

PROPOSITION 3.2. *Let* $\alpha\colon \Sigma^{s,w}\mathbf{1} \to \mathbf{1}$ *be an endomorphism. The canonical maps induce a short exact sequence*

$$0 \to \pi_{n+(\star)}\mathbf{1}/\alpha\pi_{n-s+w+(\star-w)}\mathbf{1} \xrightarrow{c_*} \pi_{n+(\star)}\mathsf{C}_\alpha \xrightarrow{d_*} {}_\alpha\pi_{n-1-s+w+(\star-w)}\mathbf{1} \to 0$$

of $\mathbf{K}^{\mathsf{MW}}$*-modules, with target the submodule of elements annihilated by* α*.*

PROOF. This follows from the homotopy cofiber sequence

$$\Sigma^{s,w}\mathbf{1} \xrightarrow{\alpha} \mathbf{1} \xrightarrow{c} \mathsf{C}_\alpha \xrightarrow{d} \Sigma^{s+1,w}\mathbf{1}$$

defining the motivic Moore spectrum. □

In particular, Morel's connectivity theorem implies with Proposition 3.2 that $\pi_{0+(\star)}\mathsf{C}_\alpha \cong \mathbf{K}^{\mathsf{MW}}/\alpha\mathbf{K}^{\mathsf{MW}}$ if $\alpha \in \mathbf{K}^{\mathsf{MW}}$ (that is, if $s = w$).

4. Toda brackets

Consider three composable maps

$$\mathsf{D} \xrightarrow{\gamma} \mathsf{E} \xrightarrow{\beta} \mathsf{F} \xrightarrow{\alpha} \mathsf{G}$$

of motivic spectra such that $\beta\gamma = 0 = \alpha\beta$. The *Toda bracket*

$$\langle \alpha, \beta, \gamma \rangle \qquad (\operatorname{mod} \alpha \circ [\Sigma^{1,0}\mathsf{D}, \mathsf{F}] + [\Sigma^{1,0}\mathsf{E}, \mathsf{G}] \circ \Sigma^{1,0}\gamma)$$

is the coset of the displayed subgroup of $[\Sigma^{1,0}\mathsf{D}, \mathsf{G}]$ given by those compositions $\Sigma^{1,0}\mathsf{D} \to \mathsf{C}(\beta) \to \mathsf{G}$ such that the obvious diagrams commute. The base scheme or field may be indicated by a subscript. The most relevant case is where all the motivic spectra involved are appropriate suspensions of the motivic sphere spectrum $\mathbf{1}$. See the classical source [**22**], as well as [**5**] and [**8**] for interesting Toda brackets in the motivic stable homotopy category. We consider a few examples and make no claim to originality. Angled brackets are used to denote both Toda brackets and quadratic forms (as in the paragraph after Theorem 2.1). The reader should be aware of this possible confusion, but context will always make the meaning unambiguous.

PROPOSITION 4.1. *The following equalities of subsets of* $\pi_{*+(\star)}\mathbf{1}$ *hold.*

$$\langle \mathsf{h}, \eta, \mathsf{h} \rangle = \{\eta\eta_{\mathsf{top}} + [\phi^2]\nu \colon \phi^2 \in 2\mathbf{K}_1^{\mathsf{M}}\{\nu\}/24\} \tag{4.1}$$

$$\langle \eta, \mathsf{h}, \eta \rangle = \{6\nu, -6\nu\} \tag{4.2}$$

$$\langle \eta, \nu, \eta \rangle = \{\nu^2\} \tag{4.3}$$

$$\langle 2 + 11\mathsf{h}, \nu, 2 + 11\mathsf{h} \rangle_{\mathbb{R}} = \{\rho^2\nu^2\} \tag{4.4}$$

PROOF. The indeterminacy in (4.1) is the subgroup $\mathsf{h} \circ \pi_{2,1}\mathbf{1} = 2\mathbf{K}_1^{\mathsf{M}}\{\nu\}/24$, using the isomorphism $\pi_{2,1}\mathbf{1} \cong \mathbf{K}_0^{\mathsf{M}}\{\eta\eta_{\mathsf{top}}\}/2 \oplus \mathbf{K}_1^{\mathsf{M}}\{\nu\}/24$ from Theorem 2.5 and the equalities $[-1]\eta(\eta\eta_{\mathsf{top}}) = [-1]\eta^2\eta_{\mathsf{top}} = [-1]12\nu = 0 = [-1]\eta\nu$ given in Lemma 2.6. That the Toda bracket contains $\eta\eta_{\mathsf{top}}$ follows from appropriate realization functors and the equality $\langle 2, \eta_{\mathsf{top}}, 2 \rangle_{\mathsf{top}} = \{\eta_{\mathsf{top}}\}$ in classical stable homotopy groups [**22**].

The indeterminacy in (4.2) is $\eta \circ \pi_{2,1} = \{0, 12\nu\}$, since Theorem 2.5 implies $\pi_{2,1}\mathbf{1} \cong \mathbf{K}_0^{\mathsf{M}}\{\eta\eta_{\mathsf{top}}\}/2 \oplus \mathbf{K}_1^{\mathsf{M}}\{\nu\}/24$, and $\eta\eta\eta_{\mathsf{top}} = 12\nu$, whereas $\eta\nu = 0$. Complex realization maps $\langle \eta, \mathsf{h}, \eta \rangle$ to $\langle \eta_{\mathsf{top}}, 2, \eta_{\mathsf{top}} \rangle_{\mathsf{top}} = \{6\nu_{\mathsf{top}}, -6\nu_{\mathsf{top}}\}$. The relevant group $\pi_{3,2}\mathbf{1}$ does not depend on the base field (at least after inverting the exponential characteristic), giving the result.

The indeterminacy in (4.3) is the subgroup $\eta \circ \pi_{4,2} = \{0\}$, since $\pi_{4,2}\mathbf{1} \cong \mathbf{K}_2^{\mathsf{M}}\{\nu^2\}/2$ by Theorem 2.7, and $\eta\nu^2 = 0$. The unique element in $\langle \eta, \nu, \eta \rangle$ has to be the unique nontrivial one in $\pi_{6,4}\mathbf{1} = \mathbb{Z}/2$ by appropriate realization.

Similar to the previous case, the indeterminacy in (4.4) is $\{0\}$ by Theorem 2.7, since $(2+11\mathsf{h})\nu^2 = 0$. Real realization sends $\langle 2+11\mathsf{h}, \nu, 2+11\mathsf{h} \rangle_{\mathbb{R}}$ to $\langle 2, \eta_{\mathsf{top}}, 2 \rangle_{\mathsf{top}} = \{\eta_{\mathsf{top}}^2\}$, whence the previous Toda bracket has to contain the unique nonzero element in $\pi_{4,2}\mathbf{1}_{\mathbb{R}}$. □

The following statement, which could be formulated in greater generality, indicates the relevance of Toda brackets for the structure of motivic Moore spectra. The notation is as in Proposition 3.2.

PROPOSITION 4.2. *Suppose* $\alpha \in \pi_{s,w}\mathbf{1}$, $\beta \in \pi_{t,x}\mathbf{1}$, *and* $\gamma \in \pi_{u,y}\mathbf{1}$ *are elements with* $\alpha\beta = 0 = \beta\gamma$. *The Toda bracket* $\langle \alpha, \beta, \gamma \rangle$ *coincides with the set of elements* $\delta \in \pi_{s+t+u+1,w+x+y}\mathbf{1}$ *for which there exists an element* $\tilde{\beta} \in \pi_{s+t+1,w+x}\mathsf{C}_\alpha$ *with* $d_*(\tilde{\beta}) = \beta$ *and* $c_*(\delta + \alpha \cdot \pi_{t+u+1,x+y}\mathbf{1}) = \tilde{\beta} \cdot \gamma$.

PROOF. The proof consists of comparing various diagrams in the motivic stable homotopy category and is left as an exercise to the reader. □

EXAMPLE 4.3. Consider $\alpha = \gamma = \eta \in \pi_{1,1}\mathbf{1}$. Then $\pi_{5,3}\mathsf{C}_\eta$ contains an element $\tilde{\nu}$ with $\tilde{\nu} \cdot \eta \neq 0$. In fact, the Toda bracket $\langle \eta, \nu, \eta \rangle = \{\nu^2\}$ given in (4.3) contains a single nonzero element, and $c_* \colon \pi_{6,4}\mathbf{1} \to \pi_{6,4}\mathsf{C}_\eta$ is injective, since multiplication with η is the zero map $\pi_{5,3}\mathbf{1} = h^{1,1}\{\nu^2\} \to h^{0,0}\{\nu^2\} = \pi_{6,4}\mathbf{1}$ by Theorem 2.7.

EXAMPLE 4.4. Let $\alpha = \gamma = 2 + 11\mathsf{h} \in \pi_{0,0}\mathbf{1}_{\mathbb{R}}$ (or any formally real field). Then $\pi_{4,2}\mathsf{C}_{2+11\mathsf{h}}$ contains an element $\tilde{\nu}$ with $\tilde{\nu} \cdot (2 + 11\mathsf{h}) \neq 0$. Indeed, the Toda bracket given in (4.4) contains a nonzero element (uniquely for $\mathbb{R}$), and $c_* \colon \pi_{4,2}\mathbf{1}_{\mathbb{R}} \to \pi_{4,2}\mathsf{C}_{2+11\mathsf{h}}$ is injective, since multiplication with $2 + 11\mathsf{h}$ induces the zero map $\pi_{4,2}\mathbf{1}_{\mathbb{R}} = h^{2,2}\{\nu^2\} \to \pi_{4,2}\mathbf{1}_{\mathbb{R}}$ by Theorem 2.7.

5. Multiplications

The "constant presheaf" functor defines a strict symmetric monoidal triangulated functor const: $\mathbf{SH} \to \mathbf{SH}(S)$ for any base scheme S. The motivic Moore spectra C_n for $n \in \mathbb{N}$ are in its image. In particular, multiplications or pairings on these motivic Moore spectra can be transferred from the corresponding topological ones [**16**].

LEMMA 5.1. *Let $\alpha\colon \Sigma^{s,w}\mathbf{1} \to \mathbf{1}$ be an endomorphism and $c\colon \mathbf{1} \to \mathsf{C}_\alpha$ the map to the homotopy cofiber. There exists a left unital pairing*

$$\mathsf{C}_\alpha \wedge \mathsf{C}_\alpha \to \mathsf{C}_\alpha$$

if and only if the identity on C_α is annihilated by α.

PROOF. Consider the following diagram:

$$\begin{array}{ccccc}
\mathbf{1}\wedge\mathbf{1} = \mathbf{1} & \xrightarrow{c} & \mathsf{C}_\alpha = \mathbf{1}\wedge\mathsf{C}_\alpha & \xrightarrow{c\wedge\mathsf{C}_\alpha} & \mathsf{C}_\alpha\wedge\mathsf{C}_\alpha \\
\big\downarrow{\scriptstyle \mathrm{id}} & & \big\downarrow{\scriptstyle \mathrm{id}} & & \\
\mathbf{1} & \xrightarrow{c} & \mathsf{C}_\alpha & \xrightarrow{\mathrm{id}} & \mathsf{C}_\alpha
\end{array}$$

There exists a left unital pairing

$$\mathsf{C}_\alpha \wedge \mathsf{C}_\alpha \to \mathsf{C}_\alpha$$

if and only if the map $c \wedge \mathsf{C}_\alpha$ admits a retraction. After smashing with C_α, the homotopy cofiber sequence

$$\Sigma^{s,w}\mathbf{1} \xrightarrow{\alpha} \mathbf{1} \xrightarrow{c} \mathsf{C}_\alpha \xrightarrow{d} \Sigma^{s+1,w}\mathbf{1}$$

induces a long exact sequence

$$\cdots \xleftarrow{d^*} [\Sigma^{s,w}\mathsf{C}_\alpha, \mathsf{C}_\alpha] \xleftarrow{\alpha} [\mathsf{C}_\alpha, \mathsf{C}_\alpha] \xleftarrow{c^*} [\mathsf{C}_\alpha \wedge \mathsf{C}_\alpha, \mathsf{C}_\alpha] \xleftarrow{d^*} [\Sigma^{s+1,w}\mathsf{C}_\alpha, \mathsf{C}_\alpha] \xleftarrow{\alpha} \cdots$$

which shows the desired statement. □

If a left unital pairing $\mu\colon \mathsf{C}_\alpha \wedge \mathsf{C}_\alpha \to \mathsf{C}_\alpha$ exists, then $\mu \circ (\mathsf{C}_\alpha \wedge c) \circ c = c$. The short exact sequence

$$0 \to \pi_{s+1,w}\mathsf{C}_\alpha \xrightarrow{d^*} [\mathsf{C}_\alpha, \mathsf{C}_\alpha] \xrightarrow{c^*} \pi_{0,0}\mathsf{C}_\alpha \to 0$$

then supplies an element $\psi \in \pi_{s+1,w}\mathsf{C}_\alpha$ with $\psi \circ d = \mu \circ (\mathsf{C}_\alpha \wedge c) - \mathrm{id}_{\mathsf{C}_\alpha}$. The short exact sequence

$$0 \to \pi_{2s+2,2w}\mathsf{C}_\alpha \xrightarrow{d^*} [\Sigma^{s+1,w}\mathsf{C}_\alpha, \mathsf{C}_\alpha] \xrightarrow{c^*} \pi_{s+1,w}\mathsf{C}_\alpha \to 0$$

provides $\theta\colon \Sigma^{s,w}\mathsf{C}_\alpha \to \mathsf{C}_\alpha$ with $\theta \circ \Sigma^{s+1,w}d = \psi$. It follows that the left unital pairing $\mu - \theta \circ (d \wedge \mathsf{C}_\alpha)$ is also right unital, because

$$\big(\mu - \theta\circ(d\wedge\mathsf{C}_\alpha)\big)\circ(\mathsf{C}_\alpha\wedge c) = \mu\circ(\mathsf{C}_\alpha\wedge c) - \theta\circ(d\wedge c) = \mu\circ(\mathsf{C}_\alpha\wedge c) - \theta\circ\Sigma^{s+1,w}c\circ d = \mathrm{id}_{\mathsf{C}_\alpha}.$$

Hence if a left unital pairing on C_α exists, a unital pairing exists as well. In the following, "multiplication" stands for "unital pairing".

LEMMA 5.2. *Let $\alpha\colon \Sigma^{s,w}\mathbf{1} \to \mathbf{1}$ be an endomorphism. Then $\alpha^2 \cdot \mathrm{id}_{\mathsf{C}_\alpha} = 0$.*

PROOF. The homotopy cofiber sequence defining C_α induces a short exact sequence

$$(5.1)\qquad 0 \to [\Sigma^{s+1,w}\mathbf{1}, \mathsf{C}_\alpha]/\alpha \xrightarrow{d^*} [\mathsf{C}_\alpha, \mathsf{C}_\alpha] \xrightarrow{c^*} {}_\alpha[\mathbf{1}, \mathsf{C}_\alpha] = [\mathbf{1}, \mathsf{C}_\alpha] = [\mathbf{1}, \mathbf{1}]/\alpha \to 0$$

for which Proposition 3.2 implies the identification of the outer terms. In particular, $c^*(\mathrm{id}_{\mathsf{C}_\alpha})$ is annihilated by α. Hence $\alpha \cdot \mathrm{id}_{\mathsf{C}_\alpha} = d^*(x)$ for some $x \in [\Sigma^{s+1,w}\mathbf{1}, \mathsf{C}_\alpha]/\alpha$. Thus $\alpha \cdot x = 0$, which proves the equality $\alpha^2 \cdot \mathrm{id}_{\mathsf{C}_\alpha} = \alpha \cdot d^*(x) = d^*(\alpha \cdot x) = 0$. $\square$

As a consequence, there exists a pairing

$$\mathsf{C}_\alpha \wedge \mathsf{C}_{\alpha^2} \to \mathsf{C}_\alpha$$

which in the classical case of the topological sphere spectrum and $\alpha = 2$ was described by Oka in [**16**]. For example, using $\mathsf{h}^2 = 2\mathsf{h}$, there results a pairing

$$\mathsf{C}_\mathsf{h} \wedge \mathsf{C}_{2\mathsf{h}} \to \mathsf{C}_\mathsf{h}$$

which realizes to Oka's pairing for subfields of the complex numbers. The specific role that squares play will be clarified by the following statement lifting a theorem of Brayton Gray [**7**, Theorem 10].

THEOREM 5.3. *Let $\alpha\colon \Sigma^{s,w}\mathbf{1} \to \mathbf{1}$ be any endomorphism. Then C_{α^2} admits a multiplication.*

PROOF. Lemma 5.1 implies it suffices to show that $\alpha^2 \cdot \mathrm{id}_{\mathsf{C}_{\alpha^2}} = 0$. In order to prove this, consider the following general construction for $\alpha\colon \mathsf{D} \to \mathsf{E}$, modeled on [**22**], and observing that the ε-graded ring structure on $\pi_{*+(\star)}\mathbf{1}$ may be equally defined via composition or via smash product. Let $R(\alpha)$ be the set of all maps $A\colon \Sigma^{1,0}\mathsf{D} \wedge \mathsf{D} \to \mathsf{E} \wedge \mathsf{E}$ such the diagram

$$\begin{array}{ccc} \Sigma^{1,0}\mathsf{D}\wedge\mathsf{D} & \xrightarrow{A} & \mathsf{E}\wedge\mathsf{E} \\ {\scriptstyle \mathsf{D}\wedge d}\uparrow & & \downarrow{\scriptstyle \mathsf{E}\wedge c} \\ \mathsf{D}\wedge\mathsf{C}_\alpha & \xrightarrow{\alpha\wedge\mathsf{C}_\alpha} & \mathsf{E}\wedge\mathsf{C}_\alpha \end{array}$$

commutes. If $R(\alpha)$ contains an element of the form $\beta \wedge \alpha$, where $\beta\colon \Sigma^{1,0}\mathsf{D} \to \mathsf{E}$, then $\alpha \wedge \mathsf{C}_\alpha = \alpha \cdot \mathrm{id}_{\mathsf{C}_\alpha}$ is the zero map. Several natural commutative diagrams show that if $\alpha\colon \mathsf{D} \to \mathsf{E}$ and $\beta\colon \mathsf{F} \to \mathsf{G}$ are two maps, then for every $A \in R(\alpha)$, the element

$$\Sigma^{1,0}\mathsf{D}\wedge\mathsf{F}\wedge\mathsf{D}\wedge\mathsf{F} \xrightarrow{\text{twist}} \Sigma^{1,0}\mathsf{D}\wedge\mathsf{D}\wedge\mathsf{F}\wedge\mathsf{F} \xrightarrow{A\wedge\beta\wedge\beta} \mathsf{E}\wedge\mathsf{E}\wedge\mathsf{G}\wedge\mathsf{G} \xrightarrow{\text{twist}} \mathsf{E}\wedge\mathsf{G}\wedge\mathsf{E}\wedge\mathsf{G}$$

lies in $R(\alpha \wedge \beta)$. In particular, for $\alpha \in \pi_{s,w}\mathbf{1}$ and $\beta \in \pi_{t,x}\mathbf{1}$, the inclusion $(-1)^{(s-w)(t-x)}\varepsilon^{wx}R(\alpha)\beta^2 \subset R(\alpha \wedge \beta)$ holds, using [**4**, Equation (2.4)]. In the special case $\alpha = \beta$, one obtains the inclusion $(-1)^{(s-w)^2}\varepsilon^{w^2}R(\alpha)\alpha^2 \subset R(\alpha \wedge \alpha)$. If $R(\alpha)$ is not empty, this implies that $\alpha^2 \wedge \mathsf{C}_{\alpha^2}$ is zero. It remains to see that $R(\alpha)$ is not empty for $\alpha \in \pi_{s,w}\mathbf{1}$. Since $\alpha \wedge \mathsf{C}_\alpha \circ \Sigma^{s,w}c = c \circ \Sigma^{s,w}\alpha = 0$, there exists $\gamma \in \pi_{2s+1,2w}\mathsf{C}_\alpha$ with $\gamma \circ \Sigma^{s,w}d = \alpha \wedge \mathsf{C}_\alpha$. If $d \circ \gamma = 0$, then $R(\alpha)$ contains a lift of γ along c. If $d \circ \gamma \neq 0$, then the equation

$$d \circ \gamma \circ \Sigma^{s,w}d = d \circ \alpha \wedge \mathsf{C}_\alpha = \Sigma^{s+1,w}\alpha \circ \Sigma^{s,w}d = 0$$

supplies $\phi \in \pi_{0,0}\mathbf{1}$ with $\phi \circ \alpha = d \circ \gamma$. Since $\alpha \circ \phi \circ \alpha = 0$, there exists $\psi \in \pi_{s+1,w}\mathsf{C}_\alpha$ with $d \circ \psi \circ \alpha = \phi \circ \alpha$ It follows that $(\gamma - \psi \circ \alpha) \circ d = \gamma \circ d = \alpha \wedge \mathsf{C}_\alpha$ and $d \circ (\gamma - \psi \circ \alpha) = 0$. A lift of $\gamma - \psi \circ \alpha \in \pi_{2s+1,2w}\mathsf{C}_\alpha$ along c provides an element in $R(\alpha)$, which concludes the proof. $\square$

Now for some negative results regarding multiplications on motivic Moore spectra. Recall that $\pi_{0,0}\mathbf{1} = GW$ identifies with the Grothendieck-Witt ring of quadratic forms by Theorem 2.1, which comes equipped with a dimension ring homomorphism $\dim\colon \mathbf{GW} \to \mathbb{Z}$.

THEOREM 5.4. *Let $\alpha\colon \mathbf{1} \to \mathbf{1}$ be an endomorphism with* $\dim(\alpha) \equiv 2(4)$. *Then* C_α *does not admit a multiplication.*

PROOF. This result follows by complex realization from [**16**] for subfields of the complex numbers. In any case, consider motivic cohomology $h^{*,*}$ with coefficients in $\mathbb{F}_2$. Since multiplication with α induces multiplication with $\dim(\alpha)$ on motivic cohomology, there results a split short exact sequence

$$0 \to h^{*,*} \to h^{*,*}(\mathsf{C}_\alpha) \to h^{*+1,*} \to 0$$

on motivic cohomology. Pick basis elements $x_0 \in h^{0,0}(\mathsf{C}_\alpha), x_1 \in h^{1,0}(\mathsf{C}_\alpha)$, and note that $\mathsf{Sq}^1(x_0) = x_1$ since $\dim(\alpha) \equiv 2(4)$. The split short exact sequence

$$0 \to h^{*,*}(\mathsf{C}_\alpha) \to h^{*,*}(\mathsf{C}_\alpha \wedge \mathsf{C}_\alpha) \to h^{*+1,*}(\mathsf{C}_\alpha) \to 0$$

induced by multiplication with α on C_α then supplies $h^{*,*}(\mathsf{C}_\alpha \wedge \mathsf{C}_\alpha)$ with basis elements $x_0 \otimes y_0, x_1 \otimes y_0, x_0 \otimes y_1, x_1 \otimes y_1$. Here $y_0 \in h^{0,0}(\mathsf{C}_\alpha)$ and $y_1 \in h^{1,0}(\mathsf{C}_\alpha)$ are basis elements in the other factor of the smash product $\mathsf{C}_\alpha \wedge \mathsf{C}_\alpha$. A Künneth theorem supplies the basis elements for the motivic cohomology of the smash product. The Cartan formula

$$\mathsf{Sq}^2(x_0 \otimes y_0) = \mathsf{Sq}^2(x_0) \otimes y_0 + x_0 \otimes \mathsf{Sq}^2(y_0) + \tau \mathsf{Sq}^1(x_0) \otimes \mathsf{Sq}^1(y_0) = \tau x_1 \otimes y_1$$

from [**23**, Proposition 9.7] then shows that Sq^2 acts nontrivially on $h^{*,*}(\mathsf{C}_\alpha \wedge \mathsf{C}_\alpha)$. If $\alpha \cdot \mathrm{id}_{\mathsf{C}_\alpha} = 0$, then $\mathsf{C}_\alpha \wedge \mathsf{C}_\alpha = \mathsf{C}_\alpha \vee \Sigma^{1,0}\mathsf{C}_\alpha$, and $h^{*,*}(\mathsf{C}_\alpha \wedge \mathsf{C}_\alpha)$ splits accordingly as a module over the motivic Steenrod algebra, implying $\mathsf{Sq}^2(x_0 \otimes y_0) = 0$. The result follows. □

LEMMA 5.5. *The motivic Moore spectrum C_{η^ℓ} admits a multiplication if and only if $\ell > 1$.*

PROOF. Let us prove first that C_η does not admit a multiplication. Again an argument via complex realization works for subfields of the complex numbers. However, one may refer to Example 4.3 and Proposition 4.2, which imply the existence of an element $\tilde{\nu} \in \pi_{3,2}\mathsf{C}_\eta$ such that $\tilde{\nu} \cdot \eta \neq 0$. In particular, $\eta \cdot \mathrm{id}_{\mathsf{C}_\eta} \neq 0$.

Consider now $\ell > 1$. The element $\eta^\ell \cdot \mathrm{id}_{\mathsf{C}_{\eta^\ell}}$ lies inside the group which sits in the middle of the short exact sequence

$$0 \to [\Sigma^{2\ell+1,2\ell}\mathbf{1}, \mathsf{C}_{\eta^\ell}]/\eta^\ell \xrightarrow{d^*} [\Sigma^{\ell,\ell}\mathsf{C}_{\eta^\ell}, \mathsf{C}_{\eta^\ell}] \xrightarrow{c^*} {}_{\eta^\ell}[\Sigma^{\ell,\ell}\mathbf{1}, \mathsf{C}_{\eta^\ell}] = [\Sigma^{\ell,\ell}\mathbf{1}, \mathbf{1}]/\eta^\ell \to 0$$

whose final term is zero, since $\eta^\ell\colon \pi_{0,0}\mathbf{1} \to \pi_{\ell,\ell}\mathbf{1}$ is surjective. The initial term is computed by Proposition 3.2. More precisely, the group $\pi_{2\ell+1,2\ell}\mathsf{C}_{\eta^\ell}$ sits inside the short exact sequence

$$0 \to \pi_{2\ell+1,2\ell}\mathbf{1}/\eta^\ell \pi_{\ell+1,\ell}\mathbf{1} \xrightarrow{c_*} \pi_{2\ell+1,2\ell}\mathsf{C}_{\eta^\ell} \xrightarrow{d_*} {}_{\eta^\ell}\pi_{\ell,\ell}\mathbf{1} \to 0$$

where ${}_{\eta^\ell}\pi_{\ell,\ell}\mathbf{1} = 0$ since $\eta^\ell\colon \pi_{\ell,\ell}\mathbf{1} \to \pi_{2\ell,2\ell}\mathbf{1}$ is an isomorphism for $\ell \geq 1$. The group $\pi_{2\ell+1,2\ell}\mathbf{1}$ vanishes by Theorem 2.2 for $\ell > 1$, whence so does $\pi_{2\ell+1,2\ell}\mathsf{C}_{\eta^\ell}$, and thus also $[\Sigma^{\ell,\ell}\mathsf{C}_{\eta^\ell}, \mathsf{C}_{\eta^\ell}]$. □

The existence of a multiplication may depend on the base field in general. Example 4.4 and Lemma 5.1 show that the motivic Moore spectrum $\mathsf{C}_{2+11\mathsf{h}}$ admits no multiplication over a formally real field F. Base change to $F(\sqrt{-1})$ produces the motivic Moore spectrum $\mathsf{C}_{2+11\mathsf{h}} = \mathsf{C}_{24}$, which admits a multiplication by lifting from topology and quoting [**16**]. Plenty of similar examples may be constructed. A systematic study on these matters remains a project for the future. Another project for the future is to enumerate possible multiplications, as well as investigate their qualitative properties like associativity and commutativity. Already the enumeration can be challenging. For example, the group $[\mathsf{C}_{\mathsf{h}^2} \wedge \mathsf{C}_{\mathsf{h}^2}, \mathsf{C}_{\mathsf{h}^2}]$ receives a nontrivial map from $[\Sigma^{1,0}\mathsf{C}_{\mathsf{h}^2}, \mathsf{C}_{\mathsf{h}^2}]$, which is injective on the contribution from the h^2-torsion in $\pi_{0,0}\mathbf{1}$ (see Proposition 3.2). The latter coincides with the fundamental ideal in the Grothendieck-Witt ring, as the following statement implies.

LEMMA 5.6. *Let $\alpha \in \mathbf{GW}(F)$ be an element with $\dim(\alpha) \neq 0$. Then*

$$_{\alpha\mathsf{h}}\mathbf{GW}(F) = {}_{\dim(\alpha)\mathsf{h}}\mathbf{GW}(F) = {}_{\mathsf{h}}\mathbf{GW}(F) = \mathbf{I}(F).$$

PROOF. Recall that $\mathsf{h} = 1 - \epsilon = \langle 1, -1\rangle$ is a form of dimension two with the property that the subgroup of $\mathbf{GW}(F)$ generated by h coincides with the ideal of $\mathbf{GW}(F)$ generated by h. The latter follows from the similarity

$$\langle u\rangle \cdot \langle 1, -1\rangle = \langle u, -u\rangle \sim \langle 1, -1\rangle$$

of quadratic forms, where $u \in F^\times$. Hence for every $\alpha \in \mathbf{GW}(F)$, the equation $\alpha \cdot \mathsf{h} = \dim(\alpha) \cdot \mathsf{h}$ follows, giving the first equation. Also the other equations follow for $\dim(\alpha) \neq 0$, since $\beta \in \mathbf{GW}(F)$ then satisfies $\beta \cdot (\alpha \cdot \mathsf{h}) = 0$ if and only if $\dim(\beta) = 0$. Here recall the short exact sequence

$$0 \to \mathbf{I}(F) \to \mathbf{GW}(F) \xrightarrow{\dim} \mathbb{Z} \to 0$$

defining the fundamental ideal in $\mathbf{GW}(F)$. □

Prompted by a recent request, this section concludes with a specific example. Let $n > 0$ be a natural number, and let $n_\varepsilon = \sum_{k=0}^{n-1} \langle (-1)^k \rangle = \langle 1, -1, 1 \ldots, \pm 1\rangle \in \mathbf{GW}$, a quadratic form of dimension n. It turns out that the motivic Moore spectrum $\mathsf{C}_{n_\varepsilon}$ admits a multiplication precisely if the topological Moore spectrum $\mathbb{S}/n$ does. Before proving this, set $e\colon \mathsf{C}_\alpha \xrightarrow{d} \Sigma^{s+1,w}\mathbf{1} \xrightarrow{\Sigma^{s+1,w}c} \Sigma^{s+1,w}\mathsf{C}_\alpha$ for $\alpha \in \pi_{s,w}\mathbf{1}$. Following [**16**], a multiplication $\mu\colon \mathsf{C}_\alpha \wedge \mathsf{C}_\alpha \to \mathsf{C}_\alpha$ is called *regular* if the equality $e \circ \mu = (\Sigma^{s+1,w}\mu) \circ (e \wedge \mathsf{C}_\alpha + \mathsf{C}_\alpha \wedge e)\colon \mathsf{C}_\alpha \wedge \mathsf{C}_\alpha \to \Sigma^{s+1,w}\mathsf{C}_\alpha$ holds.

LEMMA 5.7. *There exists a multiplication on $\mathsf{C}_{n_\varepsilon}$ if and only if $n \not\equiv 2(4)$.*

PROOF. Theorem 5.4 says that $\mathsf{C}_{n_\varepsilon}$ does not admit a multiplication if $n \equiv 2(4)$. Suppose first that $n = 2m+1$ is an odd natural number. Any element $\alpha \in \mathbf{GW}$ with $n_\varepsilon \alpha = 0$ then satisfies $\dim(\alpha) = 0$, and hence lies in the fundamental ideal. Since $n_\varepsilon = 1 + m\mathsf{h}$ and the element $\mathsf{h}\alpha$ is hyperbolic of dimension zero, one concludes $\alpha = 0$. Proposition 3.2 then provides an isomorphism $\pi_{1,0}\mathsf{C}_{n_\varepsilon} \cong \pi_{1,0}\mathbf{1}/n_\varepsilon\pi_{1,0}\mathbf{1}$. Lemma 2.6 implies that the map $\pi_{1,0}\mathbf{1} \xrightarrow{n_\varepsilon} \pi_{1,0}\mathbf{1}$ is surjective if $n \not\equiv 0(3)$. Hence if n is odd and not divisible by three, $\pi_{1,0}\mathsf{C}_{n_\varepsilon} = 0$. The short exact sequence (5.1) then reduces to $[\mathsf{C}_{n_\varepsilon}, \mathsf{C}_{n_\varepsilon}] \cong \pi_{0,0}\mathsf{C}_{n_\varepsilon}$. In particular, the identity on $\mathsf{C}_{n_\varepsilon}$ is annihilated by n_ε, which provides the existence of a multiplication by Lemma 5.1. If n is odd and divisible by three, $\pi_{1,0}\mathsf{C}_{n_\varepsilon} \cong \mathbf{K}^{\mathsf{M}}_2/3$. In order to conclude for such odd numbers as well, it suffices to prove that the short exact sequence

$$(5.2)\qquad 0 \to \pi_{1,0}\mathsf{C}_{n_\varepsilon} \cong \mathbf{K}^{\mathsf{M}}_2/3 \to [\mathsf{C}_{n_\varepsilon},\mathsf{C}_{n_\varepsilon}] \to \pi_{0,0}\mathsf{C}_{n_\varepsilon} \to 0$$

splits. Because $n_\varepsilon = 1+m\cdot\mathsf{h}$ acts as the identity on the Witt group, $\pi_{0,0}\mathsf{C}_{n_\varepsilon} \cong \mathbb{Z}/n$ for any field, generated by h. Hence if the sequence (5.2) splits over prime fields, it does so over any field, reducing the task to $F=\mathbb{Q}$. In this case, [**13**, Theorem 11.6] supplies an isomorphism $\mathbf{K}^{\mathsf{M}}_2/3(\mathbb{Q}) \cong \bigoplus_{p \text{ prime}, p\equiv 1(3)} \mathbb{Z}/3$ whose contribution at the prime $p\equiv 1(3)$ is induced by the localization sequence

$$\mathrm{Spec}(\mathbb{F}_p) \hookrightarrow \mathrm{Spec}(\mathbb{Z}_{(p)}) \hookleftarrow \mathrm{Spec}(\mathbb{Q}).$$

Since both h and the element n_ε are defined over $\mathrm{Spec}(\mathbb{Z})$, it follows that the sequence (5.2) splits.

Suppose now that $n=2^r m$ for some odd natural number m, with $r\geq 2$. Then $n_\varepsilon = m\mathsf{h}^r$. By Theorem 5.3, $\mathsf{C}_{\mathsf{h}^2}$ admits a multiplication μ. If μ is not regular, the difference $x = e\circ\mu - \mu\circ(e\wedge \mathsf{C}_{\mathsf{h}^2} + \mathsf{C}_{\mathsf{h}^2}\wedge e)$ lifts to produce a unique element $y\in\pi_{1,0}\mathsf{C}_{\mathsf{h}^2}$ such that $y\circ(d\wedge d)=x$. The latter equation shows that x is an element of order two, because $d\wedge d = -d\wedge d$, hence so is y. The vanishing $0=d\circ y\in\pi_{0,0}\mathsf{C}_{\mathsf{h}^2}$ implies that there exists $z\in\pi_{1,0}\mathbf{1}$ with $c\circ z = y$. The image of $\pi_{1,0}\mathbf{1}$ under composition with c is isomorphic, by Theorem 2.5, to $\mathbf{K}^{\mathsf{M}}_2/4\oplus\mathbf{k}^{\mathsf{M}}_1\oplus\mathbf{k}^{\mathsf{M}}_0$. The map $\lambda\colon \mathsf{C}_{\mathsf{h}^2}\to\mathsf{C}_{\mathsf{h}^4}$ induced by multiplication with h^2 on the bottom cell does not necessarily induce the zero homomorphism on $\pi_{1,0}$. More specifically, it induces multiplication by 4 as a homomorphism

$$\mathbf{K}^{\mathsf{M}}_2/4\oplus\mathbf{k}^{\mathsf{M}}_1\oplus\mathbf{k}^{\mathsf{M}}_0 \to \mathbf{K}^{\mathsf{M}}_2/8\oplus\mathbf{k}^{\mathsf{M}}_1\oplus\mathbf{k}^{\mathsf{M}}_0$$

and in particular sends the 2-torsion element y to zero. The homotopy cofiber sequence

$$\mathsf{C}_{\mathsf{h}^2} \xrightarrow{\lambda} \mathsf{C}_{\mathsf{h}^2} \xrightarrow{\rho} \mathsf{C}_{\mathsf{h}^2} \xrightarrow{e} \Sigma^{1,0}\mathsf{C}_{\mathsf{h}^2}$$

implies the existence of $w\in\pi_{2,0}\mathsf{C}_{\mathsf{h}^2}$ with $e\circ w = y$. Then $\mu - w\circ(d\wedge d)$ is a regular multiplication on $\mathsf{C}_{\mathsf{h}^2}$. Hence a regular multiplication exists on $\mathsf{C}_{\mathsf{h}^2}$.

Inductively, one may lift a regular multiplication on $\mathsf{C}_{\mathsf{h}^\ell}$ to a regular multiplication on $\mathsf{C}_{\mathsf{h}^{\ell+1}}$ via the homotopy cofiber sequence

$$\mathsf{C}_{\mathsf{h}} \xrightarrow{\lambda} \mathsf{C}_{\mathsf{h}^{\ell+1}} \xrightarrow{\rho} \mathsf{C}_{\mathsf{h}^\ell} \xrightarrow{\delta} \mathsf{C}_{\mathsf{h}}.$$

Here λ and ρ are induced by suitable multiplications with h^ℓ on the bottom cell and h on the top cell, respectively, and δ is the composition $\mathsf{C}_{\mathsf{h}^\ell} \xrightarrow{d} \Sigma^{1,0}\mathbf{1} \xrightarrow{\Sigma^{1,0}c} \mathsf{C}_{\mathsf{h}}$. The homotopy cofiber sequence

$$\mathsf{C}_{\mathsf{h}} \xrightarrow{\lambda} \mathsf{C}_m \xrightarrow{\rho} \mathsf{C}_{m\mathsf{h}^r} \xrightarrow{\delta} \mathsf{C}_{\mathsf{h}^r}$$

allows to lift a regular multiplication μ on $\mathsf{C}_{\mathsf{h}^r}$ to a regular multiplication on $\mathsf{C}_{n_\varepsilon}$. Over quadratically closed fields, $\delta=0$ (as in topology, see [**16**, Lemma 5]), but $\delta\neq 0$ for formally real fields. Nevertheless the equality $\delta\circ\mu\circ(\rho\wedge\rho)=0$ always holds. This computation follows from the fact that μ is regular and two other inputs. One input is the result from topology that m is zero on $\mathbb{S}/m$, and hence also on its image C_m in the motivic stable homotopy category. In order to apply the regularity, one uses that the composition $\mathsf{C}_{\mathsf{h}^r} \xrightarrow{d} \Sigma^{1,0}\mathbf{1} \xrightarrow{\Sigma^{1,0}c} \Sigma^{1,0}\mathsf{C}_m$ factors as

$\mathsf{C}_{\mathsf{h}^r} \xrightarrow{d} \Sigma^{1,0}\mathbf{1} \xrightarrow{\Sigma^{1,0}c} \Sigma^{1,0}\mathsf{C}_{\mathsf{h}^r} \xrightarrow{\Sigma^{1,0}\phi} \Sigma^{1,0}\mathsf{C}_m$ for some $\phi\colon \mathsf{C}_{\mathsf{h}^r} \to \mathsf{C}_m$. This provides a regular multiplication on $\mathsf{C}_{n_\varepsilon}$ for $n \not\equiv 2(4)$, which concludes the proof. □

6. Slices of motivic Moore spectra

Instead of considering slices for general motivic Moore spectra, the focus here – motivated by arguments regarding the vanishing of higher slice differentials in [**21**, Section 4] – is on $\mathsf{C}_{n\mathsf{h}}$, where $0 < n \in \mathbb{N}$. As explained in the proof of Lemma 5.6, it suffices to consider motivic Moore spectra with respect to $n\mathsf{h}$ instead of $\alpha\mathsf{h}$ for $\alpha \in \mathbf{GW}(F)$ of $\dim(\alpha) \neq 0$. Since $\mathsf{h} = 0\colon \mathsf{KW} \to \mathsf{KW}$ as an endomorphism on the motivic spectrum representing higher Witt groups, the motivic spectrum $\mathsf{KW}_{n\mathsf{h}}$ splits as $\mathsf{KW} \vee \Sigma^{1,0}\mathsf{KW}$. The same holds for its (effective or connective) covers, and also for the corresponding slices. In particular, the first slice differential for $\mathsf{KW}_{n\mathsf{h}}$ splits. For reference purposes, the explicit form is as follows.

THEOREM 6.1. *Let $0 < n \in \mathbb{N}$. The restriction of the slice d^1-differential to the summand $\Sigma^{q+j,q}\mathbf{M}\mathbb{Z}/2$ of $\mathsf{s}_q(\mathsf{KW}_{n\mathsf{h}})$ is given by*

$$\mathsf{d}^1(\mathsf{KW}_{n\mathsf{h}})(q,j) = \begin{cases} (\mathsf{Sq}^3\mathsf{Sq}^1, 0, \mathsf{Sq}^2) & j \equiv 0,1 \bmod 4 \\ (\mathsf{Sq}^3\mathsf{Sq}^1, 0, \mathsf{Sq}^2 + \rho\mathsf{Sq}^1, 0, \tau) & j \equiv 2,3 \bmod 4 \end{cases}$$

Here the ith component of the map $\mathsf{d}^1(\mathsf{KW}_{n\mathsf{h}})(q,j)$ of motivic spectra is a map $\Sigma^{q+j,q}\mathbf{M}\mathbb{Z}/2 \to \Sigma^{q+j+i,q+1}\mathbf{M}\mathbb{Z}/2$.

PROOF. This follows from the determination of the first slice differential for KW from [**19**], and the aforementioned splitting $\mathsf{KW}_{n\mathsf{h}} \simeq \mathsf{KW} \vee \Sigma^{1,0}\mathsf{KW}$. □

Here and in the following, the notation regarding the motivic Steenrod algebra is standard; for example, ρ is the class of -1 in $h^{1,1}$, and $\mathsf{Q}_1 = \mathsf{Sq}^2\mathsf{Sq}^1 + \mathsf{Sq}^1\mathsf{Sq}^2$. The slice computation is slightly more complicated for the motivic spectrum KQ representing hermitian K-theory. For comparison purposes, the case of kq, the very effective cover of KQ [**1**], [**2**], is more convenient. Set $\mathsf{kq}_{n\mathsf{h}} := \mathsf{kq} \wedge \mathsf{C}_{n\mathsf{h}}$, and similarly $\mathsf{cKW}_{n\mathsf{h}} := \mathsf{cKW} \wedge \mathsf{C}_{n\mathsf{h}} = \mathsf{cKW} \vee \Sigma^{1,0}\mathsf{cKW}$. Here $\mathsf{cKW} := \mathsf{KW}_{\geq 0} \cong \mathsf{kq}[\eta^{-1}]$ is the connective cover of KW, not its very effective cover (which might deserve the notation kw). The element η acts still invertibly on cKW.

The final set of notation concerns long exact sequences of motivic cohomology groups induced by change of coefficients in cyclic groups. Given natural numbers m, n, the inclusion $\mathbb{Z}/m \hookrightarrow \mathbb{Z}/mn$ induces a homomorphism on motivic cohomology denoted inc^m_{mn}, and the projection $\mathbb{Z}/mn \twoheadrightarrow \mathbb{Z}/n$ induces a homomorphism on motivic cohomology denoted pr^{mn}_n. The short exact sequence

$$0 \to \mathbb{Z}/m \to \mathbb{Z}/mn \to \mathbb{Z}/n \to 0$$

induces a Bockstein or boundary homomorphism ∂^n_m.

THEOREM 6.2. *Let $0 < n, 0 \leq q$. The slices of $\mathsf{kq}_{n\mathsf{h}}$ are given as follows:*

$$\mathsf{s}_{2q}(\mathsf{kq}_{n\mathsf{h}}) = \Sigma^{4q,2q}\mathbf{M}\mathbb{Z}/2n \vee \bigvee_{j=0}^{2q-1} \Sigma^{2q+j,2q}\mathbf{M}\mathbb{Z}/2$$

$$\mathsf{s}_{2q+1}(\mathsf{kq}_{n\mathsf{h}}) = \mathsf{s}_{2q+1}(\mathsf{kq}) \vee \Sigma^{1,0}\mathsf{s}_{2q+1}(\mathsf{kq}) = \bigvee_{j=0}^{2q+1} \Sigma^{2q+1+j,2q+1}\mathbf{M}\mathbb{Z}/2$$

The canonical map $\mathsf{kq}_{n\mathsf{h}} \to \mathsf{cKW}_{n\mathsf{h}}$ *induces the following map on summands of slices:*

$$\Sigma^{4q,2q}\mathbf{M}\mathbb{Z}/2n \xrightarrow{(\partial^{2n}_2,\mathrm{pr}^{2n}_2)} \Sigma^{4q+1,2q}\mathbf{M}\mathbb{Z}/2 \vee \Sigma^{4q,2q}\mathbf{M}\mathbb{Z}/2$$

$$\Sigma^{q+j,q}\mathbf{M}\mathbb{Z}/2 \xrightarrow{(1)} \Sigma^{q+j,q}\mathbf{M}\mathbb{Z}/2 \quad \text{if } j \equiv 0(2) \text{ or } n \equiv 0(2)$$

$$\Sigma^{q+j,q}\mathbf{M}\mathbb{Z}/2 \xrightarrow{(\mathsf{Sq}^1,1)} \Sigma^{q+j+1,q}\mathbf{M}\mathbb{Z}/2 \vee \Sigma^{q+j,q}\mathbf{M}\mathbb{Z}/2 \quad \text{if } j \equiv 1(2) \text{ and } n \equiv 1(2)$$

The restriction of the slice d^1*-differential for* $\mathsf{kq}_{n\mathsf{h}}$ *to the summands* $\Sigma^{q+j,q}\mathbf{M}\mathbb{Z}/2$ *or* $\Sigma^{4q,2q}\mathbf{M}\mathbb{Z}/2n$ *of* $\mathsf{s}_q(\mathsf{kq}_{n\mathsf{h}})$ *for* n *even is given by*

$$\mathsf{d}^1(\mathsf{kq}_{n\mathsf{h}})(q,j) = \begin{cases} (\mathsf{Sq}^3\mathsf{Sq}^1, 0, \mathsf{Sq}^2) & q-1 > j \equiv 0,1 \bmod 4 \\ (\mathsf{Sq}^3\mathsf{Sq}^1, 0, \mathsf{Sq}^2 + \rho\mathsf{Sq}^1, 0, \tau) & q-1 > j \equiv 2,3 \bmod 4 \end{cases}$$

$$\mathsf{d}^1(\mathsf{kq}_{n\mathsf{h}})(q,q-1) = \begin{cases} (\partial^2_{2n}\mathsf{Sq}^2\mathsf{Sq}^1, 0, \mathsf{Sq}^2) & q-1 \equiv 0 \bmod 4 \\ (\mathsf{Sq}^3\mathsf{Sq}^1, 0, \mathsf{Sq}^2) & q-1 \equiv 1 \bmod 4 \\ (\partial^2_{2n}\mathsf{Sq}^2\mathsf{Sq}^1, 0, \mathsf{Sq}^2 + \rho\mathsf{Sq}^1, 0, \tau) & q-1 \equiv 2 \bmod 4 \\ (\mathsf{Sq}^3\mathsf{Sq}^1, 0, \mathsf{Sq}^2 + \rho\mathsf{Sq}^1, 0, \tau) & q-1 \equiv 3 \bmod 4. \end{cases}$$

$$\mathsf{d}^1(\mathsf{kq}_{n\mathsf{h}})(q,q) = \begin{cases} (\mathsf{Sq}^2\partial^{2n}_2, \mathsf{Sq}^2\mathrm{pr}^{2n}_2) & q \equiv 0 \bmod 4 \\ (\mathrm{inc}^2_{2n}\mathsf{Sq}^2\mathsf{Sq}^1, \mathsf{Sq}^2) & q \equiv 1 \bmod 4 \\ (\mathsf{Sq}^2\partial^{2n}_2, \mathsf{Sq}^2\mathrm{pr}^{2n}_2, \tau\partial^{2n}_2, \tau\mathrm{pr}^{2n}_2) & q \equiv 2 \bmod 4 \\ (\mathrm{inc}^2_{2n}\mathsf{Sq}^2\mathsf{Sq}^1, \mathsf{Sq}^2 + \rho\mathsf{Sq}^1, 0, \tau) & q \equiv 3 \bmod 4. \end{cases}$$

The restriction of the slice d^1*-differential for* $\mathsf{kq}_{n\mathsf{h}}$ *to the summands* $\Sigma^{q+j,q}\mathbf{M}\mathbb{Z}/2$ *or* $\Sigma^{4q,2q}\mathbf{M}\mathbb{Z}/2n$ *of* $\mathsf{s}_q(\mathsf{kq}_{n\mathsf{h}})$ *for* n *odd is given by*

$$\mathsf{d}^1(\mathsf{kq}_{n\mathsf{h}})(q,j) = \begin{cases} (\mathsf{Sq}^3\mathsf{Sq}^1, 0, \mathsf{Sq}^2) & q-1 > j \equiv 0 \bmod 4 \\ (\mathsf{Sq}^3\mathsf{Sq}^1, Q_1, \mathsf{Sq}^2, \tau\mathsf{Sq}^1) & q-1 > j \equiv 1 \bmod 4 \\ (\mathsf{Sq}^3\mathsf{Sq}^1, 0, \mathsf{Sq}^2 + \rho\mathsf{Sq}^1, 0, \tau) & q-1 > j \equiv 2 \bmod 4 \\ (\mathsf{Sq}^3\mathsf{Sq}^1, Q_1, \mathsf{Sq}^2 + \rho\mathsf{Sq}^1, \tau\mathsf{Sq}^1 + \rho, \tau) & q-1 > j \equiv 3 \bmod 4, \end{cases}$$

$$\mathsf{d}^1(\mathsf{kq}_{n\mathsf{h}})(q,q-1) = \begin{cases} (\partial^2_{2n}\mathsf{Sq}^2\mathsf{Sq}^1, 0, \mathsf{Sq}^2) & q-1 \equiv 0 \bmod 4 \\ (\mathsf{Sq}^3\mathsf{Sq}^1, Q_1, \mathsf{Sq}^2, \tau\mathsf{Sq}^1) & q-1 \equiv 1 \bmod 4 \\ (\partial^2_{2n}\mathsf{Sq}^2\mathsf{Sq}^1, 0, \mathsf{Sq}^2 + \rho\mathsf{Sq}^1, 0, \tau) & q-1 \equiv 2 \bmod 4 \\ (\mathsf{Sq}^3\mathsf{Sq}^1, Q_1, \mathsf{Sq}^2 + \rho\mathsf{Sq}^1, \tau\mathsf{Sq}^1 + \rho, \tau) & q-1 \equiv 3 \bmod 4. \end{cases}$$

$$\mathsf{d}^1(\mathsf{kq}_{n\mathsf{h}})(q,q) = \begin{cases} (\mathsf{Sq}^2\partial^{2n}_2, \mathsf{Sq}^2\mathrm{pr}^{2n}_2) & q \equiv 0 \bmod 4 \\ (\mathrm{inc}^2_{2n}\mathsf{Sq}^2\mathsf{Sq}^1 + \partial^2_{2n}\mathsf{Sq}^2, \mathsf{Sq}^2, \tau\mathsf{Sq}^1) & q \equiv 1 \bmod 4 \\ (\mathsf{Sq}^2\partial^{2n}_2, \mathsf{Sq}^2\mathrm{pr}^{2n}_2, \tau\partial^{2n}_2, \tau\mathrm{pr}^{2n}_2) & q \equiv 2 \bmod 4 \\ (\mathrm{inc}^2_{2n}\mathsf{Sq}^2\mathsf{Sq}^1 + \partial^2_{2n}\mathsf{Sq}^2, \mathsf{Sq}^2 + \rho\mathsf{Sq}^1, \tau\mathsf{Sq}^1 + \rho, \tau) & q \equiv 3 \bmod 4. \end{cases}$$

PROOF. The description of the slices follows from the fact that the slice functors are triangulated, and [**1**, Theorem 3.2]. The effect of the canonical map $\mathsf{kq}_{n\mathsf{h}} \to \mathsf{cKW}_{n\mathsf{h}}$ on slices is readily obtained, except for the occurrence of Sq^1. The latter follows from the determination of the first slice differential for $\mathsf{cKW}_{n\mathsf{h}} = \mathsf{cKW} \vee \Sigma^{1,0}\mathsf{cKW}$, compared with possible first slice differentials for $\mathsf{kq}_{n\mathsf{h}}$ compatible with the first slice differential for kq, as described in [**1**, Theorem 3.5]. Determining the first slice differential for $\mathsf{kq}_{n\mathsf{h}}$ is then essentially straightforward. □

Let $1 < n \in \mathbb{N}$, set $\mathsf{kq}_n := \mathsf{kq} \wedge \mathsf{C}_n$, and $\mathsf{cKW}_n := \mathsf{cKW} \wedge \mathsf{C}_n$. For comparison, consider the form of the first slice differential for cKW_2, as obtained in [**9**, Theorem 4.3, Theorem 4.14]. Despite the abstract isomorphisms $\mathsf{s}_*\mathsf{cKW}_2 \cong \mathsf{s}_*\mathsf{cKW}_\mathsf{h}$ and $\mathsf{s}_*\mathsf{kq}_2 \cong \mathsf{s}_*\mathsf{kq}_\mathsf{h}$, the first slice differentials differ. Note that if n is odd, then all slices of cKW_n vanish, although cKW_n itself does not over formally real fields.

THEOREM 6.3. *The restriction of the slice d^1-differential for cKW_2 to the summand $\Sigma^{q+j,q}\mathbf{M}\mathbb{Z}/2$ of $\mathsf{s}_q(\mathsf{cKW}_2)$ is given by*

$$\mathsf{d}^1(\mathsf{cKW}_2)(q,j) = \begin{cases} (\mathsf{Sq}^3\mathsf{Sq}^1, 0, \mathsf{Sq}^2) & j \equiv 0 \bmod 4 \\ (\mathsf{Sq}^3\mathsf{Sq}^1, \mathsf{Q}_1, \mathsf{Sq}^2, \rho + \tau\mathsf{Sq}^1) & j \equiv 1 \bmod 4 \\ (\mathsf{Sq}^3\mathsf{Sq}^1, 0, \mathsf{Sq}^2 + \rho\mathsf{Sq}^1, 0, \tau) & j \equiv 2 \bmod 4 \\ (\mathsf{Sq}^3\mathsf{Sq}^1, \mathsf{Q}_1, \mathsf{Sq}^2 + \rho\mathsf{Sq}^1, \tau\mathsf{Sq}^1, \tau) & j \equiv 3 \bmod 4. \end{cases}$$

Here the ith component of $\mathsf{d}^1(\mathsf{cKW}_2)(q,j)$ is a map $\Sigma^{q+j,q}\mathbf{M}\mathbb{Z}/2 \to \Sigma^{q+j+i,q+1}\mathbf{M}\mathbb{Z}/2$.

THEOREM 6.4. *The restriction of the slice d^1-differential for kq_2 to the summand $\Sigma^{q+j,q}\mathbf{M}\mathbb{Z}/2$ of $\mathsf{s}_q(\mathsf{kq}_2)$ is given by*

$$\mathsf{d}^1(\mathsf{kq}_2)(q,j) = \begin{cases} (\mathsf{Sq}^3\mathsf{Sq}^1, 0, \mathsf{Sq}^2) & q > j \equiv 0 \bmod 4 \\ (\mathsf{Sq}^3\mathsf{Sq}^1, \mathsf{Q}_1, \mathsf{Sq}^2, \rho + \tau\mathsf{Sq}^1) & q > j \equiv 1 \bmod 4 \\ (\mathsf{Sq}^3\mathsf{Sq}^1, 0, \mathsf{Sq}^2 + \rho\mathsf{Sq}^1, 0, \tau) & q > j \equiv 2 \bmod 4 \\ (\mathsf{Sq}^3\mathsf{Sq}^1, \mathsf{Q}_1, \mathsf{Sq}^2 + \rho\mathsf{Sq}^1, \tau\mathsf{Sq}^1, \tau) & q > j \equiv 3 \bmod 4, \end{cases}$$

$$\mathsf{d}^1(\mathsf{kq}_2)(q,q) = \begin{cases} (\mathsf{Sq}^2\mathsf{Sq}^1, \mathsf{Sq}^2 + \rho\mathsf{Sq}^1) & q \equiv 0 \bmod 4 \\ (\mathsf{Q}_1, \mathsf{Sq}^2, \rho + \tau\mathsf{Sq}^1, 0) & q \equiv 1 \bmod 4 \\ (\mathsf{Sq}^2\mathsf{Sq}^1, \mathsf{Sq}^2 + \rho\mathsf{Sq}^1, \tau\mathsf{Sq}^1, \tau) & q \equiv 2 \bmod 4 \\ (\mathsf{Q}_1, \mathsf{Sq}^2 + \rho\mathsf{Sq}^1, \tau\mathsf{Sq}^1, \tau) & q \equiv 3 \bmod 4. \end{cases}$$

Here the ith component of $\mathsf{d}^1(\mathsf{kq}_2)(q,j)$ is a map $\Sigma^{q+j,q}\mathbf{M}\mathbb{Z}/2 \to \Sigma^{q+j+i,q+1}\mathbf{M}\mathbb{Z}/2$.

Note that [**9**, Theorems 4.24, 4.36] contain information about first slice differentials for kq_{2^r} and cKW_{2^r}, which could be used to analyze the first slice differentials for C_{2^r}.

THEOREM 6.5. *Let $n > 0$, and let $g := \gcd(2n, 12)$. The slices of $\mathsf{C}_{n\mathsf{h}}$ are given as follows, up to summands of simplicial suspension higher than $q+2$ for $q \geq 4$.*

$$\mathsf{s}_0(\mathsf{C}_{n\mathsf{h}}) = \mathsf{s}_0(\mathbf{1})/2n$$

$$\mathsf{s}_1(\mathsf{C}_{n\mathsf{h}}) = \mathsf{s}_1(\mathbf{1}) \vee \Sigma^{1,0}\mathsf{s}_1(\mathbf{1})$$

$$\mathsf{s}_2(\mathsf{C}_{n\mathsf{h}}) = \Sigma^{2,2}\mathbf{M}\mathbb{Z}/2\{\alpha_1^2\} \vee \Sigma^{3,2}\mathbf{M}\mathbb{Z}/2\{\overline{\alpha_1^2}\} \vee \Sigma^{3,2}\mathbf{M}\mathbb{Z}/g\{\alpha_2\} \vee \Sigma^{4,2}\mathbf{M}\mathbb{Z}/g\{\overline{\alpha_2}\}$$

$$\mathsf{s}_3(\mathsf{C}_{n\mathsf{h}}) = \mathsf{s}_3(\mathbf{1}) \vee \Sigma^{1,0}\mathsf{s}_3(\mathbf{1})$$

$$\mathsf{s}_4(\mathsf{C}_{n\mathsf{h}}) = \Sigma^{4,4}\mathbf{M}\mathbb{Z}/2\{\alpha_1^4\} \vee \Sigma^{5,4}\mathbf{M}\mathbb{Z}/2\{\overline{\alpha_1^4}\} \vee \Sigma^{6,4}\mathbf{M}\mathbb{Z}/2\{\alpha_1\alpha_3\} \vee \Sigma^{6,4}\mathbf{M}\mathbb{Z}/2\{\nu^2\} \vee \cdots$$

$$\mathsf{s}_q(\mathsf{C}_{n\mathsf{h}}) = \Sigma^{q,q}\mathbf{M}\mathbb{Z}/2\{\alpha_1^q\} \vee \Sigma^{q+1,q}\mathbf{M}\mathbb{Z}/2\{\overline{\alpha_1^q}\} \vee \Sigma^{q+2,q}\mathbf{M}\mathbb{Z}/2\{\alpha_1^{q-3}\alpha_3\} \vee \cdots$$

The unit map $\mathbf{1} \to \mathsf{kq}$ induces a map $\mathsf{C}_{n\mathsf{h}} \to \mathsf{kq}_{n\mathsf{h}}$, which induces the identity map on the slice summands

$$\mathbf{M}\mathbb{Z}/2n, \Sigma^{q,q}\mathbf{M}\mathbb{Z}/2\{\alpha_1^q\}, \Sigma^{q+1,q}\mathbf{M}\mathbb{Z}/2\{\overline{\alpha_1^q}\},$$

$$\Sigma^{q+2,q}\mathbf{M}\mathbb{Z}/2\{\alpha_1^{q-3}\alpha_3\}, \text{ and } \Sigma^{q+3,q}\mathbf{M}\mathbb{Z}/2\{\overline{\alpha_1^{q-3}\alpha_3}\},$$

and the map

$$\Sigma^{3,2}\mathbf{M}\mathbb{Z}/g \vee \Sigma^{4,2}\mathbf{M}\mathbb{Z}/g \xrightarrow{\begin{pmatrix}\partial^g_{2n} & \mathrm{inc}^g_{2n}\end{pmatrix}} \Sigma^{4,2}\mathbf{M}\mathbb{Z}/2n$$

on summands of the two-slices.

PROOF. The form of the slices follows from [**10**, Section 8] or [**21**, Theorem 2.12], since h induces multiplication by 2 on slices, and the slice functors are triangulated. The map $\mathsf{C}_{n\mathsf{h}} \to \mathsf{kq}_{n\mathsf{h}}$ is determined by the unit map $\mathbf{1} \to \mathsf{kq}$, whose behaviour on slices can be read off from [**21**, Lemmas 2.28, 2.29]. □

THEOREM 6.6. *The first slice differential has the following form on the given summands:*

$$\mathsf{d}^1(\mathsf{C}_{n\mathsf{h}})(\alpha_1^0) = (\mathsf{Sq}^2\partial_2^{2n}, \mathsf{Sq}^2\mathrm{pr}_2^{2n})$$

$$\mathsf{d}^1(\mathsf{C}_{n\mathsf{h}})(\alpha_1) = \begin{cases} (\partial_g^2\mathsf{Sq}^2\mathsf{Sq}^1, \mathrm{inc}_g^2\mathsf{Sq}^2\mathsf{Sq}^1, 0, \mathsf{Sq}^2) & n \equiv 0(4) \\ (0, \mathrm{inc}_g^2\mathsf{Sq}^2\mathsf{Sq}^1, 0, \mathsf{Sq}^2) & n \equiv 2(4) \\ (\partial_g^2\mathsf{Sq}^2\mathsf{Sq}^1, 0, 0, \mathsf{Sq}^2) & n \equiv 1(2) \end{cases}$$

$$\mathsf{d}^1(\mathsf{C}_{n\mathsf{h}})(\alpha_1^q) = (\mathsf{Sq}^3\mathsf{Sq}^1, 0, \mathsf{Sq}^2) \quad q \geq 2$$

$$\mathsf{d}^1(\mathsf{C}_{n\mathsf{h}})(\overline{\alpha_1}) = \begin{cases} (\mathrm{inc}_g^2\mathsf{Sq}^2\mathsf{Sq}^1, 0, \mathsf{Sq}^2) & n \equiv 0(2) \\ (\mathrm{inc}_g^2\mathsf{Sq}^2\mathsf{Sq}^1, \mathrm{inc}_g^2\mathsf{Sq}^2, \mathsf{Sq}^2, \tau\mathsf{Sq}^1) & n \equiv 1(2) \end{cases}$$

$$\mathsf{d}^1(\mathsf{C}_{n\mathsf{h}})(\overline{\alpha_1^q}) = \begin{cases} (\mathsf{Sq}^3\mathsf{Sq}^1, 0, \mathsf{Sq}^2) & n \equiv 0(2) \\ (\mathsf{Sq}^3\mathsf{Sq}^1, \mathsf{Q}_1, \mathsf{Sq}^2, \tau\mathsf{Sq}^1) & n \equiv 1(2) \end{cases}$$

$$\mathsf{d}^1(\mathsf{C}_{n\mathsf{h}})(\alpha_2) = (0, \mathsf{Sq}^2\partial_2^g, 0, \tau\partial_2^g)$$

$$\mathsf{d}^1(\mathsf{C}_{n\mathsf{h}})(\overline{\alpha_2}) = \begin{cases} (\mathsf{Sq}^2\partial_2^g, 0, \tau\partial_2^g, 0) & n \equiv 0(4) \\ (\mathsf{Sq}^2\partial_2^g, \mathsf{Sq}^2\mathrm{pr}_2^g, \tau\partial_2^g, \tau\mathrm{pr}_2^g) & n \not\equiv 0(4) \end{cases}$$

$$\mathsf{d}^1(\mathsf{C}_{n\mathsf{h}})(\alpha_1^{q-3}\alpha_3) = (\mathsf{Sq}^3\mathsf{Sq}^1, 0, \mathsf{Sq}^2 + \rho\mathsf{Sq}^1, 0, \tau) \quad q \geq 3$$

$$\mathsf{d}^1(\mathsf{C}_{n\mathsf{h}})(\overline{\alpha_1^{q-3}\alpha_3}) = \begin{cases} (\mathsf{Sq}^3\mathsf{Sq}^1, 0, \mathsf{Sq}^2 + \rho\mathsf{Sq}^1, 0, \tau) & n \equiv 0(2) \\ (\mathsf{Sq}^3\mathsf{Sq}^1, \mathsf{Q}_1, \mathsf{Sq}^2 + \rho\mathsf{Sq}^1, \tau\mathsf{Sq}^1 + \rho, \tau) & n \equiv 1(2) \end{cases}$$

PROOF. Based on the form of the unit map, most parts of the first slice differential are determined by Theorem 6.2. More precisely, the differentials $\mathsf{s}_0\mathsf{C}_{n\mathsf{h}} \to \Sigma^{1,0}\mathsf{s}_1\mathsf{C}_{n\mathsf{h}}$, $\mathsf{s}_2\mathsf{C}_{n\mathsf{h}} \to \Sigma^{1,0}\mathsf{s}_3\mathsf{C}_{n\mathsf{h}}$ and the listed behaviour for $q \geq 2$ is determined by Theorem 6.2. The remaining identities follow from Adem relations and the property $\mathsf{d}^2 = 0$, as well as comparison with $\mathsf{d}^1(\mathbf{1})$. □

Theorem 6.6 may be used for computations of $\pi_{1+(\star)}\mathsf{C}_{n\mathsf{h}}$ and $\pi_{2+(\star)}\mathsf{C}_{n\mathsf{h}}$. Compared with corresponding slice spectral sequence computations for $\mathbf{1}$, the absence of integral motivic cohomology groups may be viewed as an advantage. Concrete presentations for $H^{\star-k,\star}$ as $\mathbf{K}^{\mathsf{MW}}$-modules seem to be known only in very few cases, contrary to the $\mathbf{K}^{\mathsf{MW}}$-module $h^{\star-k,\star} \cong \mathbf{K}^{\mathsf{MW}}_{\star-k}/(\eta, 2)$.

References

[1] A. Ananyevskiy, O. Röndigs, and P. A. Østvær, *On very effective hermitian K-theory*, Math. Z. (2019), 14 pages, https://doi.org/10.1007/s00209-019-02302-z.

[2] Tom Bachmann, *The generalized slices of Hermitian K-theory*, J. Topol. **10** (2017), no. 4, 1124–1144, DOI 10.1112/topo.12032. MR3743071

[3] M. G. Barratt, *Track groups. II*, Proc. London Math. Soc. (3) **5** (1955), 285–329, DOI 10.1112/plms/s3-5.3.285. MR0073179

[4] Daniel Dugger and Daniel C. Isaksen, *Motivic Hopf elements and relations*, New York J. Math. **19** (2013), 823–871. MR3141814

[5] Daniel Dugger and Daniel C. Isaksen, *Low-dimensional Milnor-Witt stems over* $\mathbb{R}$, Ann. K-Theory **2** (2017), no. 2, 175–210, DOI 10.2140/akt.2017.2.175. MR3590344

[6] Bogdan Gheorghe, *The motivic cofiber of* τ, Doc. Math. **23** (2018), 1077–1127. MR3874951

[7] Brayton Gray, *Ring spectra*, Proceedings of the Advanced Study Institute on Algebraic Topology (Aarhus Univ., Aarhus, 1970), Mat. Inst., Aarhus Univ., Aarhus, 1970, pp. 143–153. Various Publ. Ser., No. 13. MR0343271

[8] Bertrand J. Guillou and Daniel C. Isaksen, *The* η*-inverted* $\mathbb{R}$*-motivic sphere*, Algebr. Geom. Topol. **16** (2016), no. 5, 3005–3027, DOI 10.2140/agt.2016.16.3005. MR3572357

[9] J. Kylling, O. Röndigs, and P. A. Østvær, *Hermitian* K*-theory, Dedekind* ζ*-functions, and quadratic forms over rings of integers in number fields*, arXiv:1811.03940.

[10] Marc Levine, *A comparison of motivic and classical stable homotopy theories*, J. Topol. **7** (2014), no. 2, 327–362, DOI 10.1112/jtopol/jtt031. MR3217623

[11] L. Mantovani, *Localizations and completions in motivic homotopy theory*, Dissertation, Universität Duisburg-Essen, 2018.

[12] John Milnor, *Algebraic* K*-theory and quadratic forms*, Invent. Math. **9** (1969/1970), 318–344, DOI 10.1007/BF01425486. MR0260844

[13] John Milnor, *Introduction to algebraic* K*-theory*, Princeton University Press, Princeton, N.J.; University of Tokyo Press, Tokyo, 1971. Annals of Mathematics Studies, No. 72. MR0349811

[14] Fabien Morel, *On the motivic* π_0 *of the sphere spectrum*, Axiomatic, enriched and motivic homotopy theory, NATO Sci. Ser. II Math. Phys. Chem., vol. 131, Kluwer Acad. Publ., Dordrecht, 2004, pp. 219–260, DOI 10.1007/978-94-007-0948-5_7. MR2061856

[15] Fabien Morel, *The stable* $\mathbb{A}^1$*-connectivity theorems*, K-Theory **35** (2005), no. 1-2, 1–68, DOI 10.1007/s10977-005-1562-7. MR2240215

[16] Shichirô Oka, *Multiplications on the Moore spectrum*, Mem. Fac. Sci. Kyushu Univ. Ser. A **38** (1984), no. 2, 257–276, DOI 10.2206/kyushumfs.38.257. MR760188

[17] O. Röndigs, *On the* η*-inverted sphere*, K-Theory, Tata Inst. Fundam. Res. Stud. Math., vol. 19, Tata Inst. Fund. Res., Mumbai, 2018, pp. 41–64.

[18] Oliver Röndigs and Paul Arne Østvær, *Rigidity in motivic homotopy theory*, Math. Ann. **341** (2008), no. 3, 651–675, DOI 10.1007/s00208-008-0208-5. MR2399164

[19] Oliver Röndigs and Paul Arne Østvær, *Slices of hermitian* K*-theory and Milnor's conjecture on quadratic forms*, Geom. Topol. **20** (2016), no. 2, 1157–1212, DOI 10.2140/gt.2016.20.1157. MR3493102

[20] O. Röndigs, M. Spitzweck, and P. A. Østvær, *The second stable homotopy groups of motivic spheres*, In preparation.

[21] Oliver Röndigs, Markus Spitzweck, and Paul Arne Østvær, *The first stable homotopy groups of motivic spheres*, Ann. of Math. (2) **189** (2019), no. 1, 1–74, DOI 10.4007/annals.2019.189.1.1. MR3898173

[22] Hirosi Toda, *Composition methods in homotopy groups of spheres*, Annals of Mathematics Studies, No. 49, Princeton University Press, Princeton, N.J., 1962. MR0143217

[23] Vladimir Voevodsky, *Reduced power operations in motivic cohomology*, Publ. Math. Inst. Hautes Études Sci. **98** (2003), 1–57, DOI 10.1007/s10240-003-0009-z. MR2031198

Institut für Mathematik, Universität Osnabrück, Germany
Email address: oliver.roendigs@uni-osnabrueck.de

Contemporary Mathematics
Volume **745**, 2020
https://doi.org/10.1090/conm/745/15027

Oriented Schubert calculus in Chow–Witt rings of Grassmannians

Matthias Wendt

Abstract. We apply the previous calculations of Chow–Witt rings of Grassmannians to develop an oriented analogue of the classical Schubert calculus. As a result, we get complete diagrammatic descriptions of the ring structure in Chow–Witt rings and twisted Witt groups. In the resulting arithmetic refinements of Schubert calculus, the multiplicity of a solution subspace is a quadratic form encoding additional orientation information. We also discuss a couple of applications, such as a Chow–Witt version of the signed count of balanced subspaces of Fehér and Matszangosz.

Contents

1. Introduction

Schubert calculus is a collection of techniques to compute intersection products in the Chow rings of Grassmannians. It can be used to answer questions about intersections of subspaces of an F-vector space V, like "how many k-dimensional subspaces of V satisfy specified intersection conditions with respect to a finite set of flags in V?" For every flag, the k-dimensional subspaces satisfying the specified intersection conditions form a Schubert variety. Over an algebraically closed field, the number of subspaces satisfying the specified conditions (assuming that the

2010 *Mathematics Subject Classification.* 14C17, 14F42, 14M15, 14N15.

Key words and phrases. Chow–Witt rings, characteristic classes, Grassmannians, oriented Schubert calculus.

flags are in general position) can be computed as degree of the intersection of the corresponding Schubert varieties (provided that intersection is zero-dimensional).

If the field F is not algebraically closed, then some of the solution subspaces may not be defined over F and then the degree of the intersection of Schubert varieties is not necessarily equal to the number of solution subspaces. For example, over $F = \mathbb{R}$, the real points of the variety of generic flags for the Schubert problem (with the analytic topology) is not path-connected and the numbers of solution subspaces in different path-components are usually different. In particular, the number of solution subspaces defined over the base field F is not invariant in general. In real enumerative geometry and real Schubert calculus, this problem has been solved by introducing signs depending on additional orientations. Counting the solution subspaces with signs produces an invariant number which can be related to intersection products in cohomology rings of real Grassmannians, cf. [**FK15**, **FM16**].

The point of the present paper is to push this idea further: we want to provide an arithmetic refinement of Schubert calculus which provides more information about the number of solution subspaces over arbitrary fields of characteristic $\neq 2$. Following the ideas of Kass and Wickelgren [**KW17**], the number of solution subspaces should be replaced by a quadratic form depending on the field of definition and additional orientation information of the solution subspaces. The right setting to realize this idea is given by the Chow–Witt rings of Grassmannians. The structure of these rings was determined in [**Wen18**]. One of the goals of the present paper is to make the intersection product more transparent by providing diagrammatical rules in terms of even Young diagrams of [**BC12a**], thus also settling the open question regarding multiplicative structure of twisted Witt groups of Grassmannians. The other main goal of the present paper is to explain how (canonical) choices of orientations on Schubert varieties yield cycles in the Chow–Witt ring and how the multiplicity of a solution subspace in the intersection of such Schubert cycles is a quadratic form encoding information on the field of definition and orientation of the solution subspace.

Now for a more detailed description of the results of the paper. While we are mostly interested in Chow–Witt rings of Grassmannians, many of the results can be reduced to statements about the cohomology of the Grassmannians $\mathrm{Gr}(k,n)$ with coefficients in the sheaves $\mathbf{I}^j$ of powers of the fundamental ideal in the Witt sheaf $\mathbf{W}$, possibly twisted by a line bundle $\mathscr{L}$. The first thing to note is that the natural morphism $\widetilde{\mathrm{CH}}^\bullet(\mathrm{Gr}(k,n),\mathscr{L}) \to \mathrm{CH}^\bullet(\mathrm{Gr}(k,n))$ from Chow–Witt groups twisted by a line bundle $\mathscr{L}$ to Chow groups is not generally surjective. The obstruction is the Steenrod square, and there is a precise formula for the Steenrod square in terms of Young diagrams. The following is only an informal statement of the combined results of Proposition 4.1, Theorem 4.2, Proposition 4.6 and Theorem 6.11. The precise formulations are given there and require quite a bit more of notation to be set up.

THEOREM 1.1.

(1) *A class* $x \in \mathrm{CH}^j(\mathrm{Gr}(k,n))$ *lifts to the Chow–Witt group* $\widetilde{\mathrm{CH}}^j(\mathrm{Gr}(k,n),\mathscr{L})$ *if and only if the Steenrod square* $\mathrm{Sq}^2_{\mathscr{L}}(x)$ *is trivial.*
(2) *For a Schubert class* $\sigma_\Lambda \in \mathrm{CH}^j(\mathrm{Gr}(k,n))$ *corresponding to a Young diagram* Λ*, the Steenrod squares* $\mathrm{Sq}^2_{\mathscr{O}}(\sigma_\Lambda)$ *and* $\mathrm{Sq}^2_{\det}(\sigma_\Lambda)$ *can be computed in*

terms of the Young diagram as follows: filling the boxes with a checkerboard pattern starting with a black box in the upper left corner, the respective Steenrod square is given as the sum of those Young diagrams with checkerboard pattern in $k \times (n-k)$-frame which can be obtained by adding a single white or black box, respectively. In particular, a class lifts if all the corners of the Young diagram have even length.

(3) *Let Σ_Λ be a Schubert variety corresponding to the Young diagram Λ, and assume that the Schubert class $\sigma_\Lambda = [\Sigma_\Lambda]$ in $\mathrm{CH}^j(\mathrm{Gr}(k,n))$ has vanishing $\mathrm{Sq}^2_{\mathscr{L}}$. Then there is a canonical orientation of the twisted relative canonical bundle $\omega_{p_\Lambda} \otimes p_\Lambda^*\mathscr{L}$ for a specific flag resolution of singularities $p_\Lambda\colon \mathrm{Fl}(\Lambda) \to \mathrm{Gr}(k,n)$ of the Schubert variety Σ_Λ. A canonical lift of the Schubert class to the Chow–Witt ring is given by the pushforward of the constant form $\langle 1\rangle$ along p_Λ (with the canonical choice of orientation).*

In particular, we can determine additive generators for the $\mathbf{I}^\bullet$-cohomology ring. Moreover, the even Young diagrams provide an additive basis of its quotient mod torsion, the $\mathbf{W}$-cohomology ring, generalizing classical results of Pontryagin [**Pon47**]. To understand the multiplication of non-torsion classes one can use an isomorphism from a suitable Chow ring, based on doubled partitions. On the other hand, multiplication with torsion classes is much simpler and can be determined by reduction to the mod 2 Chow ring $\mathrm{Ch}^\bullet(\mathrm{Gr}(k,n))$. We only state the simplest case, for $\mathrm{Gr}(2a,2b)$. For the other cases, which require more notation to set up, cf. Proposition 3.1, Proposition 3.14 and Theorem 3.16.

THEOREM 1.2.

(1) *The even Young diagrams, cf. Definition 3.8, form a $\mathrm{W}(F)$-basis for the total $\mathbf{W}$-cohomology ring $\mathrm{H}^\bullet(\mathrm{Gr}(k,n), \mathbf{W} \oplus \mathbf{W}(\det))$.*

(2) *There is a natural ring isomorphism*

$$\mathrm{Ch}^\bullet(\mathrm{Gr}(a,b)) \to \mathrm{H}^{4\bullet}(\mathrm{Gr}(2a,2b), \mathbf{W})\colon \ \mathrm{c}_i \mapsto \mathrm{p}_i\,.$$

This allows to completely describe the multiplication in $\mathbf{W}$-cohomology in terms of classical Schubert calculus for the Chow ring.

(3) *The product of any class with an $\mathrm{I}(F)$-torsion class is $\mathrm{I}(F)$-torsion and completely determined by its image in the mod 2 Chow ring $\mathrm{Ch}^\bullet(\mathrm{Gr}(k,n))$.*

As a direct consequence, we can also describe the multiplicative structure of the twisted Witt groups of Grassmannians. Their additive structure was determined in [**BC12a**], but the ring structure was left open in loc.cit.

COROLLARY 1.3. *There is a natural isomorphism*

$$\alpha\colon \mathrm{W}^i(\mathrm{Gr}(k,n),\mathscr{L}) \xrightarrow{\cong} \bigoplus_{j\equiv i \bmod 4} \mathrm{H}^j(\mathrm{Gr}(k,n), \mathbf{W}(\mathscr{L}))$$

of $\mathrm{W}(F)$-modules. Summing over all $i \in \mathbb{Z}/4\mathbb{Z}$ and $[\mathscr{L}] \in \mathrm{Pic}(\mathrm{Gr}(k,n))/2$, this induces a natural isomorphism from the total Witt ring of $\mathrm{Gr}(k,n)$ to the total $\mathbf{W}$-cohomology ring. In particular, the description of the multiplicative structure on $\mathbf{W}$-cohomology in Theorem 1.2 completely determines the multiplicative structure on the twisted Witt groups of Grassmannians.

Now we can state how "oriented" Schubert calculus works: a classical Schubert problem in the Chow ring can be lifted to the Chow–Witt ring if the Steenrod square condition from Theorem 1.1 is satisfied, and in this case, we have a canonical lift provided by canonical orientations of the Schubert varieties. This provides an "oriented Schubert problem": a collection of Schubert varieties with additional orientations. On the one hand, we can then use the above description of the Chow–Witt ring structure in Theorem 1.2 to compute the degree of the intersection of these oriented Schubert cycles in the top Chow–Witt group. On the other hand, the local contribution of a solution subspace in the intersection of the Schubert cycles is given by a quadratic form depending on the field of definition of the solution subspace as well as the additional orientation information. The following is Theorem 7.17 which describes the local contributions in the case of smooth oriented Schubert problems. There is a similar statement for the singular case, cf. Theorem 7.19, which again requires more notation for a precise statement (and is not quite as explicit).

THEOREM 1.4. *Let F be a field of characteristic $\neq 2$.*

(1) *For any oriented Schubert problem, the degree of the intersection in the top Chow–Witt group always has the form $a\langle 1\rangle + b\langle -1\rangle$ for $a, b \in \mathbb{Z}$. If one of the Schubert classes is torsion, then $a = b$.*
(2) *Fix a smooth oriented Schubert problem in the Grassmannian $\mathrm{Gr}(k,n)$. For any solution subspace $[W] \in \mathrm{Gr}(k,n)$ defined over E/F, there is a natural identification ξ_W of the tangent space of $\mathrm{Gr}(k,n)$ at $[W]$ with the direct sum of the normal spaces to the smooth Schubert varieties intersecting at $[W]$, and the oriented multiplicity of $[W]$ in the intersection cycle is $\mathrm{tr}_{E/F}\langle \det \xi_W\rangle$, i.e., it encodes the square class of the unit comparing the two natural orientations above.*

It should be pointed out clearly that the above results describe two sides of the multiplicative structure of the $\mathbf{I}^\bullet$-cohomology which are both necessary to understand enumerative consequences: on the one hand, we have the picture of characteristic classes, and a presentation of the cohomology ring in terms of Pontryagin classes which allows to compute degrees of intersection products via the folklore isomorphism of Theorem 1.2. On the other hand, we have very canonical cohomology classes of geometric origin, arising as pushforward of $\langle 1\rangle$ from resolutions of singularities for Schubert varieties corresponding to even Young diagrams; and oriented multiplicities for intersections of such geometric cycles can be determined very explicitly as in Theorem 1.4. But to get enumerative consequences, we need that both pictures coincide, i.e., that the geometric cycles are actually representatives of the respective characteristic classes. Then we can compute the degree of the intersection (i.e., the refined number of solution subspaces) using Theorem 1.2 and its relatives *and* identify the local contribution of each solution subspace in terms of orientation information encoded in the notion of type. This identification of geometric cycles and characteristic classes is provided in Proposition 8.5.

As a sample application, we have the following generalization of the signed count of balanced subspaces in [**FM16**], cf. Theorem 9.7.

THEOREM 1.5. *Let F be a field of characteristic $\neq 2$. Let $a = 2i$ and $b = 2j$ be even numbers with $a < b$. Then the following equality holds in $\mathrm{GW}(F) \cong \widetilde{\mathrm{CH}}^{4(ab-a^2)}(\mathrm{Gr}(2a, 2b))$:*

$$\sum_{\text{balanced } W} \mathrm{tr}_{F(W)/F}\langle \det \xi_W\rangle = \frac{1}{2}\left(\binom{2j}{2i} + \binom{j}{i}\right)\langle 1\rangle + \frac{1}{2}\left(\binom{2j}{2i} - \binom{j}{i}\right)\langle -1\rangle.$$

For F an algebraically closed field, this recovers the classical results from Schubert calculus. For $F = \mathbb{R}$, this recovers results of [**FM16**], and generally encodes lower bounds for the number of real solutions of a Schubert problem. For $F = \mathbb{F}_q$, this provides mod 2 congruences of solution subspaces over even-degree extensions with oriented splitting and solution subspaces over odd-degree extensions with non-oriented splitting, cf. Proposition 8.9.

Structure of the paper: We begin in Section 2 by recalling the descriptions of Chow–Witt rings of Grassmannians from [**Wen18**]. Then we describe the multiplication of torsion-free classes (in the $\mathbf{W}$-cohomology ring) in terms of even Young diagrams in Section 3, and the lifting conditions and multiplication of torsion classes are discussed in Section 4. After a short recollection of pushforwards and products in Witt groups and $\mathbf{I}^\bullet$-cohomology in Section 5, we discuss the explicit geometric representatives of lifts of Schubert classes in Section 6. Then we discuss the explicit geometric identification of oriented intersection multiplicities in Section 7. Finally, all the results are combined in Section 8, including the comparison of $\mathbf{W}$-cohomology rings and twisted Witt groups, and a couple of sample enumerative applications are given in Section 9.

Acknowledgements: First of all, many thanks to Kirsten Wickelgren; without the discussions about refined enumerative geometry and her work during the "étale and motivic homotopy 2017" conference in Heidelberg and the "motives and refined enumerative geometry" workshop in May 2018 in Essen, the present work would not exist. I would also like to thank to Thomas Hudson for discussions about resolutions of singularities for Schubert varieties. The package `genyoungtabtikz` was used to generate Young diagrams, and example computations were done using the `SchurRings` packages of `Macaulay2`. Finally, comments from Marc Levine, Kirsten Wickelgren and an anonymous referee helped improve the paper by removing a couple of minor mistakes and inaccuracies.

2. Recollection on Chow–Witt rings of Grassmannians

In this section, we recall the computations of the Chow–Witt rings resp. $\mathbf{I}^\bullet$-cohomology rings of the Grassmannians from [**Wen18**].

We fix a base field F which later will be assumed to be perfect of characteristic $\neq 2$. For a natural number k and an F-vector space V, we denote by $\mathrm{Gr}(k,V)$ the Grassmannian of k-dimensional subspaces of V. Most of the time, we actually consider $V = F^n$ and denote the Grassmannian by $\mathrm{Gr}(k,n)$.

The Grassmannian $\mathrm{Gr}(k,n)$ carries an exact sequence of vector bundles

$$0 \to \mathscr{S}_k \to \mathscr{O}^{\oplus n}_{\mathrm{Gr}(k,n)} \to \mathscr{Q}_{n-k} \to 0$$

where the fiber of $\mathscr{S}_k$ at a point $[W]$ is the subspace $W \subseteq V$, and the fiber of $\mathscr{Q}_{n-k}$ is V/W. We speak of the tautological sub- and quotient bundle, respectively. The short description of the cohomology of the Grassmannians is that the characteristic classes of these two bundles generate the cohomology rings, modulo the relations coming from the Whitney sum formula. The one exception to this rule is the case where $\dim \mathrm{Gr}(k,n) = k(n-k)$ is odd, where there is a new exterior class R. A recollection on definitions and properties of the relevant characteristic classes (Pontryagin classes p_i, Euler classes e_n, Chern classes c_i, Stiefel–Whitney classes $\overline{\mathrm{c}}_i$ and their Bocksteins $\beta_{\mathscr{L}}(\overline{\mathrm{c}}_{2i_1}\cdots\overline{\mathrm{c}}_{2i_j})$) can be found in [**HW17, Wen18**].

REMARK 2.1. *After pullback along the natural projection*

$$\pi\colon \mathrm{GL}_n / \mathrm{GL}_k \times \mathrm{GL}_{n-k} \to \mathrm{Gr}(k,n),$$

the exact sequence splits, i.e., over $\mathrm{GL}_n / \mathrm{GL}_k \times \mathrm{GL}_{n-k}$ *we have the tautological* k*-plane bundle* $\mathscr{E}_k = \pi^*\mathscr{S}_k$ *and the complementary* $n-k$*-plane bundle* $\mathscr{E}^\perp_{n-k} = \pi^*\mathscr{Q}_{n-k}$*, with* $\mathscr{E}_k \oplus \mathscr{E}^\perp_{n-k} \cong \mathscr{O}^{\oplus n}_{\mathrm{GL}_n / \mathrm{GL}_k \times \mathrm{GL}_{n-k}}$*. The projection morphism* π *is a vector bundle torsor and therefore induces isomorphisms in all cohomology theories considered in this paper. In particular, we could equivalently consider the characteristic classes of* $\mathscr{E}_k$ *and* $\mathscr{E}^\perp_{n-k}$ *as natural generators in cohomology of* $\mathrm{Gr}(k,n)$*.*

The cohomology groups we consider in the paper are always sheaf cohomology groups of a smooth scheme X over F (mostly Grassmannians and smooth Schubert varieties in flag varieties) in the Nisnevich (or equivalently the Zariski) topology. The coefficient sheaves that appear are the Milnor–Witt K-theory sheaves $\mathbf{K}^{\mathrm{MW}}_\bullet$ for Chow–Witt groups and the sheaf $\mathbf{W}$ of Witt rings as well as the sheaves $\mathbf{I}^\bullet$ of powers of the fundamental ideal sheaf in $\mathbf{W}$. In these cases, the cohomology groups are naturally modules over the coefficient sheaf evaluated at the base field F, which is the Grothendieck–Witt ring $\mathrm{GW}(F)$ for Chow–Witt groups and the Witt ring $\mathrm{W}(F)$ for cohomology with $\mathbf{W}$ or $\mathbf{I}^\bullet$-coefficients.

We also consider the cohomology of twists of these sheaves by line bundles. On the Grassmannian $\mathrm{Gr}(k,n)$, we have $\mathrm{Pic}(\mathrm{Gr}(k,n))/2 \cong \mathbb{Z}/2\mathbb{Z}$. Natural choices for the generator are the determinants $\det \mathscr{S}_k^\vee$ and $\det \mathscr{Q}^\vee_{n-k}$ of the (duals of the) tautological sub- and quotient bundle, respectively. Sometimes, the twist of a sheaf $\mathbf{A}$ by $\det \mathscr{S}_k^\vee$ will be denoted by $\mathbf{A}(\det)$. Some of the statements below are formulated for the total Chow–Witt ring or $\mathbf{W}$-cohomology ring of the Grassmannian, which has a $\mathbb{Z} \times \mathrm{Pic}(\mathrm{Gr}(k,n))/2$-grading. The technical inaccuracy of neglecting choices of isomorphisms between different representatives of isomorphism classes of line bundles can be fixed by the methods of [**BC12b**]. It should be pointed out that the canonical representatives of Schubert classes in Chow–Witt rings further below arise from choices of orientations relative to $\mathscr{O}_{\mathrm{Gr}(k,n)}$ or $\det \mathscr{S}_k^\vee$ so that most of the time we use preferred explicit representatives for the isomorphisms classes in $\mathrm{Pic}(\mathrm{Gr}(k,n))/2$.

The first relevant statement is the following, cf. [**Wen18**, Theorem 5.10].

PROPOSITION 2.2. *Let* F *be a perfect field of characteristic* $\neq 2$*, and let* $1 \leq k < n$*. Then there is a cartesian square of* $\mathbb{Z} \oplus \mathbb{Z}/2\mathbb{Z}$*-graded* $\mathrm{GW}(F)$*-algebras:*

$$\begin{array}{ccc}
\widetilde{\mathrm{CH}}^\bullet(\mathrm{Gr}(k,n), \mathscr{O} \oplus \det \mathscr{S}_k^\vee) & \longrightarrow & \ker \partial_{\mathscr{O}} \oplus \ker \partial_{\det \mathscr{S}_k^\vee} \\
\downarrow & & \downarrow \scriptstyle{\bmod 2} \\
\mathrm{H}^\bullet(\mathrm{Gr}(k,n), \mathbf{I}^\bullet \oplus \mathbf{I}^\bullet(\det \mathscr{S}_k^\vee)) & \xrightarrow[\rho]{} & \mathrm{Ch}^\bullet(\mathrm{Gr}(k,n))^{\oplus 2},
\end{array}$$

where

$$\partial_{\mathscr{L}}\colon \mathrm{CH}^\bullet(\mathrm{Gr}(k,n)) \to \mathrm{Ch}^\bullet(\mathrm{Gr}(k,n)) \xrightarrow{\beta_{\mathscr{L}}} \mathrm{H}^{\bullet+1}(\mathrm{Gr}(k,n), \mathbf{I}^{\bullet+1}(\mathscr{L}))$$

is the (twisted) integral Bockstein map.

As a direct consequence, computations in the Chow–Witt ring of $\mathrm{Gr}(k,n)$ can be done by separate computations in the Chow ring (which usually are classical and well-known, cf. [**EH17**]) and the $\mathbf{I}^\bullet$-cohomology ring. For this reason, we will focus on computations in the $\mathbf{I}^\bullet$-cohomology from now on.

We next formulate the relevant statements describing the $\mathbf{I}^\bullet$-cohomology of the Grassmannians, cf. [**Wen18**, Theorems 5.6, 5.8, 6.7]. First, we have a result which essentially allows to completely determine the torsion in the cohomology:

PROPOSITION 2.3. *Let F be a perfect field of characteristic $\neq 2$, and let $1 \leq k < n$.*

(1) *The image of the Bockstein morphism*

$$\beta_{\mathscr{L}}\colon \mathrm{Ch}^\bullet(\mathrm{Gr}(k,n)) \to \mathrm{H}^{\bullet+1}(\mathrm{Gr}(k,n), \mathbf{I}^{\bullet+1}(\mathscr{L}))$$

coincides exactly with the $\mathrm{I}(F)$-torsion. The images of the Bockstein morphisms form an ideal in $\mathrm{H}^\bullet(\mathrm{Gr}(k,n), \mathbf{I}^\bullet \oplus \mathbf{I}^\bullet(\det \mathscr{S}_k))$.

(2) *The reduction morphism $\rho_{\mathscr{L}}\colon \mathrm{H}^\bullet(\mathrm{Gr}(k,n), \mathbf{I}^\bullet(\mathscr{L})) \to \mathrm{Ch}^\bullet(\mathrm{Gr}(k,n))$ is injective on the image of $\beta_{\mathscr{L}}$, and $\rho_{\mathscr{L}} \circ \beta_{\mathscr{L}}(x) = \mathrm{Sq}^2_{\mathscr{L}}(x)$. In particular, products of Bockstein classes $\beta_{\mathscr{L}}(x)$ with other classes can be determined by computing the products of the reductions in $\mathrm{Ch}^\bullet(\mathrm{Gr}(k,n))$.*

(3) *Products in $\mathrm{Ch}^\bullet(\mathrm{Gr}(k,n))$ with elements $\mathrm{Sq}^2_{\mathscr{L}}(x)$ are computed using that the Steenrod square $\mathrm{Sq}^2_{\mathscr{O}}$ is a derivation, $\mathrm{Sq}^2_{\det}(x) = \overline{\mathrm{c}}_1 \cdot x + \mathrm{Sq}^2_{\mathscr{O}}(x)$ and the Wu formula:*

$$\mathrm{Sq}^2_{\mathscr{O}}\colon \overline{\mathrm{c}}_j \mapsto \overline{\mathrm{c}}_1\overline{\mathrm{c}}_j + (j-1)\overline{\mathrm{c}}_{j+1}; \qquad \mathrm{Sq}^2_{\det}\colon \overline{\mathrm{c}}_j \mapsto (j-1)\overline{\mathrm{c}}_{j+1}.$$

Next, to study the $\mathrm{W}(F)$-torsionfree quotient of the cohomology, there is a natural exact sequence

$$\mathrm{Ch}^j(X) \xrightarrow{\beta_{\mathscr{L}}} \mathrm{H}^{j+1}(X, \mathbf{I}^{j+1}(\mathscr{L})) \to \mathrm{H}^{j+1}(X, \mathbf{W}(\mathscr{L})) \to 0$$

by means of which the cokernel of $\beta_{\mathscr{L}}$ can be identified with the cohomology ring $\mathrm{H}^\bullet(\mathrm{Gr}(k,n), \mathbf{W}(\mathscr{L}))$ of the Grassmannians with coefficients in the sheaf of Witt groups $\mathbf{W}$. It has a description completely analogous to the integral cohomology mod torsion of the real Grassmannians. Some of these formulas already appeared in the computations of SL-oriented theories (in particular Witt groups) of Grassmannians in [**Ana15**]. The exact relation with the computations of twisted Witt groups of Grassmannians in [**BC12a**] will be discussed in Section 8.3.

PROPOSITION 2.4. *Let F be a perfect field of characteristic $\neq 2$ and let $1 \leq k < n$. The total $\mathbf{W}$-cohomology ring $\bigoplus_{i,\mathscr{L}} \mathrm{H}^i(\mathrm{Gr}(k,n), \mathbf{W}(\mathscr{L}))$ has the following presentation, as commutative $\mathbb{Z} \oplus \mathrm{Pic}(\mathrm{Gr}(k,n))/2$-graded $\mathrm{W}(F)$-algebra:*

(1) *For k, n even,*

$$\frac{\mathrm{W}(F)[\mathrm{p}_2, \dots, \mathrm{p}_{k-2}, \mathrm{e}_k, \mathrm{p}_2^\perp, \dots, \mathrm{p}_{n-k-2}^\perp, \mathrm{e}_{n-k}^\perp]}{(\mathrm{p} \cdot \mathrm{p}^\perp = 1, \mathrm{e}_k \cdot \mathrm{e}_{n-k}^\perp = 0)}$$

where $\mathrm{p} = \mathrm{p}_2 + \cdots + \mathrm{p}_{k-2}$ and $\mathrm{p}^\perp = \mathrm{p}_2^\perp + \cdots + \mathrm{p}_{n-k-2}^\perp$ denote the appropriate total Pontryagin classes.

(2) *If n is odd,*

$$\begin{cases} \dfrac{\mathrm{W}(F)[\mathrm{p}_2, \dots, \mathrm{p}_{k-2}, \mathrm{e}_k, \mathrm{p}_2^\perp, \dots, \mathrm{p}_{n-k-1}^\perp]}{(\mathrm{p} \cdot \mathrm{p}^\perp = 1)} & k \text{ even} \\ \dfrac{\mathrm{W}(F)[\mathrm{p}_2, \dots, \mathrm{p}_{k-1}, \mathrm{p}_2^\perp, \dots, \mathrm{p}_{n-k-2}^\perp, \mathrm{e}_{n-k}^\perp]}{(\mathrm{p} \cdot \mathrm{p}^\perp = 1)} & k \text{ odd} \end{cases}$$

(3) *For $k, n-k$ odd,*

$$\frac{\mathrm{W}(F)[\mathrm{p}_2,\ldots,\mathrm{p}_{k-1},\mathrm{p}_2^\perp,\ldots,\mathrm{p}_{n-k-1}^\perp]}{(\mathrm{p}\cdot\mathrm{p}^\perp=1)}\otimes\bigwedge[R]$$

Here the bidegrees of the even Pontryagin classes p_{2i} are $(4i,0)$, the bidegrees of the Euler classes e_k and $\mathrm{e}_{n-k}^\perp$ are $(k,1)$ and $(n-k,1)$, respectively, and the class R in the last case has bidegree $(n-1,0)$.

The above statements on torsion (Proposition 2.3) and the torsion-free quotient (Proposition 2.4) can be combined into a description of the $\mathbf{I}^\bullet$-cohomology $\mathrm{H}^j(\mathrm{Gr}(k,n),\mathbf{I}^j(\mathscr{L}))$: it is a direct sum of the torsion part $\mathrm{Im}\,\beta_{\mathscr{L}}$ and the torsion-free quotient $\mathrm{H}^j(\mathrm{Gr}(k,n),\mathbf{W}(\mathscr{L}))$. With the above notation, we can also describe the reduction morphism relevant for computations with the fiber square of Proposition 2.2.

PROPOSITION 2.5. *The reduction morphism*

$$\rho_{\mathscr{L}}\colon\ \mathrm{H}^\bullet(\mathrm{Gr}(k,n),\mathbf{I}^\bullet(\mathscr{L}))\to\mathrm{Ch}^\bullet(\mathrm{Gr}(k,n))$$

is given by

$$\begin{array}{rcl}
\mathrm{p}_{2i}^{(\perp)} & \mapsto & (\overline{\mathrm{c}}_{2i}^{(\perp)})^2,\\
\beta_{\mathscr{L}}(x) & \mapsto & \mathrm{Sq}^2_{\mathscr{L}}(x),\\
\mathrm{e}_k & \mapsto & \overline{\mathrm{c}}_k,\\
\mathrm{e}_{n-k}^\perp & \mapsto & \overline{\mathrm{c}}_{n-k}^\perp,\\
\mathrm{R} & \mapsto & \overline{\mathrm{c}}_{k-1}\overline{\mathrm{c}}_{n-k}^\perp=\overline{\mathrm{c}}_k\overline{\mathrm{c}}_{n-k-1}^\perp.
\end{array}$$

REMARK 2.6. *The Grassmannian $\mathrm{Gr}(k,n)$ has dimension $k(n-k)$. From the above presentation, we have*

$$\mathrm{H}^d(\mathrm{Gr}(k,n),\mathbf{I}^d)\cong\begin{cases}\mathrm{W}(F) & n \text{ even}\\ \mathbb{Z}/2\mathbb{Z} & n \text{ odd}\end{cases}$$

For twisted coefficients $\mathbf{I}^d(\det(\mathscr{S}_k))$, the situation is exactly reversed. This is a reflection of the classical fact that the real Grassmannian $\mathrm{Gr}_k(\mathbb{R}^n)$ is orientable if and only if n is even.

It follows that the image of $\beta_{\mathscr{O}}$ in $\mathrm{H}^d(\mathrm{Gr}(k,n),\mathbf{I}^d)$ is trivial for n even. This has direct consequences for products of elements which land in top degree (which is the relevant case for Schubert calculus): the factors of a product can be changed by elements from the image of β without affecting the product, and a product with an $\mathrm{I}(F)$-torsion class is always trivial.

3. Multiplicative structure via even Young diagrams

Now we want to describe the multiplicative structure of the cohomology ring $\bigoplus_{j,\mathscr{L}}\mathrm{H}^j(\mathrm{Gr}(k,n),\mathbf{I}^j(\mathscr{L}))$ more explicitly. From Proposition 2.3, this cohomology ring splits additively as the direct sum of the torsion part, given by the image of

$$\beta_{\mathscr{L}}\colon\ \mathrm{Ch}^\bullet(\mathrm{Gr}(k,n))\to\mathrm{H}^\bullet(\mathrm{Gr}(k,n),\mathbf{I}^\bullet(\mathscr{L})),$$

and the cokernel of β, which coincides with $\mathrm{H}^\bullet(\mathrm{Gr}(k,n),\mathbf{W}\oplus\mathbf{W}(\det))$. Moreover, the image of the Bockstein maps is in fact an ideal. Hence, to determine the multiplicative structure on $\mathbf{I}^\bullet$-cohomology, it suffices to determine the multiplicative structure on $\mathrm{H}^\bullet(\mathrm{Gr}(k,n),\mathbf{W}\oplus\mathbf{W}(\det))$ and the image of $\beta_{\mathscr{L}}$ separately.

The torsion part will be discussed in the next section, for now we will focus on the multiplicative structure of **W**-cohomology. Essentially, the multiplication in this part of the $\mathbf{I}^\bullet$-cohomology ring can be understood in terms of diagrammatical calculations with the even Young diagrams of [**BC12a**], similar to the computations with Young diagrams which describe computations in Chow rings of Grassmannians. This provides analogues of some relevant pieces of classical Schubert calculus: the Pieri and Giambelli formulas and the Littlewood–Richardson coefficients.

3.1. Young diagrams and multiplication in Chow rings. We shortly recall how Young diagrams for partitions are used in the description of Chow rings of Grassmannians. This is a very well-known subject which is discussed in detail e.g. in [**EH17**].[1] Similar to the discussion in Section 2, the Chow ring $\mathrm{CH}^\bullet(\mathrm{Gr}(k,n))$ is generated by the Chern classes $\mathrm{c}_1,\dots,\mathrm{c}_k$ of the tautological subbundle $\mathscr{S}_k$ and the Chern classes $\mathrm{c}_1^\perp,\dots,\mathrm{c}_{n-k}^\perp$ of the tautological quotient bundle $\mathscr{Q}_{n-k}$, with the Whitney sum formula as defining relation:

$$\mathrm{CH}^\bullet(\mathrm{Gr}(k,n)) \cong \mathbb{Z}[\mathrm{c}_1,\dots,\mathrm{c}_k,\mathrm{c}_1^\perp,\dots,\mathrm{c}_{n-k}^\perp]/(\mathrm{c}\cdot\mathrm{c}^\perp = 1).$$

Fix a complete flag $\mathcal{F}\colon 0 = V_0 \subset V_1 \subset \cdots \subset V_{n-1} \subset V_n = V$, $\dim V_i = i$ of subspaces of $V = F^n$. For a partition $\underline{a} = (a_1,\dots,a_k)$ with $n-k \geq a_1 \geq a_2 \geq \cdots \geq a_k \geq 0$, there is an associated *Schubert variety*

$$\Sigma_{\underline{a}}(\mathcal{F}) = \{U \in \mathrm{Gr}(k,V) \mid \forall i\colon \dim(V_{n-k+i-a_i} \cap U) \geq i\} \subseteq \mathrm{Gr}(k,V)$$

of dimension $\dim \Sigma_{\underline{a}}(\mathcal{F}) = \sum a_i =: |\underline{a}|$. Partitions can be visualized by Young diagrams, where the partition $\underline{a} = (a_1,\dots,a_k)$ corresponds to k left-aligned rows of boxes, with a_i boxes in the i-th row. The corresponding cycle class $\sigma_{\underline{a}} = [\Sigma_{\underline{a}}] \in \mathrm{Ch}^{|a|}(\mathrm{Gr}(k,n))$ is independent of the choice of flag $\mathcal{F}$. The partitions $(a_1,\dots,a_k)$ with $n-k \geq a_1 \geq a_2 \geq \cdots \geq a_k \geq 0$ correspond to Young diagrams fitting into a $k \times (n-k)$ rectangle, and they form an additive basis of $\mathrm{CH}^\bullet(\mathrm{Gr}(k,n))$. There are two types of *special Schubert classes*, the classes σ_{1^i} which correspond to a single column of i boxes correspond to the Chern classes c_i, and the classes σ_j with a single row of j boxes correspond to the Chern classes $\mathrm{c}_j^\perp$.

With the above additive basis, the multiplicative structure of the Chow ring $\mathrm{CH}^\bullet(\mathrm{Gr}(k,n))$ can be described by the following classical formulas:

(1) The *Pieri formula* describes multiplication with special Schubert classes: for a Schubert class $\sigma_{\underline{a}}$ and any integer b, we have

$$\sigma_b \cdot \sigma_{\underline{a}} = \sum_{\substack{|c|=|a|+b \\ a_i \leq c_i \leq a_{i-1} \forall i}} \sigma_c \qquad \text{and} \qquad \sigma_{1^b} \cdot \sigma_{\underline{a}} = \sum_{\substack{|c|=|a|+b \\ a_i \leq c_i \leq a_i+1 \forall i}} \sigma_c$$

Diagrammatically, the product is given by the sum over all Young diagrams (fitting into a $k \times (n-k)$-rectangle) which are obtained from the Young diagram for $\underline{a}$ by adding a boxes with at most one box added to each column (in case σ_b) resp. to each row (in case σ_{1^b}).

[1] Most of the notation for Schubert calculus in Chow rings will be following that reference.

(2) Arbitrary Schubert classes for partitions $\underline{a} = (a_1, \ldots, a_k)$ can be expressed in terms of special Schubert classes via the *Giambelli formula*:

$$\sigma_{\underline{a}} = \begin{vmatrix} \sigma_{a_1} & \sigma_{a_1+1} & \sigma_{a_1+2} & \cdots & \sigma_{a_1+k-1} \\ \sigma_{a_2-1} & \sigma_{a_2} & \sigma_{a_2+1} & \cdots & \sigma_{a_2+k-2} \\ \sigma_{a_3-2} & \sigma_{a_3-1} & \sigma_{a_3} & \cdots & \sigma_{a_3+k-3} \\ \vdots & \vdots & \vdots & \ddots & \vdots \\ \sigma_{a_k-k+1} & \sigma_{a_k-k+2} & \sigma_{a_k-k+3} & \cdots & \sigma_{a_k} \end{vmatrix}.$$

(3) In a slightly different formulation, if $\sigma_{\underline{a}}$ and $\sigma_{\underline{b}}$ are Schubert classes, then the product is given by

$$\sigma_{\underline{a}} \cdot \sigma_{\underline{b}} = \sum_{|c|=|a|+|b|} \gamma_{\underline{a},\underline{b};\underline{c}} \sigma_{\underline{c}}$$

where $\gamma_{\underline{a},\underline{b};\underline{c}} = \deg(\sigma_{\underline{a}}\sigma_{\underline{b}}\sigma_{\underline{c}^*})$ are the *Littlewood–Richardson coefficients.*

3.2. Chow ring vs. W-cohomology ring. Now we want to set up a similar picture for the $\mathbf{W}$-cohomology ring, i.e., the torsion-free quotient of $\mathbf{I}^\bullet$-cohomology. To understand the multiplicative structure of $\mathbf{W}$-cohomology we will make use of a ring homomorphism from a suitable Chow ring. The following result is a version for $\mathbf{W}$-cohomology of what appears to be a folklore result concerning the rational cohomology of the real Grassmannians $\mathrm{Gr}_k(\mathbb{R}^n)$. The result identifies the trivial duality part of $\mathbf{W}$-cohomology in terms of the Chow ring of a different Grassmannian via a morphism which multiplies degrees by 4. I learned about this from [**FM16**, Section 3] as well as the MathOverflow question 119273 "Real Schubert calculus" by László Fehér.

PROPOSITION 3.1. *Let $1 \le k < n$ be natural numbers.*

(1) *If $k(n-k)$ is even, then mapping $\mathrm{c}_i \mapsto \mathrm{p}_{2i}, \mathrm{c}_j^\perp \mapsto \mathrm{p}_{2j}^\perp$ induces a natural $\mathrm{W}(F)$-algebra isomorphism*

$$\omega\colon \mathrm{CH}^\bullet(\mathrm{Gr}(\lfloor k/2\rfloor, \lfloor n/2\rfloor)) \otimes_{\mathbb{Z}} \mathrm{W}(F) \to \mathrm{H}^{4\bullet}(\mathrm{Gr}(k,n), \mathbf{W}).$$

(2) *If $k(n-k)$ is odd, then mapping $\mathrm{c}_i \mapsto \mathrm{p}_{2i}, \mathrm{c}_j^\perp \mapsto \mathrm{p}_{2j}^\perp$ and $R \to \mathrm{R}$ induces a natural $\mathrm{W}(F)$-algebra isomorphism*

$$\omega\colon \mathrm{CH}^\bullet(\mathrm{Gr}(\lfloor k/2\rfloor, n/2)) \otimes_{\mathbb{Z}} \mathrm{W}(F)[R]/(R^2) \to \mathrm{H}^{4\bullet}(\mathrm{Gr}(k,n), \mathbf{W}).$$

PROOF. In the case where $k(n-k)$ is even, parts (1) and (2) of Proposition 2.4 imply that $\mathbf{W}$-cohomology (with trivial duality) has the presentation

$$\mathrm{W}(F)[\mathrm{p}_2, \ldots, \mathrm{p}_{2\lfloor k/2\rfloor}, \mathrm{p}_2^\perp, \ldots, \mathrm{p}^\perp_{2(\lfloor n/2\rfloor - \lfloor k/2\rfloor)\rfloor}]/(\mathrm{p}\cdot\mathrm{p}^\perp = 1).$$

If we replace $\mathrm{p}_{2i}^{(\perp)}$ by $\mathrm{c}_i^{(\perp)}$ and the coefficient ring $\mathrm{W}(F)$ by $\mathbb{Z}$, we get exactly the well-known presentation of $\mathrm{CH}^\bullet(\mathrm{Gr}(\lfloor k/2\rfloor, \lfloor n/2\rfloor))$. The existence of the natural ring isomorphism is then immediately clear.

Similarly, in the second case, part (3) of Proposition 2.4 implies that the $\mathbf{W}$-cohomology (with trivial duality) is an exterior algebra over the Chow ring of the appropriate Grassmannian, with coefficients extended to $\mathrm{W}(F)$. □

REMARK 3.2. *Actually, the assignment* $\mathrm{c}_i^{(\perp)} \mapsto \mathrm{p}_{2i}^{(\perp)}$ *lifts to a* $\mathrm{W}(F)$*-linear map*

$$\tilde{\omega}\colon \mathrm{CH}^\bullet(\mathrm{Gr}(\lfloor k/2\rfloor, \lfloor n/2\rfloor)) \otimes_{\mathbb{Z}} \mathrm{W}(F) \to \bigoplus_j \mathrm{H}^j(\mathrm{Gr}(k,n), \mathbf{I}^j).$$

Note, however, that this morphism will not respect the ring structures. The relations defining the Chow ring are encoded in the Whitney sum formula $\mathrm{c}\cdot\mathrm{c}^\perp = 1$*. The image of this relation under the above assignment is*

$$(1+\mathrm{p}_2+\cdots+\mathrm{p}_{2\lfloor k/2\rfloor})\cdot(1+\mathrm{p}_2^\perp+\cdots+\mathrm{p}_{2\lfloor (n-k)/2\rfloor}^\perp) = 1$$

This relation doesn't hold in the $\mathbf{I}^\bullet$*-cohomology ring. Instead, we have the Whitney sum formula* $\mathrm{p}\cdot\mathrm{p}^\perp = 1$*, which additionally contains all the odd Pontryagin classes (which are in the image of* $\beta_{\mathscr{O}}$*). The above assignment fails to be a ring homomorphism because the image of the Whitney sum formula from the Chow ring will be a non-trivial* $\mathrm{I}(F)$*-torsion element in* $\mathbf{I}^\bullet$*-cohomology.*

The next step is to identify the composition of $\tilde{\omega}$ with the reduction morphism $\mathrm{H}^j(\mathrm{Gr}(k,n),\mathbf{I}^j) \to \mathrm{Ch}^j(\mathrm{Gr}(k,n))$. It actually suffices to identify this composition modulo images of torsion classes. Using that, it will be possible to identify the composition of reduction and $\tilde{\omega}$ with the following morphism given by doubling partitions.

DEFINITION 3.3. *We denote by*

$$\gamma\colon \mathrm{CH}^\bullet(\mathrm{Gr}(\lfloor k/2\rfloor, \lfloor n/2\rfloor)) \to \mathrm{CH}^{4\bullet}(\mathrm{Gr}(k,n))$$

the morphism of abelian groups which maps a class corresponding to a partition $(a_1, a_2, \ldots, a_r)$ *to the class corresponding to the doubled partition*

$$(2a_1, 2a_1, 2a_2, 2a_2, \ldots, 2a_r, 2a_r).$$

REMARK 3.4. *In terms of Young diagrams, the doubling of partitions corresponds to replacing every box of the Young diagram by a* (2×2)*-square of boxes*

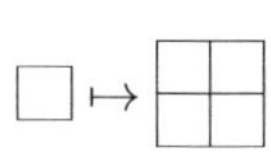

REMARK 3.5. *The map* $\gamma\colon \mathrm{CH}^\bullet(\mathrm{Gr}(k,n)) \to \mathrm{CH}^{4\bullet}(\mathrm{Gr}(2k,2n))$ *is only a homomorphism of abelian groups. It is not compatible with the ring structures. The simplest example is as follows. In* $\mathrm{Gr}(2,4)$*, we have* $\square^2 = \square\square + \begin{smallmatrix}\square\\\square\end{smallmatrix}$*. However, in* $\mathrm{Gr}(4,8)$*, we have*

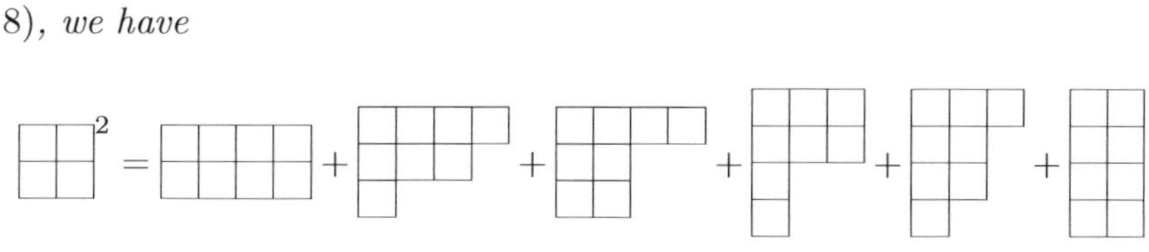

Only the first and last summand are doubled partitions, the presence of the other classes implies that partition doubling doesn't induce a ring homomorphism (even after reduction mod 2).

PROPOSITION 3.6. *Let $1 \leq k < n$ be a natural numbers. Then the difference of the two compositions*

$$\mathrm{CH}^j(\mathrm{Gr}(\lfloor k/2 \rfloor, \lfloor n/2 \rfloor)) \xrightarrow{\tilde{\omega}} \mathrm{H}^{4j}(\mathrm{Gr}(k,n), \mathbf{I}^{4j}) \xrightarrow{\rho} \mathrm{Ch}^{4j}(\mathrm{Gr}(k,n))$$

$$\mathrm{CH}^j(\mathrm{Gr}(\lfloor k/2 \rfloor, \lfloor n/2 \rfloor)) \xrightarrow{\gamma} \mathrm{CH}^{4j}(\mathrm{Gr}(k,n)) \xrightarrow{\bmod 2} \mathrm{Ch}^{4j}(\mathrm{Gr}(k,n))$$

lands in the image of $\mathrm{Sq}^2 \colon \mathrm{Ch}^\bullet(\mathrm{Gr}(k,n)) \to \mathrm{Ch}^{\bullet+1}(\mathrm{Gr}(k,n))$.

PROOF. By the Giambelli formula, we have

$$\sigma_{2i,2i} = \begin{vmatrix} \sigma_{2i} & \sigma_{2i+1} \\ \sigma_{2i-1} & \sigma_{2i} \end{vmatrix} = \sigma_{2i}^2 - \sigma_{2i+1}\sigma_{2i-1}.$$

The classes σ_{2i+1} and σ_{2i-1} correspond to odd Chern classes in the Chow ring, and consequently they have lifts $\beta(\overline{\mathrm{c}}_{2i})$ and $\beta(\overline{\mathrm{c}}_{2i-2})$ in $\mathrm{H}^\bullet(\mathrm{Gr}(k,n), \mathbf{I}^\bullet)$, cf. Proposition 2.5. Note that the odd Chern classes cannot lift to non-torsion classes since all non-torsion classes live in even degrees, by Proposition 2.4. The reduction homomorphism $\mathrm{H}^{4i}(\mathrm{Gr}(k,n), \mathbf{I}^{4i}) \to \mathrm{Ch}^{4i}(\mathrm{Gr}(k,n))$ maps the Pontryagin class p_{2i} to $\overline{\mathrm{c}}_{2i}^2$. Combined with the previous observation, we find that the reduction of $\mathrm{p}_{2i} - \beta(\overline{\mathrm{c}}_{2i-2})\beta(\overline{\mathrm{c}}_{2i})$ is exactly the class $\sigma_{2i,2i}$ of the doubled partition and this proves the claim. □

REMARK 3.7. *In particular (up to Bockstein classes) the Pontryagin class $\mathrm{p}_{2i}^{(\perp)}$ is actually a lift of the double of the partition corresponding to the Chern class $\overline{\mathrm{c}}_i^{(\perp)}$ in the mod 2 Chow ring. Combining Proposition 3.1 and 3.6, we get an additive basis of classes in $\mathrm{H}^\bullet(\mathrm{Gr}(k,n), \mathbf{W})$ whose "underlying Schubert varieties" corrrespond to doubled partitions. We will see later in Section 8 that the Pontryagin classes can be interpreted as Schubert cycle equipped with an additional choice of orientation.*

3.3. Multiplication of even Young diagrams. Now we can discuss the multiplicative structure in $\mathbf{W}$-cohomology from a diagrammatical perspective. For this, we first recall definitions of even Young diagrams from [**BC12a**, Section 2] and then explain how the previous degree-doubling morphism describes multiplication of classes in terms of even Young diagrams.

DEFINITION 3.8. *Let $1 \leq k < n$.*

(1) *A Young diagram in $k \times (n-k)$-frame is a k-tuple $\Lambda = (\Lambda_1, \Lambda_2, \ldots, \Lambda_k)$ of integers such that $n-k \geq \Lambda_1 \geq \Lambda_2 \geq \cdots \geq \Lambda_k \geq 0$.*
(2) *Such a diagram is called* even *if all segments of the boundary which are strictly inside the $k \times (n-k)$-frame have even length. More precisely, the k-tuple Λ is of the form*

$$\underbrace{\Lambda_1 = \cdots = \Lambda_{d_1}}_{d_1 \text{ terms}} > \underbrace{\Lambda_{d_1+1} = \cdots = \Lambda_{d_2}}_{d_2 - d_1 \text{ terms}} > \cdots > \underbrace{\Lambda_{d_{p-1}+1} = \cdots = \Lambda_{d_p}}_{d_p - d_{p-1} \text{ terms}} = \Lambda_k$$

and consequently, we obtain a sequence of integers $(d_1, \ldots, d_p)$. There is another sequence of integers $(e_1, \ldots, e_p)$ given by $e_i = n - k - \Lambda_{d_i}$. Now a diagram Λ is called even, if the following conditions are satisfied:
   (a) *$d_{i+1} - d_i$ is even for all $i = 1, \ldots, p-2$,*
   (b) *$e_{i+1} - e_i$ is even for all $i = 1, \ldots, p-1$,*
   (c) *when $0 < e_1 < e$ further require that d_1 is even,*
   (d) *when $0 < e_p < e$ further require that $d_p - d_{p-1}$ is even*

(3) *We call a Young diagram* completely even *if all the segments of the boundary have even length (even those which lie on the frame). In the above notation, this corresponds to requiring that d_1 and $d_p - d_{p-1}$ are even (irrespective of the parities of e_1 and e_p).*

REMARK 3.9. *The definition of the tuples $(d_1, \dots, d_p)$ and $(e_1, \dots, e_p)$ is illustrated in Figure* 1 *which is taken from* [**BC12a**, Figure 3].

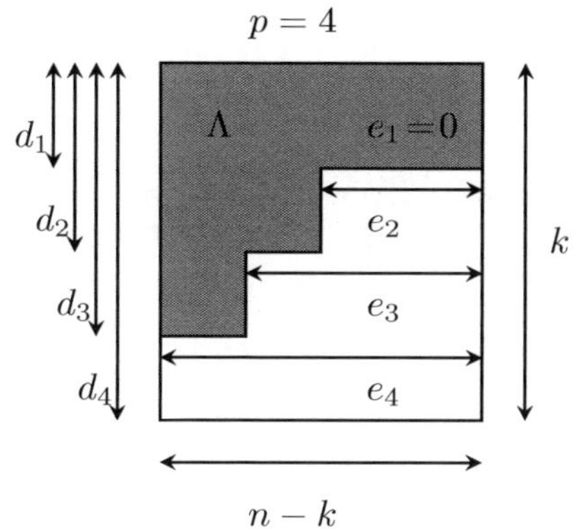

FIGURE 1. An example of a Young diagram in $k \times (n-k)$-frame with the relevant tuples $(d_1, \dots, d_4)$ and $(e_1, \dots, e_4)$.

EXAMPLE 3.10.

- *If k is even, then $\Lambda = \underbrace{(1, \dots, 1)}_{k \text{ times}}$ is an even Young diagram in $k \times (n-k)$-frame. This is the Young diagram for the top Chern class c_k of the tautological rank k subbundle on $\mathrm{Gr}(k, n)$.*
- *If $n - k$ is even, then $\Lambda = (n-k, 0, \dots, 0)$ is an even Young diagram in $k \times (n-k)$-frame. This is the Young diagram for the top Chern class $\mathrm{c}^\perp_{n-k}$ of the tautological rank $(n-k)$ quotient bundle on $\mathrm{Gr}(k, n)$.*
- *If both k and $n-k$ are odd, then $\Lambda = (n-k, 1, \dots, 1)$ is an even Young diagram in $k \times (n-k)$-frame. This is the Young diagram for the product $\mathrm{c}_{k-1}\,\mathrm{c}^\perp_{n-k} = \mathrm{c}_k\,\mathrm{c}^\perp_{n-k-1}$.*

□

DEFINITION 3.11. *Fix $1 \leq k < n$. We define the following map from the set of even Young diagrams in $k \times (n-k)$-frame to $\mathrm{H}^\bullet(\mathrm{Gr}(k, n), \mathbf{I}^\bullet \oplus \mathbf{I}^\bullet(\det))$:*

- *If the diagram is completely even, then it is the Young diagram for the double $\gamma(\underline{a})$ of some partition $\underline{a}$. Denoting by $\sigma_{\underline{a}}$ the Schubert class for the partition $\underline{a}$, the corresponding class in the $\mathbf{I}^n$-cohomology ring is given by $\tilde{\omega}(\sigma_{\underline{a}})$ from Remark* 3.2.
- *If the diagram is not completely even, then it is obtained from a completely even Young diagram by adding one of the diagrams from Example* 3.10. *The corresponding class is the product of the doubled class for the completely even part and the Euler class e_k in case (1), the Euler class $\mathrm{e}^\perp_{n-k}$ in case (2) and the class R in case (3).*

EXAMPLE 3.12. *The following Young diagram in $k \times 8$-frame with $k \geq 5$ is not completely even, since $d_1 = 1$ (and $d_4 - d_3$ is odd for even k):*

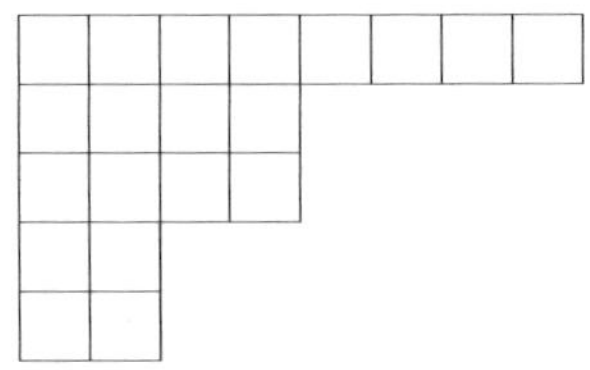

It is the product of $\mathrm{e}_8^\perp$*, corresponding to the first line, with a completely even Young diagram corresponding to the class* $\mathrm{p}_4^\perp \mathrm{p}_2^\perp - \mathrm{p}_6^\perp$*, cf. also Example* 3.20 □

REMARK 3.13. *We can exactly determine the appropriate Chow–Witt group that an even Young diagram $\Lambda = (\Lambda_1, \ldots, \Lambda_k)$ is mapped to: the cohomological degree is given by the number $\sum_i \Lambda_i$ of boxes in the diagram (this is called the area in* [**BC12a**]*). This follows easily: for a partition $\underline{a}$, the corresponding Schubert class $\sigma_{\underline{a}}$ is a class in $\mathrm{CH}^{|\underline{a}|}$, and its image under $\tilde{\omega}$ lands in cohomological degree $4|\underline{a}|$ which equals exactly the area of the doubled diagram $\gamma(\underline{a})$.*

The twist/half-perimeter $t(\Lambda)$ of an even Young diagram Λ, defined in [**BC12a**, Definition 4.3]*, also determines the twist for the $\mathbf{I}^\bullet$-cohomology: the class corresponding to Λ under the assignment in Definition* 3.11 *has coefficients in $\mathbf{I}^\bullet$ if $t(\Lambda) = 0$ and coefficients in $\mathbf{I}^\bullet(\mathrm{det})$ otherwise. In the case distinction of Definition* 3.11*, the Young diagram has odd half-perimeter if and only if it corresponds to a product of a completely even diagram with an Euler class.*

PROPOSITION 3.14. *The map of Definition* 3.11 *provides a bijection between the even Young diagrams in $k \times (n-k)$-frame and a $\mathrm{W}(F)$-basis of the total $\mathbf{W}$-cohomology ring $\mathrm{H}^\bullet(\mathrm{Gr}(k,n), \mathbf{W} \oplus \mathbf{W}(\mathrm{det}))$.*

PROOF. If $k(n-k)$ is even, the part with nontwisted coefficients follows from Proposition 3.1 and the fact that the ordinary Young diagrams provide an additive basis for the Chow ring. For the twisted coefficients, Proposition 2.4 implies that any class in $\mathrm{H}^\bullet(\mathrm{Gr}(k,n), \mathbf{W})$ is the product of a nontwisted class with an Euler class e_k or $\mathrm{e}_{n-k}^\perp$. Since $\mathrm{e}_k \cdot \mathrm{e}_{n-k}^\perp = 0$ only one Euler class factor can appear. These classes correspond to the even Young diagrams which are not completely even and have $t(\Lambda) = 1$. In particular, everything is in the image of the map from Definition 3.11. It can be checked from the presentation of Proposition 2.4 that different even Young diagrams with $t(\Lambda) = 1$ also correspond to different classes in $\mathrm{H}^\bullet(\mathrm{Gr}(k,n), \mathbf{W})$.

For the case where $k(n-k)$ is odd, Proposition 3.1 implies that an additive basis of $\mathrm{H}^\bullet(\mathrm{Gr}(k,n), \mathbf{W})$ is given by the images of an additive basis of the Chow ring, together with the images of products of such an additive basis with the class R. The former classes are again exactly the classes corresponding to completely even Young diagrams. The latter classes are even Young diagrams which are not completely even but have $t(\Lambda) = 0$, i.e., they are obtained from an even Young diagram by adding a maximal hook. Nontwisted coefficients don't appear in this case, and there are no even Young diagrams which are not of the above form. □

REMARK 3.15. *Analogous statements on additive bases for integral cohomology of real Grassmannians mod torsion are well-known, cf.* [**FK15**]*. It seems the classical topological result was first proved by Pontryagin* [**Pon47**].

Now we can formulate a complete description of the multiplication in the $\mathbf{W}$-cohomology rings of Grassmannians. To introduce appropriate notation, let $2 \le k < n$ and denote by $\mathfrak{S}_{\underline{a}}$ the image of the Schubert class $\sigma_{\underline{a}}$ under the homomorphism

$$\mathrm{CH}^\bullet(\mathrm{Gr}(\lfloor k/2\rfloor, \lfloor n/2\rfloor)) \to \mathrm{H}^{4\bullet}(\mathrm{Gr}(k,n), \mathbf{W})$$

of Proposition 3.1. By Proposition 3.14, every class in $\mathrm{H}^\bullet(\mathrm{Gr}(k,n), \mathbf{W})$ is a $\mathrm{W}(F)$-linear combination of classes of one of the four types $\mathfrak{S}_{\underline{a}}$, $\mathfrak{S}_{\underline{a}}\,\mathrm{e}_k$, $\mathfrak{S}_{\underline{a}}\,\mathrm{e}^\perp_{n-k}$ or $\mathfrak{S}_{\underline{a}}\,\mathrm{R}$. Therefore, we only need to describe the products of these four types of classes. Essentially, everything will be easily reduced to products of the form $\mathfrak{S}_{\underline{a}}\mathfrak{S}_{\underline{b}}$ which can be computed using the folklore homomorphism of Proposition 3.1.

THEOREM 3.16. *With the above notation, we have the following statements:*

(1) *The following table explains how to compute products of the four types in terms of products of the form $\mathfrak{S}_{\underline{a}}\mathfrak{S}_{\underline{b}}$:*

	$\mathfrak{S}_{\underline{b}}$	$\mathfrak{S}_{\underline{b}}\,\mathrm{e}_k$	$\mathfrak{S}_{\underline{b}}\,\mathrm{e}^\perp_{n-k}$	$\mathfrak{S}_{\underline{b}}\,\mathrm{R}$
$\mathfrak{S}_{\underline{a}}$	$\mathfrak{S}_{\underline{a}}\mathfrak{S}_{\underline{b}}$	$\mathfrak{S}_{\underline{a}}\mathfrak{S}_{\underline{b}}\,\mathrm{e}_k$	$\mathfrak{S}_{\underline{a}}\mathfrak{S}_{\underline{b}}\,\mathrm{e}^\perp_{n-k}$	$\mathfrak{S}_{\underline{a}}\mathfrak{S}_{\underline{b}}\,\mathrm{R}$
$\mathfrak{S}_{\underline{a}}\,\mathrm{e}_k$	$\mathfrak{S}_{\underline{a}}\mathfrak{S}_{\underline{b}}\,\mathrm{e}_k$	$\mathfrak{S}_{\underline{a}}\mathfrak{S}_{\underline{b}}\mathfrak{S}_{1^k}$	0	—
$\mathfrak{S}_{\underline{a}}\,\mathrm{e}^\perp_{n-k}$	$\mathfrak{S}_{\underline{a}}\mathfrak{S}_{\underline{b}}\,\mathrm{e}^\perp_{n-k}$	0	$\mathfrak{S}_{\underline{a}}\mathfrak{S}_{\underline{b}}\mathfrak{S}_{n-k}$	—
$\mathfrak{S}_{\underline{a}}\,\mathrm{R}$	$\mathfrak{S}_{\underline{a}}\mathfrak{S}_{\underline{b}}\,\mathrm{R}$	—	—	0

Note that the $\mathbf{W}$-cohomology rings in question are commutative so that the order of factors doesn't play a role in the above table. The combinations in the table marked by — are impossible: the Euler classes e_k and $\mathrm{e}^\perp_{n-k}$ require k and $n-k$ to be even, respectively, but the class R requires $k(n-k)$ to be odd.

(2) *There is an oriented version of the Pieri formula: for any Schubert class σ_a and any integer b, we have*

$$(\mathfrak{S}_b \cdot \mathfrak{S}_{\underline{a}}) = \sum_{\substack{|\underline{c}|=|\underline{a}|+b\\ a_i \le c_i \le a_{i-1} \forall i}} \mathfrak{S}_{\underline{c}}.$$

(3) *There is an oriented version of the Giambelli formula:*

$$\mathfrak{S}_{(a_1,a_2,\ldots,a_k)} = \begin{vmatrix} \mathfrak{S}_{a_1} & \mathfrak{S}_{a_1+1} & \mathfrak{S}_{a_1+2} & \cdots & \mathfrak{S}_{a_1+k-1} \\ \mathfrak{S}_{a_2-1} & \mathfrak{S}_{a_2} & \mathfrak{S}_{a_2+1} & \cdots & \mathfrak{S}_{a_2+k-2} \\ \mathfrak{S}_{a_3-2} & \mathfrak{S}_{a_3-1} & \mathfrak{S}_{a_3} & \cdots & \mathfrak{S}_{a_3+k-3} \\ \vdots & \vdots & \vdots & \ddots & \vdots \\ \mathfrak{S}_{a_k-k+1} & \mathfrak{S}_{a_k-k+2} & \mathfrak{S}_{a_k-k+3} & \cdots & \mathfrak{S}_{a_k} \end{vmatrix}.$$

(4) *The following describes the oriented Littlewood–Richardson coefficients for multiplication of classes $\mathfrak{S}_{\underline{a}}$ and $\mathfrak{S}_{\underline{b}}$: for any three Schubert classes $\sigma_{\underline{a}}$, $\sigma_{\underline{b}}$ and $\sigma_{\underline{c}}$, we have the following equality, where $\gamma_{\underline{a},\underline{b};\underline{c}} = \deg(\sigma_{\underline{a}}\sigma_{\underline{b}}\sigma^*_{\underline{c}})$ are the classical Littlewood–Richardson coefficients from the Chow ring:*

$$\mathfrak{S}_{\underline{a}} \cdot \mathfrak{S}_{\underline{b}} = \sum_{|\underline{c}|=|\underline{a}|+|\underline{b}|} \gamma_{\underline{a},\underline{b};\underline{c}}\langle 1\rangle \cdot \mathfrak{S}_{\underline{c}}.$$

PROOF. (1) The fact that all classes are $\mathrm{W}(F)$-linear combinations of classes appearing in the table follows from Proposition 2.4: the relations $\mathrm{e}_k\,\mathrm{e}^\perp_{n-k} = 0$, $\mathrm{e}_k^2 = \mathrm{p}_k$, $(\mathrm{e}^\perp_{n-k})^2 = \mathrm{p}^\perp_{n-k}$ and $\mathrm{R}^2 = 0$ imply that we always can reduce to classes appearing in the table. All the entries of the table describing the multiplication also follow from these relations.

The items (2), (3) and (4) are simply obtained by taking the classical Pieri and Giambelli formulas resp. Littlewood–Richardson coefficients and using the ring isomorphism of Proposition 3.1, and translate the result into even Young diagrams via Definition 3.11 and Proposition 3.14. □

REMARK 3.17. *In particular, while the coefficients describing the intersection products in* $\mathrm{H}^\bullet(\mathrm{Gr}(k,n),\mathbf{W})$ *are quadratic forms, they are always of the simple form* $m\langle 1\rangle$ *with* $m\in\mathbb{N}$. *Essentially, everything is defined over* $\operatorname{Spec}\mathbb{Z}$ *and there are not many unramified quadratic forms there.*

3.4. Examples. We discuss a couple of examples to illustrate the computational rules in Theorem 3.16.

EXAMPLE 3.18. *We discuss the ring structure of* $\mathrm{H}^\bullet(\mathrm{Gr}(5,10),\mathbf{W})$.[2] *Note that, since* $\dim\mathrm{Gr}(5,10)=25$ *is odd, there are no classes in cohomology with twisted coefficients. The additive generators have been described in* [**BC12a**, Figure 16, p. 641]. *The multiplicative structure is given as follows: the first row in Fig. 16, loc.cit., are the diagrams obtained by doubling Young diagrams for Schubert classes in* $\mathrm{Gr}(2,4)$. *Their products can be obtained via doubling of the corresponding computations in* $\mathrm{Gr}(2,4)$, *e.g.*

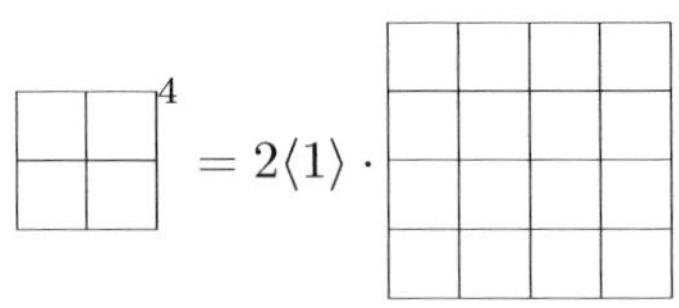

The second row in [**BC12a**, Figure 16] *contains the classes which are products of the degree 9 class* R *and a class obtained by doubling the Young diagram for a Schubert class in* $\mathrm{Gr}(2,4)$. *Since* $\mathrm{R}^2=0$, *products of classes from the second row are all zero. Products of a class from the first row with a class from the second row can be obtained as product of* R *with a product of the corresponding classes from the first row. For example,*

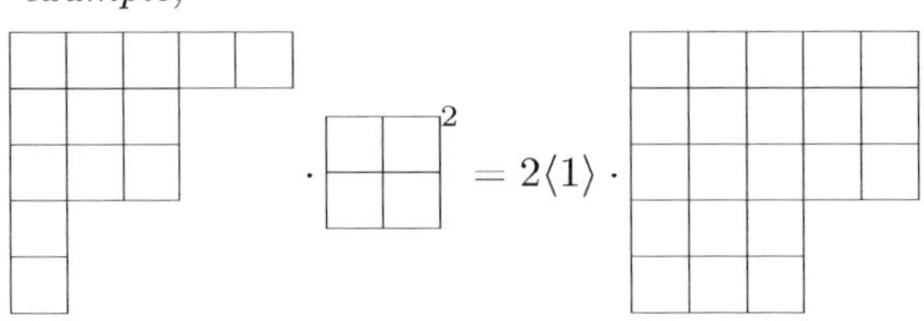

□

EXAMPLE 3.19. *The class* $\mathrm{p}_2\,\mathrm{e}_8^\perp$ *in* $\mathrm{Gr}(6,14)$ *corresponds to the following even Young diagram in* (6×8)*-frame:*

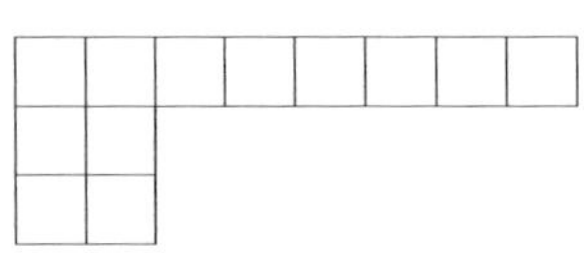

The 4-fold self-intersection of this class is $\mathrm{p}_2^4\,\mathrm{p}_8^2$. *It turns out that this equals the full* (6×8)*-rectangle with multiplicity 1: The class* p_8^2 *in* $\mathrm{Gr}(6,14)$ *is a* (4×8)*-rectangle*

[2] We will see later that this also describes the ring structure on the Witt group $\mathrm{W}^\bullet(\mathrm{Gr}(5,10))$.

and by the Pieri formula only the components of p_2^4 *which fit into a* (2×8)*-rectangle are relevant for the product* $\mathrm{p}_2^4\,\mathrm{p}_8^2$. *However, the only component of* p_2^4 *which fits into such rectangle is* p_8^2 *with multiplicity 1.* ☐

EXAMPLE 3.20. *An example of the oriented Giambelli formula is the following one: in the notation of Theorem 3.16, the doubled partition* $(2a, 2a, 2b, 2b)$ *is the class* $\mathfrak{S}_{a,b}$ *and consequently,*

$$\mathfrak{S}_{a,b} = \begin{vmatrix} \mathfrak{S}_a & \mathfrak{S}_{a+1} \\ \mathfrak{S}_{b-1} & \mathfrak{S}_b \end{vmatrix} = \mathfrak{S}_a\mathfrak{S}_b - \mathfrak{S}_{a+1}\mathfrak{S}_{b-1}.$$

The corresponding diagrammatic formula for the case $\mathfrak{S}_{2,1}$ *is the following, which is the obvious double of the usual Giambelli formula for the Schubert class* $\sigma_{2,1}$:

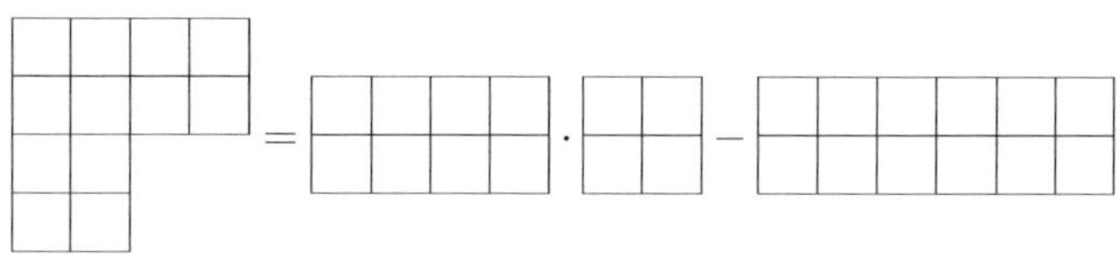

In the specific case of the class $\mathfrak{S}_{2,1}\,\mathrm{R}$ *in* $\mathrm{Gr}(5,10)$, *the formula would simplify because the second summand doesn't fit the* (5×5)*-frame:*

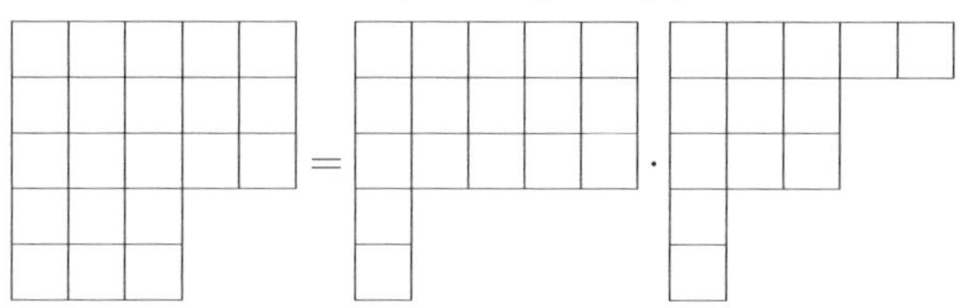

☐

4. Computations with torsion classes

In this section, we will extend the results on $\mathbf{W}$-cohomology rings from Section 3 to a discussion of the structure of the cohomology rings $\bigoplus_{j,\mathscr{L}} \mathrm{H}^j(\mathrm{Gr}(k,n), \mathbf{I}^j(\mathscr{L}))$. First, the natural reduction morphism

$$\bigoplus_{j,\mathscr{L}} \mathrm{H}^j(\mathrm{Gr}(k,n), \mathbf{I}^j(\mathscr{L})) \to \mathrm{Ch}^j(\mathrm{Gr}(k,n))$$

is not generally surjective and the obstruction is given by the Steenrod square $\mathrm{Sq}^2_{\mathscr{L}}$. So we first discuss a description of Steenrod squares in terms of Young diagrams, essentially a version of results of Lenart [**Len98**]. This allows to explicitly determine which classes in $\mathrm{Ch}^j(\mathrm{Gr}(k,n))$ lift to $\mathbf{I}^\bullet$-cohomology. The multiplication with torsion classes is then determined by reduction to the mod 2 Chow rings, and this completes the description of the ring structure of $\mathbf{I}^\bullet$-cohomology. Then we can discuss some examples of liftable and non-liftable classes, and examples of multiplication of torsion classes in terms of classical Schubert calculus.

4.1. Lifting of torsion classes and non-refinable problems. Recall that the natural projection morphisms

$$\widetilde{\mathrm{CH}}^j(\mathrm{Gr}(k,n), \mathscr{L}) \to \mathrm{CH}^j(\mathrm{Gr}(k,n)) \text{ resp. } \mathrm{H}^j(\mathrm{Gr}(k,n), \mathbf{I}^j(\mathscr{L})) \to \mathrm{Ch}^j(\mathrm{Gr}(k,n))$$

are not generally surjective. For a class $x \in \mathrm{Ch}^\bullet(\mathrm{Gr}(k,n))$, the obstruction to lifting to $\mathscr{L}$-twisted coefficients is given by the Steenrod square $\mathrm{Sq}^2_{\mathscr{L}}(x)$:

PROPOSITION 4.1. *A cycle* $\sigma \in \mathrm{CH}^i(\mathrm{Gr}(k,n))$ *lifts along the natural projection map* $\widetilde{\mathrm{CH}}^i(\mathrm{Gr}(k,n),\mathscr{L}) \to \mathrm{CH}^i(\mathrm{Gr}(k,n))$ *if and only if* $\overline{\sigma} \in \mathrm{Ch}^i(\mathrm{Gr}(k,n))$ *is in the kernel of* $\mathrm{Sq}^2_{\mathscr{L}}$.

PROOF. By the Bär sequence, cf. Proposition 2.2, a class lifts if and only if the class is in the kernel of the composition

$$\mathrm{CH}^i(\mathrm{Gr}(k,n)) \xrightarrow{\text{mod } 2} \mathrm{Ch}^i(\mathrm{Gr}(k,n)) \xrightarrow{\beta_{\mathscr{L}}} \mathrm{H}^{i+1}(\mathrm{Gr}(k,n),\mathbf{I}^{i+1}(\mathscr{L})).$$

By Proposition 2.3 (2), the reduction $\rho_{\mathscr{L}}$ is injective on the image of $\beta_{\mathscr{L}}$. Therefore, the class is in the kernel of the above composition if and only if its mod 2 reduction is in the kernel of $\mathrm{Sq}^2_{\mathscr{L}}$. □

The above result can be used in combination with Proposition 2.3 (3) to decide if classes lift: the Steenrod square $\mathrm{Sq}^2_{\mathscr{L}}$ can be computed using the Wu formula

$$\mathrm{Sq}^2(\overline{c}_j) = \overline{c}_1\overline{c}_j + (j-1)\overline{c}_{j+1}, \quad \mathrm{Sq}^2_{\det}(\overline{c}_j) = (j-1)\overline{c}_{j+1}.$$

Via the Giambelli formula, this can be used to compute arbitrary Steenrod squares in $\mathrm{Ch}^\bullet(\mathrm{Gr}(k,n))$. The following provides an easy diagrammatic description of the Steenrod squares:

THEOREM 4.2. *Fix* $1 \leq k < n$. *For a Young diagram* Λ, *fill the boxes with a checkerboard pattern, starting with a black box in the upper left corner.*

(1) *The ordinary Steenrod square* $\mathrm{Sq}^2_{\mathscr{O}}(\Lambda)$ *in* $\mathrm{Ch}^\bullet(\mathrm{Gr}(k,n))$ *is given as the sum of those Young diagrams with checkerboard pattern in* $k \times (n-k)$*-frame which can be obtained from* Λ *by adding a single white box.*
(2) *The twisted Steenrod square* $\mathrm{Sq}^2_{\det}(\Lambda)$ *in* $\mathrm{Ch}^\bullet(\mathrm{Gr}(k,n))$ *is given as the sum of those Young diagrams with checkerboard pattern in* $k \times (n-k)$*-frame which can be obtained from* Λ *by adding a single black box.*

PROOF. Since the Chow ring $\mathrm{Ch}^\bullet(\mathrm{Gr}(k,n))$ is generated by the Stiefel–Whitney classes $\overline{c}_1,\dots,\overline{c}_k$, the Steenrod squares $\mathrm{Sq}^2_{\mathscr{O}}$ and $\mathrm{Sq}^2_{\det}$ are determined by their values on $\overline{c}_i$ and the fact that $\mathrm{Sq}^2_{\mathscr{O}}$ is a derivation (resp. that $\mathrm{Sq}^2_{\det}$ has a derivation-like property). To prove the theorem, it suffices to show that the checkerboard rules agree with the Wu formula on the $\overline{c}_i$, and to show that the checkerboard rules have the appropriate multiplicative behaviour.

We first consider the Stiefel–Whitney classes $\overline{c}_i$, which correspond to the Young diagram with i boxes in a single column. The checkerboard pattern ends in a white box at the bottom if i is even, and a black box otherwise. To get a Young diagram, we can only add a box in the first row or at the bottom of the column. For the twisted Steenrod square, the box in the first row cannot be black, so the only possibility when we can add a black box (at the bottom of the column) would be when i is even or $i = k$. This corresponds exactly to the Wu formula $\mathrm{Sq}^2_{\det}(\overline{c}_i) = (i-1)\overline{c}_{i+1}$. For the non-twisted Steenrod square, we can always add a white box in the first row, and a white box at the bottom of the column can be added if and only if i is odd. The sum of these two possibilities corresponds exactly to $\overline{c}_1\overline{c}_i$ by the Pieri formula. Consequently, the checkerboard description agrees exactly with the Wu formula $\mathrm{Sq}^2_{\mathscr{O}}(\overline{c}_i) = \overline{c}_1\overline{c}_i + (i-1)\overline{c}_{i+1}$. This finishes the discussion of the Stiefel–Whitney classes $\overline{c}_i$, the Stiefel–Whitney classes $\overline{c}_i^\perp$ are handled similarly.

Now we show that the checkerboard rules have the right multiplicative behaviour: $\mathrm{Sq}^2_{\mathscr{O}}$ is a derivation and $\mathrm{Sq}^2_{\det}(a \cdot b) = a \cdot \mathrm{Sq}^2_{\det}(b) + b \cdot \mathrm{Sq}^2_{\mathscr{O}}(a)$. Actually, using the Giambelli formula, we only need to show that the derivation property is satisfied for products of an arbitrary Young diagram with one for the $\overline{c}_i$. The two multiplication statements for $\mathrm{Sq}^2_{\mathscr{O}}(\Lambda \cdot \overline{c}_i)$ and $\mathrm{Sq}^2_{\det}(\Lambda \cdot \overline{c}_i)$ are equivalent, because both sides of the equation differ by $\overline{c}_1 \cdot \Lambda \cdot \overline{c}_i$. Therefore, it suffices to prove the formula for $\mathrm{Sq}^2_{\det}$; this is easier, because the Wu formula for $\mathrm{Sq}^2_{\det}(\overline{c}_i)$ is the easier one. We only sketch the combinatorial argument: on the left-hand side, the diagrammatic description of $\mathrm{Sq}^2_{\det}(\Lambda \cdot \overline{c}_i)$ is obtained as sum of all possibilities of adding i boxes according to the Pieri formula, and then adding a black box. Note that the boxes added via the Pieri formula cannot be stacked vertically, but the last black box can be added below one of the boxes added before. On the right-hand side of the diagrammatic description, one of the contributions is

$$\Lambda \cdot \mathrm{Sq}^2_{\det}(\overline{c}_i) = \begin{cases} 0 & i \text{ odd} \\ \Lambda \cdot \overline{c}_{i+1} & i \text{ even} \end{cases}$$

and $\Lambda \cdot \overline{c}_{i+1}$ is obtained by adding a column of $i+1$ boxes to Λ such that no two boxes are in the same row. The other contribution on the right-hand side is $\mathrm{Sq}^2_{\mathscr{O}}(\Lambda) \cdot \overline{c}_i$ which is obtained by first adding a white square to Λ and then adding i boxes according to the Pieri formula.

If on the left-hand side (i boxes according to Pieri formula plus one black box), the black box is in a different row from all the i boxes added before, then this contribution is in the $\Lambda \cdot \overline{c}_{i+1}$-part of the right-hand side. If it is to the right of a box added before, the previous box must be white, and this contributes to the $\mathrm{Sq}^2_{\mathscr{O}}(\Lambda) \cdot \overline{c}_i$-part of the right-hand side. So we have the same contributions on either side, proving the required equality. □

REMARK 4.3. *Here is an alternative argument: we first consider the real Grassmannians $\mathrm{Gr}_k(\mathbb{R}^n)$. It is known, cf. e.g.* [**CK13**] *that the coefficients for the maps in the complex computing integral cellular cohomology are either* 0 *or* ± 2. *By definition of the Bockstein map, the Bockstein $\beta(x)$ of a class $x \in \mathrm{H}^j(\mathrm{Gr}_k(\mathbb{R}^n), \mathbb{Z}/2\mathbb{Z})$ is non-trivial if and only if the boundary of a representative of x in the chain complex for $\mathrm{Gr}_k(\mathbb{R}^n)$ is non-trivial; in that case the Bockstein is the reduction of half the boundary. The criterion in Theorem* 4.2 *then follows directly from the description of the boundary map in* [**CK13**]. *The result follows since the real cycle class map $\mathrm{H}^j(\mathrm{Gr}(k,n), \overline{\mathbf{I}}^j) \cong \mathrm{Ch}^j(\mathrm{Gr}(k,n)) \to \mathrm{H}^j(\mathrm{Gr}_k(\mathbb{R}^n), \mathbb{Z}/2\mathbb{Z})$ takes the Steenrod square $\mathrm{Sq}^2_{\mathscr{L}}$ to the topological Bockstein operation $\beta^{\mathrm{top}}_{\mathscr{L}}$, a fact that will be established in forthcoming joint work with Jens Hornbostel, Heng Xie and Marcus Zibrowius.*

On the other hand, the description of $\mathrm{Sq}^1 = \beta$ on $\mathrm{H}^j(\mathrm{Gr}_k(\mathbb{R}^n), \mathbb{Z}/2\mathbb{Z})$ agrees with the description of Sq^2 on $\mathrm{H}^j(\mathrm{Gr}_k(\mathbb{C}^n), \mathbb{Z}/2\mathbb{Z})$: essentially, Sq^2 in the complex setting is nontrivial if the attaching map is the Hopf map η, and in the real case Sq^1 is nontrivial if the attaching map is the natural projection $\mathrm{S}^n \to \mathbb{RP}^n$. Therefore, the diagrammatic description above also describes Sq^2 for $\mathrm{Gr}_k(\mathbb{C}^n)$. Then the cycle class map $\mathrm{Ch}^j(\mathrm{Gr}(k,n)) \to \mathrm{H}^{2j}(\mathrm{Gr}_k(\mathbb{C}^n), \mathbb{Z}/2\mathbb{Z})$ is compatible with Sq^2 (essentially because Voevodsky's description of the Steenrod operations is a version of the topological definition in the $\mathbb{A}^1$-homotopy category). This is another way to see that the description of Sq^2 on $\mathrm{Ch}^\bullet(\mathrm{Gr}(k,n))$ agrees with the description of attaching maps in the cell complex for $\mathrm{Gr}_k(\mathbb{R}^n)$. Note that this agrees with the description in [**Len98**].

EXAMPLE 4.4. *To illustrate the conditions in Theorem 4.2, consider the following computation of Steenrod squares with checkerboard rule:*

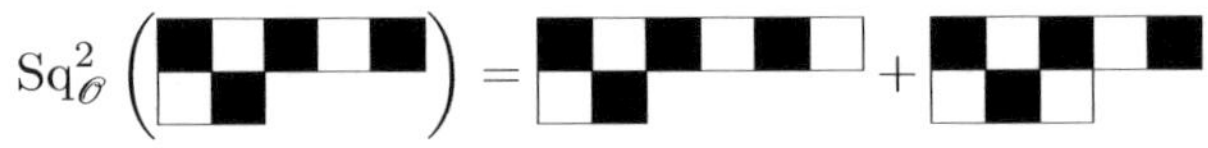

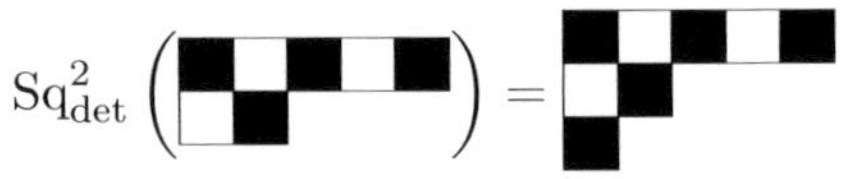

□

EXAMPLE 4.5. *We work out the Steenrod square of the class $\sigma_{a,b}$. By the Giambelli formula, we have*

$$\sigma_{a,b} = \det \begin{pmatrix} \sigma_a & \sigma_{a+1} \\ \sigma_{b-1} & \sigma_b \end{pmatrix} = \sigma_a\sigma_b - \sigma_{a+1}\sigma_{b-1}.$$

Now we have a case distinction

$$\mathrm{Sq}^2_{\mathscr{O}}\,\sigma_{a,b} = \begin{cases} -\sigma_{a+1}\sigma_b - \sigma_{b-1}\sigma_{a+2} = \sigma_{a+1,b} & a,b \text{ odd} \\ \sigma_a\sigma_{b+1} - \sigma_{b-1}\sigma_{a+2} = \sigma_{a,b+1} + \sigma_{a+2,b} & a, b-1 \text{ odd} \\ 0 & a-1, b \text{ odd} \\ \sigma_a\sigma_{b+1} + \sigma_{a+1}\sigma_b & a,b \text{ even} \end{cases}$$

The simplifications are obtained by expanding the product $\sigma_i\sigma_j$ as $\sigma_{i+j} + \sigma_{i+j-1,1} + \cdots + \sigma_{i+j-\min(i,j),\min(i,j)}$ and cancelling corresponding terms. Note that in case a, b even, the result can be simplified to $\sigma_{a,b+1}$ if $a > b$ and 0 if $a = b$. This fits exactly with the geometric description: in the first row a white square can be added if and only if a is odd, and in the second row a white square can be added if and only if b is even. In the third, we can never add a white square. □

Actually, the triviality of the Steenrod squares can be identified by means of an elementary size condition on the Young diagrams. The condition is exactly the one discussed in [**BC12a**, Section 4]; in particular, the vanishing of the Steenrod square $\mathrm{Sq}^2_{\mathscr{L}}$ for a Schubert variety is equivalent to the relative canonical bundle of a resolution of singularities being isomorphic to the pullback of the corresponding line bundle $\mathscr{L}$ from the Grassmannian.

PROPOSITION 4.6. *Fix $1 \leq k < n$ and let Λ be a Young diagram. In the notation from Definition 3.8 and Figure 1, we have the following:*

(1) *Then $\mathrm{Sq}^2_{\mathscr{O}}(\Lambda) = 0$ if and only if*
   (a) *for all $1 \leq i < p$ we have $d_{i+1} - d_i + e_{i+2} - e_{i+1}$ is even, and*
   (b) *either ($n-k-e_1$ and $d_1 + e_2 - e_1$ are even) or ($e_1 = 0$ and both $n-k$ and $d_1 + e_2 - e_1$ are odd).*

(2) *Then $\mathrm{Sq}^2_{\det}(\Lambda) = 0$ if and only if*
   (a) *for all $1 \leq i < p$ we have $d_{i+1} - d_i + e_{i+2} - e_{i+1}$ is even, and*
   (b) *either ($n-k-e_1$ is odd and $d_1 + e_2 - e_1$ is even) or ($e_1 = 0$, $n-k$ is even and $d_1 + e_2 - e_1$ is odd).*

PROOF. We prove the first case, the second is similar. The Steenrod square $\mathrm{Sq}^2_{\mathscr{O}}(\Lambda)$ vanishes if, after filling the Young diagram Λ with a checkerboard pattern, there is no way to add a white box and still get a Young diagram in $k \times (n-k)$-frame which has a checkerboard pattern. The places where can possibly add a white box to get a Young diagram would be at the end of the first row, or in the corners of hooks; in the notation of Figure 1, such a box would be at the coordinates $(d_i+1, n-k-e_{i+1}+1)$. We cannot add a box in the first row if and only if the length is either even (thus ending in a white square) or has length $n-k$ (so that adding a square would leave the $k\times(n-k)$-frame). We cannot add boxes in the hook corners if and only if the boxes at the coordinates $(d_i, n-k-e_{i+1}+1)$ resp. $(d_i+1, n-k-e_{i+1})$ are white. Now we note that the boxes $(d_i+1, n-k-e_{i+1})$ and $(d_{i+1}, n-k-e_{i+2}+1)$ have the same color if and only if $d_{i+1}-d_i+e_{i+2}-e_{i+1}$ is even. Moreover, the box at the end of the first row has the same color as the one in the rightmost hook if and only if $d_1+e_2-e_1$ is even. Therefore, if the first row has even length, then it suffices to require $d_{i+1}-d_i+e_{i+2}-e_{i+1}$ and $d_1+e_2-e_1$ to be even. However, if the first row has odd length $n-k$ (equivalently $e_1=0$), then we need to require $d_1+e_2-e_1$ to be odd. □

PROPOSITION 4.7. *Fix $1\leq k<n$. The Steenrod squares $\mathrm{Sq}^2_{\mathscr{L}}(\sigma_\Lambda)$ of Schubert classes $\sigma_\Lambda\in \mathrm{Ch}^\bullet(\mathrm{Gr}(k,n))$ form a set of additive generators for the $\mathrm{I}(F)$-torsion in $\bigoplus_{j,\mathscr{L}}\mathrm{H}^j(\mathrm{Gr}(k,n),\mathbf{I}^j(\mathscr{L}))$.*

PROOF. This follows directly from Proposition 2.3. □

At this point, getting an additive generating set as above is the best statement we can make (but the previous discussion of the Steenrod squares allows to make the generating set fairly explicit). It seems more difficult to provide a $\mathbb{Z}/2\mathbb{Z}$-basis for the torsion part of the $\mathbf{I}^\bullet$-cohomology. The following example shows that we cannot expect to get an additive basis simply by taking lifts of Schubert classes.

EXAMPLE 4.8. *One of the easiest such examples appears in $\mathrm{H}^2(\mathrm{Gr}(2,5),\mathbf{I}^2)$. The Chow group $\mathrm{Ch}^2(\mathrm{Gr}(2,5))$ is generated by the two classes $\overline{\mathrm{c}}_2$ and $\overline{\mathrm{c}}_2^\perp$. Both have nontrivial $\mathrm{Sq}^2_{\mathscr{O}}$ and hence do not lift to $\mathbf{I}^2$-cohomology. However, their non-twisted Steenrod squares are both equal to the hook $\sigma_{2,1}$ and consequently $\overline{\mathrm{c}}_2+\overline{\mathrm{c}}_2^\perp=\overline{\mathrm{c}}_1^2$ lifts, namely to the element $\beta_{\mathscr{O}}(\overline{\mathrm{c}}_1)$.* □

4.2. Multiplication of torsion classes. We finally discuss the multiplicative structure of the $\mathbf{I}^\bullet$-cohomology ring. By Proposition 3.14, the even Young diagrams provide a $\mathrm{W}(F)$-basis for the free part of $\bigoplus_{j,\mathscr{L}}\mathrm{H}^j(\mathrm{Gr}(k,n),\mathbf{I}^j(\mathscr{L})$. The remaining part is $\mathrm{I}(F)$-torsion, given exactly as the image of $\beta_{\mathscr{L}}$, cf. Proposition 2.3. In particular, any class x in $\mathbf{I}^\bullet$-cohomology has a unique decomposition $x=x_{\mathrm{tor}}+x_{\mathrm{ev}}$ as a sum of an element x_{ev} in the $\mathrm{W}(F)$-subalgebra generated by even Young diagrams and an element x_{tor} which is $\mathrm{I}(F)$-torsion.

Now if we have two classes $x=x_{\mathrm{tor}}+x_{\mathrm{ev}}$ and $y=y_{\mathrm{tor}}+y_{\mathrm{ev}}$ in $\mathbf{I}^\bullet$-cohomology, we can determine their product $x\cdot y$ as follows. The product $x\cdot y$ is given by $x_{\mathrm{ev}}\cdot y_{\mathrm{ev}}+x_{\mathrm{tor}}\cdot y_{\mathrm{tor}}+x_{\mathrm{tor}}\cdot y_{\mathrm{ev}}+x_{\mathrm{ev}}\cdot y_{\mathrm{tor}}$, and all summands except possibly the first one are torsion. Non-torsion contributions can only arise from $x_{\mathrm{ev}}\cdot y_{\mathrm{ev}}$ but as the example in 3.5 shows, $x_{\mathrm{ev}}\cdot y_{\mathrm{ev}}$ can also contain torsion contributions. The non-torsion contributions in $x_{\mathrm{ev}}\cdot y_{\mathrm{ev}}$ can be determined using Theorem 3.16. The overall class of the product can then be determined by reduction:

PROPOSITION 4.9. *Fix $1\leq k<n$, and let x,y be two classes in $\mathbf{I}^\bullet$-cohomology. Then the reduction of $x\cdot y$ is given by the product of the reductions of x and y in $\mathrm{Ch}^\bullet(\mathrm{Gr}(k,n))$.*

PROOF. This follows directly from Proposition 2.3. □

In particular, products with torsion classes are determined completely on the reduction mod 2, and these can be evaluated by classical Young diagram calculations. As a relevant consequence for Schubert calculus (where the intersection product lands in top degree), we have the following:

COROLLARY 4.10. *Let F be a perfect field of characteristic $\neq 2$, fix $1 \leq k < n$ and assume n is even. Any product $x_1 \dots x_l \in \mathrm{H}^{k(n-k)}(\mathrm{Gr}(k,n), \mathbf{I}^{k(n-k)})$ involving an $\mathrm{I}(F)$-torsion class is trivial.*

PROOF. Under the assumption that n is even, we have

$$\mathrm{H}^{k(n-k)}(\mathrm{Gr}(k,n), \mathbf{I}^{k(n-k)}) \cong \mathrm{W}(F),$$

i.e., the top $\mathbf{I}^\bullet$-cohomology group is $\mathrm{W}(F)$-torsion-free. In particular, the image of $\beta_{\mathscr{O}}$ is trivial. Any product involving $\mathrm{I}(F)$-torsion is again $\mathrm{I}(F)$-torsion, lies in the image of $\beta_{\mathscr{O}}$ and is consequently trivial. □

As an example how to compute products in $\mathbf{I}^\bullet$-cohomology, we resume the discussion of the example in Remark 3.5.

EXAMPLE 4.11. *We consider the square of the class $\sigma_{2,2}$ in $\mathrm{Ch}^8(\mathrm{Gr}(4,8))$. We will see in Section 6 that the class $\sigma_{2,2}$ has a canonical lift to $\mathrm{H}^4(\mathrm{Gr}(4,8), \mathbf{I}^4)$ given by a canonical orientation on the normal bundle of the smooth Schubert variety $\Sigma_{2,2}$. The corresponding class $\sigma_{2,2}^2$ in $\mathrm{H}^8(\mathrm{Gr}(4,8), \mathbf{I}^8)$ has two non-torsion contributions (computed using Theorem 3.16): the classes $\sigma_{4,4}$ and $\sigma_{2,2,2,2}$ corresponding to even Young diagrams with trivial twist. The torsion contribution to $\sigma_{2,2}^2$ can be computed in the mod 2 Chow ring, cf. Remark 3.5; it is given by subtracting from $\sigma_{2,2}^2$ in the mod 2 Chow ring the above even Young diagrams representing the non-torsion part. The class that remains is $\sigma_{4,3,1}+\sigma_{4,2,2}+\sigma_{3,3,1,1}+\sigma_{3,2,2,1}$, and this is a torsion class with non-twisted coefficients. Note that none of the individual Schubert classes in this sum lifts because they all have non-trivial Steenrod square. However, the full sum lifts because*

$$\begin{aligned}
& \mathrm{Sq}^2_{\mathscr{O}}(\sigma_{4,3,1}) + \mathrm{Sq}^2_{\mathscr{O}}(\sigma_{4,2,2}) + \mathrm{Sq}^2_{\mathscr{O}}(\sigma_{3,3,1,1}) + \mathrm{Sq}^2_{\mathscr{O}}(\sigma_{3,2,2,1}) \\
= \; & \sigma_{4,3,2} + \sigma_{4,3,1,1} + \sigma_{4,3,2} + \sigma_{4,2,2,1} + \sigma_{4,3,1,1} + \sigma_{3,3,2,1} + \sigma_{4,2,2,1} + \sigma_{3,3,2,1} \\
= \; & 0
\end{aligned}$$

The torsion lift is uniquely determined by its reduction mod 2, the lift is given as $\beta_{\mathscr{O}}(\sigma_{4,2,1} + \sigma_{3,2,1,1})$ □

5. Recollection on Witt groups and cohomology

In this section, we provide a short recollection on Witt groups as well as $\mathbf{I}^n$-cohomology. Most important for our later discussions will be the definition of push-forwards (and the dependence on choices of orientations) and the intersection products. We discuss the natural morphism from Witt groups to $\mathbf{I}^\bullet$-cohomology and its compatibility with push-forwards and intersection products. The compatibility with the pushforwards implies that the results from Section 3 also describe the multiplication on the twisted Witt groups, and the compatibility with intersection products is an oriented version of Serre's Tor-formula for intersection multiplicities.

5.1. Witt groups and comparison morphism. We shortly recall some relevant definitions from the theory of Witt groups of schemes, cf. in particular [**Bal05**]. For a category $\mathcal{C}$ with duality (given by an involutive endofunctor $(-)^\vee\colon \mathcal{C}^{\mathrm{op}} \to \mathcal{C}$ with an explicit identification $\varpi\colon \mathrm{id}_{\mathcal{C}} \xrightarrow{\cong} (-)^{\vee\vee}$), a symmetric space in $\mathcal{C}$ is a pair (X, ϕ) with X an object from $\mathcal{C}$ and an isomorphism $\phi\colon X \to X^\vee$ such that

$$\begin{array}{ccc} X & \xrightarrow{\phi} & X^\vee \\ {\scriptstyle \varpi_X}\downarrow{\scriptstyle\cong} & & \downarrow{\scriptstyle =} \\ X^{\vee\vee} & \xrightarrow[\phi^\vee]{} & X^\vee \end{array}$$

The Witt group of a category with duality is given as the quotient of the isometry classes of symmetric spaces modulo metabolic spaces. There is a similar notion of triangulated categories with duality, together with the corresponding notion of Witt group, called triangular Witt groups, cf. [**Bal05**, Section 1.4].

For the present work, we are primarily interested in the following situation. For a smooth scheme X over a field F of characteristic $\neq 2$ and a line bundle $\mathscr{L}$ over X, one can consider the category of vector bundles on X equipped with the duality given by

$$\mathscr{E} \mapsto \mathrm{Hom}_{\mathscr{O}_X}(\mathscr{E}, \mathscr{O}_X) \otimes_{\mathscr{O}_X} \mathscr{L} \cong \mathrm{Hom}_{\mathscr{O}_X}(\mathscr{E}, \mathscr{L}).$$

This exact category with duality produces the classical *Witt groups of schemes.* One can also consider the *derived Witt groups*, which arise from the triangulated category of perfect complexes with the derived version of the above duality. It is also possible to consider the category of bounded complexes of quasi-coherent sheaves with coherent cohomology, with the same duality as above; this gives rise to the *coherent Witt groups* of X. In the case of smooth schemes, all these constructions produce isomorphic Witt groups because coherent sheaves have finite resolutions by vector bundles.

A morphism of categories with duality is a functor $G\colon \mathcal{C} \to \mathcal{D}$ with an explicit isomorphism $G \circ (-)^{\vee_{\mathcal{C}}} \xrightarrow{\cong} (-)^{\vee_{\mathcal{D}}} \circ G$. The image of a symmetric space $\phi\colon X \to X^\vee$ is then given by $G(X) \xrightarrow{G(\phi)} G(X^\vee) \cong G(X)^\vee$, where the latter isomorphism is the one from the definition of morphism of categories with duality. Under apppropriate assumptions, a morphism of schemes $f\colon X \to Y$ gives rise to functors $\mathrm{L}f^*$ and $\mathrm{R}f_*$ of categories of quasi-coherent sheaves or perfect complexes, which preserves the duality. This gives rise to pullback and push-forward morphisms for (ordinary, derived, coherent) Witt groups of schemes, cf. [**CH11**].

There is also a base-change formula, cf. [**CH11**, Theorem 5.4]: for a pullback diagram of schemes

$$\begin{array}{ccc} W & \xrightarrow{g'} & X \\ {\scriptstyle f'}\downarrow & & \downarrow{\scriptstyle f} \\ Y & \xrightarrow[g]{} & Z, \end{array}$$

there is a natural morphism of functors $\epsilon\colon \mathrm{L}f^* \circ \mathrm{R}g_* \to \mathrm{R}g'_* \circ \mathrm{L}(f')^*$ which is an isomorphism if the diagram is homologically transversal, cf. the discussion in Section 7. This will be relevant for computation of intersection products.

For coherent and derived Witt groups, the derived tensor product of complexes gives rise to duality-preserving functors and consequently to pairings in triangular Witt groups, cf. [**GN03**]. This is usually called the $\star$-product. For complexes concentrated in a single degree, the product on Witt groups reduces to the ordinary tensor product of symmetric forms as defined by Knebusch, cf. the discussion in [**BG05**, Section 2].

We recall the natural morphisms from Witt-groups to $\mathbf{I}^\bullet$-cohomology from [**Fas07**, Section 7]. Let $Z \subset X$ be a closed subset of pure codimension i. Denoting by $\mathrm{GW}^i_Z(X)$ the i-th Grothendieck–Witt group of perfect complexes on X with support in Z, there is a homomorphism

$$\mathrm{GW}^i_Z(X) \to \mathrm{GW}^i(\mathrm{Der}^{\mathrm{b}}(X)^{(i)}) \to \mathrm{GW}^i(\mathrm{Der}^{\mathrm{b}}_i(X))$$

which is obtained as composition of the inclusion of $\mathrm{Der}^{\mathrm{b}}_Z(X)$ in the category $\mathrm{Der}^{\mathrm{b}}(X)^{(i)}$ of complexes with support in codimension $\geq i$ followed by the projection to $\mathrm{Der}^{\mathrm{b}}_i(X) = \mathrm{Der}^{\mathrm{b}}(X)^{(i)}/\mathrm{Der}^{\mathrm{b}}(X)^{(i+1)}$. The composition

$$\mathrm{GW}^i(\mathrm{Der}^{\mathrm{b}}(X)^{(i)}) \to \mathrm{GW}^i(\mathrm{Der}^{\mathrm{b}}_i(X)) \to \mathrm{W}^{i+1}(\mathrm{Der}^{\mathrm{b}}(X)^{(i+1)})$$

is zero, and consequently, there is a homomorphism $\mathrm{GW}^i_Z(X) \to \widetilde{\mathrm{CH}}^i(X)$. The same argument also furnishes a homomorphism

$$\alpha\colon \mathrm{W}^i_Z(X, \mathscr{L}) \to \mathrm{H}^i(X, \mathbf{I}^i(\mathscr{L})).$$

To accomodate the additional line bundle twist by $\mathscr{L}$, simply use the derived categories above with the duality given by $\mathrm{Hom}_{\mathscr{O}_X}(-, \mathscr{L})$.

5.2. Pushforwards and orientations. We shortly discuss push-forwards in cohomology with coefficients in $\mathbf{I}^\bullet$ and $\mathbf{W}$.

For a proper morphism $f\colon X \to Y$ between smooth schemes over F, there are push-forward maps, cf. [**Fas07**, Corollaire 10.4.5] or [**CF14**, Section 3]:

$$\mathrm{H}^n(X, \mathbf{I}^n(\omega_{X/Y} \otimes f^*\mathscr{L})) \to \mathrm{H}^{n-d}(Y, \mathbf{I}^{n-d}(\mathscr{L}))$$

where $d = \dim X - \dim Y$ and $\omega_{X/Y} = \omega_{X/F} \otimes f^*\omega^\vee_{Y/F}$ is the relative canonical bundle. The push-forward is defined in [**Fas08**, Section 8] on the level of Gersten–Rost–Schmid complexes $\mathrm{C}^n(X, \mathbf{I}^n(\omega_{X/F})) \to \mathrm{C}^{n-d}(Y, \mathbf{I}^{n-d}(\omega_{Y/F}))$: for a point $x \in X$ with image $y = f(x)$ such that the residue field extension $F(y) \subseteq F(x)$ is finite, the corresponding morphism on the Witt group coefficients is given by the transfers $\mathrm{tr}^y_x\colon \mathrm{W}(F(x), \omega_{F(x)/F}) \to \mathrm{W}(F(y), \omega_{F(y)/F})$.

The line bundle twists in the above are really necessary. Moreover, for isomorphic line bundles $\mathscr{L}_1$ and $\mathscr{L}_2$, the identification $\mathrm{H}^n(X, \mathbf{I}^n(\mathscr{L}_1)) \cong \mathrm{H}^n(X, \mathbf{I}^n(\mathscr{L}_2))$ depends on the choice of isomorphism $\mathscr{L}_1 \cong \mathscr{L}_2$. In particular, to get a pushforward $\mathrm{H}^n(X, \mathbf{I}^n) \to \mathrm{H}^{n-d}(Y, \mathbf{I}^{n-d}(\mathscr{L}))$ for a proper map $f\colon X \to Y$, one needs to choose an isomorphism $\omega_{X/Y} \otimes f^*\mathscr{L} \cong \mathscr{O}_X$. Actually, since everything is well-defined up to squares (in the sense of the discussion of lax similitude in [**BC12b**]), it suffices to choose an identification $\omega_{X/Y} \otimes f^*\mathscr{L} \cong \mathscr{N}^2$ for some line bundle $\mathscr{N}$, i.e., the pushforward maps for $\mathbf{I}^\bullet$-cohomology depend on the choice of orientations, cf. [**KW17**, Definition 16]:

DEFINITION 5.1. *Let X be a smooth scheme and let $\mathscr{L}$ be a line bundle over X. An* orientation *of $\mathscr{L}$ is a pair $(\mathscr{N}, \phi)$ where $\mathscr{N}$ is a line bundle over X and $\phi\colon \mathscr{N} \otimes \mathscr{N} \cong \mathscr{L}$ is an isomorphism. Two orientations $(\mathscr{N}_1, \phi_1)$ and $(\mathscr{N}_2, \phi_2)$ are equivalent if there exists an isomorphism $\alpha\colon \mathscr{N}_1 \to \mathscr{N}_2$ such that $\phi' \circ (\alpha \otimes \alpha) = \phi$.*

EXAMPLE 5.2. *If the line bundle $\mathscr{L}$ has a global invertible section $\sigma\colon \mathscr{O}_X \to \mathscr{L}$, then we get an orientation $\mathscr{O}_X \otimes \mathscr{O}_X \cong \mathscr{O}_X \xrightarrow{\sigma} \mathscr{L}$. Two such orientations are equivalent if their quotient is the square of a global invertible section. For a finite-dimensional vector space V over F, we recover the classical concept of orientation as a choice of generator of $\bigwedge^{\dim V} V$. Two orientations are equivalent if the corresponding generators differ by the square of a unit in $F^\times$.* □

A choice of orientation $\mathscr{N} \otimes \mathscr{N} \xrightarrow{\cong} \mathscr{L}$ provides an isomorphism $\mathrm{W}(X, \mathscr{O}) \cong \mathrm{W}(X, \mathscr{L})$ which maps a form $P_\bullet \otimes P_\bullet \to \mathscr{O}$ (alternatively written as $P_\bullet \xrightarrow{\cong} \mathrm{Hom}(P_\bullet, \mathscr{O})$) to the form $(P_\bullet \otimes \mathscr{N}) \otimes (P_\bullet \otimes \mathscr{N}) \to \mathscr{L}$ (alternatively written as $P_\bullet \otimes \mathscr{N} \xrightarrow{\cong} \mathrm{Hom}(P_\bullet \otimes \mathscr{N}, \mathscr{O}) \otimes \mathscr{L}$). The identifications for $\mathbf{I}^\bullet$-cohomology and $\mathbf{W}$-cohomology can be made similarly precise.

Finally, we discuss the compatibility of the natural comparison homomorphism $\alpha\colon \mathrm{W}^j_Z(X, \mathscr{L}) \to \mathrm{H}^j(X, \mathbf{I}^j(\mathscr{L}))$ with basic maps. The compatibility of the comparison homomorphism with pullbacks is established in [**Fas07**, Remark 7.5]. To check compatibility with pushforwards, we trace through the definitions:

PROPOSITION 5.3. *Let F be a perfect field of characteristic $\neq 2$, let $f\colon X \to Y$ be a proper morphisms of smooth schemes over F, and set $d = \dim X - \dim Y$. Then there is a commutative diagram*

$$\begin{array}{ccc} \mathrm{GW}^i(X, \omega_{X/Y} \otimes f^*\mathscr{L}) & \xrightarrow{f_*} & \mathrm{GW}^{i-d}(Y, \mathscr{L}) \\ \downarrow{\scriptstyle \alpha_X} & & \downarrow{\scriptstyle \alpha_Y} \\ \widetilde{\mathrm{CH}}^i(X, \omega_{X/Y} \otimes f^*\mathscr{L}) & \xrightarrow[f_*]{} & \widetilde{\mathrm{CH}}^{i-d}(Y, \mathscr{L}). \end{array}$$

PROOF. For a finite morphism $f\colon X \to Y$ (between smooth schemes, say), the push-forward on Chow–Witt-rings and $\mathbf{I}^\bullet$-cohomology is induced from the push-forward on Witt groups, cf. [**Fas08**, Corollaries 5.3.5–5.3.7] resp. [**Fas08**, Corollary 6.3.10] for the case of smooth schemes which is more relevant for us. In particular, it follows from the definition of push-forwards on Chow–Witt groups in [**Fas08**] that the comparison morphism between Grothendieck–Witt groups and Chow–Witt rings are compatible with push-forwards along finite morphisms. Since the push-forward on Chow–Witt groups for general proper morphisms $f\colon X \to Y$ is also defined by reduction to the case of finite morphisms, cf. [**Fas08**, Section 8], we get the claim. □

5.3. Intersection products and oriented Tor formula. We shortly recall the definition of the intersection products in Chow–Witt rings and $\mathbf{I}^\bullet$-cohomology from [**Fas07**]: first, there is an exterior product

$$\boxtimes\colon \widetilde{\mathrm{CH}}^i(X, \mathscr{L}) \times \widetilde{\mathrm{CH}}^j(X, \mathscr{M}) \to \widetilde{\mathrm{CH}}^{i+j}(X \times X, \mathrm{pr}_1^*(\mathscr{L}) \otimes \mathrm{pr}_2^*(\mathscr{M})).$$

On the other hand, we have a pullback morphism

$$\Delta^!\colon \widetilde{\mathrm{CH}}^{i+j}(X \times X, \mathrm{pr}_1^*(\mathscr{L}) \otimes \mathrm{pr}_2^*(\mathscr{M})) \to \widetilde{\mathrm{CH}}^{i+j}(X, \mathscr{L} \otimes \mathscr{M})$$

for the embedding $\Delta\colon X \hookrightarrow X \times X$ (which is a closed immersion between smooth schemes). The intersection product is then given as the composition of these two operations: it maps a pair $x_i \in \widetilde{\mathrm{CH}}^{n_i}(X, \mathscr{L}_i)$, $i = 1, 2$, to $\Delta^!(x_1 \boxtimes x_2)$.

A different way to view the construction of the pullback is given in [**AF16a**, Section 2], in particular the discussion leading up to Theorem 2.3.4. There is also a base-change formula for Chow–Witt rings and $\mathbf{I}^\bullet$-cohomology, cf. [**AF16a**, Theorem 2.4.1], which can be used to compute some intersection products, cf. the proof of Theorem 7.17.

The relation between the $\star$-product on (Grothendieck–)Witt groups and the intersection product on Chow–Witt-rings or $\mathbf{I}^\bullet$-cohomology is given by the following commutative diagram which can be interpreted as an oriented analogue of Serre's Tor-formula, cf. [**Fas07**, Theorem 7.6]:

$$\begin{array}{ccc}
\mathrm{GW}^n_V(X) \times \mathrm{GW}^m_W(X) & \xrightarrow{\ \star\ } & \mathrm{GW}^{m+n}_{V\cap W}(X) \\
\Big\downarrow{\scriptstyle \alpha_V \times \alpha_W} & & \Big\downarrow{\scriptstyle \alpha_{V\cap W}} \\
\widetilde{\mathrm{CH}}^n(X) \times \widetilde{\mathrm{CH}}^m(X) & \xrightarrow[\times]{} & \widetilde{\mathrm{CH}}^{n+m}(X).
\end{array}$$

We shortly recall the classical Tor formula in intersection theory. Assume Y and Z are irreducible closed subschemes of a smooth scheme X intersecting properly. If C is an irreducible component of the intersection $Y \cap Z$ then the multiplicity of C in the intersection of Y and Z is given by

$$\mu(Y,Z;C) = \sum_{j\geq 0} (-1)^j \operatorname{len}_{\mathscr{O}_{X,C}} \left(\operatorname{Tor}_j^{\mathscr{O}_{X,C}}(\mathscr{O}_{Y,C}, \mathscr{O}_{Z,C}) \right).$$

Here, we take the Tor-groups of the local rings of C in Y and Z over $\mathscr{O}_{X,C}$, and $\operatorname{len}_{\mathscr{O}_{X,C}}$ is the length of the respective Tor-group as module over the local ring $\mathscr{O}_{X,C}$. In particular, multiplicities in intersection products can be computed by means of the product structure on K-groups which is given in terms of derived tensor products of complexes of coherent sheaves. In this setting, it is clear why homological transversality is useful: the Tor-formula simply reduces to the term in degree 0 which is the length of the intersection subscheme.

The oriented Tor-formula, as expressed in the commutative diagram above, now allows to compute some intersection products in the Chow–Witt ring: the lower composition is the intersection product of two cycles which are in the image of the natural comparison map $\alpha\colon \mathrm{GW}^i_Z(X,\mathscr{L}) \to \widetilde{\mathrm{CH}}^i_Z(X,\mathscr{L})$, and the upper composition is the cycle associated to the $\star$-product, which is itself given in terms of the derived tensor products of forms on perfect complexes over X, with $\mathscr{L}$-twisted duality. Again, in the presence of homological transversality, the oriented Tor-formula simplifies: the corresponding derived tensor products are ordinary tensor products, and in the simplest case, when we only consider structure sheaves of closed subvarieties, the $\star$-product is then simply a form on the structure sheaf of the intersection subscheme which encodes a "refined length" containing additional orientation information.

6. Oriented Schubert classes

In this section, we can now write down geometric representatives for the classes in the $\mathbf{I}^\bullet$-cohomology ring. As in [**BC12a**], the representatives are obtained by push-forward of the form $\langle 1\rangle$ from resolutions of singularities of Schubert varieties. Of course, the push-forward depends on choices of relative orientations, but we will actually show that the relative canonical bundles of these resolutions have canonical (relative) orientations. This implies that Schubert classes for even Young diagrams have canonical lifts to the Chow–Witt ring.

6.1. Resolutions of Schubert varieties. We first discuss the relevant resolutions of singularities for Schubert varieties in Grassmannians, including some geometric facts concerning the descriptions of relative canonical bundles. Most of this is a recollection of statements from [**BC12a**]. These statements will be required for the discussion of canonical orientations in the next subsection.

Let F be a field of characteristic $\neq 2$. Fix $1 \leq k < n$, and let V be an n-dimensional F-vector space. Let Λ be a Young diagram in $k \times (n-k)$-frame corresponding to the partition $\underline{a} = (a_1, \dots, a_k)$, then we have associated tuples $(d_1, \dots, d_p = k)$ and $(e_1, \dots, e_p)$ as in Definition 3.8 and Figure 1. For a given full flag

$$\mathcal{F}\colon 0 \leq V_1 \leq \cdots \leq V_{n-1} \leq V$$

with $\dim_F V_i = i$, the Schubert variety

$$\Sigma_\Lambda(\mathcal{F}) = \{W \in \mathrm{Gr}(k, V) \mid \dim_F(V_{n-k+i-a_i} \cap W) \geq i\}$$

corresponding to Λ is a representative of the Schubert class corresponding to the Young diagram Λ.

Definition 6.1. *For the partial flag $0 \leq V_{d_1+e_1} \leq \cdots \leq V_{d_p+e_p} \leq V$, we can consider the following Schubert variety in the flag variety $\mathrm{Fl}(d_1, \dots, d_p = k; n)$:*

$$\mathrm{Fl}(\underline{d}, \underline{e}, \mathcal{F}) := \left\{P_{d_1} \leq P_{d_2} \leq \cdots \leq P_{d_p} \in \mathrm{Fl}(\underline{d}; n) \mid P_{d_i} \leq V_{d_i+e_i}\right\}$$

Alternatively, we use the notation $\mathrm{Fl}(\Lambda)$ for this closed subvariety of $\mathrm{Fl}(\underline{d}; n)$. There is a natural projection morphism

$$p_\Lambda\colon \mathrm{Fl}(\Lambda) \to \mathrm{Gr}(k, n)\colon (P_{d_1}, \dots, P_{d_p}) \mapsto P_{d_p}.$$

Remark 6.2. *The variety $\mathrm{Fl}(\Lambda)$ is smooth and its image is the Schubert variety $\Sigma_\Lambda(\mathcal{F})$. The morphism p_Λ is proper and birational, and hence is a resolution of singularities of $\Sigma_\Lambda(\mathcal{F})$.*

On the flag variety $\mathrm{Fl}(d_1, \dots, d_p; n)$, we have the tautological bundles $\mathscr{P}_{d_i}$. The fiber of the bundle $\mathscr{P}_{d_i}$ at the point corresponding to the flag $P_{d_1} \leq \cdots \leq P_{d_p}$ is P_{d_i}. These bundles form a filtration $\mathscr{P}_{d_1} \leq \cdots \leq \mathscr{P}_{d_p}$ of the trivial bundle $\mathscr{O}_{\mathrm{Fl}}^{\oplus n}$. We also have the corresponding tautological subquotient bundles $\mathscr{S}_i = \mathscr{P}_{d_i}/\mathscr{P}_{d_{i-1}}$.

Next, we need to describe the tangent bundles of flag varieties as well as the relative canonical bundles for the resolutions $p_\Lambda\colon \mathrm{Fl}(\Lambda) \to \mathrm{Gr}(k, n)$. In the special case of the Grassmannian, we have an exact sequence $0 \to \mathscr{S} \to \mathscr{O}_{\mathrm{Gr}}^{\oplus n} \to \mathscr{Q} \to 0$ which decomposes the trivial rank n bundle as extension of the tautological rank k subbundle $\mathscr{S}$ by the tautological rank $n-k$ quotient bundle $\mathscr{Q}$. It is a standard fact that the tangent space of the Grassmannian is canonically identified with $\mathrm{Hom}(\mathscr{S}, \mathscr{Q})$, cf. [**FM16**, Section 4.1] or [**EH17**, Theorem 3.5].

To describe the tangent bundles of the flag varieties $\mathrm{Fl}(\underline{d}, \underline{e})$, we can use the fact that these flag varieties have a fibration by Grassmannians, cf. [**BC12a**, Lemma 1.11]. For two tuples $(d_1, \dots, d_p)$ and $(e_1, \dots, e_p)$ (as the ones arising from even Young diagrams, cf. Figure 1), $\mathrm{Fl}(\underline{d}, \underline{e})$ can be identified as Grassmannian of $d_p - d_{p-1}$-dimensional subspaces in the tautological quotient bundle $\mathscr{O}_{\mathrm{Fl}}^{\oplus d_p+e_p}/\mathscr{P}_{d_{p-1}}$ of rank $d_p - d_{p-1} + e_p$ on $\mathrm{Fl}((d_1, \dots, d_{p-1}), (e_1, \dots, e_{p-1}))$. Hence, we have a fibration

$$\mathrm{Gr}(d_p - d_{p-1}, d_p - d_{p-1} + e_p) \to \mathrm{Fl}(\underline{d}, \underline{e}) \to \mathrm{Fl}((d_1, \dots, d_{p-1}), (e_1, \dots, e_{p-1}))$$

With these identifications, the flag of subbundles on $\mathrm{Fl}(\underline{d},\underline{e})$ is given as follows: the first $k-1$ subbundles are the pullbacks of the subbundles $\mathscr{P}_i$ for $1\leq i\leq k-1$, and the last two subquotients $\mathscr{P}_{d_p}/\mathscr{P}_{d_{p-1}}$ and $\mathscr{O}_{\mathrm{Fl}(\underline{d},\underline{e})}^{\oplus d_p+e_p}/\mathscr{P}_{d_p}$ identified with the tautological sub- and quotient bundle of the relative Grassmannian $\mathrm{Fl}(\underline{d},\underline{e})\to\mathrm{Fl}((d_1,\ldots,d_{p-1}),(e_1,\ldots,e_{p-1}))$, respectively. In particular, the relative tangent bundle of $\mathrm{Fl}(\underline{d},\underline{e})/\mathrm{Fl}((d_1,\ldots,d_{p-1}),(e_1,\ldots,e_{p-1}))$ is canonically identified with $\mathrm{Hom}(\mathscr{P}_{d_p}/\mathscr{P}_{d_{p-1}},\mathscr{O}_{\mathrm{Fl}(\underline{d},\underline{e})}^{\oplus d_p+e_p}/\mathscr{P}_{d_p})$. The tangent bundle for $\mathrm{Fl}(\underline{d},\underline{e})$ is then an extension of the relative tangent bundle by the pullback of the tangent bundle of $\mathrm{Fl}((d_1,\ldots,d_{p-1}),(e_1,\ldots,e_{p-1}))$. Inductively, we obtain the following description:

PROPOSITION 6.3. *The tangent bundle of* $\mathrm{Fl}(\underline{d},\underline{e})$ *is canonically identified as an iterated extension of pullbacks of the bundles*

$$\mathrm{Hom}_{\mathrm{Fl}((d_1,\ldots,d_j),(e_1,\ldots,e_j))}(\mathscr{P}_{d_j}/\mathscr{P}_{d_{j-1}},\mathscr{O}^{\oplus d_j+e_j}/\mathscr{P}_{d_j})$$

with $j=1,\ldots,p$. *In particular, the rank of the tangent bundle is* $\sum_{i=1}^p(d_i-d_{i-1})e_i$.

6.2. Relative orientations of relative normal bundles. Now we discuss how to provide canonical relative orientations on the relative canonical bundles for the flag resolutions of Schubert varieties.

PROPOSITION 6.4. *Fix* $1\leq k<n$ *and consider a Young diagram* Λ *in* $k\times(n-k)$*-frame. If* Λ *satisfies the conditions of Proposition* 4.6 *for the line bundle* $\mathscr{L}$ *but is not even, then the class* $(p_\Lambda)_*(\langle 1\rangle)\in\mathrm{H}^{|\Lambda|}(\mathrm{Gr}(k,n),\mathbf{I}^{|\Lambda|}(\mathscr{L}))$ *is independent of the choice of relative orientation of* $\omega_{p_\Lambda}\otimes p_\Lambda^*(\mathscr{L})$.

PROOF. If Λ satisfies one of the conditions of Proposition 4.6 for the line bundle $\mathscr{L}$, then the corresponding class $\sigma_\Lambda\in\mathrm{Ch}^{|\Lambda|}(\mathrm{Gr}(k,n))$ lifts to the cohomology group $\mathrm{H}^{|\Lambda|}(\mathrm{Gr}(k,n),\mathbf{I}^{|\Lambda|}(\mathscr{L}))$. If Λ is not even, then the corresponding lift is necessarily $\mathrm{I}(F)$-torsion. In particular, by Proposition 2.3, it is uniquely determined by requiring that its reduction is the class σ_Λ. Since the reduction morphism $\mathrm{H}^\bullet(X,\mathbf{I}^\bullet(\mathscr{L}))\to\mathrm{Ch}^\bullet(X)$ is compatible with pushforwards, the class $(p_\Lambda)_*(\langle 1\rangle)$ has reduction σ_Λ (equal to the pushforward of 1 along p_Λ on the level of mod 2 Chow rings), by definition. The claim now follows from the uniqueness statement above. □

In particular, for the torsion Schubert classes in $\mathrm{H}^\bullet(\mathrm{Gr}(k,n),\mathbf{I}^\bullet(\mathscr{L}))$, we don't need to worry about choices of orientations; we can always consider $(p_\Lambda)_*(\langle 1\rangle)$ to be a geometric representative of the corresponding class, independent of the orientation for the relative canonical bundle. Therefore, in the subsequent discussion, we can now focus on even Young diagrams. These correspond to non-torsion classes, and the class $(p_\Lambda)_*(\langle 1\rangle)$ really depends on the choice of orientations.

Before we can discuss the canonical orientations for relative canonical bundles, we need a couple of preliminary statements on determinants of vector bundles. First, the following provides a canonical identification of determinants for tensor products, cf. [**OT14**, Lemma 13].

LEMMA 6.5. *Let* V,W *be two finite-dimensional* F*-vector spaces. Then there is a canonical identification:*

$$\det(V\otimes_F W)\cong\det(V)^{\otimes\dim_F W}\otimes_F\det(W)^{\otimes\dim_F V}.$$

As a direct consequence, we get a variation of [**FM16**, Lemma 4.4]: if $\dim_F V$ and $\dim_F W$ are both even, the tensor product $V \otimes_F W$ has a canonical orientation given by $\mathscr{N} \cong \det(V)^{\otimes(\dim_F W/2)} \otimes_F \det(W)^{\otimes(\dim_F V/2)}$ and the above identification; this can also be applied to $\mathrm{Hom}_F(V,W) \cong V^\vee \otimes_F W$.

PROPOSITION 6.6. *For integers $1 \le d_1 < d_2 < \cdots < d_p = k < n$, consider the flag variety $\mathrm{Fl}(d_1,\ldots,d_p;n)$. If*

$$d_1 \equiv d_2 - d_1 \equiv d_3 - d_2 \equiv \cdots \equiv d_p - d_{p-1} \equiv n-k \mod 2$$

then the flag variety $\mathrm{Fl}(d_1,\ldots,d_p;n)$ has a canonical orientation.

PROOF. The tangent bundle of $\mathrm{Fl}(d_1,\ldots,d_p;n)$ has a filtration whose subquotients are Hom-bundles $\mathrm{Hom}(\mathscr{S}_i,\mathscr{S}_j)$ where $i<j$ and $\mathscr{S}_i$ is the tautological subquotient bundle of rank $d_i - d_{i-1}$. Using this together with Lemma 6.5, we obtain canonical isomorphisms

$$\begin{aligned}\omega^\vee_{\mathrm{Fl}(d_1,\ldots,d_p;n)} &\cong \bigotimes_{i<j} \det \mathrm{Hom}(\mathscr{S}_i,\mathscr{S}_j)\\ &\cong \bigotimes_{i<j}\left(\det \mathscr{S}_i^{\otimes(-\mathrm{rk}\mathscr{S}_j)} \otimes \det \mathscr{S}_j^{\otimes(\mathrm{rk}\mathscr{S}_i)}\right).\end{aligned}$$

This can be rearranged to a tensor product of powers $\det \mathscr{S}_i^{\otimes m_i}$ of the subquotient bundles, with the multiplicity given by

$$m_i = \sum_{j<i} \mathrm{rk}\,\mathscr{S}_j - \sum_{j>i} \mathrm{rk}\,\mathscr{S}_j = \sum_{j<i}(d_j - d_{j-1}) - \sum_{j>i}(d_j - d_{j-1}) = d_{i-1} + d_i - n.$$

Therefore, we have a canonical identification $\omega^\vee_{\mathrm{Fl}} \cong \det \mathscr{S}_i^{\otimes(d_{i-1}+d_i-n)}$.

By the above computation, if all the $d_j - d_{j-1}$ are even, then so are all the m_i. In this case, the canonical bundle ω_{Fl} has a canonical orientation. If all $d_j - d_{j-1}$ are odd, there are two further cases: if p is odd, then we always add an even number of odd summands, therefore m_i is even and the bundle ω_{Fl} has a canonical orientation. Finally, if p is even, then all m_i are odd. Since the subquotient bundles arise from a filtration of the trivial bundle, we have a canonical identification $\bigotimes_i \det \mathscr{S}_i \cong \mathscr{O}_{\mathrm{Fl}}$. In particular, with this identification, ω_{Fl} is again a tensor product of even tensor powers of $\mathscr{S}_i$ and hence canonically oriented. □

REMARK 6.7. *In the special case $\mathrm{Gr}(k,n)$, we have an extension*

$$0 \to \mathscr{S} \to \mathscr{O}_{\mathrm{Gr}}^{\oplus n} \to \mathscr{Q} \to 0$$

where $\mathscr{S}$ is the tautological rank k subbundle of the trivial bundle and $\mathscr{Q}$ is the tautological rank $n-k$ quotient bundle. The tangent bundle is isomorphic to $\mathrm{Hom}(\mathscr{S},\mathscr{Q})$ and consequently Lemma 6.5 provides a canonical identification

$$\omega^\vee_{\mathrm{Gr}(k,n)} \cong (\det \mathscr{S}^\vee)^{\otimes(n-k)} \otimes (\det \mathscr{Q})^{\otimes k}.$$

From the above extension we also have a canonical isomorphism $\det \mathscr{S} \otimes \det \mathscr{Q} \cong \mathscr{O}_{\mathrm{Gr}}$. We can use this canonical isomorphism $\det \mathscr{S}^\vee \cong \det \mathscr{Q}$ to write $\omega^\vee_{\mathrm{Gr}(k,n)} \cong \det \mathscr{Q}^{\otimes n}$. In particular, if n is even, the Grassmannian $\mathrm{Gr}(k,n)$ has a canonical orientation.

We now can describe the canonical orientations for the relative canonical bundles of the flag resolutions.

PROPOSITION 6.8. *For an even Young diagram Λ, consider the associated tuples $\underline{d} = (d_1, \dots, d_p = k)$ and $\underline{e} = (e_1, \dots, e_p)$ with $0 < d_1 < \cdots < d_p < n$ and $e_1 + d_1 \leq \cdots \leq e_p + d_p$, cf. Figure 1 and Definition 3.8.*

(1) *If the Young diagram Λ has twist 0, then the relative canonical bundle ω_{p_Λ} for the resolution $p_\Lambda \colon \mathrm{Fl}(\underline{d}, \underline{e}) \to \mathrm{Gr}(k, V)$ is canonically orientable.*

(2) *If the Young diagram Λ has twist 1, then the twisted relative canonical bundle $\omega_{p_\Lambda} \otimes p_\Lambda^* \det \mathscr{E}_k^\vee$ is canonically orientable.*

PROOF. We compute the canonical bundle of $\mathrm{Fl}(\underline{d}, \underline{e})$, this is essentially the same argument as in [**BC12a**, Proposition 1.13]. By Proposition 6.3, the tangent bundle is canonically identified as iterated extension of Hom-bundles. This translates into a canonical identification of $\omega_{\mathrm{Fl}(\underline{d},\underline{e})}$ as tensor product of exterior powers of Hom-bundles, and the latter can be canonically identified via Lemma 6.5. Inductively, we obtain a canonical identification of the canonical bundle:

$$\omega_{\mathrm{Fl}(\underline{d},\underline{e})} \cong \bigotimes_{i=1}^{p-1} \det \mathscr{P}_{d_i}^{\otimes(d_i - d_{i-1} + e_i - e_{i+1})} \otimes \det \mathscr{P}_{d_p}^{\otimes(d_p - d_{p-1} + e_p)}.$$

The relative canonical bundle ω_{p_Λ} is obtained from this by tensoring with the dual of the pullback of the canonical bundle of the Grassmannian, which by Proposition 6.6 is $\det \mathscr{P}_{d_p}^{d_p+n}$, cf. also [**BC12a**, Corollary 1.14].

(1) As in the proof of Proposition 6.6, if the twist of Λ is 0, then all the determinants of subquotient bundles in the relative canonical bundle identified above appear with powers which all have the same parity. We get a canonical orientation if all the powers are even. If all powers are odd, we can again use the identification $\otimes_i \det \mathscr{S}_i \cong \mathscr{O}_{\mathrm{Fl}}$ to reduce to the case where all powers are even, hence we also have a canonical orientation.

(2) If the twist of Λ is 1, we can remove even powers of line bundles as in (1). What remains is a single copy of $\det \mathscr{P}_{d_p}$ which is the pullback of a representative of the generator of $\mathrm{Pic}(\mathrm{Gr}(k,n))/2$. In particular, we get a canonical orientation on $\omega_{p_\Lambda} \otimes p_\Lambda^* \mathscr{L}$. □

REMARK 6.9. *Note that the orientability of ω_{p_Λ} resp. $\omega_{p_\Lambda} \otimes p_\Lambda^* \mathscr{L}$ is only possible if the class of the relative canonical bundle ω_{p_Λ} belongs to the image of the restriction morphism $p_\Lambda^* \colon \mathrm{Pic}(\mathrm{Gr}(k,V))/2 \to \mathrm{Pic}(\mathcal{F}l(\Lambda))/2$. By [**BC12a**], this is the case if and only if the following conditions are satisfied:*

(1) *$d_i - d_{i-1} + e_{i+1} - e_i$ is even for every $i = 2, \dots, k-1$ (for $k \geq 3$)*

(2) *when $0 < e_1 < e$ and $k \geq 2$, $d_1 + e_2 - e_1$ is also even*

In turn, these conditions are satisfied if and only if the corresponding Schubert class lifts to the Chow–Witt ring, i.e., one of the conditions in Proposition 4.6 must be satisfied.

REMARK 6.10. *Schubert varieties are generally singular, but they are normal and Cohen–Macaulay. In particular, the canonical bundle ω_{X_w} of a Schubert variety X_w can be obtained as push-forward of the canonical bundle from the smooth locus. This implies that restricting a choice of orientation for the canonical bundle of a singular Schubert variety to the smooth locus is the same thing as an orientation of the canonical bundle of the smooth locus.*

Now we can discuss our natural choices for fundamental classes of Schubert varieties in $\mathbf{I}^\bullet$-cohomology. We consider an appropriate resolution, a (canonical) choice of orientation, and then the fundamental class of the Schubert variety in the $\mathbf{I}^\bullet$-cohomology ring is given by pushforward of $\langle 1\rangle$ along the resolution.

THEOREM 6.11. *Fix $1 \leq k < n$, let Λ be an even Young diagram in $k\times(n-k)$-frame and denote by $p_\Lambda\colon \mathrm{Fl}(\Lambda)\to \mathrm{Gr}(k,n)$ the resolution of singularities for the Schubert variety Σ_Λ. Then $(p_\Lambda)_*(\langle 1\rangle) \in \mathrm{H}^{|\Lambda|}(\mathrm{Gr}(k,n),\mathbf{I}^{|\Lambda|}(\mathscr{L}))$ is a lift of the Schubert class $\sigma_\Lambda \in \mathrm{Ch}^{|\Lambda|}(\mathrm{Gr}(k,n))$. A canonical choice for such lift is given by the canonical orientations of Proposition* 6.8.

PROOF. The morphism $p_\Lambda\colon \mathrm{Fl}(\Lambda)\to\mathrm{Gr}(k,n)$ is a proper morphism between smooth F-schemes, hence we have a push-forward map, cf. Section 5.2:

$$\mathrm{H}^0(\mathrm{Fl}(\Lambda),\mathbf{I}^0(\omega_{p_\Lambda}\otimes p_\Lambda^*\mathscr{L})) \xrightarrow{(p_\Lambda)_*} \mathrm{H}^d(\mathrm{Gr}(k,n),\mathbf{I}^d(\mathscr{L})).$$

Here $d = \dim \mathrm{Gr}(k,n) - \mathrm{Fl}(\Lambda) = |\Lambda|$, cf. [**BC12a**, Propositions 1.3, 1.13]. For an even Young diagram, the appropriate choice of twisting line bundle $\mathscr{L}$ is a generator of $\mathrm{Pic}(\mathrm{Gr}(k,n))/2$ if one of the two possible boundary components of Λ is odd; the general twist is determined by the half-perimeter [**BC12a**, Definition 4.3].

Since the push-forward is compatible with the reduction morphism $\mathrm{H}^j(X,\mathbf{I}^j)\to \mathrm{Ch}^j(X)$, the image of $(p_\Lambda)_*(\langle 1\rangle)$ in $\mathrm{Ch}^{|\Lambda|}(\mathrm{Gr}(k,n))$ is the class of the subscheme given by the image of p_Λ, which is exactly the Schubert class σ_Λ.

It remains to note that the push-forward morphism depends on the choice of a relative orientation, i.e., a line bundle $\mathscr{M}$ and an isomorphism $\mathscr{M}^{\otimes 2}\cong \omega_{p_\Lambda}\otimes p_\Lambda^*\mathscr{L}$. But we can use the canonical orientations from Proposition 6.8 to get canonical lifts of Schubert classes. □

REMARK 6.12. *The canonical lifts in the above theorem are in fact generators of the corresponding $\mathrm{W}(F)$-summand in $\mathbf{I}^{|\Lambda|}(\mathscr{L})$-cohomology. In* [**BC12a**]*, inductive arguments involving localization sequences and explicit computations of images of generators under boundary morphisms were used to show that the classes $(p_\Lambda)_*(\langle 1\rangle)$ are generators of $\mathrm{W}^{\mathrm{tot}}(\mathrm{Gr}(k,n))$. The same arguments can now also be used to show that the classes $(p_\Lambda)_*(\langle 1\rangle)$ are generators for $\bigoplus_{j,\mathscr{L}}\mathrm{H}^j(\mathrm{Gr}(k,n),\mathbf{I}^j(\mathscr{L}))$, using the compatibility of the comparison morphism $\alpha\colon \mathrm{W}^j(X,\mathscr{L})\to\mathrm{H}^j(X,\mathbf{I}^j(\mathscr{L}))$ with pullbacks, pushforwards and localization sequences (possibly up to sign).*

7. Geometric interpretation of oriented intersection multiplicities

In this section, we can now describe the oriented intersection multiplicity of a solution subspace in an intersection of Schubert varieties. Essentially, we can use the classical homological transversality results to make the Schubert varieties transverse and then the oriented intersection multiplicity is given by the length of the reduced subscheme plus additional relative orientation information.

7.1. Transversality. Before we can provide explicit geometric identifications of the oriented intersection multiplicities, we need to discuss how to make the Schubert varieties in the Schubert problem transversal. Classically, one uses (homological) transversality results to make the Schubert cycles transverse, and we will discuss how to take care of the additional orientations that are floating around.

We first recall the very classical Kleiman–Bertini transversality theorem, cf. [**Kle74**]:

THEOREM 7.1. *Let F be a field of characteristic* 0*, let G be a geometrically connected algebraic group such that the set $G(F)$ of F-points is Zariski dense in G. Assume that $G \looparrowright X$ is a smooth variety over F with geometrically transitive G-action, with integral closed subschemes $Y, Z \subseteq X$. Then there exists a dense open subset U of G such that for each point $g \in U(F)$, the intersection $gY \cap Z$ is proper and each component appears with multiplicity one, i.e., there exists a dense open in $gY \cap Z$ which is regular and has pure dimension* $\dim Y + \dim Z - \dim X$.

REMARK 7.2. *The results in* [**Kle74**] *are only stated for algebraically closed fields. However, requiring the geometrically transitive G-action above implies that the conditions for Kleiman's results from* [**Kle74**] *are satisfied. The condition that $G(F)$ is Zariski dense then implies that the open set U produced by Kleiman's result contains some F-point. The conditions are of course satisfied for the* GL_n*-action on the Grassmannian* $\mathrm{Gr}(k,n)$.

In particular, at a general point of the intersection $gY \cap Z$, the tangent space to X will decompose as direct sum of the normal spaces to gY and Z and the tangent space to $gY \cap Z$. (Note that this is only a statement about vector spaces; if the intersection has positive dimension, the relevant restricted tangent and normal bundles are only related by exact sequences.) Moreover, if we have $Y, Z \subseteq X$ of complementary dimensions, the intersection $gY \cap Z$ is a zero-dimensional reduced scheme which is contained in the smooth loci of Y and Z, respectively. In positive characteristic, this strong version of transversality fails, but there are some homological transversality results that can be used (in conjunction with Serre's Tor-formula) to compute intersection products in Chow rings.

DEFINITION 7.3. *Let F be a field and let X be a variety over F. Two quasi-coherent sheaves $\mathscr{E}$ and $\mathscr{F}$ on X are called* homologically transverse *if the higher Tor-sheaves over $\mathscr{O}_X$ vanish:*

$$\mathscr{T}\!or_j^X(\mathscr{E}, \mathscr{F}) = 0 \text{ for all } j \geq 1.$$

If $\mathscr{E} = \mathscr{O}_Y$ for a closed subvariety $Y \subseteq X$, we say that $\mathscr{F}$ is homologically transverse to Y.

We consider a scheme F defined over a field X, fix an algebraic closure $\overline{F}$ of F and denote $\overline{X} = X \times \operatorname{Spec} \overline{F}$. For a coherent sheaf $\mathscr{F}$ on X, we also denote by $\overline{\mathscr{F}}$ the pullback of $\mathscr{F}$ to $\overline{X}$. The following criterion for homological transversality being generic is proved in [**Sie09**, Theorem 1.2]:

THEOREM 7.4. *Let F be a field with algebraic closure $\overline{F}$. Let X be an F-scheme with a left action of a smooth algebraic group G, and let $\mathscr{F}$ be a coherent sheaf on X. If for all closed points $x \in \overline{X}$ the sheaf $\overline{\mathscr{F}}$ is homologically transverse to the closure of the $G(\overline{F})$-orbit of x, then for all coherent sheaves $\mathscr{E}$ on X there is a Zariski open and dense subset $U \subseteq G$ such that for all $g \in U(F)$ the sheaf $g\mathscr{F}$ is homologically transverse to $\mathscr{E}$.*

REMARK 7.5. *We can apply this to $X = \mathrm{Gr}(k,n)$ as variety with GL_n-action over F and $\mathscr{E} = \mathscr{O}_Y$, $\mathscr{F} = \mathscr{O}_Z$ for two Schubert varieties $Y, Z \subseteq X$. Then the conditions of Sierra's theorem are satisfied, and the two Schubert varieties Y and*

Z can be made homologically transversal by taking a generic translate gY for $g \in \mathrm{GL}_n(F)$. Note that this requires F to be infinite; for a finite field $F = \mathbb{F}_q$, the dense open subset $U \subseteq G$ need not contain $\mathbb{F}_q$-rational points. Note also that this result can be used to make Y transversal to finitely many other Schubert varieties: the intersection of the relevant open dense subsets $U_1, \ldots, U_l$ will be open and dense.

REMARK 7.6. *There are also other criteria that could be applied to deduce homological transversality for intersections of Schubert varieties in Grassmannians. For example,* [**Bri02**, Lemma 1] *states that any proper intersection of equidimensional Cohen–Macaulay subschemes Y and Z in a smooth variety X is homologically transverse.*

Changing a Schubert variety $\Sigma_\Lambda(\mathscr{F})$ to a GL_n-translate $g\Sigma_\Lambda(\mathscr{F})$ doesn't change the corresponding class in the Chow ring. The same is true for the Chow–Witt ring:

PROPOSITION 7.7. *Let F be a perfect field of characteristic $\neq 2$. Let $Y = \Sigma_\Lambda(\mathscr{F}) \subseteq \mathrm{Gr}(k,n)$ be a Schubert variety corresponding to a Young diagram satisfying one of the conditions of Proposition 4.6 for the line bundle $\mathscr{L}$. For any $g \in \mathrm{SL}_n(F)$, denote by $p_Y\colon \mathrm{Fl}(\Lambda, \mathscr{F}) \to \mathrm{Gr}(k,n)$ and $p_{gY}\colon \mathrm{Fl}(\Lambda, g\mathscr{F}) \to \mathrm{Gr}(k,n)$ the corresponding resolutions of singularities of Y and gY, respectively. Then we have*

$$(p_Y)_*(\langle 1\rangle) = (p_{gY})_*(\langle 1\rangle) \in \mathrm{H}^{|\Lambda|}(\mathrm{Gr}(k,n), \mathbf{I}^{|\Lambda|}(\mathscr{L}))$$

PROOF. Since $g \in \mathrm{SL}_n(F)$, it has an elementary factorization. Therefore, we have a chain of $\mathbb{A}^1$-families of translates of Y connecting Y and gY. It suffices to prove the claim for a single $\mathbb{A}^1$-family $Y \times \mathbb{A}^1 \to \mathrm{Gr}(k,n)$. This $\mathbb{A}^1$-family gives rise to a family of resolutions $q\colon \mathrm{Fl}(\Lambda) \times \mathbb{A}^1 \to \mathrm{Gr}(k,n)$. We view this as a proper morphism $\mathrm{Fl}(\Lambda) \times \mathbb{A}^1 \to \mathrm{Gr}(k,n) \times \mathbb{A}^1$ (over $\mathbb{A}^1$), and we want to compare the pushforwards of $\langle 1\rangle$ from the two fibers, for Y and gY. The form $\langle 1\rangle$ obviously extends to the whole family $q\colon \mathrm{Fl}(\Lambda) \times \mathbb{A}^1 \to \mathrm{Gr}(k,n) \times \mathbb{A}^1$. Since $\mathrm{Fl}(\Lambda)$ is smooth, the bundle $\omega_q \otimes q^*\mathscr{L}$ is isomorphic to the pullbacks of $\omega_{p_Y} \otimes p_Y^*\mathscr{L}$ and $\omega_{p_{gY}} \otimes p_{gY}^*\mathscr{L}$ along the respective $\mathbb{A}^1$-projections. Moreover, the corresponding induced orientations on $\omega_q \otimes q^*\mathscr{L}$ agree. In particular, we have a pushforward $(p_q)_*(\langle 1\rangle)$ in the Chow–Witt ring of $\mathrm{Gr}(k,n) \times \mathbb{A}^1$. By the base change result of [**AF16a**, Theorem 2.4.1], the restriction of this class to the fibers over 0 and 1 (corresponding to the Schubert varieties Y and gY) agrees with the classes $(p_Y)_*(\langle 1\rangle)$ and $(p_{gY})_*(\langle 1\rangle)$, thus proving the claim. □

REMARK 7.8. *More generally, one would expect for a smooth variety X with an action of an $\mathbb{A}^1$-connected group G, that translation of $\mathbf{I}^\bullet$-cohomology cycles by $g \in G(F)$ doesn't change the cohomology class. In the corresponding Borel–Moore homology picture, cf.* [**DF17**]*, the above argument shows that this is true because of $\mathbb{A}^1$-invariance for singular varieties. However, in the $\mathbf{I}^\bullet$-cohomology picture, this is not so clear, but the above special case is enough for our purposes.*

REMARK 7.9. *Note also that the above result can be applied to arbitrary Schubert varieties. The homological transversality result of Sierra, applied to the SL_n-action on $\mathrm{Gr}(k,n)$, implies that we can make Schubert varieties homologically transversal by elements from SL_n. In particular, the canonical fundamental classes of Schubert varieties in $\mathbf{I}^\bullet$-cohomology discussed in Section 6 are independent of the choice of flag defining the Schubert variety. Hence, to compute oriented intersection products* *of such fundamental classes, we can always assume that the corresponding Schubert varieties are homologically transversal, provided the base field is infinite.*

Finally, we want to mention Vakil's generic smoothness theorem [**Vak06**, Theorem 1.5] which provides all the relevant consequences of the Kleiman–Bertini theorem in positive characteristic, for zero-dimensional intersections.[3] As a particular consequence of Vakil's generic smoothness result, for a given Schubert problem over a finite field, there is a positive density of flags (possibly defined over finite field extensions $E/\mathbb{F}_q$) where the intersection subscheme is just a union of reduced E-rational points. A fortiori, there will be a positive density of flags where the corresponding intersection of Schubert varieties is a reduced subscheme. In particular, while it's not quite clear if Schubert varieties can be made transversal over finite fields, this will always be possible after a finite field extension.

7.2. The type of subspaces. Finally, we need to interpret the local contribution of each solution subspace to the degree of the oriented zero-cycle computed previously. This will be done via comparison of the canonical orientations of the Schubert varieties involved, generalizing the notion of type of subspaces in [**FM16**] similar to the type of lines in [**KW17**].

We first discuss the geometric interpretation of intersection multiplicities in the case where the relevant Schubert varieties are smooth (or intersect in points where the resolutions of singularities from Section 6 are isomorphisms) and there are no twists involved. Let F be a field of characteristic $\neq 2$, let V be an n-dimensional F-vector space, and let $\mathfrak{S}_1, \ldots, \mathfrak{S}_l$ be smooth Schubert varieties in $\operatorname{Gr}(k, V)$ with $\sum_i \dim \mathfrak{S}_i = k(n-k)$, intersecting properly in a closed point $[W] \in \operatorname{Gr}(k, n)$. Then $[W]$ is defined over a finite extension field E/F, and corresponds to a k-dimensional E-vector space $W \subseteq V \otimes_F E$. Since the Schubert varieties are smooth, we have normal spaces $\mathrm{N}_{[W]}\,\mathfrak{S}_i$ of dimension $k(n-k) - \dim \mathfrak{S}_i$, and the normal space can be explicitly identified as $\operatorname{Hom}(W \cap U, V/\langle U, W\rangle)$ with $U \leq V$ a subspace of suitable dimension.[4] Moreover, we have a canonical quotient map $\mathrm{T}_{[W]} \operatorname{Gr}(k, n) \to \mathrm{N}_{[W]}\,\mathfrak{S}_i$ which can be explicitly identified as

$$\begin{array}{rcl} \mathrm{T}_{[W]} \operatorname{Gr}(k, n) \cong \operatorname{Hom}(W, V/W) & \to & \operatorname{Hom}(W \cap U, V/\langle U, W\rangle) \cong \mathrm{N}_{[W]}\,\mathfrak{S}_i \\ \psi\colon W \to V/W & \mapsto & \left(W \cap U \hookrightarrow W \xrightarrow{\psi} V/W \twoheadrightarrow V/\langle U, W\rangle\right) \end{array}$$

cf. e.g. the discussion in [**FM16**, Section 4.1]. Since the Schubert varieties intersect properly at $[W]$, the natural map

$$\xi_W\colon\ \mathrm{T}_{[W]} \operatorname{Gr}(k, n) \to \bigoplus_{i=1}^{l} \mathrm{N}_{[W]}\,\mathfrak{S}_i$$

obtained from the above projections to the normal spaces is an isomorphism, called the *splitting map*. The splitting map induces an isomorphism of determinants,

[3] A comment after the statement of Theorem 1.5 in loc.cit. suggests that we might hope for a Kleiman–Bertini type generic smoothness theorem in arbitrary characteristic if we assume generic transversality with respect to Schubert varieties. Such a result would solve all the transversality issues left in the present discussion, but this seems open at the moment.

[4] The Schubert variety being smooth is equivalent to the Young diagram being a rectangle. In particular, the Schubert variety is determined by intersection conditions w.r.t. a flag $0 \leq U \leq V$.

which we can rewrite as a canonical orientation

$$\mathscr{O}_{\mathrm{Gr}(k,n),[W]} \cong \omega_{\mathrm{Gr}(k,n)}|_{[W]} \otimes \bigotimes_{i=1}^{l} \omega_{\mathfrak{S}_i/\mathrm{Gr}(k,n)}|_{[W]}.$$

On the other hand, the canonical orientations of the tangent bundle of $\mathrm{Gr}(k,n)$ and the normal bundles of $\mathfrak{S}_i$, cf. Propositions 6.6 and 6.8, provide another such orientation of the right-hand side tensor product. Note that the right-hand side is a one-dimensional E-vector space, and each of the two orientations is a generator. Since an orientation is only well-defined up to multiplication by squares, the two orientations differ by a unit $u \in E^\times$ whose square residue in $E^\times/(E^\times)^2$ only depends on the orientations and not on the specific generators representing them. It makes sense to denote this unit u by $\det \xi_W$: if we choose bases in $\mathrm{T}_{[W]}\mathrm{Gr}(k,n)$ and $\bigoplus_{i=1}^{l} \mathrm{N}_{[W]}\mathfrak{S}_i$ which induce the orientations, then the unit u is given as the square class of the determinant of the matrix representing the splitting map in the chosen bases.

We can now define the type of the solution subspace W which will turn out to coincide with the oriented intersection multiplicity of the point $[W]$ in the intersection product $[\mathfrak{S}_1]\cdots[\mathfrak{S}_l]$ in the Chow–Witt ring.

DEFINITION 7.10. *In the above notation, if W is a solution subspace in the intersection of the Schubert varieties $\mathfrak{S}_1,\dots,\mathfrak{S}_l$, defined over E/F, then the* type *of W is defined to be*

$$\mathrm{tr}_{E/F}\langle \det \xi_W\rangle \in \mathrm{GW}(F),$$

i.e., it is the Scharlau trace of the one-dimensional quadratic form given by the determinant of the splitting map.

REMARK 7.11. *The same procedure works if we consider non-orientable Schubert varieties. In this case, the normal bundles to some of the $\mathfrak{S}_i$ may not be orientable. However, the twisting by $\mathscr{L}_i \cong \det \mathscr{E}_k^\vee|_{\mathfrak{S}_i}$ makes the bundles orientable. Since we want to consider intersection multiplicities, we have to look at Schubert problems in Grassmannians with n even, and then we must necessarily have an even number of non-orientable Schubert varieties involved in the intersection problem. The type of a subspace in the intersection is then as in Definition* 7.10 *above: the type is given by the square class of units which compares the canonical orientations on $\omega_{\mathrm{Gr}(k,n)}|_{[W]} \otimes \bigotimes_{i=1}^{l} \omega_{\mathfrak{S}_i/\mathrm{Gr}(k,n)}|_{[W]}(\mathscr{L}_i)$, one given by the natural morphisms and the other one given by the canonical orientations.*

We also need to discuss a notion of types of solution subspaces for the case of intersections of singular Schubert varieties. While this definition is really applicable to all intersection problems, it will usually be difficult to actually evaluate in practice.

Let F be a field of characteristic $\neq 2$, let V be an n-dimensional F-vector space, and let $\mathfrak{S}_1,\dots,\mathfrak{S}_l$ be Schubert varieties in $\mathrm{Gr}(k,V)$ with $\sum_i \dim \mathfrak{S}_i = k(n-k)$, intersecting properly in a closed point $[W] \in \mathrm{Gr}(k,n)$. Consider the resolutions of singularities $p_i\colon \mathrm{Fl}(\Lambda_i) \to \mathrm{Gr}(k,V)$ of Section 6, where Λ_i is the Young diagram for the Schubert variety $\mathfrak{S}_i$. For every i, we have a line bundle $\mathscr{L}_i$ (either equal to $\mathscr{O}_{\mathrm{Gr}(k,n)}$ or $\det \mathscr{E}_k^\vee$) such that $\omega_{p_i} \otimes p_i^*\mathscr{L}_i$ has a canonical orientation, cf. Proposition 6.8, i.e., there is a line bundle $\mathscr{M}_i$ such that $\mathscr{M}_i \otimes \mathscr{M}_i \xrightarrow{\cong} \omega_{p_i} \otimes p_i^*\mathscr{L}_i$. We can now consider the fiber $p_i^{-1}([W])$ of the resolution $p_i\colon \mathrm{Fl}(\Lambda_i) \to \mathrm{Gr}(k,V)$ over the point $[W]$. The push-forward of the orientation form $\mathscr{M}_i \otimes \mathscr{M}_i \xrightarrow{\cong} \omega_{p_i} \otimes p_i^* \mathscr{L}_i$ is a form $\mathrm{R}(p_i)_*(\mathscr{M}_i) \otimes \mathrm{R}(p_i)_*(\mathscr{M}_i) \to \mathscr{L}_i$, cf. [**CH11**] for the exact definition of the push-forward of forms. Restricted to the point $[W]$, we get an intersection form on coherent cohomology

$$\cup\colon \mathrm{H}^r(p_i^{-1}([W]), \mathscr{M}_i) \otimes \mathrm{H}^s(p_i^{-1}([W]), \mathscr{M}_i) \to \mathscr{O}_{[W]}$$

DEFINITION 7.12. *In the above notation, if W is a solution subspace in the intersection of the Schubert varieties $\mathfrak{S}_1, \dots, \mathfrak{S}_l$, then the* type *of W is the image of the product in $\mathrm{GW}_{[W]}(\mathrm{Gr}(k,V))$ of the intersection forms*

$$\left(\bigoplus_{j=0}^{\dim p_i^{-1}([W])} \mathrm{H}^j(p_i^{-1}([W]), \mathscr{M}_i)[-j], \cup \right)$$

for all $i = 1, \dots, l$ under the identification $\mathrm{GW}_{[W]}(\mathrm{Gr}(k,V)) \cong \mathrm{GW}(F)$ given by the canonical orientation of the tangent space $\mathrm{T}_{[W]}\,\mathrm{Gr}(k,V)$.

REMARK 7.13. *This is very much inspired by the Levine–Serre formula for the motivic Euler classes of smooth varieties in terms of trace forms on $\bigoplus \mathrm{H}^i(X, \Omega^j)$, cf.* [**Lev18**].

REMARK 7.14. *Note that Definition* 7.12 *reduces to the simpler Definition* 7.10 *when the resolutions $p\colon \mathrm{Fl}(\Lambda) \to \mathrm{Gr}(k,V)$ are isomorphisms at the intersection point. In this case, both forms are simply given by the canonical orientations of the normal bundle of the Schubert variety at the (smooth) intersection point.*

7.3. Oriented intersection multiplicities.

REMARK 7.15. *In the following, we want to identify the multiplicity of a closed point $[W] \in \mathrm{Gr}(k,V)$ in the intersection product (in the Chow–Witt ring resp. the $\mathbf{I}^\bullet$-cohomology) of oriented Schubert classes $\mathfrak{S}_1, \dots, \mathfrak{S}_l$. This will be done using the oriented Tor formula of Section* 5.3 *and computations of $\star$-products in Witt rings. Of course, since $\mathrm{W}^{\mathrm{tot}}(\mathrm{Gr}(k,V))$ is a free $\mathrm{W}(F)$-module generated by the even Young diagrams, this procedure doesn't apply to intersections involving torsion classes. However, the product with torsion classes is a torsion class and can be identified with the corresponding intersection product in the mod 2 Chow ring. In these cases, the only information contained in the intersection multiplicity is the length of the subscheme mod 2, and this can be computed with classical methods. We can therefore restrict to intersections involving only even Young diagrams, where we have canonical lifts to $\mathrm{W}(\mathrm{Gr}(k,V))$, cf. Section* 8.3.

REMARK 7.16. *In the below results, we have restrictions on the twists of the Schubert classes. These arise because we don't have a pushforward from the top Chow–Witt group to the Grothendieck–Witt ring $\mathrm{GW}(F)$ of the base field if the Grassmannian $\mathrm{Gr}(k,n)$ is not oriented, i.e., if n is odd. Of course, one can ask about the corresponding intersection problems in the orientation cover $\widetilde{\mathrm{Gr}}(k,n)$, and basically the same Schubert calculus applies to such computations.*

The first result is a special case in which we can apply the simpler notion of types of Definition 7.10.

THEOREM 7.17. *Let F be a field of characteristic $\neq 2$. Let $V = F^n$, and fix an orientation on $V = F^n$. Let $[\mathfrak{S}_1], \dots, [\mathfrak{S}_l]$ be oriented Schubert classes in $\mathrm{Gr}(k,V)$*

such that $\sum_i \dim \Sigma_i = k(n-k)$. *Assume that* F *is of characteristic* 0, *or that the Schubert varieties are smooth. Let* x *be a point in the intersection of general translates of* $\mathfrak{S}_1, \ldots, \mathfrak{S}_l$, *corresponding to a subspace* $W \subset V \otimes_F E$ *over a finite extension field* E/F. *Then the multiplicity of* x *in the intersection product*

$$[\mathfrak{S}_1] \cdots [\mathfrak{S}_l] \in \widetilde{\mathrm{CH}}^{k(n-k)}(\mathrm{Gr}(k,n))$$

is given by $\mathrm{type}(W) = \mathrm{tr}_{E/F}\langle \det \xi_W \rangle \in \mathrm{GW}(F)$.

Proof. If $\operatorname{char} F = 0$, we can use Kleiman's transversality, cf. Theorem 7.1, to see that general translates of the Schubert varieties $\mathfrak{S}_1, \ldots, \mathfrak{S}_l$ are transverse. If $\operatorname{char} F = p > 0$, we can use Sierra's homological transversality, cf. Theorem 7.4, to see that general translates of the Schubert varieties are homologically transverse, and by the smoothness assumption, the intersection points in the Schubert varieties are smooth. Summing up, for the arguments below we can assume that the intersection of Schubert varieties is a reduced zero-dimensional subscheme of $\mathrm{Gr}(k,n)$ and the direct sum of the normal spaces of the Schubert varieties is isomorphic to the tangent space.

The relevant classes $[\sigma_j]$ for the intersection problem are given as $(\iota_j)_*(\langle 1 \rangle)$ where $\iota_j \colon \mathfrak{S}_j \to \mathrm{Gr}(k,n)$ is the inclusion of the relevant Schubert variety. Note that the pushforward map $(\iota_j)_*$ depends on the choice of orientation of V, but we take the canonical relative orientation of the Schubert varieties from Proposition 6.8.

The intersection product is now computed by restricting the exterior product

$$(\iota_1)_*(\langle 1 \rangle) \boxtimes \cdots \boxtimes (\iota_l)_*(\langle 1 \rangle) \in \widetilde{\mathrm{CH}}^{k(n-k)}(\mathrm{Gr}(k,n)^{\times l})$$

along the diagonal inclusion $\Delta \colon \mathrm{Gr}(k,n) \hookrightarrow \mathrm{Gr}(k,n)^{\times l}$. By [**CF14**], the proper pushforward commutes with the exterior product; in the case at hand there are no signs because the pushforwards start at $\widetilde{\mathrm{CH}}^0$. In particular, we have

$$(\iota_1)_*(\langle 1 \rangle) \boxtimes \cdots \boxtimes (\iota_l)_*(\langle 1 \rangle) = (\iota_1 \times \cdots \times \iota_l)_*(\langle 1 \rangle).$$

Note that the data determining the pushforward $(\iota_1 \times \cdots \times \iota_l)_*$ is exactly given by the orientations of the exterior product of the normal bundles of the Schubert varieties $\mathfrak{S}_j$.

Now we consider the following pullback diagram

$$\begin{array}{ccc} \mathfrak{S}_1 \cap \cdots \cap \mathfrak{S}_l & \xrightarrow{\Delta'} & \mathfrak{S}_1 \times \cdots \times \mathfrak{S}_l \\ {\scriptstyle \iota}\downarrow & & \downarrow{\scriptstyle \iota_1 \times \cdots \times \iota_l} \\ \mathrm{Gr}(k,n) & \xrightarrow[\Delta]{} & \mathrm{Gr}(k,n)^{\times l} \end{array}$$

In this case, the map $\iota_1 \times \cdots \times \iota_l$ is a regular immersion of smooth schemes, and the diagram is transversal in the sense of the base-change result [**AF16a**, Theorem 2.4.1]. Consequently, we have

$$\Delta^*(\iota_1 \times \cdots \times \iota_l)_*(\langle 1 \rangle) = \iota_* \Delta'^*(\langle 1 \rangle) = \iota_*(\langle 1 \rangle).$$

In particular, for any point $x \in \mathfrak{S}_1 \cap \cdots \cap \mathfrak{S}_l$ the corresponding form is $(\iota_x)_*(\langle 1 \rangle) \in \widetilde{\mathrm{CH}}^{k(n-k)}(\mathrm{Gr}(k,n))$. Note that as usual the pushforward $(\iota_x)_*$ depends on choices, and in this case the relevant choices are given by the orientations of the normal spaces of the Schubert varieties intersecting at x. The relevant multiplicity of the point x in the intersection product depends on the choice of generator of

$\widetilde{\mathrm{CH}}^{k(n-k)}(\mathrm{Gr}(k,n))$. The natural choice is given by the canonical orientation of the tangent space of $\mathrm{Gr}(k,n)$, cf. Proposition 6.6; the corresponding generator of the top Chow–Witt group is the pushforward of $\langle 1\rangle$ along ι_x, with the tangent-space orientation.[5] With this choice of isomorphism $\widetilde{\mathrm{CH}}^{k(n-k)}(\mathrm{Gr}(k,n))\cong \mathrm{GW}(F)$ the class of $(\iota_x)_*(\langle 1\rangle)$ (here with the orientation coming from the normal spaces of the Schubert varieties) is mapped exactly to $\mathrm{type}(W)$ as claimed. □

REMARK 7.18. *This result can also be used to show that the classes $(p_\Lambda)_*(\langle 1\rangle)$ are generators, cf. Remark 6.12. Take an even Young diagram Λ and its complementary Young diagram $\overline{\Lambda}$ (obtained as complement of Λ in the $k\times(n-k)$-frame, rotated by 180 degrees). By classical Schubert calculus, the degree of their intersection in the top Chow group is 1, i.e., generic translates of the corresponding Schubert varieties intersect in a single rational point. By the above, the oriented intersection multiplicity of this rational point is the square class of the unit comparing the two natural orientations (from the tangent space of the Grassmannian and the direct sum of the two normal spaces for the Schubert varieties). This is a unit in the Grothendieck–Witt ring and therefore the canonical orientations on the two Schubert varieties for Λ and $\overline{\Lambda}$ must be generators of the corresponding* $\mathrm{GW}(F)$*-summands in the Chow–Witt ring.*

THEOREM 7.19. *Let F be a field of characteristic $\neq 2$, let n be an even number and set $V=F^n$. Consider oriented Schubert classes $\mathfrak{S}_i\in\widetilde{\mathrm{CH}}^\bullet(\mathrm{Gr}(k,V),\mathscr{L}_i)$, $i=1,\dots,l$ such that the underlying Schubert varieties are in general position, $\sum_i\dim\mathfrak{S}_i=k(n-k)$ and $\bigotimes_i\mathscr{L}_i=0$ in $\mathrm{Pic}(\mathrm{Gr}(k,V))/2$. For a subspace $[W]\in\mathfrak{S}_1\cap\dots\cap\mathfrak{S}_l$, the oriented multiplicity of $[W]$ in the intersection product $[\mathfrak{S}_1]\cdots[\mathfrak{S}_l]$ is given by the type of W in the sense of Definition* 7.12.

PROOF. The assumptions imply that the underlying Schubert varieties of the cycles $\mathfrak{S}_i$ are homologically transverse.

Now we want to apply the oriented Tor formula, cf. Section 5.3 which provides a commutative diagram

$$\begin{array}{ccc}
\mathrm{GW}^{n_1}_{\mathfrak{S}_1}(\mathrm{Gr}(k,V),\mathscr{L}_1)\times\cdots\times\mathrm{GW}^{n_l}_{\mathfrak{S}_l}(\mathrm{Gr}(k,V),\mathscr{L}_l) & \xrightarrow{\star} & \mathrm{GW}^{k(n-k)}_{\mathfrak{S}_1\cap\cdots\cap\mathfrak{S}_l}(\mathrm{Gr}(k,V)) \\
\downarrow{\scriptstyle \alpha_{\mathfrak{S}_1}\times\cdots\times\alpha_{\mathfrak{S}_l}} & & \downarrow{\scriptstyle \alpha_{(\mathfrak{S}_1\cap\cdots\cap\mathfrak{S}_l)}} \\
\widetilde{\mathrm{CH}}^{n_1}(\mathrm{Gr}(k,V),\mathscr{L}_1)\times\cdots\times\widetilde{\mathrm{CH}}^{n_l}(\mathrm{Gr}(k,V),\mathscr{L}_l) & \xrightarrow[\times]{} & \widetilde{\mathrm{CH}}^{k(n-k)}(\mathrm{Gr}(k,V)).
\end{array}$$

Actually, we want to work with the corresponding diagram involving Witt groups (instead of Grothendieck–Witt groups) and $\mathbf{I}^\bullet$-cohomology (instead of Chow–Witt groups). The relevant class in $\mathrm{W}^{n_1}_{\mathfrak{S}_1}(\mathrm{Gr}(k,V),\mathscr{L}_1)\times\cdots\times\mathrm{W}^{n_l}_{\mathfrak{S}_l}(\mathrm{Gr}(k,V),\mathscr{L}_l)$ is the exterior product of the classes $p_*(\langle 1\rangle)\in\mathrm{W}^{n_i}_{\mathfrak{S}_i}(\mathrm{Gr}(k,V),\mathscr{L}_i)$ where $p\colon \mathrm{Fl}(\Lambda)\to\mathrm{Gr}(k,V)$ is the resolution of singularities of the Schubert variety $\mathfrak{S}_i$ (corresponding to the Young diagram Λ).

To determine an explicit representative of the form

$$(p_\Lambda)_*(\langle 1\rangle)\in\mathrm{W}^{n_i}_{\mathfrak{S}_i}(\mathrm{Gr}(k,V),\mathscr{L}_i),$$

[5] This is exactly the class in the Chow–Witt ring corresponding to the full Young diagram in $k\times(n-k)$-frame.

we first note that the form $\langle 1\rangle$ in $\mathrm{W}^0(\mathrm{Fl}(\Lambda),\mathscr{O})$ is given by $\mathrm{id}\colon \mathscr{O}_{\mathrm{Fl}(\Lambda)} \xrightarrow{\cong} \mathscr{O}_{\mathrm{Fl}(\Lambda)}$. Next, we use the identification $\mathrm{W}^0(\mathrm{Fl}(\Lambda),\mathscr{O}) \cong \mathrm{W}^0(\mathrm{Fl}(\Lambda),\omega_{p_\Lambda}\otimes p_\Lambda^*(\mathscr{L}_i))$ which is induced by the explicit choice of (canonical) isomorphism $\mathscr{N}^{\otimes 2} \xrightarrow{\cong} \omega_{p_\Lambda}\otimes p_\Lambda^*(\mathscr{L}_i)$: it maps a form $Q \xrightarrow{\cong} \mathrm{Hom}(Q,\mathscr{O})$ corresponding to $Q\otimes Q\to\mathscr{O}$ to $(Q\otimes\mathscr{N})\otimes(Q\otimes\mathscr{N})\to \omega_{p_\Lambda}\otimes p_\Lambda^*\mathscr{L}_i$ by tensoring with the given orientation. In the next step, the relevant push-forward is

$$(p_\Lambda)_*\colon\ \mathrm{W}^0(\mathrm{Fl}(\Lambda),\omega_{p_\Lambda}\otimes p_\Lambda^*(\mathscr{L}_i)) \to \mathrm{W}^{|\Lambda|}(\mathrm{Gr}(k,V),\mathscr{L}),$$

and the form we want to push forward is simply the orientation $\mathscr{N}\otimes\mathscr{N}\to\omega_{p_\Lambda}\otimes p_\Lambda^*(\mathscr{L}_i)$, cf. Section 6. From the definitions in [**CH11**, Theorem 4.4], the push-forward on Witt groups is induced from the derived pushforward $\mathrm{R}p_*$ on the level of triangulated categories of perfect complexes and the morphism $\zeta_K\colon \mathrm{R}p_*\circ\sharp'_{\omega_{\mathrm{Fl}(\Lambda)}}\to \sharp'_{\omega_{\mathrm{Gr}(k,V)}}\circ\mathrm{R}p_*$ exchanging push-forward and dualities. In our case, the push-forward is given by

$$\begin{aligned}
\mathrm{R}(p_\Lambda)_*\mathscr{N} &\to \mathrm{R}(p_\Lambda)_*\left(\mathrm{Hom}(\mathscr{N},\mathscr{O}_{\mathrm{Fl}(\Lambda)})\otimes\omega_{p_\Lambda}\otimes p_\Lambda^*(\mathscr{L}_i)\right)\\
&\to \mathrm{Hom}\left(\mathrm{R}(p_\Lambda)_*\mathscr{N},\mathrm{R}(p_\Lambda)_*(\omega_{p_\Lambda}\otimes p_\Lambda^*(\mathscr{L}_i))\right)\\
&\to \mathrm{Hom}\left(\mathrm{R}(p_\Lambda)_*\mathscr{N},\mathscr{O}\right)\otimes\mathscr{L}_i,
\end{aligned}$$

i.e., it is a form on $\mathrm{R}(p_\Lambda)_*\mathscr{N}$ with values in the twisting bundle $\mathscr{L}_i$ (which we can assume is either $\mathscr{O}_{\mathrm{Gr}(k,V)}$ or the determinant of the tautological sub- or quotient bundle).

Now we can apply a base-change argument similar to the proof of Theorem 7.17. We want to apply the base-change formula for Witt groups, cf. [**CH11**, Theorem 5.4] and the discussion in Section 5, to the following diagram

$$\begin{array}{ccc}
\prod_{i=1}^{l} p_i^{-1}(\mathfrak{S}_1\cap\cdots\cap\mathfrak{S}_l) & \longrightarrow & \mathrm{Fl}(\Lambda_1)\times\cdots\times\mathrm{Fl}(\Lambda_l)\\
{\scriptstyle q}\big\downarrow & & \big\downarrow{\scriptstyle \prod p_i}\\
\mathrm{Gr}(k,V) & \xrightarrow[\Delta]{} & \mathrm{Gr}(k,V)\times\cdots\times\mathrm{Gr}(k,V)
\end{array}$$

This diagram is cartesian and the upper left entry is the product of all the fibers of the resolutions $p_i\colon \mathrm{Fl}(\Lambda_i)\to\mathrm{Gr}(k,V)$ of the respective Schubert varieties $\mathfrak{S}_i$ over the intersection subscheme $\mathfrak{S}_1\cap\cdots\cap\mathfrak{S}_l$. Note that by assumption, this diagram is homologically transverse so that the base-change formula for Witt groups can be applied. Consequently, the restriction of the push-forward $\mathrm{R}(p_1)_*(\langle 1\rangle)\boxtimes\cdots\boxtimes\mathrm{R}(p_l)_*(\langle 1\rangle)$ along the diagonal $\Delta\colon \mathrm{Gr}(k,V)\hookrightarrow\mathrm{Gr}(k,V)^{\times l}$ is isomorphic to the push-forward of $\langle 1\rangle$ along the morphism $q\colon \prod_{i=1}^{l} p_i^{-1}(\mathfrak{S}_1\cap\cdots\cap\mathfrak{S}_l)\to\mathrm{Gr}(k,V)$. Note also that q factors as structural morphism of the product of fibers over the intersection points, followed by the closed immersion of the intersection points into $\mathrm{Gr}(k,V)$. By the compatibility of exterior product with push-forwards, the pushforward of $\langle 1\rangle$ along the product morphism can alternatively be computed as $\star$-product of the pushforwards of $\langle 1\rangle$ along the individual structure maps $p_i\colon p_i^{-1}(\mathfrak{S}_i)\to\mathrm{Gr}(k,V)$. Note again that the pushforward depends on the choice of orientation, and we use the canonical ones from Proposition 6.8. For a point $[W]\in\mathfrak{S}_1\cap\cdots\cap\mathfrak{S}_l$, the $\star$-product of the pushforwards of $\langle 1\rangle$ along the corresponding structure maps is then, by Definition 7.12, the type of the corresponding solution subspace. This finishes the argument. $\square$

REMARK 7.20. *Note that the transversality result of Sierra, cf. Theorem 7.4, together with Proposition 7.7 implies that over infinite fields the situation in Theorem 7.19 above can always be achieved by taking generic translates.*

8. Main results and consequences

Now we finally come to formulate the main results in oriented Schubert calculus. Fix a perfect field F of characteristic $\neq 2$, and let V be an n-dimensional F-vector space. An *oriented Schubert problem* for the Grassmannian $\mathrm{Gr}(k, V)$ is a collection of oriented Schubert classes $[\mathfrak{S}_i] \in \widetilde{\mathrm{CH}}^\bullet(\mathrm{Gr}(k,V), \mathscr{L}_i)$, $i = 1, \ldots, l$ with $\sum_i \dim \mathfrak{S}_i = k(n-k)$, i.e., Schubert varieties equipped with the canonical orientations given by Theorem 6.11. The transversality results imply that, up to translation by GL_n, we can assume that the intersection of the Schubert varieties $\mathfrak{S}_i$ is proper and homologically transversal, in particular, the intersection $\mathfrak{S}_1 \cap \cdots \cap \mathfrak{S}_l$ is a zero-dimensional subscheme. A *solution subspace* is a closed point $[W] \in \mathrm{Gr}(k, V)$ in the intersection. The answer to an oriented Schubert problem is a quadratic form in $\mathrm{GW}(F)$ which not only encodes the number of solution subspaces (and the degrees of the corresponding closed points, i.e., the degrees of the field extensions over which the solution is defined) but also additional information on the compatibility of the canonical orientations of the intersecting Schubert varieties, encoded in the types of Definitions 7.10 and 7.12.

8.1. Geometric cycles vs characteristic classes. We still need to identify the geometric generators in terms of the characteristic classes used to describe the multiplication in Section 3.

First, we note that there is some stabilization for the Grassmannians: fix a field F and an F-vector space V of dimension n. Mapping a k-dimensional subspace $\iota\colon W \subset V$ to $W \subseteq V \oplus F^m$ via $\iota \oplus 0^m$ induces a morphism $\mathrm{Gr}(k,n) \to \mathrm{Gr}(k, n+m)$. By construction, the tautological subbundle of $\mathrm{Gr}(k, n+m)$ restricts to the tautological subbundle of $\mathrm{Gr}(k,n)$ along this inclusion, and consequently the restriction maps the Pontryagin classes p_i of $\mathrm{Gr}(k, n+m)$ to the ones for $\mathrm{Gr}(k,n)$. Moreover, the tautological quotient bundle of $\mathrm{Gr}(k, n+m)$ restricts to the Whitney sum of the tautological quotient bundle for $\mathrm{Gr}(k,n)$ and the trivial bundle of rank m and the Pontryagin classes $\mathrm{p}_j^\perp$ restrict accordingly. Similarly, mapping a k-dimensional subspace $\iota\colon W \subset V$ to $W \oplus F^m \subseteq V \oplus F^m$ via $\iota \oplus \mathrm{id}^m$ induces a morphism $\mathrm{Gr}(k,n) \to \mathrm{Gr}(k+m, n+m)$. In this case, the tautological quotient bundle of $\mathrm{Gr}(k+m, n+m)$ restricts to the tautological quotient bundle for $\mathrm{Gr}(k,n)$ and the restriction maps the Pontryagin classes $\mathrm{p}_j^\perp$ of $\mathrm{Gr}(k+m, n+m)$ to the ones for $\mathrm{Gr}(k,n)$. Note that smooth Schubert varieties are smaller Grassmannians embedded as above.

The statements concerning Pontryagin classes above imply that the restriction morphism on $\mathbf{W}^\bullet$-cohomology is the canonical quotient projection modulo the ideal generated by the Pontryagin classes of the bigger Grassmannian which vanish on the smaller Grassmannian. A direct consequence of the above, we have the following:

LEMMA 8.1. *Let F be a field of characteristic $\neq 2$ and fix $1 \leq k < n$. The Pontryagin class p_j in $\mathrm{H}^{2j}(\mathrm{Gr}(k,n), \mathbf{W})$ is uniquely determined by its restriction to $\mathrm{Gr}(j, n+j-k)$. The Pontryagin class $\mathrm{p}_j^\perp \in \mathrm{H}^{2j}(\mathrm{Gr}(k,n), \mathbf{W})$ is uniquely determined by its restriction to $\mathrm{Gr}(k, k+j)$.*

In particular, to compare Pontryagin classes and the explicit geometric cycles, it then suffices to consider the top Pontryagin class p_j or $\mathrm{p}_j^{\perp}$ for a smaller Grassmannian. If we can show that the Pontryagin class p_j on $\mathrm{Gr}(j,n)$ is given by pushforward of $\langle 1\rangle$ from the corresponding smooth Schubert variety (for the canonical orientation), then this will be true for $\mathrm{Gr}(k,m)$ for all $k\geq j$ and $m\geq n+k-j$: this follows from the base-change in the cartesian diagram

$$\begin{array}{ccc} \Sigma_j & \xrightarrow{=} & \Sigma_j \\ \downarrow & & \downarrow \\ \mathrm{Gr}(j,n) & \longrightarrow & \mathrm{Gr}(k,m). \end{array}$$

Similar statements hold for $\mathrm{p}_j^{\perp}$: if the statement holds for $\mathrm{Gr}(k,k+j)$, then it holds for $\mathrm{Gr}(k',m)$ with $k'\geq k$ and $m\geq k'+j$. The following establishes the relevant claim for the Euler classes, and this implies the corresponding statements for Pontryagin classes.

LEMMA 8.2. *Let F be a perfect field of characteristic $\neq 2$ and fix $1\leq k<n$.*

(1) *Assume k is even. Denoting by $\iota\colon \Sigma_{1^k}\hookrightarrow \mathrm{Gr}(k,n)$ the closed immersion of the Schubert variety into the Grassmannian, we have $\iota_*(\langle 1\rangle)=\mathrm{e}_k$ in $\mathrm{H}^k(\mathrm{Gr}(k,n),\mathbf{I}^k(\det\mathscr{E}_k^{\vee}))$.*
(2) *Assume $n-k$ is even. Denoting by $\iota\colon \Sigma_{n-k}\hookrightarrow \mathrm{Gr}(k,n)$ the closed immersion of the Schubert variety into the Grassmannian, we have $\iota_*(\langle 1\rangle)=\mathrm{e}_{n-k}^{\perp}$ in $\mathrm{H}^{n-k}(\mathrm{Gr}(k,n),\mathbf{I}^{n-k}(\det\mathscr{E}_{n-k}^{\vee}))$.*

PROOF. We prove (1), (2) is proved similarly. First, we note that the vanishing locus of a generic section s of the tautological subspace bundle $\mathscr{S}$ is the Schubert variety Σ_{1^k}. The Euler class of $\mathscr{S}$ can be concretely represented as Koszul complex of the generic section s of $\mathscr{S}$, by the computations in [**Fas08**, Sections 13, 14]. On the other hand, the pushforward of $\langle 1\rangle$ along the inclusion (for the canonical relative orientation) is also identified with the Koszul complex for the section, cf. the discussion in [**CH11**, Section 7.2]. In particular, both $\iota_*(\langle 1\rangle)$ and e_k are both represented by the Koszul complex for a generic section of $\mathscr{S}$, which proves the claim. □

REMARK 8.3. *Alternatively, as discussed before the statement of the proposition, we can reduce (1) above to the claim about $\mathrm{Gr}(k,k+1)\cong\mathbb{P}^k$ and use* [**Fas13**] *to deduce that the Euler class $\mathrm{H}^k(\mathbb{P}^k,\mathbf{I}^k(\det\mathscr{Q}_k))$ is given by the pushforward of $\langle 1\rangle$ along the inclusion of a point, with the canonical orientation of $\mathbb{P}^k$ relative to $\mathscr{Q}_k$.*

REMARK 8.4. *We discuss an analogue of the complementary intersections in* [**EH17**, Section 4.2.2]. *Let F be a field of characteristic $\neq 2$, and fix $1\leq k<n$. Fix two flags $\mathcal{V}$ and $\mathcal{W}$ in V which are transversal in the sense of* [**EH17**, Definition 4.4]. *In particular, we can choose a basis $e_1,\dots,e_n$ of V such that $V_i=\langle e_1,\dots,e_i\rangle$ and $W_j=\langle e_{n+1-j},\dots,e_n\rangle$. For two complementary (even) Young diagrams $\Lambda=(a_1,\dots,a_k)$ and $\Lambda'=(b_1,\dots,b_k)$ with $|\Lambda|+|\Lambda'|=k(n-k)$, the corresponding Schubert varieties Σ_Λ and $\Sigma_{\Lambda'}$ intersect transversely in a single F-rational point which is the subspace $U\subset V$ generated by the basis vectors $e_{n-k+i-a_i}$ for $i=1,\dots,k$.*

The tangent space of the Grassmannian at the point $[U]$ is canonically identified with $\mathrm{Hom}(U,V/U)$. Taking the basis $f_i:=e_{n-k+i-a_i}$ for U and the U-cosets of the *other basis vectors* e_j, *denoted as* g_j, *as basis for* V/U, *the corresponding maps* $v \mapsto f_i^\vee(v) \cdot g_j$ *form a basis for* $\mathrm{Hom}(U, V/U)$ *which induces the canonical orientation.*

The tangent space of the Schubert variety Σ_Λ *at the point* $[U]$ *is canonically identified with the subspace of* $\mathrm{Hom}(U, V/U)$ *which maps* $V_{n-k+i-a_i} \cap U \subset U$ *into* $(V_{n-k+i-a_i} + U)/U \subset V/U$, *cf.* [**EH17**, Proposition 4.1]. *In particular, a basis of the tangent space is given by the maps* $v \mapsto f_i^\vee(v) \cdot g_j$ *such that* $g_j \in V_{n-k+l-a_l}$ *whenever* $f_i \in V_{n-k+l-a_l}$. *The other maps form a basis for the normal space to* Σ_Λ *at* $[U]$, *and this basis induces the canonical orientation of the normal space. For the Schubert variety* $\Sigma_{\Lambda'}$, *the situation is exactly reversed, i.e., the basis for the tangent space to* Σ_Λ *above is a basis for the normal space to* $\Sigma_{\Lambda'}$. *Consequently, the decomposition of the tangent space to the Grassmannian at* $[U]$ *as direct sum of the normal spaces to* Σ_Λ *and* $\Sigma_{\Lambda'}$ *is compatible with the canonical orientations. Reformulated in the* $\mathbf{I}^\bullet$*-cohomology ring, the intersection product* $[\Sigma_\Lambda]\cdot[\Sigma_{\Lambda'}]$ *is equal to* $\langle 1\rangle$ *times the canonical generator of the top cohomology (which is the class given by pushforward of* $\langle 1\rangle$ *along the inclusion of an* F*-rational point, with the canonical orientation of the tangent space).*

PROPOSITION 8.5. *Let* F *be a perfect field of characteristic* $\neq 2$, *and fix* $1 \leq k < n$. *Let* Λ *be a rectangular even Young diagram with twist* t. *In the cohomology group* $\mathrm{H}^{|\Lambda|}(\mathrm{Gr}(k,n), \mathbf{I}^{|\Lambda|}(\det^t))$, *the class associated to* Λ *by Definition* 3.11 *agrees with the class* $(\iota_\Lambda)_*(\langle 1\rangle)$ *of Theorem* 6.11.

PROOF. By Lemma 8.1, it suffices to establish the cases where the longer side of the rectangle Λ is k or $n-k$. We deal with the first case, the second is similar. In the first case, the class of Definition 3.11 is a power of the Euler class e_k. By multiplicativity of the Euler classes, it is the Euler class of the appropriate tensor power of the tautological subbundle $\mathscr{S}_k$. Then the argument of Lemma 8.2 works: the class from Definition 3.11 is represented by the Koszul complex for a generic section of the appropriate tensor power of $\mathscr{S}_k$. Similarly, the class $(\iota_\Lambda)_*(\langle 1\rangle)$ of Theorem 6.11 is also represented by the Koszul complex for a generic section of the appropriate tensor power of $\mathscr{S}_k$ (whose generic section is a Schubert variety for the Young diagram Λ). This proves the identification. □

REMARK 8.6. *In the spirit of Remark* 8.4, *it is also possible to extend the Proposition* 8.5 *to show that the classes of Definition* 3.11 *agree with the classes* $(p_\Lambda)_*(\langle 1\rangle)$ *of Theorem* 6.11. *For an even Young diagram* Λ *and a Schubert variety* $\Sigma_{2a,2a}$, *denote by* Λ' *the Schubert variety for a Young diagram complementary to an even Young diagram appearing in the intersection of* Σ_Λ *and* $\Sigma_{2a,2a}$. *The intersection of* Σ_Λ, $\Sigma_{2a,2a}$ *and* $\Sigma_{\Lambda'}$ *(for transversal flags) consists of a single* F*-rational point. Then it can be checked that the decomposition of the tangent space as direct sums of the normal spaces is compatible with the canonical orientations in the definition of the classes* $(p_\Lambda)_*(\langle 1\rangle)$. *This is similar to the proof of the Pieri formula in* [**EH17**, 4.2.4] *and is a version of the oriented Pieri formula for the geometric cycles* $(p_\Lambda)_*(\langle 1\rangle)$.

THEOREM 8.7. *Let* F *be a perfect field of characteristic* $\neq 2$, *let* n *be an even number. Consider smooth oriented Schubert classes* $\mathfrak{S}_i \in \widetilde{\mathrm{CH}}^\bullet(\mathrm{Gr}(k,n), \mathscr{L}_i)$, $i = 1, \ldots, l$ *such that the underlying Schubert varieties are in general position,* $\sum_i \dim \mathfrak{S}_i = k(n-k)$ *and* $\bigotimes_i \mathscr{L}_i = 0$ *in* $\mathrm{Pic}(\mathrm{Gr}(k,n))/2$. *Then there exist natural numbers* a, b *such that we have the following equality in the top cohomology*

$\widetilde{\mathrm{CH}}^{k(n-k)}(\mathrm{Gr}(k,n)) \cong \mathrm{GW}(F)$:

$$\sum_{W \in \mathfrak{S}_1 \cap \cdots \cap \mathfrak{S}_l} \mathrm{tr}_{F(W)/F}\langle \det \xi_W \rangle = a\langle 1 \rangle + b\langle -1 \rangle$$

The numbers a and b are determined by:

(1) *$a + b$ is the number of solution spaces over an algebraically closed field, equal to the degree of the intersection product $[\mathfrak{S}_1] \cdots [\mathfrak{S}_l]$ in the Chow group $\mathrm{CH}^{k(n-k)}(\mathrm{Gr}(k,n))$.*

(2) *$a - b$ is the signed count of solution subspaces over $\mathbb{R}$, equal to the degree of the intersection product $[\mathfrak{S}_1] \cdots [\mathfrak{S}_l]$ in the integral cohomology of the real Grassmannian $\mathrm{H}^{k(n-k)}(\mathrm{Gr}_k(\mathbb{R}^n), \mathbb{Z})$.*

PROOF. This is a consequence of Proposition 8.5: the two sides of the equation correspond to two ways of computing the intersection product. The left-hand side is the sum of all the local contribution of the solution subspaces in terms of types as done in Theorem 7.17. Note that this uses the geometric representatives of characteristic classes from Section 6. On the right-hand side, we have the degree of the intersection, and this is computed in terms of the formulas from Section 3, which are formulated in terms of the Pontryagin-class presentation of the **W**-cohomology ring. The fact that the geometric representatives are exactly identified with the images of Schubert classes under the folklore map from Proposition 3.1 is exactly Proposition 8.5. Hence the left-hand side is equal to the degree of the intersection of the oriented Schubert classes $[\mathfrak{S}_i]$ (where the generator of the top Chow–Witt group is induced from the canonical orientation on the tangent space of the Grassmannian).

It remains to identify the degree of the oriented intersection as a quadratic form. The class in $\mathrm{GW}(F)$ is determined by its images in $\mathrm{W}(F)$ and the rank. First, the rank is the image under the reduction morphism

$$\widetilde{\mathrm{CH}}^{k(n-k)}(\mathrm{Gr}(k,n)) \to \mathrm{CH}^{k(n-k)}(\mathrm{Gr}(k,n)),$$

i.e., it is the degree of the intersection in the Chow ring. Second, the image in $\mathrm{W}(F)$ is the image under the reduction morphism

$$\widetilde{\mathrm{CH}}^{k(n-k)}(\mathrm{Gr}(k,n)) \to \mathrm{H}^{k(n-k)}(\mathrm{Gr}(k,n), \mathbf{I}^{k(n-k)}).$$

As a consequence of Theorem 3.16, the image in $\mathrm{W}(F)$ is of the form $m\langle 1 \rangle$. Since the reduction morphism is the quotient modulo $\mathbb{Z} \cdot [\mathbb{H}]$ with $\mathbb{H} = \langle 1 \rangle + \langle -1 \rangle$, the Chow–Witt degree of the intersection is necessarily of the form $a\langle 1 \rangle + b\langle -1 \rangle$, as claimed.

To identify a and b, note that the reduction morphism to the Chow ring maps both $\langle 1 \rangle$ and $\langle -1 \rangle$ to 1, in particular, the image in the Chow ring will be $a+b$. The reduction to $\mathrm{W}(F)$ maps $\langle -1 \rangle = -\langle 1 \rangle$, in particular the image will be $a - b$. This proves the remaining claims. □

REMARK 8.8. *In particular, for all oriented Schubert problems, the oriented degree of the intersection in the Chow–Witt ring is a simple quadratic form $a\langle 1 \rangle + b\langle -1 \rangle$ whose rank and signature can be determined by complex and real realization.*

8.2. Refined counts over special fields. We discuss a couple of consequences of the above result over various fields. Over algebraically closed fields, Theorem 8.7 simply recovers the classical results from Schubert calculus. For any real embedding $F \hookrightarrow \mathbb{R}$ the above result recovers signed counts.

In the case where $F = \mathbb{F}_q$ is a finite field of odd characteristic, there are only two possibilities for the class $\det \xi_W \in \mathbb{F}_{q^d}^\times/(\mathbb{F}_{q^d}^\times)^2$ of a solution subspace W defined over $\mathbb{F}_{q^d}$: if the determinant is a square, the splitting map ξ_W is orientation-preserving, otherwise it is orientation-reversing. The following result is an analogue of [**KW17**, Theorem 1].

PROPOSITION 8.9. *Let $F = \mathbb{F}_q$ be a finite field of odd characteristic. In the situation of Theorem 8.7, we have the following statement for the types of solution subspaces W:*

$$\#(W/\mathbb{F}_{q^d}, d \textit{ even}, \xi_W \textit{ oriented}) + \#(W/\mathbb{F}_{q^d}, d \textit{ odd}, \xi_W \textit{ non-oriented}) \equiv b \bmod 2.$$

PROOF. The discriminant of the right-hand side form in Theorem 8.7, viewed as an element of $\mathbb{F}_q^\times/(\mathbb{F}_q^\times)^2$ is b. Now we need to understand which subspaces contribute to the discriminant of the left-hand side. By [**KW17**, Lemma 58], the discriminant of $\operatorname{tr}_{\mathbb{F}_{q^d}/\mathbb{F}_q}\langle\det \xi_W\rangle$ is nonzero if d is even and $\det \xi_W$ is a square or if d is odd and $\det \xi_W$ is a nonsquare. This proves the claim. □

PROPOSITION 8.10. *Let F be a local field with odd residue characteristic. In the situation of Theorem 8.7, we have*

$$\sum_{W \in \mathfrak{S}_1 \cap \cdots \cap \mathfrak{S}_l} \operatorname{res}\left(\operatorname{tr}_{F(W)/F}\langle \det \xi_W\rangle\right) = 0,$$

i.e., the sum of the residues of the type-forms for solution subspaces is 0.

PROOF. The right-hand side of the formula in Theorem 8.7 is an unramified form, i.e., has trivial residue. □

COROLLARY 8.11. *Let F be a number field. Then for every finite place v, the sum of the Hasse invariants at v of the types for the solution subspaces must be* 0.

REMARK 8.12. *If $a\langle 1\rangle + b\langle -1\rangle$ is the degree of the intersection of an oriented Schubert problem over the field F, then for every embedding of F into a real closed field, the signed count of solution subspaces is $a - b$.*

If $a\langle 1\rangle + b\langle -1\rangle$ is the degree of the intersection of an oriented Schubert problem over $\mathbb{Q}$, and there is a single solution subspace defined over a number field of degree $[F : \mathbb{Q}] = a+b$, then the number of real embeddings of F must be equal to one of the possible numbers of real solutions to the Schubert problem over $\mathbb{R}$. In particular, $a - b$ is a lower bound on the number of real embeddings of F.

Note that even in the above case of a single solution over a number field, the refined count doesn't restrict the possible field extension. Of course, if the splitting isomorphism between the tangent space to $\operatorname{Gr}(k, V)$ and the direct sum of normal spaces preserves orientations, then the refined count would be the trace form of the number field. But this is no restriction: if the trace form of the number field is complicated (and has a lot of non-trivial local Hasse invariants), then this only means that the determinant of the splitting map must be similarly complicated.

8.3. Comparison with twisted Witt groups. Now we are ready to compare the previous computations of $\mathbf{W}$-cohomology in Section 3 with the computations of total Witt groups in [**BC12a**].

THEOREM 8.13. *Let F be a perfect field of characteristic $\neq 2$, and let $1 \leq k < n$. The natural morphisms $\alpha\colon \mathrm{W}^i(\mathrm{Gr}(k,n),\mathscr{L}) \to \mathrm{H}^i(\mathrm{Gr}(k,n),\mathbf{W}(\mathscr{L}))$ induce an isomorphism of $\mathrm{W}(F)$-modules:*

$$\mathrm{W}^i(\mathrm{Gr}(k,n),\mathscr{L}) \xrightarrow{\cong} \bigoplus_{j\equiv i \bmod 4} \mathrm{H}^j(\mathrm{Gr}(k,n),\mathbf{W}(\mathscr{L})).$$

Summing over all $i \in \mathbb{Z}/4\mathbb{Z}$ and $[\mathscr{L}] \in \mathrm{Pic}(\mathrm{Gr}(k,n))/2$, this induces an isomorphism of commutative graded $\mathrm{W}(F)$-algebras from the total Witt ring to the total $\mathbf{W}$-cohomology ring.

PROOF. By Proposition 5.3, we know that the morphisms α are compatible with pushforwards. In particular, the generators of $\mathrm{W}^i(\mathrm{Gr}(k,n),\mathscr{L})$ discussed in [**BC12a**, Theorem 6.1] are mapped exactly to the generators of the $\mathbf{W}$-cohomology group $\mathrm{H}^i(\mathrm{Gr}(k,n),\mathbf{W}(\mathscr{L}))$, cf. Propositions 3.14 and 8.5. Consequently, we have a $\mathrm{W}(F)$-linear homomorphism between finitely generated free $\mathrm{W}(F)$-modules which is induced by a bijection of total bases, hence we get an isomorphism of $\mathrm{W}(F)$-modules. Compatibility with the intersection products follows from the oriented Tor-formula discussed in Section 5. □

REMARK 8.14. *The above $\mathrm{W}(F)$-algebra homomorphism provides a complete description of the multiplication in the total Witt groups $\bigoplus_{i,\mathscr{L}} \mathrm{W}^i(\mathrm{Gr}(k,n),\mathscr{L})$. Products of classes corresponding to even Young diagrams can be evaluated using the results in Section 3. This resolves the open question of determining Littlewood–Richardson coefficients for the multiplication in Witt groups, cf.* [**BC12a**, Remark 7.8].

9. Sample enumerative applications

In this section, we discuss a couple of applications to refined enumerative geometry which demonstrate the type of results that can be obtained using oriented Schubert calculus. What we provide are classical problems, either refined by lifting to the Chow–Witt (resp. $\mathbf{I}^\bullet$-cohomology) ring or modified by replacing Chern classes by Pontryagin classes. The classical problems (like lines on hypersurfaces, balanced subspaces etc.) can all be found in the book [**EH17**], and some of the refinements have appeared in the literature on real Schubert calculus before.

9.1. Intersections with torsion classes. We first note a simple consequence for intersection products involving torsion classes. In the orientable cases, the refined degree of the intersection is always a multiple of the hyperbolic plane. More precisely, we have:

THEOREM 9.1. *Let F be a perfect field of characteristic $\neq 2$, and consider oriented Schubert classes $[\mathfrak{S}_i] \in \widetilde{\mathrm{CH}}^\bullet(\mathrm{Gr}(k,n),\mathscr{L}_i)$, $i = 1,\ldots,l$ such that $\sum_i \dim \mathfrak{S}_i = k(n-k)$.*

(1) *Assume that n is even and the class of $\bigotimes_i \mathscr{L}_i$ in $\mathrm{Pic}(\mathrm{Gr}(k,n))/2$ is trivial. If the corresponding Young diagram for one of the oriented Schubert*

classes fails to be even, then in the notation of Theorem 8.7, we have

$$\sum_{W \in \mathfrak{S}_1 \cap \cdots \cap \mathfrak{S}_l} \mathrm{tr}_{F(W)/F} \langle \det \xi_W \rangle = a\mathbb{H}$$

where a is half the degree of the intersection product $[\mathfrak{S}_1] \cdots [\mathfrak{S}_l]$ in the Chow ring $\mathrm{CH}^{k(n-k)}(\mathrm{Gr}(k,n))$.

(2) *Assume that $n \bmod 2$ is different from the class of $\bigotimes_i \mathscr{L}_i$ in $\mathbb{Z}/2\mathbb{Z} \cong \mathrm{Pic}(\mathrm{Gr}(k,n))/2$. Then we have*

$$\widetilde{\mathrm{CH}}^{k(n-k)}(\mathrm{Gr}(k,n), \bigotimes_i \mathscr{L}_i) \cong \mathrm{CH}^{k(n-k)}(\mathrm{Gr}(k,n))$$

and the degree of the intersection $[\mathfrak{S}_1] \cdots [\mathfrak{S}_l]$ is the one in the Chow group.

PROOF. By the computations of Chow–Witt rings of Grassmannians, cf. Section 2, we have two cases for the top Chow–Witt group: if $n \bmod 2$ equals the class of $\bigotimes_i \mathscr{L}_i$ in $\mathbb{Z}/2\mathbb{Z} \cong \mathrm{Pic}(\mathrm{Gr}(k,n))/2$, then $\widetilde{\mathrm{CH}}^{k(n-k)}(\mathrm{Gr}(k,n), \bigotimes_i \mathscr{L}_i) \cong \mathrm{GW}(F)$. Otherwise, the natural projection

$$\widetilde{\mathrm{CH}}^{k(n-k)}(\mathrm{Gr}(k,n), \bigotimes_i \mathscr{L}_i) \to \mathrm{CH}^{k(n-k)}(\mathrm{Gr}(k,n))$$

is an isomorphism, since the top $\mathbf{I}^\bullet$-cohomology group maps isomorphically to the top mod 2 Chow group. This already establishes (2).

To prove (1), note that the assumption that one Young diagram is non-even implies that the corresponding class in $\mathbf{I}^\bullet$-cohomology is $\mathrm{I}(F)$-torsion. Therefore, the whole oriented intersection product is $\mathrm{I}(F)$-torsion and thus trivial in the top $\mathbf{I}^\bullet$-cohomology group (since that is $\mathrm{I}(F)$-torsion-free resp. the image of β is trivial). In particular, the intersection product in $\widetilde{\mathrm{CH}}^{k(n-k)}(\mathrm{Gr}(k,n), \bigotimes_i \mathscr{L}_i) \cong \mathrm{GW}(F)$ is a class whose image in $\mathrm{W}(F)$ is trivial, hence it must be a multiple of the hyperbolic plane. The appropriate multiple is determined by the requirement that the rank equals the degree of the intersection product in the top Chow group. This proves the claim. □

REMARK 9.2. *In particular, we can only expect quadratic forms different from hyperbolic planes (and thus non-trivial signed counts over the real numbers) if the Young diagrams in the Schubert problem are all even. In particular, the corresponding Schubert varieties will either have dimensions divisible by 4 (i.e., completely even) or are products of such with Euler classes or the hook class* R.

EXAMPLE 9.3. *We can determine the refined degree of the Grassmannian in the Plücker embedding $\mathrm{Gr}(2, n+1) \hookrightarrow \mathbb{P}(\bigwedge^2 F^{n+1})$. Classically, this degree can be computed as $\deg(\mathrm{c}_1^{2n-2})$. For the class $\overline{\mathrm{c}}_1$, we have $\mathrm{Sq}^2_{\mathscr{O}}(\overline{\mathrm{c}}_1) = \overline{\mathrm{c}}_1^2 = \overline{\mathrm{c}}_2 + \sigma_{1,1}$ and $\mathrm{Sq}^2_{\det}(\overline{\mathrm{c}}_1) = 0$. In particular, for $k(n-k) \geq 2$, the class $\overline{\mathrm{c}}_1$ is not in the image of $\mathrm{H}^1(\mathrm{Gr}(2,n+1), \mathbf{I}^1)$, but it is in the image of $\mathrm{H}^1(\mathrm{Gr}(2,n+1), \mathbf{I}^1(\det))$. The unique lift of $\overline{\mathrm{c}}_1$ to $\mathbf{I}^1$-cohomology is given by $\beta_{\det}(\langle 1 \rangle)$.*

We want to compute $\beta_{\det}(\langle 1 \rangle)^{2n-2}$. First note that the class $\beta_{\det}(\langle 1 \rangle)^{2n-2}$ is *in the image of* β. *In the case where* n *is odd, the Grassmannian* $\mathrm{Gr}(2, n+1)$ *is orientable, and the image of* $\beta_{\mathscr{O}}$ *in* $\mathrm{H}^{2n-2}(\mathrm{Gr}(2,n+1), \mathbf{I}^{2n-2}) \cong \mathrm{W}(F)$ *is trivial. Consequently,* $\beta_{\det}(\langle 1\rangle)^{2n-2} = 0$. *In the case where* n *is even, we have* $\mathrm{H}^{2n-2}(\mathrm{Gr}(2,n+1), \mathbf{I}^{2n-2}) \cong \mathbb{Z}/2\mathbb{Z}$ *and the image of* β *is detected on* $\mathrm{Ch}^{2n-2}(\mathrm{Gr}(2,n+1))$. *Then* $\beta_{\det}(\langle 1\rangle)^{2n-2}$ *is trivial if and only if* $\overline{\mathrm{c}}_1^{2n-2}$ *is. Now*

$$\deg \mathrm{c}_1^{2n-2} = \frac{(2n-2)!}{n!(n-1)!}$$

is the $(n-1)$*-th Catalan number which is odd if and only if* $n = 2^k$. *Consequently,* $\beta_{\det}(\langle 1\rangle)^{2n-2}$ *is nontrivial if and only if* $n \neq 2^k$.

For the Chow–Witt ring, we have the following consequences, obtained by combining this computation of the $\mathbf{I}^\bullet$*-cohomological degree with the classical computation, cf.* [**EH17**, Section 4.3]. *If* n *is odd, then* $\widetilde{\mathrm{CH}}^{2n-2}(\mathrm{Gr}(2,n+1)) \cong \mathrm{GW}(F)$, *and denoting the unique lift of* c_1 *by* $\widetilde{\mathrm{c}}_1$ *we have* $\widetilde{\mathrm{c}}_1^{2n-2} = \frac{(2n-2)!}{2n!(n-1)!} \cdot \mathbb{H}$*; the refined degree is half the* $(n-1)$*-th Catalan number times the hyperbolic plane. If* n *is even, then* $\widetilde{\mathrm{CH}}^{2n-2}(\mathrm{Gr}(2,n+1)) \cong \mathrm{CH}^{2n-2}(\mathrm{Gr}(2,n+1)) \cong \mathbb{Z}$*; in this case, there is no refinement and the degree simply agrees with* $\deg(\mathrm{c}_1^{2n-2})$ *in the Chow ring.*

In the case where n *is odd, this provides a refinement of the enumerative problem of how many lines in* $\mathbb{P}^n$ *meet each of* $2n-2$ *general* $(n-2)$*-planes. Even though the quadratic form providing the refined count is only a copy of hyperbolic planes whose rank is the classical degree, this provides more information than the simple number. In* [**SW18**], *Srinivasan and Wickelgren have associated to each of the lines meeting the* $(n-2)$*-planes a quadratic form (or better a class in* $\mathrm{GW}(F)$*). The refined enumerative information is that the sum of those classes in* $\mathrm{GW}(F)$ *must be* $\mathrm{C}_{n-1}/2$ *copies of the hyperbolic plane* $\mathbb{H}$. □

EXAMPLE 9.4. *As another example, we can discuss the intersection problem underlying the question how many* $(n-1)$*-planes are contained in the intersection of two general quadrics in* $\mathbb{P}^{2n}$. *By* [**EH17**, Proposition 4.15], *the corresponding cycle is trivial in* $\mathrm{Ch}^\bullet(\mathrm{Gr}(n-1, 2n+1))$. *Consequently, the cycle* Φ *from loc.cit. lifts to the Chow–Witt ring. Its self-intersection in* $\mathbf{I}^\bullet$*-cohomology will be zero. The self-intersection of the class in the Chow–Witt ring will be* $4^n/2 \cdot \mathbb{H}$, *a sum of copies of the hyperbolic plane whose rank equals the classical count* 4^n. *In this case, however, there is no enumerative interpretation: as pointed out by Kirsten Wickelgren, the relevant vector bundles whose Euler class would be the above intersection class fails to be relatively orientable. Not all computations of intersection multiplicities done in Chow–Witt rings have enumerative interpretations.* □

9.2. Counting balanced subspaces. Next we discuss an enumerative problem which can be refined to a problem involving non-torsion classes, the arithmetic count of balanced subspaces. This generalizes the signed count of balanced subspaces over $\mathbb{R}$ which was established by Fehér and Matszangosz in [**FM16**].

Let F be a field of characteristic $\neq 2$. We consider the following enumeration problem which is a higher-dimensional analogue of the classical problem of finding the number of lines in $\mathbb{P}^3$ intersecting 4 given lines in general position.

> Given four b-dimensional subspaces in general position in a $2b$-dimensional space, what is the number of $2a$-dimensional subspaces having a-dimensional intersection with all the given subspaces?

DEFINITION 9.5. *Let F be a field of characteristic $\neq 2$, and let $a < b$ be two natural numbers. Let U be an F-vector space of dimension $2b$, and let $V_1, \dots, V_4 \subset U$ be four F-subspaces of dimension b. A $2a$-dimensional subspace $W \subset U$ is* balanced *(relative to the subspaces V_i) if the intersection $W \cap V_i$ has dimension a for all i.*

In the case where F is algebraically closed, the number of balanced subspaces is determined by a theorem of Vakil [**Vak06**]:

PROPOSITION 9.6. *Let F be an algebraically closed field, and let $a < b$ be two natural numbers. The number of $2a$-dimensional balanced subspaces of a $2b$-dimensional space is $\binom{b}{a}$.*

The classical case is $a = 1$, $b = 2$: there are 2 lines meeting four lines in $\mathbb{P}^3$ in general position. The following result is now a generalization of the signed count of balanced subspaces in [**FM16**].

THEOREM 9.7. *Let F be a perfect field of characteristic $\neq 2$. Let $a = 2i$ and $b = 2j$ be even numbers with $a < b$. Then the following equality holds in $\mathrm{GW}(F) \cong \widetilde{\mathrm{CH}}^{4a(b-a)}(\mathrm{Gr}(2a, 2b))$:*

$$\sum_{\text{balanced } W} \mathrm{tr}_{F(W)/F}\langle \det \xi_W \rangle = \frac{\binom{2j}{2i} + \binom{j}{i}}{2}\langle 1\rangle + \frac{\binom{2j}{2i} - \binom{j}{i}}{2}\langle -1\rangle.$$

PROOF. For a fixed b-dimensional subspace $V \subset U$, the set of $2a$-dimensional subspaces $W \subset U$ with $\dim W \cap V \geq a$ is the Schubert variety Σ_Λ whose corresponding Young diagram is a rectangle with side lengths a and $b-a$. Note that this is a completely even Young diagram by the assumptions on a and b. In particular, the relevant oriented Schubert problem is given by the 4-fold self-intersection of the lifts of the Schubert class σ_Λ given by the canonical orientation of the normal bundle of Σ_Λ, cf. Proposition 6.8. Then we can apply Theorem 8.7.

The multiplicity can be computed separately: the dimension of the class in $\mathrm{GW}(F)$ is the degree of the 4-fold self-intersection of $[\Sigma_\Lambda]$ in the Chow ring, which by Proposition 9.6 equals $\binom{b}{a}$. The image of the class in $\mathrm{W}(F)$ is the degree of the 4-fold self-intersection in $\mathrm{H}^n(\mathrm{Gr}(2a, 2b), \mathbf{I}^n)$. By Proposition 3.1, this multiplicity is the same as the degree of the 4-fold self-intersection of the rectangular Young diagram with side lengths i and j in $\mathrm{CH}^\bullet(\mathrm{Gr}(a, b))$. Again, by Proposition 9.6, this degree is $\binom{j}{i}$. The degree in the Chow–Witt ring is then the class of the quadratic form in $\mathrm{GW}(F)$ of dimension $\binom{2j}{2i}$ whose image in $\mathrm{W}(F)$ is $\binom{j}{i}$. This class must necessarily be of the form $m\langle 1\rangle + n\langle -1\rangle$, and satisfy

$$m + n = \binom{2j}{2i}, \qquad m - n = \binom{j}{i}.$$

This provides the degree on the right-hand side of the equation. □

EXAMPLE 9.8. *The simplest case is $a = 2$, $b = 4$, in which the degree of the intersection problem in* $\mathrm{Gr}(4,8)$ *is* $4\langle 1\rangle + 2\langle -1\rangle$. *For $a = 4$, $b = 8$, the degree of the intersection problem in* $\mathrm{Gr}(8,16)$ *is* $38\langle 1\rangle + 32\langle -1\rangle$, *recovering the 70 solution subspaces from classical Schubert calculus and the signed count of* 6 *over the reals.* □

9.3. 3-planes and highest power of p_1. We discuss another enumerative problem involving non-torsion classes, the arithmetic count of 3-planes whose intersection with $2n$ subspaces in $\mathbb{P}^{2n+3}$ of dimension $(2n-1)$ in general position is a line. This generalizes the results of Section 5.2 of [**FK15**].

For $\mathbb{P}^{2n-1} \subseteq \mathbb{P}^{2n+3}$, the set of 3-planes intersecting $\mathbb{P}^{2n-1}$ in a line is the Schubert variety $\Sigma_{2,2}$ in $\mathrm{Gr}(4, 2n+4)$. By Proposition 6.8, there is a unique lift of the class $\sigma_{2,2}$ to the Chow–Witt ring, given by the canonical orientation of the normal bundle for the Schubert variety $\Sigma_{2,2}$. Note also that, as discussed in Proposition 3.6, the lift of the Schubert class $\sigma_{2,2}$ to the Chow–Witt ring is – up to the image of the Bockstein – the first Pontryagin class p_1.

For every 3-plane W in the $2n$-fold self-intersection of $\sigma_{2,2}$, we have a quadratic form given by the type of the corresponding subspace, cf. Definition 7.10, which compares the orientations of the Grassmannian with the canonical orientations of the normal spaces to the translates of $\Sigma_{2,2}$. The following consequence of Theorem 8.7 describes the refined count of 3-planes in the intersection:

THEOREM 9.9. *Let F be a perfect field of characteristic $\neq 2$. Then the following equality holds in* $\mathrm{GW}(F) \cong \widetilde{\mathrm{CH}}^{8n}(\mathrm{Gr}(4, 2n+4))$:

$$\sum_{3\text{-planes}[W]\in\sigma_{2,2}^{2n}} \mathrm{tr}_{F(W)/F}\langle \det \xi_W\rangle = \frac{\mathrm{D}_n + \mathrm{C}_n}{2}\langle 1\rangle + \frac{\mathrm{D}_n - \mathrm{C}_n}{2}\langle -1\rangle,$$

where D_n denotes the degree of $\sigma_{2,2}^{2n}$ in $\mathrm{CH}^{8n}(\mathrm{Gr}(4, 2n+4))$ *and* $\mathrm{C}_n = \frac{(2n)!}{n!(n+1)!}$ *is the n-th Catalan number.*

PROOF. The dimension of $\mathrm{Gr}(4, 2n+4)$ is $8n$, and we want to compute the degree of $\mathrm{p}_1^{2n} \in \widetilde{\mathrm{CH}}^{8n}(\mathrm{Gr}(4, 2n+4)) \cong \mathrm{GW}(F)$. We apply Theorem 8.7, then the claim follows if D_n is the degree of the intersection in the Chow ring and C_n is the degree of the intersection in the $\mathbf{I}^\bullet$-cohomology ring. So it suffices to show that the degree of the intersection in $\mathbf{I}^\bullet$-cohomology equals the n-th Catalan number. The easiest way to go is to use Proposition 3.1. By [**EH17**, Proposition 4.12], $\deg \mathrm{c}_1^{2n} \in \mathrm{CH}^{2n}(\mathrm{Gr}(2, n+2))$ is the n-th Catalan number C_n. Consequently, p_1^{2n} is the quadratic form $\mathrm{C}_n\langle 1\rangle$. □

REMARK 9.10. *It seems that no closed-form expression for the numbers D_n is known, cf. MathOverflow-question 302535 "Closed form for integer series from enumerative problem?", and the series also doesn't appear in the OEIS database. Using the* `SchurRings` *package of* `Macaulay2` *provides the following first terms of the series D_n, starting from $n = 2$:*

$$6, 145, 8806, 830622, 100317140, 14342519633, 2325250316950, 415755865974454, \ldots$$

Consequently, the first instances of the quadratic form in Theorem 9.9 are

n	$\frac{\mathrm{D}_n+\mathrm{C}_n}{2}\langle 1\rangle + \frac{\mathrm{D}_n-\mathrm{C}_n}{2}\langle -1\rangle$
2	$4\langle 1\rangle + 2\langle -1\rangle$
3	$75\langle 1\rangle + 70\langle -1\rangle$
4	$4410\langle 1\rangle + 4396\langle -1\rangle$
5	$415332\langle 1\rangle + 415290\langle -1\rangle$
6	$50158636\langle 1\rangle + 50158504\langle -1\rangle$
7	$7171260031\langle 1\rangle + 7171259602\langle -1\rangle$
8	$1162625159190\langle 1\rangle + 1162625157760\langle -1\rangle$

References

[AF16a] A. Asok and J. Fasel, *Comparing Euler classes*, Q. J. Math. **67** (2016), no. 4, 603–635. MR3609848

[Ana15] Alexey Ananyevskiy, *The special linear version of the projective bundle theorem*, Compos. Math. **151** (2015), no. 3, 461–501, DOI 10.1112/S0010437X14007702. MR3320569

[Bal05] Paul Balmer, *Witt groups*, Handbook of K-theory. Vol. 1, 2, Springer, Berlin, 2005, pp. 539–576, DOI 10.1007/978-3-540-27855-9_11. MR2181829

[BC12a] Paul Balmer and Baptiste Calmès, *Witt groups of Grassmann varieties*, J. Algebraic Geom. **21** (2012), no. 4, 601–642, DOI 10.1090/S1056-3911-2012-00613-4. MR2957690

[BC12b] Paul Balmer and Baptiste Calmès, *Bases of total Witt groups and lax-similitude*, J. Algebra Appl. **11** (2012), no. 3, 1250045, 24, DOI 10.1142/S0219498811005725. MR2928113

[BG05] Paul Balmer and Stefan Gille, *Koszul complexes and symmetric forms over the punctured affine space*, Proc. London Math. Soc. (3) **91** (2005), no. 2, 273–299, DOI 10.1112/S0024611504015199. MR2167088

[Bri02] Michel Brion, *Positivity in the Grothendieck group of complex flag varieties*, J. Algebra **258** (2002), no. 1, 137–159, DOI 10.1016/S0021-8693(02)00505-7. Special issue in celebration of Claudio Procesi's 60th birthday. MR1958901

[CF14] B. Calmès and J. Fasel. Finite Chow–Witt correspondences. Preprint, arXiv:1412.2989.

[CH11] Baptiste Calmès and Jens Hornbostel, *Push-forwards for Witt groups of schemes*, Comment. Math. Helv. **86** (2011), no. 2, 437–468, DOI 10.4171/CMH/230. MR2775136

[CK13] L. Casian and Y. Kodama. On the cohomology of real Grassmann manifolds. arXiv:1309.5520v1.

[DF17] F. Déglise and J. Fasel. The Milnor–Witt motivic ring spectrum and its associated theories. Preprint, arXiv:1708.06102v1.

[EH17] David Eisenbud and Joe Harris, *3264 and all that—a second course in algebraic geometry*, Cambridge University Press, Cambridge, 2016. MR3617981

[Fas07] Jean Fasel, *The Chow-Witt ring*, Doc. Math. **12** (2007), 275–312. MR2350291

[Fas08] Jean Fasel, *Groupes de Chow-Witt* (French, with English and French summaries), Mém. Soc. Math. Fr. (N.S.) **113** (2008), viii+197, DOI 10.24033/msmf.425. MR2542148

[Fas13] Jean Fasel, *The projective bundle theorem for* $\mathbf{I}^j$*-cohomology*, J. K-Theory **11** (2013), no. 2, 413–464, DOI 10.1017/is013002015jkt217. MR3061003

[FK13] Sergey Finashin and Viatcheslav Kharlamov, *Abundance of real lines on real projective hypersurfaces*, Int. Math. Res. Not. IMRN **16** (2013), 3639–3646, DOI 10.1093/imrn/rns135. MR3090704

[FK15] S. Finashin and V. Kharlamov, *Abundance of 3-planes on real projective hypersurfaces*, Arnold Math. J. **1** (2015), no. 2, 171–199, DOI 10.1007/s40598-015-0015-5. MR3370064

[FM16] László M. Fehér and Ákos K. Matszangosz, *Real solutions of a problem in enumerative geometry*, Period. Math. Hungar. **73** (2016), no. 2, 137–156, DOI 10.1007/s10998-016-0122-7. MR3564561

[GN03] Stefan Gille and Alexander Nenashev, *Pairings in triangular Witt theory*, J. Algebra **261** (2003), no. 2, 292–309, DOI 10.1016/S0021-8693(02)00669-5. MR1966631

[HW17] J. Hornbostel and M. Wendt. Chow–Witt rings of classifying spaces for symplectic and special linear groups. J. Topol. 12, no.3, (2019), 915–965.

[Kle74] Steven L. Kleiman, *The transversality of a general translate*, Compositio Math. **28** (1974), 287–297. MR0360616

[KW17] Kirsten Wickelgren, *An arithmetic count of the lines on a smooth cubic surface*, Notices Amer. Math. Soc. **65** (2018), no. 4, 404–405. MR3752384

[Len98] Cristian Lenart, *The combinatorics of Steenrod operations on the cohomology of Grassmannians*, Adv. Math. **136** (1998), no. 2, 251–283, DOI 10.1006/aima.1998.1731. MR1626852

[Lev18] M. Levine. Motivic Euler characteristics and Witt-valued characteristic classes. Preprint, arXiv:1806.10108v1.

[OT14] Christian Okonek and Andrei Teleman, *Intrinsic signs and lower bounds in real algebraic geometry*, J. Reine Angew. Math. **688** (2014), 219–241, DOI 10.1515/crelle-2012-0055. MR3176620

[Pon47] L. S. Pontryagin, *Characteristic cycles on differentiable manifolds* (Russian), Mat. Sbornik N. S. **21(63)** (1947), 233–284. MR0022667

[Sie09] Susan J. Sierra, *A general homological Kleiman-Bertini theorem*, Algebra Number Theory **3** (2009), no. 5, 597–609, DOI 10.2140/ant.2009.3.597. MR2578891

[SW18] P. Srinivasan and K. Wickelgren. An arithmetic count of lines meeting 4 lines in $\mathbb{P}^3$. Preprint, arXiv:1810.03503.

[Vak06] Ravi Vakil, *Schubert induction*, Ann. of Math. (2) **164** (2006), no. 2, 489–512, DOI 10.4007/annals.2006.164.489. MR2247966

[Wen18] M. Wendt. Chow–Witt rings of Grassmannians. Preprint, arXiv:1805.06142.

Matthias Wendt, Fachgruppe Mathematik/Informatik, Bergische Universität Wuppertal, Gaussstrasse 20, 42119 Wuppertal, Germany

Email address: m.wendt.c@gmail.com

Selected Published Titles in This Series

745 **Federico Binda, Marc Levine, Manh Toan Nguyen, and Oliver Röndigs, Editors,** Motivic Homotopy Theory and Refined Enumerative Geometry, 2020

740 **Alina Bucur and David Zureick-Brown, Editors,** Analytic Methods in Arithmetic Geometry, 2019

739 **Yaiza Canzani, Linan Chen, and Dmitry Jakobson, Editors,** Probabilistic Methods in Geometry, Topology and Spectral Theory, 2019

738 **Shrikrishna G. Dani, Surender K. Jain, Jugal K. Verma, and Meenakshi P. Wasadikar, Editors,** Contributions in Algebra and Algebraic Geometry, 2019

737 **Fernanda Botelho, Editor,** Recent Trends in Operator Theory and Applications, 2019

736 **Jane Hawkins, Rachel L. Rossetti, and Jim Wiseman, Editors,** Dynamical Systems and Random Processes, 2019

735 **Yanir A. Rubinstein and Bernard Shiffman, Editors,** Advances in Complex Geometry, 2019

734 **Peter Kuchment and Evgeny Semenov, Editors,** Differential Equations, Mathematical Physics, and Applications, 2019

733 **Peter Kuchment and Evgeny Semenov, Editors,** Functional Analysis and Geometry, 2019

732 **Samuele Anni, Jay Jorgenson, Lejla Smajlović, and Lynne Walling, Editors,** Automorphic Forms and Related Topics, 2019

731 **Robert G. Niemeyer, Erin P. J. Pearse, John A. Rock, and Tony Samuel, Editors,** Horizons of Fractal Geometry and Complex Dimensions, 2019

730 **Alberto Facchini, Lorna Gregory, Sonia L'Innocente, and Marcus Tressl, Editors,** Model Theory of Modules, Algebras and Categories, 2019

729 **Daniel G. Davis, Hans-Werner Henn, J. F. Jardine, Mark W. Johnson, and Charles Rezk, Editors,** Homotopy Theory: Tools and Applications, 2019

728 **Nicolás Andruskiewitsch and Dmitri Nikshych, Editors,** Tensor Categories and Hopf Algebras, 2019

727 **André Leroy, Christian Lomp, Sergio López-Permouth, and Frédérique Oggier, Editors,** Rings, Modules and Codes, 2019

726 **Eugene Plotkin, Editor,** Groups, Algebras and Identities, 2019

725 **Shijun Zheng, Marius Beceanu, Jerry Bona, Geng Chen, Tuoc Van Phan, and Avy Soffer, Editors,** Nonlinear Dispersive Waves and Fluids, 2019

724 **Lubjana Beshaj and Tony Shaska, Editors,** Algebraic Curves and Their Applications, 2019

723 **Donatella Danielli, Arshak Petrosyan, and Camelia A. Pop, Editors,** New Developments in the Analysis of Nonlocal Operators, 2019

722 **Yves Aubry, Everett W. Howe, and Christophe Ritzenthaler, Editors,** Arithmetic Geometry: Computation and Applications, 2019

721 **Petr Vojtěchovský, Murray R. Bremner, J. Scott Carter, Anthony B. Evans, John Huerta, Michael K. Kinyon, G. Eric Moorhouse, and Jonathan D. H. Smith, Editors,** Nonassociative Mathematics and its Applications, 2019

720 **Alexandre Girouard, Editor,** Spectral Theory and Applications, 2018

719 **Florian Sobieczky, Editor,** Unimodularity in Randomly Generated Graphs, 2018

718 **David Ayala, Daniel S. Freed, and Ryan E. Grady, Editors,** Topology and Quantum Theory in Interaction, 2018

717 **Federico Bonetto, David Borthwick, Evans Harrell, and Michael Loss, Editors,** Mathematical Problems in Quantum Physics, 2018

For a complete list of titles in this series, visit the
AMS Bookstore at **www.ams.org/bookstore/conmseries/**.